MULTI-C
OXI

MULTI-COPPER OXIDASES

editor

Albrecht Messerschmidt

Max-Planck-Institute for Biochemistry, Germany

World Scientific
Singapore • New Jersey • London • Hong Kong

Published by

World Scientific Publishing Co. Pte. Ltd.

P O Box 128, Farrer Road, Singapore 912805

USA office: Suite 1B, 1060 Main Street, River Edge, NJ 07661

UK office: 57 Shelton Street, Covent Garden, London WC2H 9HE

British Library Cataloguing-in-Publication Data
A catalogue record for this book is available from the British Library.

MULTI-COPPER OXIDASES

ISBN 981-02-2711-6

Printed in Singapore.

To my wife Beate
and daughter Nadja

Preface

The biological activation of dioxygen is a key reaction in biological systems. Enzymes involved in direct oxygen activation are oxidases and oxygenases. Multi-copper oxidases are an important class of oxidases reducing dioxygen in a four-electron reduction to water with concomitant one-electron oxidation of the reducing substrate. The progress in the characterization and understanding of the structure and function of these enzymes has advanced so tremendously over the last ten years that the publication of a book documenting these achievements has been overdue.

The recent discovery of a key role of the FET3 protein of *Saccharomyces cerevisae*, a multi-copper oxidase, in iron metabolism of this eukaryote has underpinned the function of the plasma multi-copper oxidase ceruloplasmin in vertebrate iron transport. The lately determined X-ray structure of human ceruloplasmin confirms its close structural relatedness to the plant multi-copper oxidase ascorbate oxidase and plant or fungal multi-copper oxidase laccase and due to strong amino-acid sequence similarities has allowed to construct a useful model of the more distantly related blood-clotting factor VIII. Furthermore, a possible physiological implication of ceruloplasmin in NO-mediated events has been disclosed.

The blue multi-copper oxidases are enzymes that have optimized intramolecular electron transfer rates suitable for the catalytic dioxygen reduction. Therefore, they are relevant and interesting systems for investigating the parameters that control the intramolecular electron transfer rates.

This book contains review articles from experts in the field, dealing with modern spectroscopy, enzyme kinetics, bioinorganic chemistry, X-ray crystallography, electron transfer reactions, molecular biology, medical aspects and potential industrial applications of the three main members of multi-copper oxidases, i.e., laccase, ascorbate oxidase and ceruloplasmin.

The various chapters cover a broad spectrum of aspects of research in the field. On the first view there may be some overlap between several chapters but it takes into account the different conclusions deduced from the data of the various applied experimental techniques. Despite the knowledge of the X-ray structures of two members of the blue multi-copper oxidases, ascorbate oxidase and ceruloplasmin, several points are subject of a controversial debate. This mainly concerns the interpretation of the spectroscopy of the trinuclear copper site and the chemistry taking place at this multi-copper species. Spectroscopy predicts a central oxygen ligand coordinated to all three copper atoms, whereas such a ligand cannot be seen in the high resolution X-ray structure of ascorbate oxidase. Azide and peroxide binding has been interpreted by spectroscopy as bridging between the type-2 copper and one of the type-3 coppers compared to the X-ray derivative structures where both molecules bind to one of the type-3 coppers in an end-on fashion. Different results have also been obtained for the type-2 depleted form. Here, the X-ray structure of the type-2 depleted form of ascorbate oxidase indicates a depletion of copper from all three copper sites of the trinuclear cluster, whereas spectroscopy assigns the depletion to one site, the type-2 site only. The authors of the different chapters have taken the chance to present their own results and interpretations and have made it possible to assemble a comprehensive and up-to-date overview of the multi-copper oxidase research.

I would like to thank Prof. Robert Huber for his continuous interest and support of this project. I am grateful to World Scientific Publishing Co. and its editors Dr. G. Tan, T. Yugarani and J. M. Tan for initiating this project and the very good cooperation.

Albrecht Messerschmidt
March 1997

CONTENTS

CHAPTER 1

EARLY AND MORE RECENT HISTORY IN THE RESEARCH ON MULTI-COPPER OXIDASES

Bo G. Malmström

Department of Biochemistry and Biophysics, Göteborg University
Medicinaregatan 9C, S-413 90 Göteborg, Sweden
E-mail: bgm@bcbp.gu.se

ABSTRACT

The early history of oxidase research is reviewed, starting with Lavoisier (1789). Bertrand's pioneering work on laccase is described, and it is pointed out that he introduced the term "oxidase" and suggested for the first time that a metal constitutes an integral part of an enzyme. The identification of the metal in laccase as Cu by Keilin and Mann is mentioned. In the period 1900–1950 ascorbate oxidase and ceruloplasmin were discovered. Early EPR investigations established the presence of three forms of Cu in blue oxidases in the period 1957–1968. The trinuclear nature of the O_2-reducing site was already suggested in 1972 but not established until 1989 with the determination of the crystal structure of ascorbate oxidase. Oxygen intermediates in blue oxidases were demonstrated both optically and by EPR during the 1970s. The concept of rack-induced bonding, introduced in 1964, is discussed in some detail. Finally, blue Cu^{2+} is compared to Cu_A of cytochrome oxidase, which has been shown by crystal structure determinations and by multifrequency EPR to be dinuclear. The two strong optical bands around 500 nm arise from a splitting due to an interaction between the two Cu monomers, and the band close to 800 nm is a charge-transfer transition between the two Cu atoms.

Contents

Abbreviations: CD–circular dichroism; EPR–electron paramagnetic resonance; LF–ligand field; MCD–magnetic circular dichroism.

1. Introduction

The blue multi-copper oxidases are laccase (benzenediol: oxygen oxidoreductase, EC 1.10.3.2), ceruloplasmin (FE(II): oxygen oxidoreductase, EC 1.16.3.1) and ascorbate oxidase (L-ascorbate: oxygen oxidoreductase, EC 1.10.3.3). They are among the few enzymes that catalyze the 4-electron reduction of O_2 to $2H_2O$. This occurs in a type 2/type 3 cluster of 3 Cu ions.[1] The primary acceptor of electrons from the reducing substrate is a type 1 or blue Cu^{2+} center, however, and from this the electron is transferred intramolecularly to the O_2-reducing cluster. In this article I will review the history of research on blue oxidases. Since the knowledge of blue Cu centers is largely derived from investigations of small blue proteins, such as azurin and plastocyanin, some discussion on these proteins will also be included. In addition, the dinuclear Cu_A site in cytochrome oxidase is derived from two blue centers,[2,3] so it appears relevant to treat certain aspects of this site as well. The review will not be comprehensive, and references will be limited to papers important in the development of key concepts concerning blue oxidases. The historic presentation will be introduced by a brief review of very early oxidase research. Some more recent results, which will be discussed in detail in later chapters of this volume, will also be mentioned to indicate which, in my opinion, are the major breakthroughs in the field.

2. Early History of Copper Oxidase Research

2.1. *From Lavoisier to Bertrand*

Antoine Lavoisier (1743–1794) can be considered as the first oxidase investigator, since he suggested in 1789 that "respiration is nothing but a slow combustion of carbon and hydrogen, which is entirely similar to that which occurs in a lighted lamp or candle...". (For documentation of the very early work, see Ref. 4.)

This idea received additional support in the early 19th century, when several investigators demonstrated that animal and plant tissues consume oxygen and release carbon dioxide. In 1836 Jöns Jacob Berzelius (1779–1848) proposed that chemical reactions in living systems are catalytic in nature, and this idea was applied to biological oxidations by Mauritz Traube (1826–1894). He summarized his investigations of respiration in a period starting in 1858 in the monograph *Oxidationsfermente* (1874). In this he expressed the view that biological oxidation involves the activation of molecular oxygen by intracellular enzymes, i.e., enzymes which we would nowadays call oxidases.

Traube's concept was strongly supported by work of Hikorukoro Yoshida (1859–1928) in Japan and Gabriel Bertrand (1867–1962) in France on enzymes involved in the darkening and hardening of the latex from Oriental lacquer trees. They showed that the color change is due to the oxidation of two phenolic substances, urushiol and laccol. Bertrand, in particular,

Fig. 1. The laccase samples of G. Bertrand, kept at Collège de France in Paris.

carried out a remarkable series of investigations on tree laccase as well as laccase from mushrooms in the period 1894–1897.[5–8] He also discovered tyrosinase in 1895–1896.[9] In the course of these studies, he coined the term oxidase ("oxydase") for enzymes activating molecular oxygen. In addition, he suggested for the first time that a metal constitutes an integral part of an enzyme. He identified the metal as Mn, but, almost half a century later, Keilin and Mann[10–12] showed that both laccase and tyrosinase are Cu enzymes.

In 1985 Jean-Marie Lehn discovered that Bertrand's laccase samples are still kept in Collège de France in Paris, as shown in Fig. 1, and he let me have samples for analysis by a non-destructive X-ray fluorescence method.[13] We found that, indeed, the preparations contained 840 ppm Mn but only 35 ppm Cu. When we analyzed an acetone powder prepared from Vietnamese lacquer and used as a starting material for purifying tree laccase in our laboratory, this was shown to contain 1480 ppm Mn and 45 ppm Cu, whereas our pure enzyme had 2310 ppm Cu and as little as 5 ppm Mn. Thus, Bertrand's mistake was caused by an insuffient purity of his enzyme, which is not surprising for protein studies carried out a century ago.

2.2. *Work on Copper Oxidases 1900–1950*

The first important event in the research on multi-copper oxidases in this century was the discovery of ascorbate oxidase by Szent-Györgyi in 1928,[14] but this was not shown to be a copper enzyme until 1940.[15] Keilin and Mann in Cambridge published several important papers in the 1930s. As already mentioned, they demonstrated that laccase is a Cu enzyme,[10,11] and they also showed that its strong blue color disappears reversibly on addition of substrate. A decade later, Tissière, working in their laboratory, managed to remove Cu reversibly from laccase by dialysis against cyanide.[17] Nutritional studies during 1934–1941 had indicated that Cu is a constituent of cytochrome oxidase,[18] and in 1938–1939 Keilin and Mann[19,20] found Cu in their preparations of this enzyme. The final key development in this half-century was the work by Holmberg and Laurell, which led to the discovery and isolation of ceruloplasmin.[21,22]

3. The Discovery of Three Forms of Copper: Early EPR Investigations

3.1. *The First EPR Studies of Blue Oxidases*

The EPR technique was introduced by the Russian physicist Zavoisky in 1945, and about a decade later the first biological and biochemical applications appeared. Many of them were concerned with free radicals, and the first study of a transition metal is probably the investigation of the binding of Mn^{2+} to a kinase by Cohn and Townsend (see Ref. 23 for documentation of these early studies). I was studying the metal-activation of enolase at the time, and my own interest in the EPR method was aroused by Cohn's work. In fact, my first EPR paper, with Vänngård and Larsson, was concerning the binding of Mn^{2+} to yeast enolase. At the same time I was approached by a microbiologist in Uppsala, Gösta Fåhraeus, who asked me if I could help him purify a fungal laccase that he was investigating. Together with a student, Rolf Mosbach, we developed a purification method for *Polyporus* laccase, and this was published in early 1958, together with a basic characterization of the enzyme.[24]

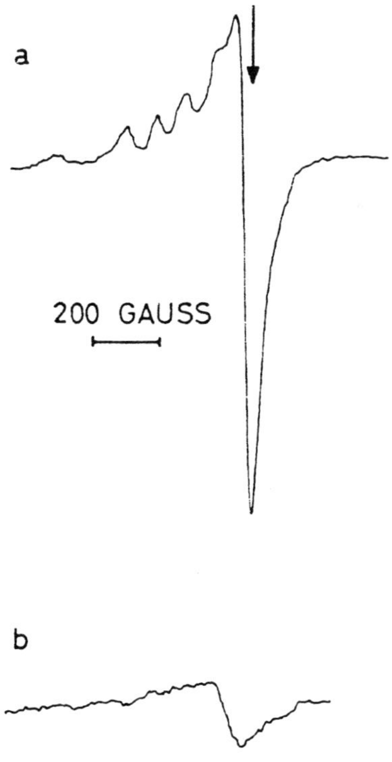

Fig. 2. EPR spectrum of fungal laccase (a) before and (b) after the addition of catechol.

Vänngård and I did our first EPR study of laccase in the autumn of 1957, and this was published in *Nature* in January of 1959[25] (some EPR results were briefly mentioned in the purification paper[24]). We showed that the enzyme gave an EPR signal with a g-value at maximum absorption (approximately equal to $g_\perp$) of 2.06, which we interpreted as the presence of covalent Cu(II). We also noted that there was an anisotropic hyperfine structure associated with the copper resonance. In addition, we observed that the Cu^{2+} ion was reduced by a substrate, catechol. These results are illustrated in Fig. 2, which contains spectra recorded in the same year. These show the presence of a low-field line, which was not included in the published spectra. We now know[26] that this stems from type 2 Cu^{2+}, but at the time we thought it represented extraneous copper.

In the *Nature* paper[25] we stated that we hoped that a detailed study of simpler Cu^{2+} complexes in progress would permit a more exact interpretation of the laccase spectrum. The publication in 1960 of our results[27] included 12 organic and inorganic complexes as well as a number of proteins, in addition to laccase, ceruloplasmin, denatured laccase and ceruloplasmin, Cu^{2+}-carboxypeptidase and erythrocuprein (now known as superoxide dismutase). This was the first demonstration that blue copper proteins have unique spectral properties, not only the strong blue color, observed already

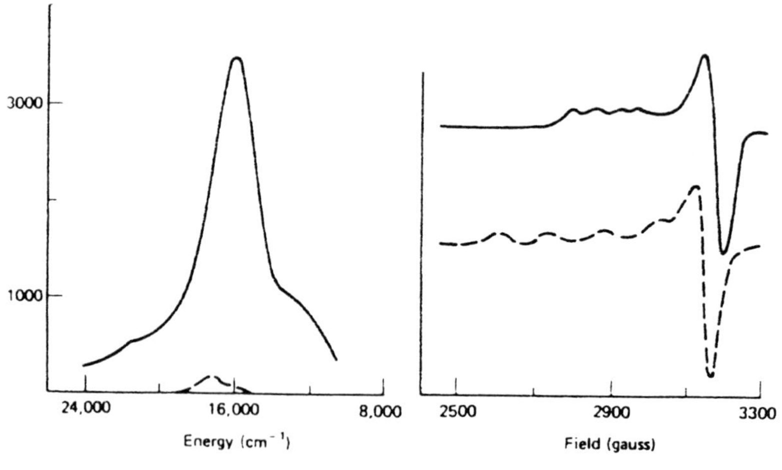

Fig. 3. Comparison of the optical and EPR spectra of a blue Cu^{2+} site (solid lines) and a normal tetragonal Cu^{2+} center(dashed lines).

by Keilin and Mann,[10] but also uniquely narrow hyperfine coupling constants, as illustrated in Fig. 3. We concluded from this, on the basis of an analysis which was actually erroneous (a possibility explicitly stated in the paper), that there was a strong degree of covalency and that the most likely ligand atoms were N and S. In addition, we suggested that a high degree of delocalization of the unpaired hole on Cu^{2+} would favor a valence-change mechanism. Finally, we proposed that the unique bonding was imposed on the Cu^{2+} ion by the protein structure, and I will return to this question in some detail later (Sec. 4.3).

When I first presented[28] the results just described at a conference on "Biological Aspects of Metal-Binding", held at Penn State in September of 1960, I was attacked by Williams, who claimed that it could not be the same copper ion that gave the EPR signal and the strong blue color, since only Cu^{1+} complexes could be strongly colored (see the printed discussion after my paper). Since I also reported that only half of the total copper

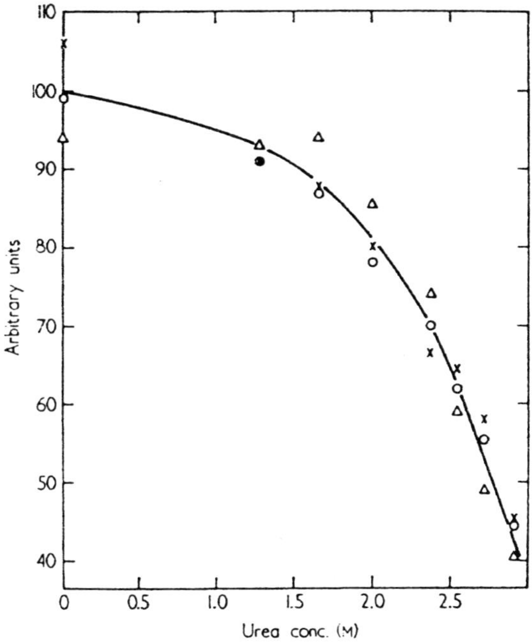

Fig. 4. The effect of urea denaturation on the absorbance at 610 nm (o), the type 1 EPR signal (Δ) and the catalytic activity (×) of cerulopasmin.

could be detected by EPR, it was possible that the other half was present as Cu^{1+}, and that this half was responsible for the color. We, therefore, devised experiments to decide whether the color originated from the EPR-detectable Cu^{2+}. One way we approached this problem was to progressively denature the enzymes with urea and measure the absorbance at 610 nm, the (type 1) EPR intensity and the catalytic activity. It was reported[29] for ceruloplasmin in 1962 that all three parameters fell on the same curve, as shown in Fig. 4. Furthermore, these properties were lost at a much lower urea concentration than that leading to gross denaturation, demonstrating that the loss was not caused by a general disruption of the protein tertiary structure. Thus, we concluded that the strong color is due to the unique bonding of Cu^{2+}, reflected in the EPR spectrum, and that this bonding is necessary for the catalytic activity.

Blumberg *et al.* confirmed in the following year (1963)[30] that it is the Cu^{2+} ions, and only the Cu^{2+} ions, that are chromophoric; they had namely, like us, found that a fraction of the total copper is EPR-undetectable. They also validated our finding of the narrow hyperfine splitting in the ceruloplasmin EPR spectrum. In 1964, Mason[31] as well as we[32] showed that azurin, which contains a single copper atom, not only has a strong blue color but also a narrow hyperfine splitting in the EPR spectrum. Consequently, it was by now beyond any doubt that the strong color of blue copper proteins is the result of a specific bonding of Cu^{2+}.

Beinert and Sands[33] reported in 1959 the first EPR spectrum of Cu_A in cytochrome oxidase, and they also showed that this ion is reducible by substrates. Three years later they published[34] an extension of this investigation, in which they showed that the *g*-value at maximum absorption is 2.03, which is lower than any value found for copper proteins or small complexes by Vänngård and myself[27] and more like that of a free radical. Remarkably, they concluded that two Cu^{2+} ions are close to each other, probably forming a mixed-valence dinuclear site. Since we now know that this is exactly how it is,[2,3] it is unfortunate that they subsequently[35] abandoned this concept because it could not be supported by power-saturation experiments.

Nakamura[36] had measured the magnetic susceptibility of *Rhus* laccase already in 1958 and concluded that all copper was in the Cu^{2+} state and contributed to the strong absorption band around 600 nm in the optical spectrum, which he had recorded in the same year.[37] Blumberg *et al.*[38]

reported in 1964 the EPR spectrum of tree laccase, which they interpreted as originating from a single rhombic component. They did find, however, that not all copper was detectable by EPR, as they had earlier[29] described for ceruloplasmin, and as Vänngård and I[28] had found for both ceruloplasmin and fungal laccase. With ceruloplasmin this was also confirmed by the Beinert group,[39] and, in addition, susceptibility measurements by Ehrenberg *et al.*[40] demonstrated that only about half of the copper is paramagnetic.

3.2. *The Establishment of the Presence of Type 2 Cu^{2+} in Blue Oxidases*

As mentioned earlier, the low-field line observed in all our spectra of fungal laccase and ceruloplasmin was originally attributed to extraneous copper. At a conference on "The Biochemistry of Copper" in 1965, Nakamura presented, in a paper[41] together with Ogura, his first EPR spectra of tree laccase and ceruloplasmin, which clearly displayed low-field lines. He maintained, however, that there is only one form of Cu^{2+}, a position that he still defended as late as 1973.[42] At the same conference, I pointed out in the discussion following Nakamura's paper that the low-field line in fungal laccase corresponded to about 40 % of the EPR-detectable copper. I also suggested that the laccase spectra of Blumberg *et al.*[38] contained a component corresponding to our "extraneous" copper, and Blumberg replied that it was possible.

The fact that the second component of Cu^{2+} in both ceruloplasmin[29] and laccase (see above) constituted about half of the EPR-detectable copper made us wonder if, indeed, this was not a specific form of Cu^{2+} associated with these enzymes. In 1968 we published an extensive investigation[43] of his question with fungal laccase, using EPR at 9 and 35 GHz, proton relaxation measurements, denaturation, reduction by quinol, and reactions with azide and cyanide. We found that the two species were present in equal amounts, as demonstrated by the experimental and simulated spectra in Fig. 5. The type 2 Cu^{2+}, as the second component came to be called,[26] showed the same saturation behavior as the signal from type 1 Cu^{2+}, both being more easily saturated than added Cu^{2+}. Type 2 Cu^{2+} was more slowly reduced by quinol compared to type 1 Cu^{2+} but still more rapidly than added Cu^{2+}. The proton relaxation measurements indicated that type 2 Cu^{2+} is more available than the type 1 ion. From our experiments we concluded the

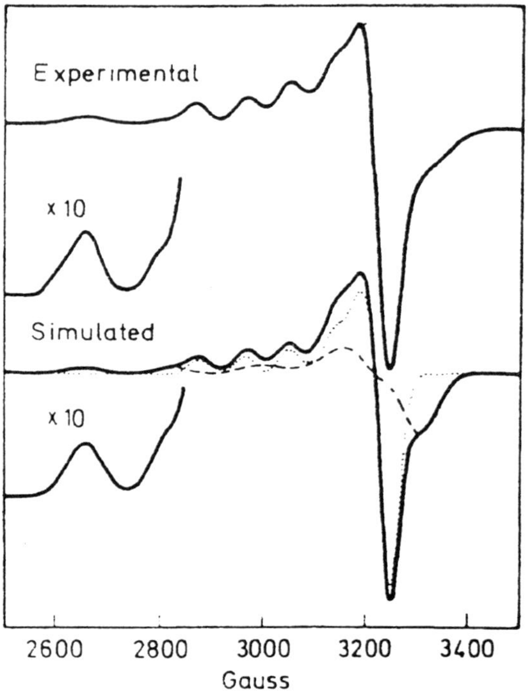

Fig. 5. Experimental and simulated EPR spectra of fungal laccase, showing the presence of type 1 (dotted line) and type 2 Cu^{2+} (dashed line).

type 2 Cu is an integral component of laccase, but we were unable to suggest any function for it.

The year after the publication of this investigation, we described[44] experiments in which we reversibly removed the type 2 Cu^{2+}. This resulted in a complete loss of catalytic activity, but the spectral properties of the type 1 Cu^{2+} were completely unchanged. Thus, type 2 Cu^{2+} is a requirement for function, which was also demonstrated[45] by the inhibition of the catalytic activity by anions, notably F^-, coordinating to type 2 Cu^{2+}. We were still, however, unable to suggest a specific function of this Cu^{2+} in the overall catalytic reaction. It is now known, of course, that it constitutes a part of a trinuclear O_2-reducing unit of these oxidases,[1] but, because of its intimate relation with the type 3 coppers, we will consider the functional properties of the type 2 ion in a later section (Sec. 4.2).

Even if the type 1 and type 2 Cu^{2+} ions are distinguished by the size of their hyperfine coupling constants, not all Cu^{2+} ions of a given type have the same EPR parameters. In a review published in 1972, Vänngård[46] showed that copper proteins can be classified in a plot of $A_{\|}$ versus $g_{\|}$, as shown in Fig. 6. In such a plot, the two types of Cu^{2+} ions fall in separate regions. Later, Peisach and Blumberg[47] developed Vänngård's idea further, showing that given proteins fall in specific areas of the plot, not only depending on which type they belong to but also on the nature of the ligand atoms.

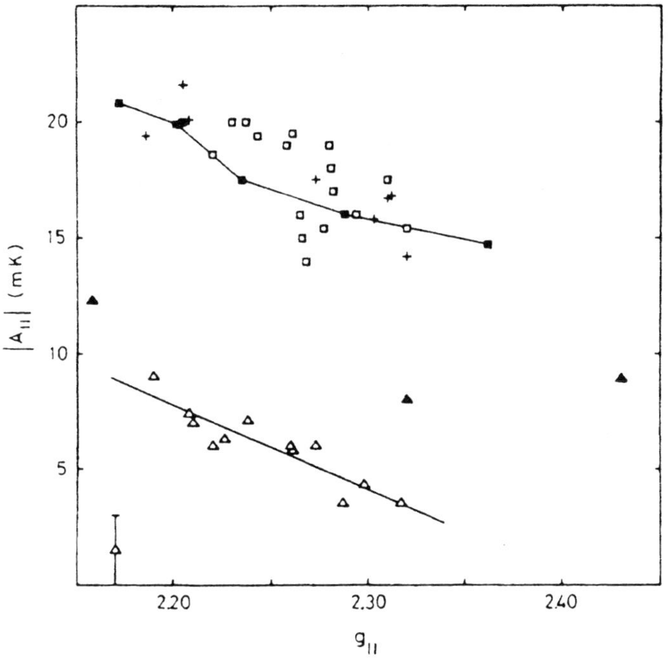

Fig. 6. A plot of $A_{\|}$ versus $g_{\|}$ for copper proteins and some model complexes. Type 1 sites are represented by (Δ), except the point in the lower left-hand corner, which represents Cu_A of cytochrome oxidase, and type 2 sites by ($\square$).

Vänngård[46] also noticed that in his plot, Cu_A of cytochrome oxidase is found in a region of its own, so that it cannot be classified as a type 1 ion. Subsequently it was found that in such a plot, the point for N_2O reductase coincides with that for cytochrome oxidase.[48,49] In the N_2O reductase case, however, a seven-line hyperfine splitting could be observed in the X-band spectrum, which is not found in native cytochrome oxidase. Because of the almost identical EPR parameters, it was suggested on this basis that in both enzymes Cu_A is dinuclear,[48] as will be discussed in greater detail in Sec. 5.

3.3. *The Discovery of Type 3 Copper*

Type 3 copper consists of a spin-coupled Cu^{2+}–Cu^{2+} pair,[26] and is hence nonparamagnetic. Its discovery involved two quite separate lines of research: (1) anaerobic redox titrations of fungal laccase[50]; and (2) stopped-flow studies of the anaerobic reduction of the enzyme by substrates.[51] In the anaerobic titrations, it was found that 3.7 electron equivalents were required to reduce the blue Cu^{2+}, and the color decreased linearly with the amount of added titrant. EPR measurements confirmed this but also showed that the type 2 Cu^{2+} behaved heterogeneously, which depended on some F^- being bound to this ion and on its slow reduction. The important finding, however, was that two reducing equivalents were taken up by a site, which did not display any EPR signal even at intermediate stages of reduction. From this we postulated the presence of a cooperative $2e^-$-acceptor in the form of an exchange-coupled Cu^{2+}–Cu^{2+} pair. A few years later, the existence of three types of electron-acceptors was also established with tree laccase[52] and ceruloplasmin.[53]

The linear decrease of the blue color in the redox titrations is a result of the fact that the type 1 Cu^{2+} and the type 3 pair have the same reduction potentials.[54] It was also shown that the binding of F^- splits the potentials in fungal laccase, which explains why we[55] could differentiate the reduction of the three types of copper in the presence of excess F^-. An important result of this was that we could show that a 330 nm band present in the oxidized-reduced difference spectrum can be associated with the type 3 unit. This band had first been observed in ceruloplasmin by Blumberg *et al.*[30] and later shown to be present in tree laccase by Nakamura and Ogura.[41,56] The latter is somewhat ironic, since, as mentioned earlier (Sec. 3.2), Nakamura[41] claimed that all copper can be seen by EPR, and, in addition, Nakamura and Ogura[56] showed that only enzymes with a 330 nm band can react rapidly with O_2. It is now established[1] that the O_2-reducing unit is a trinuclear center involving the type 2 and type 3 ions, something which I already suggested in 1972,[57] as illustrated in Fig. 7 (which unfortunately

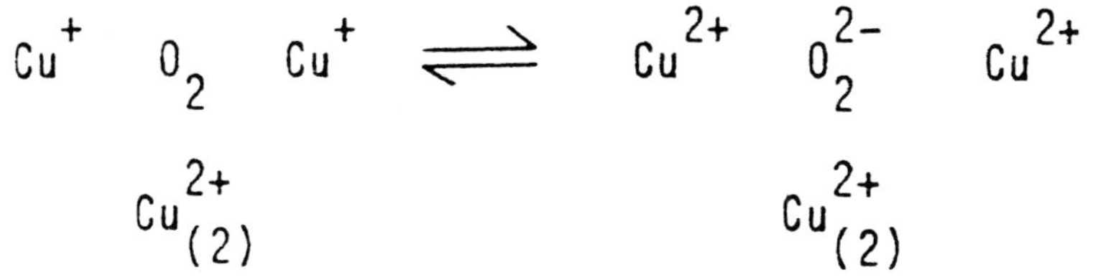

Fig. 7. The trinuclear O_2-reducing site of blue oxidases, as depicted in 1972.[57]

was published in an obscure place only, the annals of a local academy in Uppsala).

The presence of type 3 copper was independently discovered[51] in stopped-flow experiments involving anaerobic reduction that I carried out with Eraldo Antonini in Rome on my first sabbatical in 1968. The substrate we used was $Fe(CN)_6{}^{4-}$, which had an advantage; we could follow the formation of the product at 420 nm, where the enzyme has very little absorption. We found that 2.9 molecules of $Fe(CN)_6{}^{3-}$ formed per molecule of enzyme, and concluded that laccase contains two electron-accepting sites other than type 1 and type 2 Cu^{2+}, the latter not being reduced in these experiments. Thus, we obtained an independent confirmation of the presence of a type 3 center.

4. Some Later Milestones in Blue Oxidase Research

4.1. *Oxygen Intermediates*

The first direct evidence for an oxygen intermediate in a blue oxidase came from work on ceruloplasmin by Manabe *et al.*[58] These authors observed a transient species with maximum absorption at 420 nm, which was formed synchronously with the rapid reoxidation of the 610- and 330-nm chromophores. Similar reaction intermediates were later observed with both fungal and tree laccase,[59,60] suggesting that they are a species common to all blue oxidases. The optical intermediate was initially believed to be a peroxide compound,[59,61] but later it was shown with *Rhus* laccase that it forms simultaneously with the development of the 615- and 330-nm absorptions.[60] This suggests that O_2 rapidly receives three electrons from the type 1 and type 3 sites.

Since three electrons are rapidly transferred to O_2 in the formation of the optical intermediate, one would expect this to be paramagnetic.

No new EPR signal could be observed at 77 K, but at 10 K a new paramagnetic species has been reported[62,63] in both tree and fungal laccase as well as in ceruloplasmin, as illustrated in Fig. 8. These signals really represent an oxygen intermediate as this is proven by the line broadening effect of substituting ^{17}O for ^{16}O (Fig. 8). This intermediate can also be seen under turnover conditions,[64] indicating that it is a participant in the normal catalytic reaction.

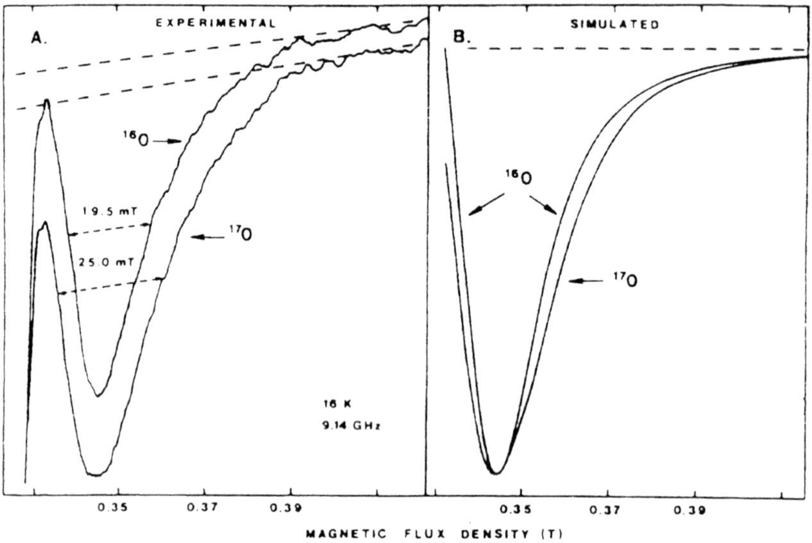

Fig. 8. Isotope effect on the experimental and simulated EPR spectra of a paramagnetic oxygen intermediate of fungal laccase.

4.2. *The Trinuclear O_2-Reducing Center of Blue Oxidases*

By reoxidizing fully reduced laccase with $^{17}O_2$, Brändén and Deinum[65] demonstrated that one of the H_2O molecules formed in the reaction becomes coordinated to type 2 Cu^{2+}. Since the type 3 copper pair is part of the O_2-reducing site, these results suggested that this site is actually a trinuclear copper site, as I had already proposed in 1972 (see Sec. 3.3). By

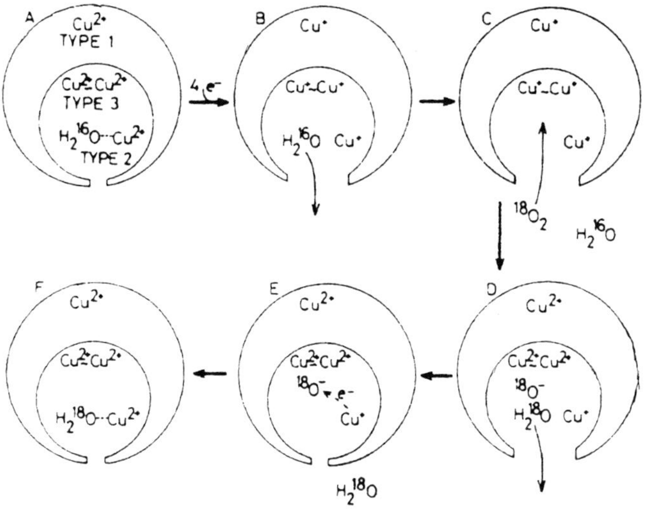

Fig. 9. The O_2-reducing cavity of laccase, showing a redox-liked conformational change leading to the trapping of one of the H_2O molecules formed on type 2 Cu^{2+}.

mass spectrometry, it was later shown[66] that the water molecule which is not trapped at the type 2 Cu^{2+} has been released into the H_2O phase. On the basis of these results,[65,66] a scheme was formulated according to which the O_2-reducing site is situated in a cavity, which can be either open or closed, as illustrated in Fig. 9. The conformational change was assumed to be induced by a change in the redox state of the enzyme.

The trinuclear nature of the O_2-reducing site in blue oxidases was also advocated by the Solomon group,[67,68] before the advent of the crystal structure of ascorbate oxidase established it,[1,69] largely on the basis of the spectroscopic properties of N_3^- derivatives of laccase. Even if the X-ray structure of the azide derivatives showed[70] that there is no bridging N_3^- between the type 2 and type 3 ions, as proposed by the Solomon group,[68,71] so that the assignment of the spectral signals had to be reinterpreted, there is no doubt that the trinuclear site is a mechanistic requirement for the reduction of O_2 to $2H_2O$.

4.3. *Rack-Induced Bonding in Blue Copper Proteins*

In Sec. 3.1, I mentioned that, together with Vänngård, I suggested[27] that the unique bonding of blue Cu^{2+} is the result of a structure forced upon the metal ion by the protein. I expanded on this idea at a conference in 1964 and give two quotes here from its printed proceedings[72]:

"The cooperative effect of a large number of weak forces stabilizing the tertiary structure of a protein could ... create a chelating site with very little flexibility."
"The energy for the distortion would then be derived from the tertiary folding of the protein, and these enzymes would be examples of the 'rack' mechanism of Lumry ...."

These ideas were at the time purely qualitative, but subsequent experimental developments made it possible (about 20 years later) to put them

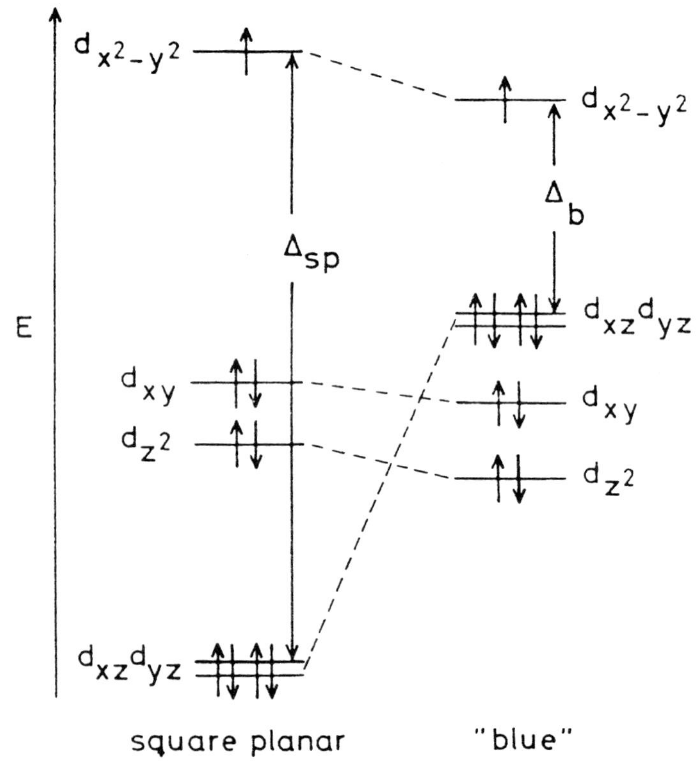

Fig. 10. The energies of the *d* electron orbitals in the ligand field of a square planar and a blue Cu^{2+} complex.

on a more quantitative basis. First, the 3-dimensional structure of a blue copper protein, plastocyanin, was published by Colman *et al.*[73] in 1977, and this established that the coordination geometry is distorted from that preferred by Cu^{2+}. Second, extensive spectroscopic investigations by CD, MCD and EPR had allowed Gray and Solomon[74] to give a complete description of the electronic structure of blue Cu^{2+} centers. With the aid of this knowledge, Gray and I[75] made an estimate of the rack energy from the decreased ligand field in a blue site compared to a square-planar Cu^{2+} complex, as shown in Fig. 10 (the assignment of the *d* orbital levels has been modified by recent theoretical work,[76,77] but this does not invalidate our analysis, which depends on the energies of the *d–d* transitions alone). Our calculation showed that the rack energy was at least 70 $kJmol^{-1}$, and we added the following comments:

"... the lowest energy of a protein conformation that would accomodate a planar CuN_2SS^* unit must be 70 $kJmol^{-1}$ ($\approx$ 6000 cm^{-1}) above that of the observed conformation. Put in another way, the energy that would be released upon formation of an optimally LF[ligand field]-stabilized CuN_2SS^* unit falls well below that required to twist the protein accordingly, with the result that the copper is forced to accept the geometrical structure that is presented to it."

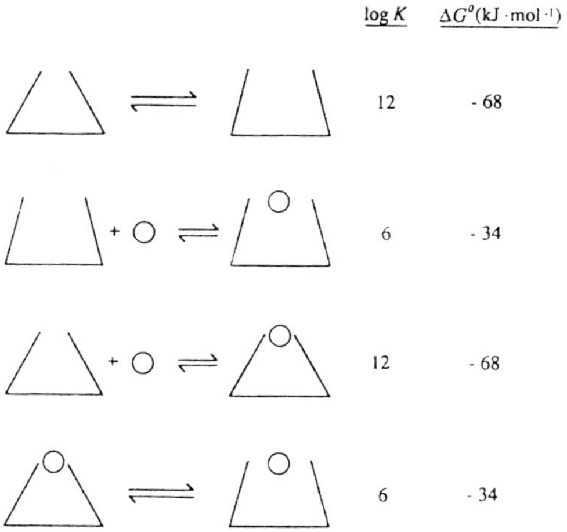

Fig. 11. Schematic representation of the conformational changes involved in rack-induced bonding.

The conformational equilibria involved in rack-induced bonding are illustrated in Fig. 11, taken from one of my recent reviews.[78] The protein is assumed to have two discrete conformations with different positions of the ligand atoms. The equilibrium is far towards the more open conformation, which binds the metal more weakly than the closed conformation. The affinity is still considerable (10^6 M^{-1}) in the more weakly binding form, however, so that the conformational equilibrium forces the metal ion into a strained conformation. The conformational free energy change has in this example been chosen to correspond to the estimated rack energy in blue copper proteins.[75]

5. The Relationship between Blue Cu^{2+} and Cu_A of Cytochrome Oxidase

The application of the technique of site-directed mutagenesis to blue copper proteins[79,80] has led to major advances in the understanding of structure-function relationships, for example, the tuning of the reduction potentials by the protein structure.[81] So far, no study of blue oxidases by this technique has been published, but it has been used extensively to investigate cytochrome oxidase,[82] despite the greater complexity of this enzyme. Here I will summarize some recent work on the Cu_A site of cytochrome oxidase, because of its close relationship to blue copper proteins, as evidenced, among others, by sequence homologies.[83,84]

Cu_A is bound to subunit II of cytochrome oxidase, which is anchored to the membrane by two transmembrane helices.[83] The copper site is, however, located in a domain outside the membrane, and recently it has been possible to express this domain of subunit II as soluble proteins.[85,86] These soluble domains have the same secondary structure[87] as blue copper proteins, a Greek key β-barrel named the cupredoxin fold,[88] and they contain a Cu_A site with EPR parameters identical to those of the native oxidase.[85,86] The relationship between Cu_A and blue copper proteins is further illustrated by the fact that it has been possible to introduce a Cu_A site into a blue protein, amicyanin.[89] It has already been mentioned that Cu_A has been suggested to be dinuclear,[48] and this has now been unequivocally established by the determination of the crystal structure of the *Paracoccus* oxidase[90] and of a soluble Cu_A domain[91] as well as by multifrequency EPR measurements[92] with ^{63}Cu and ^{65}Cu.

The preparation of soluble Cu_A domains has made possible for the first time the recording of the complete optical spectra, free from the heme absorption bands, associated with Cu_A sites.[85,86] Whereas blue protein spectra are dominated by one strong band close to 600 nm, Cu_A has two closely spaced bands of nearly equal, but lower, intensity around 500 nm. This split into two bands is a result of the dinuclear nature of the site, which leads to an interaction between the two copper monomers.[3] This is consistent with the observation that the corresponding CD bands have opposite signs. The near-IR band close to 800 nm, which is seen also in the whole oxidase, is a charge-transfer transition between the two Cu atoms. The dinuclear nature of the Cu_A site has two functional advantages: it makes it possible for an electron to enter and leave the site by different pathways, which is a necessity because the membrane-bound protein cannot rotate, and it lowers the reorganization energy associated with the electron transfer.[3]

Acknowledgments

My own work has been supported by the Swedish Natural Science Research Council and the European Commission.

References

1. A. Messerschmidt, A. Rossi, R. Ladenstein, R. Huber, M. Bolognesi, G. Gatti, A. Marchesini, R. Petruzelli and A. Finazzi-Agrò, *J. Mol. Biol.* **206**, 513 (1989).
2. P. Lappalainen and M. Saraste, *Biochim. Biophys. Acta* **1187**, 222 (1994).
3. S. Larsson, B. Källebring, P. Wittung and B. G. Malmström, *Proc. Natl. Acad. Sci. USA* **92**, 7167 (1995).
4. J. S. Fruton, *Molecules and Life* (Wiley-Interscience, New York, 1972).
5. G. Bertrand, *Compt. Rend. Acad. Sci.* **T118**, 1215 (1894).
6. G. Bertrand, *Compt. Rend. Acad. Sci.* **T120**, 266 (1895).
7. G. Bertrand, *Compt. Rend. Acad. Sci.* **T122**, 1132 (1896).
8. G. Bertrand, *Compt. Rend. Acad. Sci.* **T124**, 1032 (1897).
9. G. Bertrand, *Compt. Rend. Acad. Sci.* **T123**, 463 (1896).
10. D. Keilin and T. Mann, *Nature* **143**, 23 (1939).
11. D. Keilin and T. Mann, *Nature* **144**, 304 (1939).
12. D. Keilin and T. Mann, *Proc. Roy. Soc. London* **B125**, 187 (1938).
13. J.-M. Lehn, B. G. Malmström, E. Selin and M. Öblad, *Trends Biochem. Sci.* **11**, 228 (1986).
14. A. Szent-Györgyi, *Biochem. J.* **22**, 1387 (1928).

15. E. Stotz, *J. Biol. Chem.* **133**, C (1940).
16. A. Tissière, *Nature* **162**, 340 (1948).
17. E. Cohen and C. A. Elvehjem, *J. Biol. Chem.* **107**, 97 (1934).
18. M. O. Schultze, *J. Biol. Chem.* **138**, 219 (1941).
19. D. Keilin and E. F. Hartree, *Nature* **141**, 870 (1938).
20. D. Keilin and E. F. Hartree, *Proc. Roy. Soc. London* **B127**, 167 (1939).
21. C. G. Holmberg and C.-B. Laurell, *Acta Chem. Scand.* **1**, 944 (1947).
22. C. G. Holmberg and C.-B. Laurell, *Acta Chem. Scand.* **2**, 550 (1948).
23. H. Beinert and G. Palmer, *Advan. Enzymol.* **27**, 105 (1965).
24. B. G. Malmström, G. Fåhraeus and R. Mosbach, *Biochim. Biophys. Acta* **28**, 652 (1958).
25. B. G. Malmström, R. Mosbach and T. Vänngård, *Nature* **183**, 321 (1959).
26. R. Malkin and B. G. Malmström, *Advan. Enzymol.* **33**, 177 (1970).
27. B. G. Malmström and T. Vänngård, *J. Mol. Biol.* **2**, 118 (1960).
28. B. G. Malmström, *Fed. Proc. Suppl.* **10**, 60 (1961).
29. L. Broman, B. G. Malmström, R. Aasa and T. Vänngård, *J. Mol. Biol.* **5**, 301 (1962).
30. W. E. Blumberg, J. Eisinger, P. Aisen, A. G. Morell and I. H. Scheinberg, *J. Biol. Chem.* **238**, 1675 (1963).
31. H. S. Mason, *Biochem. Biophys. Res. Commun.* **10**, 11 (1963).
32. L. Broman, B. G. Malmström, R. Aasa and T. Vänngård, *Biochim. Biophys. Acta* **75**, 365 (1963).
33. R. H. Sands and H. Beinert, *Biochim. Biophys. Acta* **1**, 175 (1959).
34. H. Beinert, D. E. Griffiths, D. C. Wharton and R. H. Sands, *J. Biol. Chem.* **237**, 2337 (1962).
35. H. Beinert and G. Palmer, in *Oxidases and Related Redox Systems,* eds. T. E. King, H. S. Mason and M. Morrison (Wiley, New York, 1965) p. 567.
36. T. Nakamura, *Biochim. Biophys. Acta* **30**, 640 (1958).
37. T. Nakamura, *Biochim. Biophys. Acta* **30**, 44 (1958).
38. W. E. Blumberg, W. G. Levine, S. Margolis and J. Peisach, *Biochem. Biophys. Res. Commun.* **15**, 277 (1964).
39. C. B. Kasper, H. F. Deutsch and H. Beinert, *J. Biol. Chem.* **238**, 2338 (1963).
40. A. Ehrenberg, B. G. Malmström, L. Broman and R. Mosbach, *J. Mol. Biol.* **5**, 450 (1962).
41. T. Nakamura and Y. Ogura, in *The Biochemistry of Copper*, eds. J. Peisach, P. Aisen and W. E. Blumberg (Academic Press, New York, 1966) p. 389.
42. T. Nakamura and Y. Ogura, in *Oxidases and Related Redox Systems*, eds. T. E. King, H. S. Mason and M. Morrison (University Park Press, Baltimore, 1973) p. 96.
43. B. G. Malmström, B. Reinhammar and T. Vänngård, *Biochim. Biophys. Acta* **156**, 67 (1968).
44. R. Malkin, B. G. Malmström and T. Vänngård, *Eur. J. Biochem.* **7**, 253 (1969).

45. R. Brändén, B. G. Malmström and T. Vänngård, *Eur. J. Biochem.* **36**, 195 (1973).
46. T. Vänngård, in *Biological Applications of Electron Spin Resonance*, eds. H. M. Swartz, J. R. Bolton and D.C. Borg (Wiley, New York, 1972) p. 411.
47. J. Peisach and W. E. Blumberg, *Arch. Biochem. Biophys.* **165**, 691 (1974).
48. W. E. Antholine, D. H. W. Kastrau, G. C. M. Steffens, G. Buse, W. G. Zumft and P. Kroneck, *Eur. J. Biochem.* **209**, 875 (1992).
49. B. G. Malmström and R. Aasa, *FEBS Lett.* **325**, 49 (1993).
50. J. A. Fee, R. Malkin, B. G. Malmström and T. Vänngård, *J. Biol. Chem.* **244**, 4200 (1969).
51. B. G. Malmström, A. Finazzi Agrò and E. Antonini, *Eur. J. Biochem.* **9**, 383 (1969).
52. B. G. Malmström, B. Reinhammar and T. Vänngård, *Biochim. Biophys. Acta* **205**, 48 (1970).
53. R. J. Carrico, B. G. Malmström and T. Vänngård, *Eur. J. Biochem.* **20**, 518 (1971).
54. B. R. M. Reinhammar, *Biochim. Biophys. Acta* **275**, 245 (1972).
55. R. Malkin, B. G. Malmström and T. Vänngård, *Eur. J. Biochem.* **10**, 324 (1969).
56. T. Nakamura and Y. Ogura, *J. Biochem.* (Tokyo) **59**, 449 (1966).
57. B. G. Malmström, in *Kungl. Vetenskaps-Societetens Årsbok* (1973) p. 33.
58. T. Manabe, N. Manabe, K. Hiromi and H. Hatano, *FEBS Lett.* **23**, 268 (1972).
59. L.-E. Andréasson, R. Brändén, B. G. Malmström and T. Vänngård, *FEBS Lett.* **32**, 187 (1973).
60. L.-E. Andréasson, R. Brändén and B. Reinhammar, *Biochim. Biophys. Acta* **438**, 370 (1976).
61. R. Brändén, B. G. Malmström and T. Vänngård, *Eur. J. Biochem.* **18**, 238 (1971).
62. R. Aasa, R. Brändén, J. Deinum, B. G. Malmström, B. Reinhammar and T. Vänngård, *Biochem. Biophys. Res. Commun.* **70**, 1204 (1976).
63. R. Aasa, R. Brändén, J. Deinum, B. G. Malmström, B. Reinhammar and T. Vänngård, *FEBS Lett.* **61**, 115 (1976).
64. L.-E. Andréasson and B. Reinhammar, *Biochim. Biophys. Acta* **568**, 145 (1979).
65. R. Brändén and J. Deinum, *FEBS Lett.* **73**, 144 (1977).
66. R. Brändén, J. Deinum and M. Coleman, *FEBS Lett.* **89**, 180 (1978).
67. M. D. Allendorf, D. J. Spira and E. I. Solomon, *Proc. Natl. Acad. Sci. USA* **82**, 3063 (1985).
68. D. J. Spira-Solomon, M. D. Allendorf and E. I. Solomon, *J. Am Chem. Soc.* **108**, 5318 (1986).
69. A. Messerschmidt, R. Ladenstein, R. Huber, M. Bolognesi, L Avigliano, R. Petruzelli, A. Rossi and A. Finazzi-Agrò, *J. Mol. Biol.* **224**, 179 (1992).
70. A. Messerschmidt, H. Luecke and R. Huber, *J. Mol. Biol.* **230**, 997 (1993).

71. J. M. Cole, L Avigliano, L. Morpurgo and E. I. Solomon, *J. Am Chem. Soc.* **113**, 9080 (1991).
72. B. G. Malmström, in *Oxidases and Related Redox Systems*, eds. T. E. King, H. S. Mason and M. Morrison (Wiley, New York, 1965), p. 207.
73. P. M. Colman, H. C. Freeman, J. M. Guss, M. Murata, V. A. Norris, J. A. M. Ramshaw and M. P. Venkatappa, *Nature* **272**, 319 (1978).
74. H. B. Gray and E. I. Solomon, in *Copper Proteins*, ed. T. G. Spiro (Wiley, New York, 1981) p. 1.
75. H. B. Gray and B. G. Malmström, *Comments Inorg. Chem.* **2**, 203 (1983).
76. E. I. Solomon, M. D. Lowery, L. B. LaCroix and D. E. Root, *Methods Enzymol.* **226**, 1 (1993).
77. S. Larsson, A. Broo and L. Sjölin, *J. Phys. Chem.* **99**, 4860 (1995).
78. B. G. Malmström, *Eur. J. Biochem.* **223**, 711 (1994).
79. G. W. Canters and G. Gilardi, *FEBS Lett.* **325**, 39 (1993).
80. D. S. Wuttke and H. B. Gray, *Curr. Opin. Struct. Biol.* **3**, 555 (1993).
81. T. Pascher, B. G. Karlsson, M. Nordling, B. G. Malmström and T. Vänngård, *Eur. J. Biochem.* **212**, 289 (1993).
82. J. P. Hosler, S. Ferguson-Miller, M. W. Calhoun, J. W. Thomas, J. Hill, L. Lemieux, J. Ma, C. Georgiou, J. Fetter, J. Shapleigh, M. M. J. Tecklenburg, G. T. Babcock and R. B. Gennis *J. Bioenerg. Biomembr.* **25**, 121 (1993).
83. M. Saraste, *Q. Rev. Biophys.* **23**, 331 (1990).
84. B. L. Trumpower and R. B. Gennis, *Annu. Rev. Biochem.* **63**, 675 (1994).
85. P. Lappalainen, R. Aasa, B. G. Malmström and M. Saraste, *J. Biol. Chem.* **268**, 26416 (1993).
86. C. von Wachenfeldt, S. de Vries and J. van der Oost, *FEBS Lett.* **340**, 109 (1994).
87. P. Wittung, B. Källebring and B. G. Malmström, *FEBS Lett.* **349**, 286 (1994).
88. E. T. Adman, *Adv. Protein Chem.* **42**, 145 (1991).
89. C. Dennison, E. Vijgenboom, S. de Vries, J. van der Oost and G. W. Canters, *FEBS Lett.* **365**, 92 (1995).
90. S. Iwata, C. Ostermeier, B. Ludwig and H. Michel, *Nature* **376**, 660 (1995).
91. M. Wilmanns, P. Lappalainen, M. Kelly, E. Sauer-Eriksson and M. Saraste, *Proc. Natl. Acad. Sci. USA* **92**, 11955 (1995).
92. J. A. Fee, D. Sanders, C. E. Slutter, P. E. Doan, R. Aasa, M. Karpefors and T. Vänngård, *Biochem. Biophys. Res. Commun.* **212**, 77 (1995).

CHAPTER 2

SPATIAL STRUCTURES OF ASCORBATE OXIDASE, LACCASE AND RELATED PROTEINS: IMPLICATIONS FOR THE CATALYTIC MECHANISM

Albrecht Messerschmidt

Max-Planck-Institut für Biochemie, Am Klopferspitz 18A
D-82152 Martinsried bei Muenchen, Federal Republic of Germany
E-mail: messersc@.biochem.mpg.de

ABSTRACT

Ascorbate oxidase was the first blue oxidase whose spatial structure was determined by X-ray crystallography. The three-dimensional structure of ascorbate oxidase showed the nature and spatial arrangement of the copper centers and the three-domain structure. The domains are of the β-barrel type and are distantly related to the small blue copper proteins plastocyanin or azurin.

A structurally-based amino-acid sequence alignment strongly suggests a three-domain structure for laccase closely related to ascorbate oxidase and a six-domain structure for ceruloplasmin. The relationship suggests that laccase, like ascorbate oxidase, has a mononuclear blue copper in domain 3 and a trinuclear copper between domains 1 and 3 and ceruloplasmin has mononuclear copper ions in domains 2, 4 and 6 and a trinuclear copper between domains 1 and 6. The six-domain structure and the arrangement of the copper ions for cerulopasmin has recently been confirmed by an X-ray structure analysis.

X-ray structure of functional derivatives of ascorbate oxidase provide pictures of intermediate states, which will probably be passed during the catalytic cycle. A catalytic

---

Abbreviations used: $ArNO_2^{\cdot -}$ – one electron-reduced nitroaromatic; ASC – ascorbate; CD – circular dichroism; DMG – dimethyl-glyoxime; E^0 – biological standard redox potential; EDTA – ethyl diamine tetra acetate; ET – electron transfer; EPR – electron paramagnetic resonance; k_{cat} – turnover number; FC – calculated structure factor amplitude; FO – observed structure factor amplitude; K_1, K_2, K_3 – reaction constants; K_{init} – initial bimolecular reation constant; K_m – Michaelis constant; MPD – methyl pentane diol; $MV^{\cdot +}$ – methyl viologen radical; NATI – native form; PEOX – peroxide form; REDU – reduced form; 3D – three-dimensional; T1Hg – type-1 copper substituted by mercury; T2D – type-2 depleted; UV/vis – ultraviolet-visable.

mechanism that is based on the available mechanistic data and these new results has been proposed. Electron transfer properties of ascorbate oxidase are discussed.

Contents

1. Introduction

The biological functions and spectroscopical, catalytic and chemico-physical properties of the blue multi-copper oxidases are described and discussed in the other chapters of this book. The X-ray structure of human cerulopasmin is dealt with in the next chapter.

This chapter is devoted to the spatial structure of ascorbate oxidase and conclusions drawn from it in relation to the spatial structures for the other blue oxidases, laccase and ceruloplasmin, to the catalytic mechanism of the blue oxidases and to their evolutionary relatedness among each other and to other proteins. In this context it is useful to include relevant informations about amino-acid sequences, oxidation-reduction potentials, and kinetic properties of laccase and ascorbate oxidase.

2. X-Ray Structure of Ascorbate Oxidase

2.1. *Crystallization*

The first crystals of a blue copper oxidase were obtained from human ceruloplasmin in 1962.[1] Since that time several successful crystallizations of ceruloplasmin from different species have been reported.[2] Crystals of human native ceruloplasmin were crystallographically characterized by Magdoff–Fairchild *et al.*[3] and later by Zaytsev *et al.*[4] The crystals are tetragonal, space group I4, and diffract to a resolution of about 3.5 Å. The unit cell has dimensions of $a = b = 268$ Å and $c = 129$ Å and contains two molecules per asymmetric unit. Also from native human ceruloplasmin, Zaytsev *et al.*[5] could obtain a trigonal crystal form, space group $P3_121$ or $P3_221$ ($a = b = 214$ Å, $c = 84.5$ Å) with only one molecule per asymmetric unit. These crystals diffract to 3.5 Å resolution in contrast to an isomorphous trigonal form from asialo-ceruloplasmin[4] which diffract only to 6 Å. The trigonal crystal form has recently been improved to diffract to 3.1 Å resolution and the successful X-ray structure analysis is described in the next chapter as already mentioned.

There have been several attempts to crystallize laccase from different species by various groups without any encouraging results and they are therefore not documented in the literature. The main reason for this failure will be the heterogeneity in glycosylation of laccase.

The first crystals of ascorbate oxidase were grown by Ladenstein *et al.*[6] in a final 1.2 M sodium-potassium phosphate buffer, pH 7.0. They were orthorhombic, space group $P2_12_12_1$ with $a = 190.7$ Å, $b = 125.2$ Å, $c = 112.3$ Å and two molecules (4 subunits) per asymmetric unit. Some years later Bolognesi *et al.*[7] obtained a different crystal form of ascorbate oxidase with 2-methyl-2,4-pentane-diol (MPD) as precipitant. The crystals were orthorhombic as well, space group $P2_12_12$ with $a = 106.7$ Å,

$b = 105.1$ Å, $c = 113.5$ Å and one molecule (2 subunits) per asymmetric unit. In both cases, the protein material was prepared from the peels of green zucchini squash (*Cucurbita pepo medullosa*). The preliminary three-dimensional X-ray structure of ascorbate oxidase based on the analysis of both crystal forms has been published in 1989 by Messerschmidt *et al.*[8] and the polypeptide fold and the coordination of the mononuclear blue copper site and the unprecedented trinuclear copper site have been described. The structure has now been refined to 1.9 Å resolution and its detailed description has been the subject of a further publication.[9]

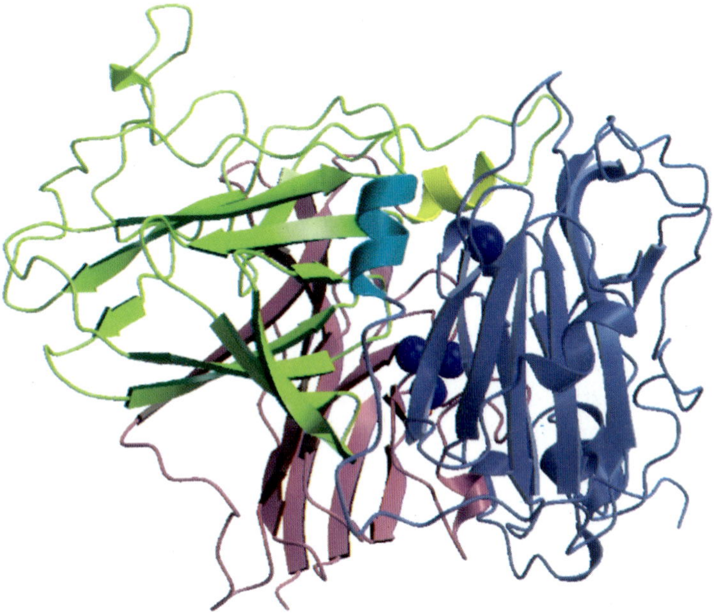

(a)

Fig. 1. A schematic representation of the monomer structure of ascorbate oxidase. (a) Monomer plus copper ions, (b) assignment of the secondary structure elements. β-sheets are represented by arrows, α-helices by helical ribbons. Figure 1(a) was made with programs MOLSCRIPT[8a] and RASTER3D.[8b] Figure 1(b) was produced by the program RIBBON.[8c]

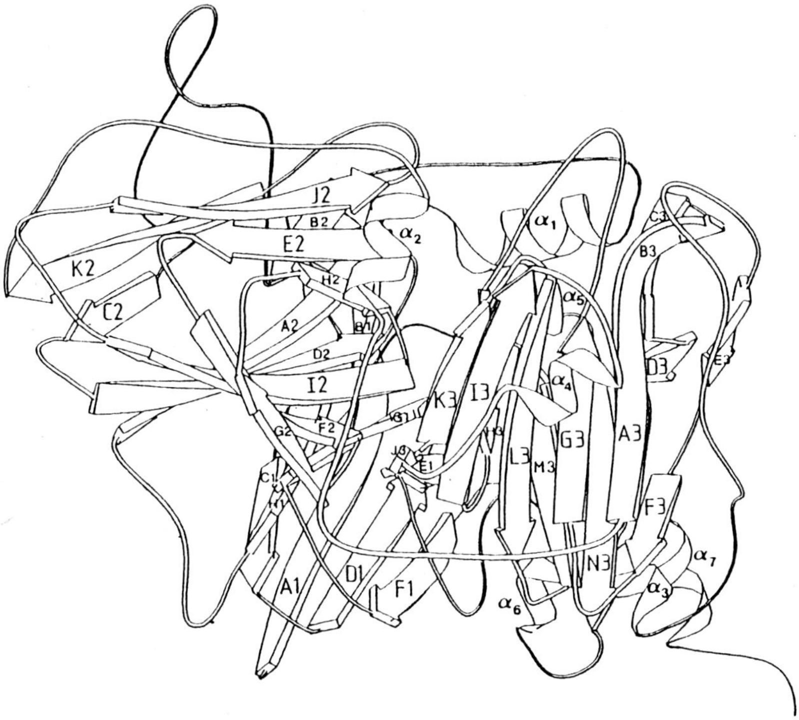

(b)

Fig. 1. (*Continued*)

2.2. *Overall Description of the Structure*

The subunits are arranged in the crystals as homo-tetramers with D2 symmetry. The structure of a subunit is shown schematically in Fig. 1. Each subunit of 552 amino acid residues has a globular shape with dimensions of 49×53×65 Å and is built up by three domains arranged sequentially on the polypeptide chain tightly associated in space. The folding of all three domains is of a similar β-barrel type. It is distantly related to the small blue copper proteins like plastocyanin or azurin. Domain 1 is made up of two four-stranded β-sheets (Fig. 1b) which form a β-sandwich structure. Domain 2 consists of a six-stranded and a five-stranded β-sheet. Finally, domain 3 is built of two five-stranded β-sheets which form the β-barrel structure and a four-stranded β-sheet which is an extension at the N-terminal part of this domain. A topology diagram of ascorbate oxidase for all three domains and of the related structures of plastocyanin and azurin is shown in Fig. 2. Ascorbate oxidase contains ten helices. Three of them are very short 3_{10}-helices and are therefore not indicated in Fig. 1b. Domain 2 has a short α-helix (α_1) between strands A2 and B2. Domain 3 exhibits four short α-helices which are located between strands D3 and E3 (α_3), M3 and N3 (α_5) and at the C-terminus (α_6, α_7) as well as a short 3_{10}-helix between I3 and J3 (α_4 in Fig. 1b). Helix α_2 connects domain 2 and domain 3.

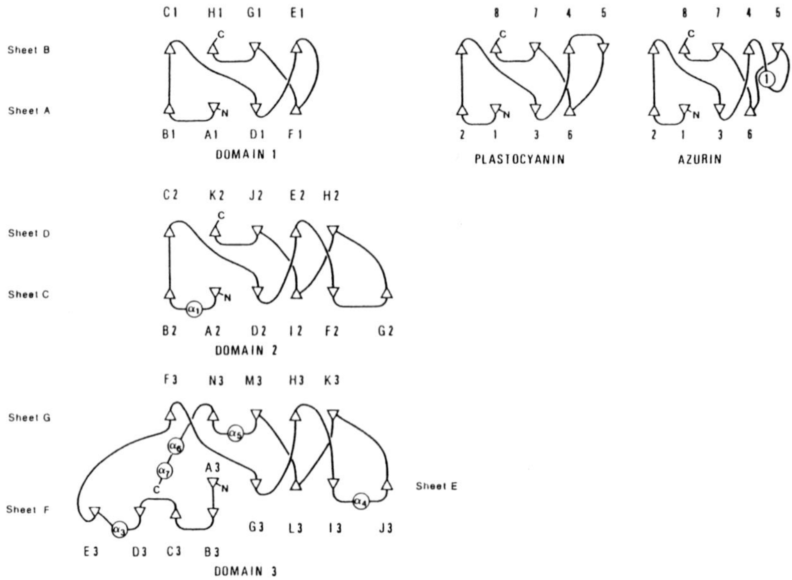

Fig. 2. A topology/packing diagram of the domains of the ascorbate oxidase monomer compared with plastocyanin and azurin. Each β-strand is represented by a triangle whose apex points up or down depending on whether the strand is viewed from the C or N terminus. α-Helices are represented by circles.

A comparison of the different variants of the β-barrel domain structure in Fig. 1 shows that domain 1 of ascorbate oxidase has the simplest β-barrel with only two four-stranded β-sheets. Plastocyanin and azurin are quite similar but between strands 4 (E1) and 6 (F1) they have insertions of one strand (plastocyanin) or one strand and an α-helix (azurin). Domain 2 has one additional strand H2 in sheet D next to strand E2 (sheet B and strand E1 in domain 1) and two additional strands, F2 and G2, in sheet C next to strand I2 (sheet A and strand F1 in domain 1). Domain 3 resembles domain 2 except for the insertion of the short α-helices and the addition of the four-standed β-sheet at its N-terminus.

The mononuclear copper site is located in domain 3 and the trinuclear copper species is bound between domain 1 and domain 3 (see Fig. 1a). The copper site geometries will be discussed later.

Each monomer exhibits three disulfide bridges. These are between domain 1 and domain 2 (Cys19-Cys201), domain 1 and domain 3 (Cys81-Cys538) and within domain 2 (Cys180-Cys193). Three putative attachment sites for N-glycosidicly-linked carbohydrate moieties are present in the amino acid sequence of ascorbate oxidase from zucchini (Asn92-Phe93-Thr94; Asn325-Phe326-Thr327; Asn440-Leu441-Ser442), but only Asn92 shows density for an N-acetyl-glucosamine group.

2.3. *Secondary Structure and Tetramer Contact Surface Areas*

Secondary structure was assigned on the basis of main-chain hydrogen bonding using the algorithm of Kabsch and Sander[10] with a cut-off energy of 0.5 kcal/mol for the definition of a hydrogen bond interaction. Ramachandran plots of the refined models for subunit A and B show that all non-glycine residues lie in or close to energetically allowed regions. All residues with positive Φ-values are located in turns of the polypeptide chain.

There are six α-helices in the monomer of ascorbate oxidase involving a total of 37 residues or 6.7 % of the polypeptide. In addition there are one longer and three short 3_{10}-helices. The average conformation angles of these secondary structures for subunit A and B have been calculated and are for both α-helices and 3_{10}-helices close to those reported by Barlow and Thornton.[11] Exceptions are α_3 and α_5, which have one $i \rightarrow i+5$ bond at their C-terminal end to give a short length π-helix.

All turns have been classified according to Crawford *et al.*[12] and Richardson.[13] One subunit exhibits 26 reverse turns (10 type I, 5 type I′, 10 type II, 1 type II′), 3 reverse turns associated with α-helices (3 type I), 3 near reverse turns (1 type 1, 2 type 2), and 3 open turns. The preference

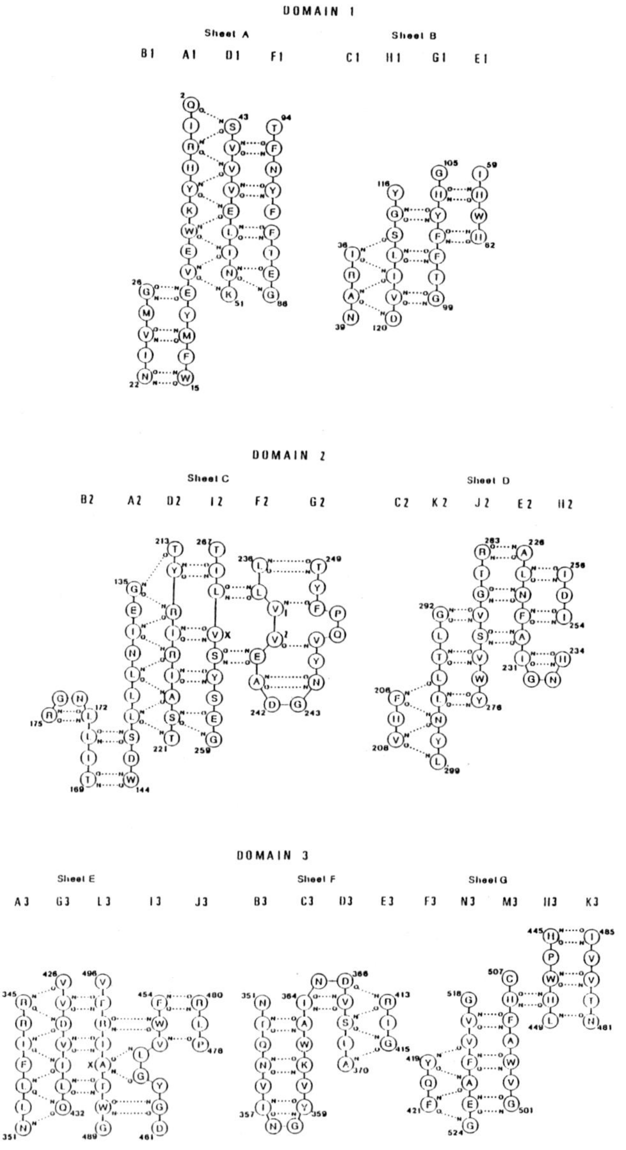

Fig. 3. Main-chain hydrogen bonding scheme of the 7 β-sheets in ascorbate oxidase. (a) Domain 1. (b) Domain 2; wide β-buldge: $x \rightarrow 264$; $1 \rightarrow 238$; $2 \rightarrow 239$. (c) Domain 3; classic β-buldge: $x \rightarrow 492$; $1 \rightarrow 457$; $2 \rightarrow 458$. The amino acid residues are indicated by the one-letter code.

of glycine at position 3 in the turn is very pronounced. There are three cis-proline turns[14] and one Asx turn[15] which displays conventional geometry. The cis-prolines have no unusual function in the structure. The residues involved in the seven β-sheet structures and the hydrogen bonding patterns are shown in Fig. 3. All sheets exhibit the characteristic right-handed twist when viewed along the strand direction. There is a "wide" β-bulge[13] in β-sheet C of domain 2 (Fig. 3b) at residues Val 238 and Val 239. β-Sheet E of domain 3 (Fig. 3c) contains a classic β-bulge[13] at residues Leu 457 and Gly 458.

Contact surface areas for the monomer–monomer interactions within the homo-tetramer present in both crystal forms were calculated using the algorithm of Lee and Richards.[16] It is evident from these calculations[9] that the contact surface areas between the two dimers related about the crystallographic dyad are by far largest. Ascorbate oxidase from zucchini occurs in solution as a dimer, which is likely the dimer mediated by the crystallographic dyad. The other dimers are related by the local dyad and nearly identical. The tetramer is a dimer of dimers, which is related by a local dyad.

2.4. *Copper Site Geometries*

The mononuclear copper site is located in domain 3 and has the four canonical type 1 copper ligands (His, Cys, His, Met) also found in plastocyanin and azurin. It is coordinated to the ND1 atoms of His445 and 512, the SG atom of the Cys507 and the SD atom of Met517 in a distorted trigonal pyramidal geometry. The SD atom is at the long apex (see Fig. 4). Bond lengths of the type-1 copper for both subunits are displayed in Table 1. They are compared with oxidized poplar plastocyanin[17] and azurin from *Pseudomonas aeruginosa*.[18] Figure 5 shows an overlay of the type-1 copper site in azurin, plastocyanin, and ascorbate oxidase.

The copper is penta coordinated in azurin (glycine) with the main-chain carbonyl oxygen of the residue preceding the first histidine copper ligand on the polypeptide chain. This carbonyl oxygen forms the fifth ligand. In poplar plastocyanin, the homologous peptide is a proline which is the beginning of a turn, causing the carbonyl oxygen to be removed to a distance of 3.8 Å from the type-1 copper ion. In ascorbate oxidase, the corresponding strand is extended, moving the carbonyl oxygen to a distance of 4.8 Å away from the type-1 copper ion. This extended strand contributes Glu 443 to the formation of the binding site of the reducing substrate.[9]

Table 1. Copper–copper and copper–ligand distances in ascorbate oxidase compared with copper–ligand distances for the type-1 copper site in plastocyanin and azurin.

		Distance Å		Azurin[a]		
Atom 1	Atom 2	A	B	wtp5	wtp9	Plastocyanin[b]
CU1	CU2	12.20	12.20			
CU1	CU3	12.73	12.65			
CU1	CU4	14.87	14.86			
		Type-1-copper site				
CU1	ND1 445[c]	2.10	2.12	2.11	2.09	2.04
CU1	SG 507	2.13	2.03	2.25	2.26	2.13
CU1	ND1 512	2.05	2.11	2.03	2.04	2.10
CU1	SD 517	2.90	2.83	3.15	3.12	2.90
CU1	O 444	4.78	4.83	2.97	2.95	3.82
		Trinuclear copper site				
CU2	CU3	3.68	3.73			
CU2	CU4	3.78	3.90			
CU3	CU4	3.66	3.69			
CU2	NE2 106	2.16	2.17			
CU2	NE2 450	2.06	2.09			
CU2	NE2 506	2.07	2.09			
CU3	ND1 62	1.98	2.13			
CU3	NE2 104	2.19	2.14			
CU3	NE2 508	2.14	2.08			
CU4	NE2 60	2.00	2.02			
CU4	NE2 448	2.09	2.05			
CU2	OH1	2.00	2.00			
CU3	OH1	2.06	2.00			
CU4	OH3	2.02	2.03			

[a] The wtp5 and wtp9 recombinant wild-type structures are from *Pseudomonas aeruginosa* azurin at pH 5.5 and 9.0. The data are from Nar *et al.*[18]
[b] Plastocyanin is from poplar leaves. The data are from Guss and Freeman.[17]
[c] Ascorbate oxidase numbering.

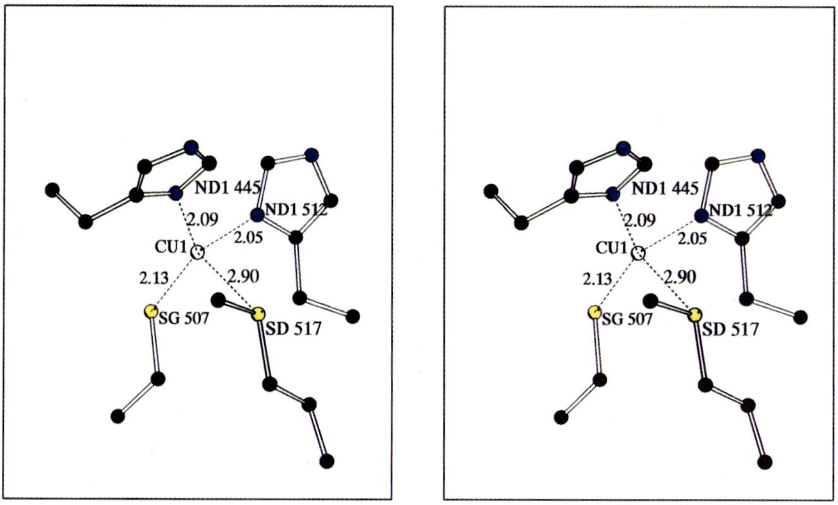

Fig. 4. Stereo drawing of the type-1 copper site in domain 3. The displayed bond distances are for subunit A. The figure was made with program MOLSCRIPT.[8a]

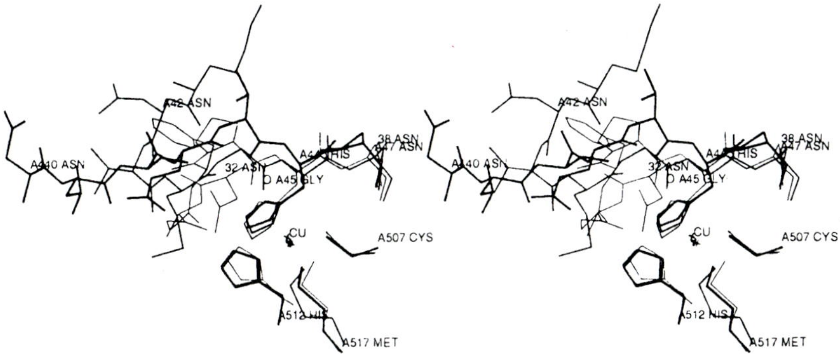

Fig. 5. Stereo plot of the overlay of the type-1 copper sites for poplar plastocyanin (thin line), *Pseudomonas aeruginosa* azurin (medium thick line), and ascorbate oxidase (thick line).

The trinuclear copper site (see Fig. 6) has eight histidine ligands symmetrically supplied by domain 1 and domain 3 and two oxygen ligands. Seven histidines are ligated by their NE2 atoms to the copper ions, whereas His62 is ligated to CU3 by its ND1 atom. In the preliminary structural report on ascorbate oxidase[8] based on lower resolution data, all eight histidine

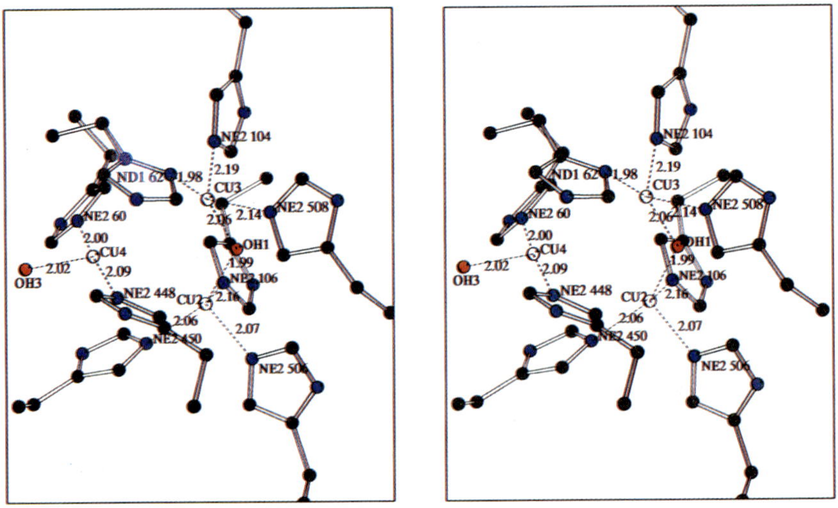

Fig. 6. Stereo drawing of the trinuclear copper site. The displayed bond distances are for subunit A. The figure was made with program MOLSCRIPT.[8a]

residues were modeled with their NE2 atoms as ligands to the copper atoms of the trinuclear site. The high resolution data allow an unequivocal interpretation. His62 is part of a β-sheet, whereas His450 is not. An overlay of the relevant parts of domain 3 onto domain 1 show that His62 comes closer with its main-chain atoms to the copper ion CU3 than the corresponding atoms of His450. As consequence, the side chain of His62 has to adopt a conformation where ND1 of the imidazole ring is ligated to the copper ion CU3.

The trinuclear copper site may be subdivided into a pair of copper (CU2, CU3) with six histidine ligands in a trigonal prismatic arrangement. The pair is bridged by an OH^- or O^{2-}, which leads to a strong antiferromagnetic coupling and makes this copper pair EPR silent. The pair very likely represents the spectroscopic type-3 copper. The remaining copper (CU4) has two histidine ligands and an OH^- or H_2O ligand and probably represents the spectroscopic type-2 copper. Both oxygen ligands could be clearly detected from difference Fourier maps. An oxygen ligand in the center of the three copper ions could not be detected.

Cole *et al.*[19] studied the electronic structure of the laccase trinuclear copper active site by the use of absorption, circular dichroism and low-temperature magnetic circular dichroism spectroscopies. The assigned ligand field transition energies indicated that all three coppers have tetragonal geometries and that the two type-3 coppers are inequivalent.

The latter interpretation fits well with our structural model as CU3 of the copper pair is ligated to one ND1 and two NE2 atoms, but the former interpretation, tetragonal coordination geometries for all three coppers, is not consistent with the structure. The coppers of the pair are both tetrahedrally coordinated, whereas the type-2 copper has three ligands. The existence of a central oxygen ligand would give rise to a penta coordination of both copper pair atoms (but not a tetragonal-pyramidal coordination) and a square-planar coordination for the spectroscopic type-2 copper. Messerschmidt *et al.*[9] examined by modeling whether, by side chain torsion angle rotations, His62 and His450 could become bridging ligands between CU4 and CU2 or CU3 with both ring nitrogens as ligating atoms but found this stereochemically unreasonable.

The copper–copper and copper-ligand distances of the trinuclear copper site are displayed in Table 1 for both subunits. Their mean values (Cu-N, 2.09 Å ($\sigma = 0.06$ Å); Cu-O, 2.02 Å ($\sigma = 0.02$ Å)) are comparable to those of binuclear copper model compounds with nitrogen and oxygen copper ligands[20,21] in the range of the accuracy of 0.1 Å for the copper ligand bonds. There is no long Cu-N bond (about 2.7 Å) as found in the binuclear copper center of hemocyanin from *Panulirus interruptus* determined by X-ray crystallography at 3.2 Å resolution.[22] The binuclear copper center in hemocyanin has a trigonal antiprismatic arrangement of the six histidine ligands, whereas the type-3 coppers in ascorbate oxidase show the trigonal prismatic coordination. The presence of the type-2 copper and its ligands would not allow the former arrangement.

The average copper–copper distance in the trinuclear copper site of ascorbate oxidase is 3.74 Å ($\sigma = 0.08$ Å) and the individual distances do not deviate by more than 0.16 Å from this mean value. The average copper–copper distance in hemocyanin is 3.54 Å.[22] The copper–copper distances are too long for copper–copper bonds but magnetic interactions are possible.

The shortest distance between the type-1 copper center and the trinuclear copper center is 12.2 Å. The His506-Cys507 His508 amino-acid sequence segment links the type-1 copper center and the type-3 copper center as a bridging ligand (see Fig. 7). Aspects of the intramolecular electron transfer between the two redox centers will be discussed in Sec. 10.

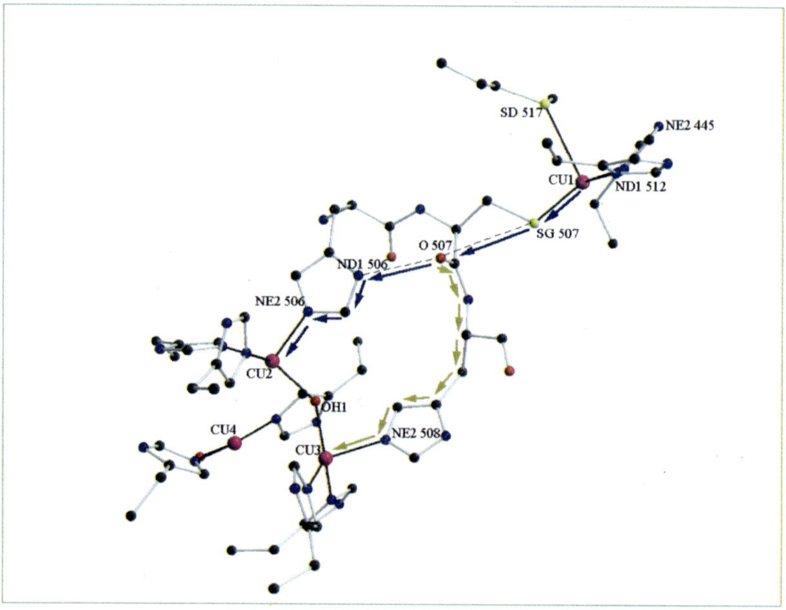

Fig. 7. Drawing of the region of the atomic model containing the type-1 copper center in domain 3 and the trinuclear copper center between domain 1 and domain 3. The figure was made with programs MOLSCRIPT[8a] and RASTER3D.[8b]

3. Structural Relationship Between the Blue Copper Oxidases

3.1. *Amino Acid Sequence Alignment*

Knowledge of the three-dimensional structure of ascorbate oxidase and of amino acid sequences of all members of the blue oxidases made it possible to carry out a structurally based amino acid sequence alignment[23] involving *Neurospora crassa* laccase, cucumber ascorbate oxidase and human ceruloplasmin. The structural basis of this alignment was the spatial superposition of the three domains of ascorbate oxidase, which allowed recognition of the conserved and variable structural elements within the single domains.

The amino acid sequence similarity of the blue oxidases supports a domain structure of ceruloplasmin and laccase. Ceruloplasmin has six and laccase three domains, which are related to the domains of ascorbate oxidase. Three domains have variable lengths over 125–200 residues but show a common β-barrel motif in which β-strands and turns may be superimposed in all domains. The occurrence of β-barrel motifs in laccase and ceruloplasmin are solely based on sequence homology with ascorbate oxidase. The variable parts of the ascorbate oxidase domains are mainly in connecting segments and correspond to the regions of weak similarity in the aligned sequences. Similarity is high in segments of the conserved secondary structure. These conserved elements include the copper ligands.

The comparison of the amino acid sequences of laccase and ascorbate oxidase shows that they are more closely related among each other than to ceruloplasmin. Fungal *Neurospora crassa* laccase contains all copper ligands of ascorbate oxidase, except methionine of type-1 copper. It forms a disulphide bridge (116-537) between domain 1 and domain 3 as ascorbate oxidase (disulphide bridge 83-541). The spatial structure of laccase is expected to be very similar to that of ascorbate oxidase, including the type-1 copper in domain 3 and the trinuclear copper site with its ligands between domain 1 and domain 3. Ascorbate oxidase is a symmetrical dimer of two laccase-like subunit. Dimerization is promoted by a metal ion, whose nature is not yet established. This metal is located on the molecular dyad and has two histidine ligands (His286) supplied symmetrically from each subunit.[8] The existence of six structurally related domains in ceruloplasmin is supported by the known disulphide bridges. SS-bridges 155-188, 515-541, and 855-881 are similar to 100-124, 273-302, and 503-527 in ascorbate oxidase according to the amino acid sequence alignment.[23] The corresponding positions in ascorbate oxidase in domains 1, 2, and 3 are in distances of about 5.0 Å so that disulphide bridges may be formed. A similar disulphide bridge (300-334) can be formed in domain 2 of *Neurospora crassa* laccase.[23] Two further SS-bridges in ceruloplasmin, 257-338 and 618-699, are similar to 41-124 and 424-525 of ascorbate oxidase. Also these disulphide bridges can be formed easily.

The tryptic cleavage sites for ceruloplasmin are at the end of the domains but approximately follow the internal triplication. A fourth cleavage site is in the middle of domain 3 in an extended and variable region of the secondary structure. The tryptic cleavage sites are in accord with the six domain structure. The X-ray structure of human ceruloplasmin described in the next chapter shows that this prediction is correct.

Human ceruloplasmin displays sequence similarities to the human blood clotting factors VIII[24] and V.[25] The amino acid sequences of the mature proteins of human factors VIII and V consist of 2332 and 1995 residues, respectively, and are similar (33 % identity and 53 % similarity when aligned with program BESTFIT.[26] Both coagulation factors exhibit three distinct types of structure domains, including a triplicated region of ~ 330 amino acid (A domain), a unique region of ~ 950 amino acid (the B domain) and a C-terminal duplicated region of ~ 150 amino acids (C domain), which are arranged in the order A1-A2-B-A3-C1-C2. The A domain of factor VIII and V show significant similarity to ceruloplasmin. The A domains can be subdivided each in two domains as in ceruloplasmin and may form similar disulphide-bridges as in ceruloplasmin because the corresponding cysteines are present. In factor VIII, A1 and A3 contain the canonical ligands for a type-1 copper at the homologous positions in the amino acid sequence. Factor V does not have the canonical type-1 copper ligands in none of the three A domains. The four His-X-His sequences, ligands to the trinuclear copper site, are not found in both factors, however. On the other hand, cysteine residues are present in domain A3 of factor V at the homologous position of ceruloplasmin. They form a disulphide bridge in this structure.

3.2. *Structural Model of the Copper Sites for Ceruplasmin*

Ceruloplasmin also contains the trinuclear copper site established for ascorbate oxidase. The 4 His-X-His sequences, providing the ligands of this copper site, are each two in domains 1 and 6 of ceruloplasmin.[23] Ligands for type-1 copper in suitable spacings on the polypeptide chain are in domains 2, 4, 6, assuming the nearly isosteric methionine and leucine as equivalent. *Neurospora crassa* laccase apparently forms a blue copper site with His, Cys, His, Leu at the canonical positions for the ligands of the type-1 copper site. The copper has three protein ligands, two His and one Cys, at this site, because the leucine, isosteric to methionine, cannot be considered as a copper ligand. It has a very high redox potential of 780 mV. Recently, it has been shown, using a recombinant Met-121-Leu mutant of azurin from *Pseudomonas aeruginosa*, that the methionine of the type-1 copper site is not required to generate the properties of this metal site.[27] The putative type-1 copper site in domain 2 of ceruloplasmin also has a leucine and is expected to have a high redox potential.

Redox titrations of ceruloplasmin, indeed, show two different standard redox potentials with values of 490 mV and 580 mV.[28] The reoxidation of the completely reduced enzyme by dioxygen as measured at 630 nm proceeds in two steps. About half of the blue colour returns fast (within milliseconds), whereas the complete reaction requires several hours. Therefore, it was proposed[29] that only one of the type-1 coppers is involved in the four-electron transfer to oxygen and thus forms a part of the active site. This copper has been called the "fast" type-1. The type-1 copper in domain 6, which may interact with the trinuclear copper, has been interpreted to be a "fast"species, explaining most of the fast reoxidation amplitude. The slight quantitative discrepancy may be due to experimental errors.[23] The assignment of the three type-1 copper centers was already proposed by Ortel *et al.*[30] and later by Moshkov *et al.*,[31] but the assignment of the ligands for the type-2 and type-3 coppers by Rydén[32] and later Moshkov *et al.*[31] is incompatible with the 3D-structure of the related blue oxidase ascorbate oxidase.

The six-domain structure and copper site model for ceruloplasmin received further support from the elucidation of the X-ray structure of a copper-containing nitrite reductase from *Achromobacter cycloclastes.*[33] This nitrite reductase occurs in solution and in the crystal as homotrimer. The trimer shows C3 symmetry. One monomer is built up by two domains of a similar β-barrel fold as found in the small blue copper proteins or in ascorbate oxidase (see Fig. 8a). The enzyme contains two mononuclear copper centers. A type-1 copper site with the canonical ligands (His, Cys, His, Met) is located in domain 1 and a type-2 copper with three histidines and one water molecule as ligands is bound between domain 1 and domain 2 of the adjacent symmetry-related molecule (see Fig. 8a, b). The copper-containing domains of ascorbate oxidase can be superimposed on the domains of nitrite reductase, which are displayed in Fig. 8b. The type-1 copper and its ligands in nitrite reductase fall at the type-1 center of ascorbate oxidase, and all three histidines of the type-2 site, as well as the copper, fall at the same site as the trinuclear cluster of ascorbate oxidase[33] (Fig. 8b, c). Sequence comparison[34] shows that two histidines from domain 1 (residues 100 and 135) and two from domain 2 (residues 255 and 306) correspond to four of the eight histidine ligands of the trinuclear copper cluster in ascorbate oxidase. In nitrite reductase only histidine residues 100, 135, and 306 are ligands. These comparisons suggest that the type-2 copper of nitrite reductase corresponds to one copper of the type-3 copper pair in ascorbate oxidase. In the trimer of nitrite reductase a six-domain structure is realized, which is reminiscent of the six-domain structure of ceruloplasmin,[35] which was deduced from the amino acid sequence alignment with the other blue oxidases.[23] However, the arrangement of the six gene segments in ceruloplasmin is not simply a triplication of an ancestral nitrite reductase gene coding for two domains, but the triplication of a gene in which the segments coding for the individual domains were inverted. This is due to the finding that the type-1 copper sites in ceruloplasmin are found in domains 2, 4 and 6 rather than in domain 1 and symmetrical counterparts in nitrite reductase.

Fig. 8. (a) Drawing of the trimer of nitrite reductase from *Achromobacter cycloclastes.* (b) Drawing of the interface between domain 1 (subunit A) and domain 2 of the adjacent symmetry-related molecule (subunit C) of nitrite reductase from *Achromobacter cycloclastes.* (c) Drawing of domain 1 and 3 of ascorbate oxidase. The type-1 copper is in domain 3 and the trinuclear copper center is between domains 1 and 3. The domains have an orientation similar to that of the corresponding domains of nitrite reductase shown in (b). The figures were produced by the program RIBBON.[8c]

4. The Blue Oxidases and Related Proteins

The genes of a number of fungal laccases and of a plant laccase have been cloned and sequenced. These comprise the sequences of the basidiomycetes *Phlebia radiata*,[36] *Coriolus hirsutus*,[37] the oyster mushroom *Pleurotus ostreatus*,[38] the mushroom *Agaricus bisporus*[39] and the ligninolytic basiomycete PM1,[40] of the ascomycetes *Neurospora crassa*[41] and the chestnut blight fungus *Cryphonectria parasitica*,[42] of the deuteromycete *Aspergillus nidulans*[43] and of the yeast-like fungus *Filobasidiella neoformans*[44] as well as of the sycamore maple tree *Acer pseudoplatanus*.[45] Primary structures of ascorbate oxidase from cucumber, *Cucumis sativus*,[46] pumpkin, *Cucurbita* sp. Ebisu Nankin,[47] and zucchini, *Cucurbita pepo medullosa*,[48] have been reported. Furthermore, amino acid sequences of five different proteins that are related to the blue oxidases have been reported. The first one is phenoxazinone synthase from *Streptomyces antibioticus*.[49] This enzyme is a copper-containing oxidase that catalyzes the coupling of 2-aminophenols to form the 2-aminophenoxazinone chromophore.[50] This reaction constitutes the final step in the biosynthesis of the potent antineoplastic agent actinomycin. The next two are the gene products of a copper resistance gene of *Pseudomonas syringae* pv. *tomato*[51] and *Xanthomonas campestris* pv. *juglandis*.[52]

These proteins are believed to function as copper-binding proteins, thus making a contribution to the copper resistance. The next one is a product of the Bp10 gene of *Brassica napus* (rape).[53] The expression of the Bp10 gene family is pollen-specific and developmentally regulated. Finally, the FET3 gene of *Saccharomyces cerevisae*[54] encodes a multi-copper oxidase that is required for ferrous iron uptake and exhibits ferroxidase activity.

A multiple amino-acid sequence alignment of these related proteins including the sequences of the domains 1, 2, and 6 of human ceruloplasmin and omitting the product of the copper resistance gene of *Xanthomonas campestris* pv. *juglandis*, which is very similar to that of *Pseudomonas syringae* pV. *tomato*, has been carried out and is shown in Table 2. The basis of this alignment was the three-dimensional structure of ascorbate oxidase. All protein copper ligands characteristic for the blue oxidases are conserved with two principal exceptions. The rape Bp10 gene product has only one histidine of the copper ligands conserved, meaning that this protein will bind no copper and therefore will have no oxidase activity.[53] The second exception concerns the position for a possible fourth ligand of the

Table 2. Structurally based amino-acid sequence alignment of the blue oxidases and related proteins.

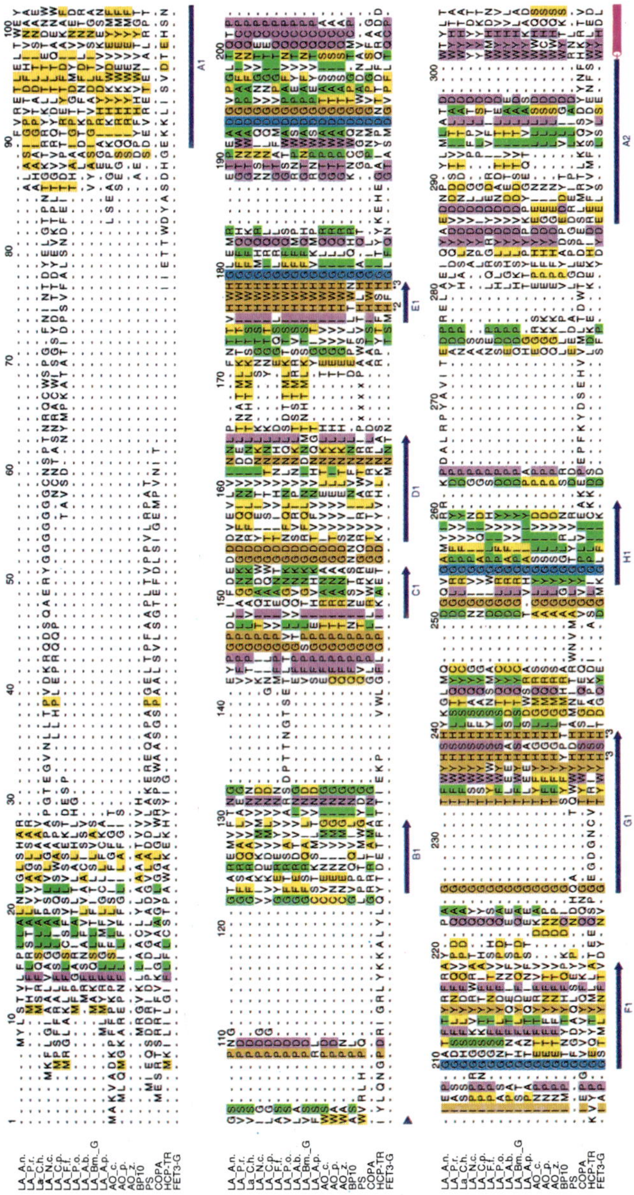

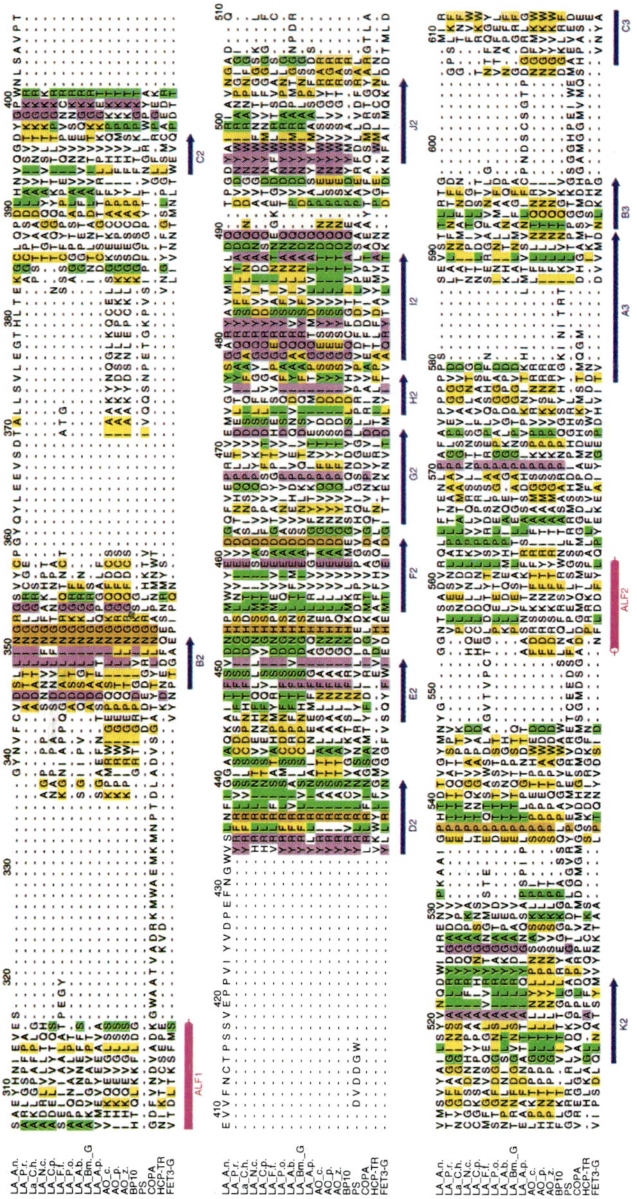

Table 2. (*Continued*)

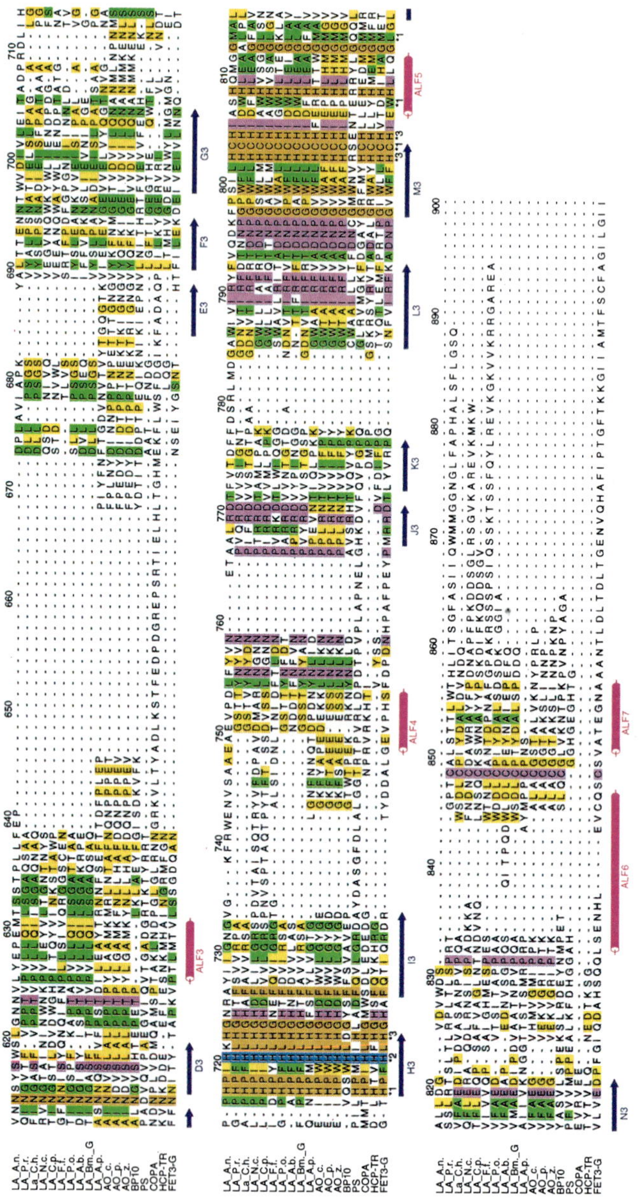

Table 2. *(Continued)*

Abbreviations for Table 2: LA_A.n.–laccase from *Aspergillus nidulands*;[43] LA_P.r.–laccase from *Phlebia radiata*;[36] LA_C.–laccase from *Coriolus hirsutus*;[37] LA_N.c.–laccase from *Neurospora crassa*;[41] LA_C.p.–laccase from *Cryphonectria parasitica*;[42] LA_F.f.–laccase from *Filobasidiella neoformans*;[44] LA_P.o.–laccase from *Pleurotus ostreatus*;[38] LA_A.b.–laccase from *Agaricus bisporus*;[39] LA_Bm._G.–laccase from ligninolytic basidiomycete PM1;[40] LA_A.p.–laccase from *Acer pseudoplatanus*;[45] AO_c.–ascorbate oxidase from cucumber–*Cucumis sativus*;[46] AO_p.–ascorbate oxidase from pumpkin, *Cucurbita* sp. Ebisu Nankin; [47] AO_z.–ascorbate oxidase from zucchini, *Cucurbita pepo medullosa*;[48] BP10–*Brassica napus* Bp10 gene product;[53] PS–phenoxazinone synthase from *Streptomyces antibioticus*;[49] COPA–copper binding protein from the copper resistance gene from *Pseudomonas syringae* pv *tomato*;[51] HCP-TR–human ceruloplasmin, [53a] domains 1, 2 and 6; FET3-G–multi-copper oxidase product of the FET-3 gene from *Saccharomyces cerevisae*.[54] The secondary structure elements, known from the three-dimensional structure of ascorbate oxidase, are given at the bottom of each alignment block. *1, ligand to type-1 copper; *2, ligand to type-2 copper; *3, ligand to type-3 copper. Color code: blue, 18 identities; brown, 15–17 identities; turquise, 10–15 identities; green, 6–9 identities; yellow, 4–5 identities. The alignment was formatted by the program ALSCRIPT.[54a]

type-1 copper. As mentioned above, there is very often a methionine at this place in small blue copper proteins as well as in blue oxidases. This methionine serves as a weak ligand to the type-1 copper. In *Aspergillus nidulans* and the maple laccases, the three ascorbate oxidases, phenoxazinone synthase, the copper-binding protein from the *Pseudomonas syringae* copper resistance gene and human ceruloplasmin, there is a methionine at this position. Similar to *Neurospora crassa* laccase, *Phlebia radiata*, *Cryphonectria parasitica*, *Pleurotus ostreatus*, *Agaricus bisporus* laccases and the product of the *Saccharomyces cerevisae* FET-3 gene also contain a leucine at this place. The explanations of the *Neurospora crassa* laccase related to the spectral and electrochemical properties of such a type-1 site should be valid as well. *Coriolus hirsutus* and the basidiomycete PM1 laccases hold a phenylalanine at this position. This residue also cannot serve as a copper ligand. The role of the residue located in this position seems to tune the redox potential of the type-1 copper site which is necessary for the function of the enzyme. Very unusual is the lysine at this position in the laccase of the yeast-like fungus *Filobasidiella neoformans*, but lysine can be a metal ligand in its neutral form and the necessary pK-shift (pK_A of lysine ~ 10.5) can easily be caused by the protein environment.

An inspection of this amino acid sequence alignment shows that the sequences align very well in the segments of the conserved copper ligands and quite well in segments of defined secondary structure elements of the ascorbate oxidase structure.

5. Evolution of the Blue Oxidases and Related Proteins from a Common Ancestor

Dwulet and Putnam[55] proposed that ceruloplasmin had evolved from a fused azurin gene and, more recently, Ryden[56] speculated about an evolution of the blue oxidases from a common ancestor, suggesting that an ancestral gene was first duplicated. The simple oxidases like laccase or ascorbate oxidase acquired than a third genetic element while a triplication of the duplicated gene lead to ceruloplasmin. This evolution scheme can be checked by the calculation of similarity matrices based on the amino acid sequence alignment given in Ref. 23 and Table 2. The program ALNED[57] was used to calculate the similarity matrices. Table 3 displays a similarity matrix computed for the amino-sequence alignment shown in Table 2. The calculations included all residues from alignment positions 83 to 888 and represent the pairwise differences in percent between the amino-acid sequences.

The similarity matrix calculated by Messerschmidt and Huber[23] (not shown here) indicates clearly the 6-domain structure of ceruloplasmin and 3-domain structures for laccase and ascorbate oxidase. The internal triplication within the ceruloplasmin amino acid sequence is reflected by values of about 60 percent difference. The comparison of both the N-terminal domains and the C-terminal domains of the blue oxidases indicates, respectively, a relationship that is closer and relevant values of percentage difference that are significantly lower than those for other comparisons. This might reflect the requirements for the trinuclear copper site. The lowest values of about 70 to 73 % difference are observed for both N-terminal and C-terminal domains of laccase and ascorbate oxidase, showing that the two oxidases are more closely related than either of them to ceruloplasmin.

Table 3 supplies insight into the relatedness within the blue oxidases and to the other included proteins. The laccases of the narrowly related basiodomycetes *Phlebia radiata*, *Coriolus hirsutus*, *Pleurotus ostreatus*, the basidiomycete PM1 and *Agaricus bisporus* show differences from about 18 to 38 % for the first four of them and of about 55 % to the *Agaricus bisporus* laccase. The differences for the first four of them below 40 % reflect that these fungi belong to the same taxonomic order of *Aphylophorales*.

Table 3. Similarity matrix for the amino acid sequence in Table 2.

		1	2	3	4	5	6	7	8	9	10	11	12	13	14	15	16	17	18
1	LA_A.n.	0.00	74.48	74.38	77.05	77.73	74.95	76.20	76.15	75.47	80.98	76.81	77.14	77.18	82.86	87.82	81.29	83.84	78.17
2	LA_P.r.		0.00	32.49	73.04	70.51	72.52	39.15	55.86	32.68	75.60	72.39	72.63	72.43	79.11	88.24	81.11	83.59	73.54
3	LA_C.h.			0.00	69.46	69.84	70.75	33.59	53.39	19.34	75.15	72.08	72.62	72.46	81.35	88.42	80.37	83.62	71.99
4	LA_N.c.				0.00	42.32	77.18	70.12	72.01	70.20	78.04	77.00	76.92	76.57	82.50	86.81	81.29	83.37	76.46
5	LA_C.p.					0.00	75.63	71.37	73.35	69.58	75.39	73.43	72.94	73.29	81.08	85.80	81.25	83.16	73.78
6	LA_F.f.						0.00	71.46	71.31	71.43	77.31	71.48	71.72	71.96	81.75	86.02	80.38	81.34	74.32
7	LA_P.o.							0.00	53.29	35.42	74.70	72.42	73.90	72.82	82.65	88.56	79.12	83.30	70.83
8	LA_A.b.								0.00	54.89	70.22	71.94	72.89	72.50	80.71	87.07	81.61	81.43	73.19
9	LA_Bm._G									0.00	74.10	72.76	72.51	72.35	80.04	88.19	80.28	83.37	71.69
10	LA_A.p.										0.00	72.01	72.90	72.76	77.93	86.69	82.75	83.37	77.35
11	AO_c.											0.00	21.42	18.33	71.54	83.66	81.20	82.19	73.76
12	AO_p.												0.00	2.55	71.05	81.48	80.33	82.29	74.49
13	AO_z.													0.00	70.70	80.70	79.82	82.57	74.89
14	BP10														0.00	87.73	84.20	86.85	81.29
15	PS															0.00	84.33	87.97	86.60
16	COPA																0.00	85.88	78.87
17	HCP-TR																	0.00	83.95
18	FET3-G																		0.00

The calculations include all residues from alignment positions 83 to 888 and represent the pairwise differences in percent between the amino-acid sequences.

The low difference of about 18 % between the *Coriolus hirsutus* and the PM1 laccases indicates that they should belong to the same taxonomic family. The higher values for the *Agaricus bisporus* laccase demonstrates its membership to another taxonomic order, namely to the *Agaricales*. The differences for the laccases from the ascomycetes *Neurospora crassa* and *Cryphonectria parasitica* are about 40 %. The laccase of the deuteromycete *Aspergillus nidulans* shows differences between 74 and 80 % to all fungal, fungal-like and plant oxidases. The differences are even greater to the bacterial copper oxidases and the Bp10 rape gene product.

The maple tree laccase is related to the fungal, fungal-like and plant proteins in a range of about 70 to 80 % differences and to the bacterial proteins with higher values. As expected, the ascorbate oxidases are very closely related to each other. The bacterial proteins phenoxazinone synthase and the copper resistance gene product from *Pseudomonas antibioticus* are very distant from the other proteins included in Table 3. The rape Bp10 gene product exhibits significant similarity to ascorbate oxidase (about 72 % differences). The relation of the copper oxidase product of the yeast FET3 gene to the other partners of the alignment is comparable to that of the deuteromycete *Aspergillus nidulans* to the others. Human ceruloplasmin exhibits approximately the same differences of about 85 % to all other alignment partners reflecting the far phylogenetic distance of this mammalian protein to the other compared gene products.

The structurally based amino-acid sequence alignments and the analysis of the published[23] and presented similarity matrices suggest an evolutionary scheme for this group of proteins, which is shown in Fig. 9. The existence of a quasi-symmetrical arrangement of the N-terminal and C-terminal domains strongly suggests evolution by a duplication of a gene encoding a single protein domain of an ancestral small blue copper protein. The two domains in tandem might have been functioned as a quasisymmetrical oxidase with two type-1 copper sites after acquisition of a trinuclear copper site. The insertion of a third "single domain" genetic element and the subsequent loss of type-1 coppers in domains 1 and 1′ may have led to the bacterial proteins phenoxazinone synthase and the copper resistance gene product from *Pseudomonas syringae*, to the fungal laccases and yeast FET3 gene copper oxidase product, to the plantal ascorbate oxidases, laccases and the rape Bp10 gene product. The rape Bp10 gene product lost all coppers during its evolution.

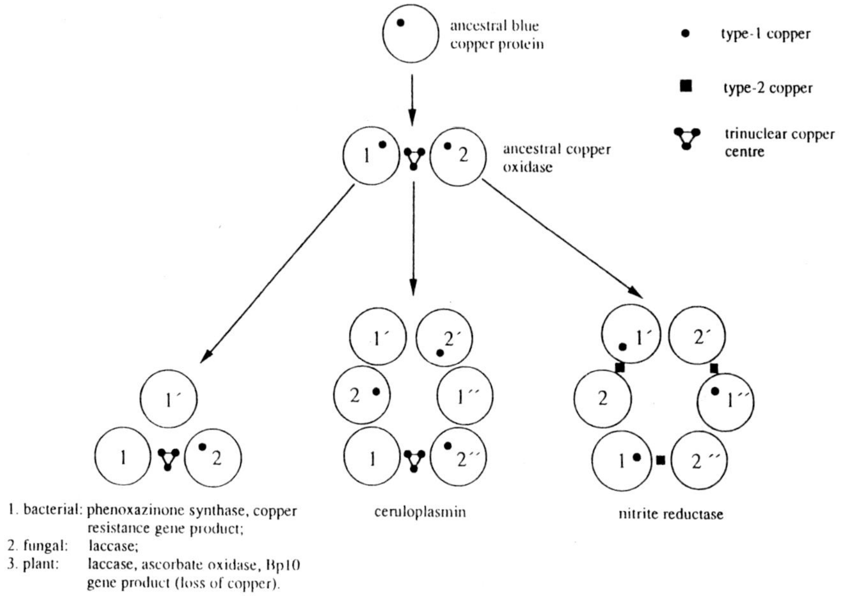

Fig. 9. Proposal of an evolutionary scheme of the blue oxidases and related proteins.

The pronounced three-fold repeat in the six-domain molecule ceruloplasmin contradicts evolution by a duplication of an ascorbate oxidase-like molecule and suggests evolution from the tandem two-domain molecule by triplication possibly before acquisition of the trinuclear copper site. This might have occurred later between the N- and C-terminal domains.

The copper-containing nitrite reductase from *Achromobacter cycloclastes* may also have evolved from this ancestral oxidase. Nitrite reductase is a two-domain protein that functions as trimeric molecule. During its evolution from the ancestral copper oxidase, domain 2 of the ancestral oxidase lost its type-1 copper and one copper of the trinuclear copper site has been conserved only.

6. Oxidation-Reduction Potentials

Redox potentials for different copper centers in the blue oxidases have been determined for all members of the group but in each case only for a limited number of species. They will be discussed in another chapter of this

Table 4. Selection of oxidation-reduction potentials of the copper center of blue copper oxidases.

Protein	pH	Potential, E^0 (mV)			Reference
		Type-1	Type-2	Type-3	
Laccase					
Fungal					
Polyporus	5.5	785	—	782	57a
+1 mM NaF	5.5	780	—	570	
Tree					
Rhus	7.5	394	365	434	57a, 57b
+10 mM NaF	7.5	390	390	390	
$+Fe(CN)^{4-}$	7.5	434	—	483	
Ascorbate oxidase					
Cucurbita pepo medullosa	7.0	344[a]	—	344[a]	59
	7.0	327[b]	—	357[b]	
Cucumis sativus	6.0	350	—	—	59a
Ceruloplasmin					
Human	5.5	490, 580	—	—	28

[a], bei 25° C.
[b], bei 10° C.

book. Some data, which is of relevance for the material of this chapter, is summarized in Table 4. The redox potentials for the type-1 copper of tree laccase and ascorbate oxidase are in the range of 330–400 mV and comparable to the values determined for the small blue copper proteins plastocyanin, azurin and cucumber basic protein (for redox potentials of small blue copper protein see the review of Sykes[58]). The high potential for the fungal *Polyporus* laccase is probably due to a leucine or phenylalanine residue at the fourth coordination position, which have been observed in the amino acid sequences of fungal laccase from other species (see Table 2 and discussion in Sec. 3.2). Two different redox potentials for the type-1 copper were observed for human ceruloplasmin.[28] The 490 mV potential can be assigned to the two type-1 copper sites with methionine ligand and the 580 mV potential to the type-1 center with the isosteric leucine at this position (see Sec. 3.2). The redox potentials for the type-3 centers are equal to or somewhat higher than the respective type-1 potentials. For ascorbate oxidase a difference of +30 mV between the type-3 and type-1 potentials was observed, when they were determined at 10°C in contrast to 25°C, where no distinction was possible.[59] It is intriguing that the type-3 potential in fungal laccase is also greater, which is necessary for its enzymatic function (the reason for this will be discussed later).

It will be interesting to learn what structural changes, compared with tree laccase or ascorbate oxidase, are required to give this elevation of redox potential once the X-ray structures of a fungal and tree laccase have been determined. Redox potentials for the type-2 site have been reported only for tree laccase. They had to be determined from a redox titration of the type-2 EPR signal. The values obtained do not differ much from those of the other copper centers. Taking into account that an integral trinuclear copper site is present in the blue oxidases and that there are no separated type-2 and type-3 centers makes it more plausible to define a unique redox potential for this trinuclear copper species. The type-2 EPR signal and the 330-nm band are both spectroscopic signals of this copper species. Redox titrations using one of these signals should give the same results for the redox potential.

7. Kinetic Properties of Laccase and Ascorbate Oxidase

Numerous studies concerning the catalytic and binding properties of the blue oxidases are well documented in reviews for laccase,[60] for ascorbate oxidase[61] and for ceruloplasmin.[2] Important data, concentrating on laccase and ascorbate oxidase only, are summarized here.

Results from steady-state kinetics for laccase and ascorbate oxidase are given in Table 5. Listed are $k_{\rm cat}$, $K_{\rm m}$ and $k_{\rm cat}/K_{\rm m}$ values at different pH values. The $k_{\rm cat}/K_{\rm m}$ values indicate the low limit of the bimolecular enzyme-substrate reaction rate. The most reactive reducing substrate for tree laccase is hydroquinone with $k_{\rm cat} = 2.4 \times 10^2$ sec^{-1}, $K_{\rm m} = 2.0 \times 10^{-1}$ M and $k_{\rm cat}/K_{\rm m} = 1.2 \times 10^3$ M^{-1} sec^{-1} and for ascorbate with $k_{\rm cat} = 7.5 \times 10^3$ sec^{-1}, $K_{\rm m} = 2.0 \times 10^{-4}$ M and $k_{\rm cat}/K_{0m} = 3.8 \times 10^7$ M^{-1} sec^{-1}. The data do not show remarkable pH dependence.

A "ping-pong di Theorell-Chance" mechanism has been deduced for tree laccase from steady-state kinetics.[62] This mechanism is characterized by the sequential entry of the two substrates and the immediate release of the respective product. The actual catalytic cycle will consist of very

Table 5. Steady-state kinetics of laccase and ascorbate oxidase.

Enzyme	Substrate	pH	k_{kat} (sec^{-1})	K_m (M)	k_{cat}/K_m (M^{-1} sec^{-1})	Ref.
Laccase	Ascorbate	6.0	1.5	1.0×10^{-2}	1.5×10^{2}	65
Rhus	$Fe(CN)_6^{4-}$	6.0	2.0×10^{1}	5.0×10^{-2}	4.0×10^{2}	65
vernicifera	Ascorbate	7.5	8.2	5.9×10^{-2}	1.4×10^{2}	65
	Hydroquinone	7.5	2.4×10^{2}	2.0×10^{-1}	1.2×10^{3}	62
	Oxygen	7.5	1.3×10^{2}	2.1×10^{-5}	6.1×10^{6}	62
Rhus succedanea	Ascorbate	6.0	3.0×10^{1}	1.7×10^{-2}	1.8×10^{3}	65
	$Fe(CN)_6^{4-}$	6.0	2.2×10^{2}	2.2×10^{-2}	1.0×10^{4}	65
	Ascorbate	7.5	8.8×10^{1}	6.0×10^{-3}	1.5×10^{4}	65
	$Fe(CN)_6^{4-}$	7.5	7.8×10^{1}	1.1×10^{2}	7.1×10^{3}	65
Ascorbate oxidase	Ascorbate	6.0	7.5×10^{3}	2.0×10^{-4}	3.8×10^{7}	61,65
Cucurbita pepo	$Fe(CN)_6^{4-}$	6.0	2.5×10^{1}	3.0×10^{-3}	8.3×10^{3}	61,65
medullosa	Ascorbate	7.5		unaffected by pH		61,65
	$Fe(CN)_6^{4-}$	7.5	1.7	5.0×10^{-3}	8.5×10^{2}	61,65

different steps but the finding that the two substrates react in a ping-pong mechanism is important.

Transient kinetics such as stopped-flow, pulse radiolysis or laser flash photolysis has been applied to monitor the different steps during the catalytic cycle. In the presence of both substrates, complex patterns that are difficult to interpret have been obtained. Therefore, transient kinetics was carried out in the presence of only one substrate in each case. One process that has been investigated is the anaerobic reduction of the enzyme. A second is the reoxidation of the reduced enzyme by dioxygen. Results of the anaerobic reduction of laccase and ascorbate oxidase are given in Table 6. The reduction of the type-1 copper, monitored by the bleaching of the 600-nm band, is a bimolecular second-order reaction. The reaction is biphasic for tree laccase and ascorbate oxidase with hydroquinone and reductate as substrates, respectively, and it is monophasic in the other cases. The rate constants for tree laccase with hydroquinone as substrate are in the range of 2.0×10^2 to 2×10^4 M^{-1} sec^{-1} with ascending values from pH 6.5 to pH 8.5.[63] The values for fungal laccase are three to four orders of magnitudes higher.[64] The rates for ascorbate oxidase with reductate as substrate are about 10^4 M^{-1} sec^{-1} and exhibit no remarkable pH dependence.[59]

Table 6. Anaerobic reduction of laccase and ascorbate oxidase.

	Kind of type-1 copper reduction and rate (~610 nm)					Kind of type-2 and type-3 copper reduction and rate (~330 nm and/or reappearance of the 610-nm band)						
Enzyme	Substrate	pH	K_{init} (M^{-1} sec^{-1})	K_2 (M^{-1} sec^{-1})	Ref.	Substrate	pH	K_{init} (M^{-1} sec^{-1})	K_1 (sec^{-1})	K_2 (sec^{-1})	K_3 (sec^{-1})	Ref.
Laccase												
Fungal	Hydroquinone	5.5	1.7×10^7	—	64	Hydroquinone	5.5	—	—	—	1.0	64
Polyporus versicolor	$Fe(CN)_6^{-4}$	5.5	1.5×10^6	—	64	Ascorbate	5.5	—	—	—	1.0	64
						Inactive form already at pH 5.5; OH^- binds to type-2 copper						
Tree	Hydroquinone	6.5	3.6×10^2	1.6×10^2	63	Hydroquinone	6.5	4.0×10^2	—	—	—	63
Rhus vernicifera	Hydroquinone	7.4	1.6×10^3	8.0×10^2	63	Hydroquinone	7.4	1.5×10^3	—	—	0.4	63
	Hydroquinone	8.5	1.8×10^4	4.0×10^3	63	Hydroquinone	8.5	8.5×10^3	—	—	0.25	63
	$ArNO_2^{\cdot -}$	7.4	$< 2 \times 10^6$	—	65	$CO_2^{\cdot -}$	7.0	—	—	—	1.0	71
						Inactive form at high pH; OH^- binds to type-2 copper						
Ascorbate oxidase	Reductate	6.1	2.2×10^4	1.2×10^4	59	Reductate	6.1	—	100	—	—	59
Cucurbita perpo	Reductate	7.0	1.7×10^4	1.0×10^4	59	Reductate	7.0	—	100	—	—	59
medullosa	Reductate	7.8	1.4×10^4	0.8×10^4	59	Reductate	7.8	—	100	—	—	59
	Ascorbate	Much faster, reaction within the deadtime of the instrument			59	Rapid freeze-quench EPS experiments indicate that type-2 copper is reduced more slowly.						59
	$ArNO_2^{\cdot -}$	7.5	$1\text{–}3 \times 10^7$	—	66							
	Lumiflavin[a]	7.0	2.7×10^7	—	67							
		7.0	3.8×10^7	—	57a	Lumiflavin[a]	7.0	—	97	—	2.4	68
	$CO_2^{\cdot -}$	7.0	1.2×10^9	—	69	$CO_2^{\cdot -}$	7.0	—	201	20	2.3	69
		7.0	1.1×10^9	—	68		7.0	—	120	—	2.0	68
	$MV^{\cdot +}$	7.0	2.6×10^7	—	68	$MV^{\cdot +}$	7.0	—	127	—	2.3	68
	Deazaflavin[a]	7.0	1.5×10^8	—	68	Deazaflavin[a]	7.0	—	121	—	2.5	68

[a]Radical form

Reaction of one electron-reduced nitroaromatics ($ArNO_2^{\cdot-}$) gives rate contants of $< 2 \times 10^6$ M^{-1} sec^{-1} for tree laccase[65] and $1–3 \times 10^7$ M^{-1} sec^{-1} for ascorbate oxidase.[66] Ascorbate as a substrate for ascorbate oxidase should be at least as fast as the given rate constants since the reaction is completed within the deadtime of the stopped flow instrument, 3 to 5 ms.[59] Recent studies on ascorbate oxidase using laser flash photolysis[67] or pulse radiolysis[68,69] reveal rate contants for the radicals of lumiflavin, deazaflavin, $CO_2^{\cdot-}$ and $MV^{\cdot+}$ between 10^7 and 10^9 M^{-1} sec^{-1}. The pH dependence was not determined. The data suggests that the type-1 copper is the primary electron receptor from the reducing substrate. In the case of ascorbate oxidase with ascorbate as substrate it has been shown that one electron is transferred from ascorbate and that the generated semidehydroascorbate spontaneously dismutates in solution.[70]

The anaerobic reduction of the trinuclear copper species monitored by the bleaching of the 330-nm band, the reappearance of the 600-nm band and the disappearance of the type-2 EPR signal appear to be a multiphasic process in most cases. For fungal laccase, the process is monophasic with a unimolecular rate constant of 1.0 sec^{-1}.[64] Tree laccase with hydroquinone as a substrate[63] displays a pH dependence with a bimolecular reaction ($K_{init} = 4.0 \times 10^2$ M^{-1} sec^{-1}) at pH 6.5 and a two-phasic behavior at pH 7.4 and 8.5. The first reaction is a bimolecular second order reaction with increasing rate constants at ascending pH. The second one is a unimolecular reaction with rate constants of about 0.3 sec^{-1}. Pulse radiolysis studies on tree laccase with the $CO_2^{\cdot-}$ radical as a substrate at pH 7.0 reported a monophasic reaction ($K = 1.0$ sec^{-1}).[71] The disappearance of the type-2 EPR-signal has been determined for tree laccase by rapid freeze-quench experiments.[72] It is as fast as the reduction of the type-1 copper. Tree laccase is inactive at high pH. Because the type-2 EPR-signal is altered, it has been concluded that an OH^- binds to the type-2 copper, causing inactivation of the enzyme.

The anaerobic reduction of the trinuclear copper center for ascorbate oxidase with different substrates presents a distinct picture. The reaction with reductate is monophasic with a unimolecular rate constant of 100 sec^{-1} (Ref. 59) independent of pH. Rapid freeze-quench EPR experiments indicate that the type-2-EPR signal vanishes more slowly.[59] The pulse radiolysis studies of the radicals of lumiflavin, deazaflavin, $CO_2^{\cdot-}$ and MV^+ at pH 7.0[68,69] showed a biphasic behavior with an initial faster reaction ($K_1 = 97$–$127\ sec^{-1}$) and a final, slower reaction ($K_3 \sim 2\ sec^{-1}$).[68] Different results have been obtained by Farver and Pecht[69] with $CO_2^{\cdot -}$ radical as a substrate. They found a triphasic reaction with unimolecular rate constants $K_1 = 201\ sec^{-1}$, $K_2 = 20\ sec^{-1}$ and $K_3 = 2.3\ sec^{-1}$. The first constant is twice that in a study by Kyritsis *et al.*[68] whereas the third constant is identical. The second constant was not observed in the study.

The unimolecular reduction of the trinuclear copper center is caused by intramolecular electron transport from the type-1 copper to the trinuclear copper species. The turnover numbers represented by the k_{cat} values (see Table 5) are one to two orders of magnitudes greater than the observed intramolecular electron transfer rate constants. They should be at least as large as the turnover numbers. The low rates seem to imply that the enzymes are in an inactive form under the conditions of the anaerobic reduction experiments. It will be shown later that there is experimental evidence for this assumption.

The reoxidation studies on laccase and ascorbate oxidase are listed in Table 7. The reoxidation of the type-1 copper and of the trinuclear copper site occurs at a rate of $5 \times 10^6\ M^{-1}\ sec^{-1}$ both for tree laccase[73] and for ascorbate oxidase.[74] During reoxidation with H_2O_2, an O_2^{2-} intermediate is formed in several minutes, which is documented for tree laccase by changes in the CD-spectrum[75] and for ascorbate oxidase in the formation of an absorption band at 350 nm.[76] The intermediates are stable for several hours in both enzymes. An interesting intermediate was found during the reoxidation of fully reduced tree laccase by dioxygen.[73] This intermediate caused the rapid formation of an absorption maximum at 330 nm and also affected the type-2 copper EPR-signal. It decayed in a first-order reaction with a half-time of 20 sec. A new EPR-signal at low tempetature (10 K) due to this radical could be detected. This radical has been described as an $O^{\cdot -}$ radical. Recently, Solomon and coworkers[19,77,77a] have identified and spectroscopically characterized an oxygen intermediate during the reaction of either fully reduced native tree laccase or T1Hg-laccase with dioxygen. The intermediate has been described as hydroperoxide binding as 1,1-μ between either CU2 and CU4 or CU3 and CU4. From these data, it follows that the dioxygen binds to the trinuclear copper site. This species may store three electrons and transfer them to the bound dioxygen followed by a final one-electron transfer.

Table 7. Reoxidation of reduced enzyme.

Enzyme	Oxidant	Signal of intermediate	Putative intermediate	Rate of reoxidation	Rate of decay of intermediates	Reference
Laccase *Polyporus versicolor*	O_2	Absorption maximum at 360 nm, new EPR-signal at low temperature (10 K)	$O_2^{\cdot -}$-radical	615-nm band, 5×10^6 $M^{-1}sec^{-1}$; 330-nm-band 5×10^6 $M^{-1}sec^{-1}$; 360-nm absorption maximum, 5×10^6 $M^{-1}sec^{-1}$; type-2 copper EPR signal, 20s halftime	absorption maximum at 360 nm, 20 s halftime, first-order reaction	73
T1Hg-derivative	O_2	Absorption bands at 340 and 470 nm, CD changes in 500- to 1000-nm range, disappearance of MCD signal at 730 nm	HO_2^-	340- and 470-nm-bands,[a] 2×10^6 $M^{-1}sec^{-1}$	Formation is irreversible	77
	H_2O_2	CD changes in 250- to 330-nm range	O_2^{2-}	CD changes, formation in < 10 min	New CD spectrum, stable at > 6 hr	75
Ascorbate oxidase *Cucurbita pepo medullosa*	O_2	no data	no data	610-nm band, 5×10^6 $M^{-1}sec^{-1}$; 330-nm band, 5×10^6 $M^{-1}sec^{-1}$	no data	74
	H_2O_2	Absortion band at 350 nm, pH 7.6	O_2^{2-}	350-nm band formation in several minutes	350-nm band decay over several hours	76

[a] Measured at 3°C.
[b] Addition to oxidized enzyme in equimolar amounts.

From the smoothed van der Waals surface (Conolly surface) calculated for the X-ray structure model of ascorbate oxidase, a binding pocket for the reducing substrate near the type-1 copper and two channels providing access to the trinuclear copper species could be identified.[9] The binding pocket for the reducing substrate is complementary to an ascorbate molecule. The channel giving access to copper atoms CU2 and CU3 of the trinuclear site is much broader than the channel leading to CU4. The former is the main channel for the entrance of the dioxygen and protons and the release of the water molecules. Figures and a detailed description of the pocket and the channels are given by Messerschmidt *et al.*[9]

8. Functional Derivatives

8.1. *General Remarks*

The blue oxidases may be inhibited by a number of different substances. For ceruloplasmin, there are systematic studies showing that two groups of more specific inhibitors exist: (1) halides and inorganic anions and (2) carboxylate ions.[78–81] The rank of inhibition effect of anions with ceruloplasmin is $CN^- > N_3^- > OCN^- > SCN^- > SeCN^- > F^- > I^- \simeq NO_3^- \simeq Cl^- > Br^- >$ (ClO_4^-, tetraborate, borate, phosphate, sulfate, and cacodylate).[79] Inhibition by CN^-, N_3^- and F^- has been investigated predominantly with laccase[72,82,83] and ascorbate oxidase.[84] The order of inhibitory action is the same as that observed for ceruloplasmin. Inhibition by these anions appears to be understood quite satisfactory. Strong binding to the spectroscopic type-2 copper is indicated by a perturbation of the EPR signal of this site.

Laccase also becomes inhibited at higher pH values. Tree laccase is inhibited above pH 6.5. It appears that at pH 7.4 ~ 50 % of the enzyme molecules are inhibited due to the binding of OH^- to the type-2 copper. The nature of this OH^- binding to the resting form of ascorbate oxidase will be demonstrated below.

The reaction of nitric oxide with laccase[85] and ascorbate oxidase[86] has been studied as well. Nitric oxide fully reduces fungal and tree laccase when added to the oxidized enzyme under anaerobic conditions. In addition the binding of one NO molecule to laccase could be detected. This is characterized by a new EPR signal and has been described as coordinated with the type-2 copper.[85] Only the reduction of the type-1 copper has been observed when NO has been added to ascorbate oxidase under anaerobic conditions.

Other functional derivatives have been prepared. Two of them deserve special mention, namely the type-2 copper depleted (T2D) enzyme and a derivative where the type-1 copper is replaced by mercury (T1Hg). The T2D derivative will be dealt with in a special paragraph. The T1Hg-form was originally prepared by McMillin and coworkers[87] from tree laccase. This derivative lacks the type-1 copper EPR signal and makes possible the recording of the type-2 EPR signal alone. It has been used to study the reaction of dioxygen with the reduced derivative.[77] As already mentioned, the formation of an oxygen intermediate could be observed. This intermediate was bound irreversibly to the trinuclear copper site due to abortion of further reduction of the dioxygen caused by the lack of the type-1 copper redox center. The preparation of the apoenzyme and reconstitution are possible for all three blue oxidases (see for laccase,[60] for ascorbate oxidase[61] and for ceruloplasmin[2]). X-ray crystal structures of four functional derivatives of ascorbate oxidase have been determined.[88,89] The results of these investigations and implications for the catalytic mechanism of the blue oxidases will be outlined in the next section.

8.2. *X-ray Structure of the Type-2 Depleted (T2D) Form of Ascorbate Oxidase*

It has been demonstrated by several groups that copper can be selectively removed from the blue oxidases, causing a disappearance of the type-2 EPR signal. Several methods have been described for laccase from *Polyporus versicolor*[90,91] and from the Japanese lacquer tree, *Rhus vernicifera*,[92–95] and for ascorbate oxidase from zucchini.[95–97] All procedures except one involve working under reducing conditions with metal chelating reagents, such as EDTA, dimethyl glyoxime, bathocuproine disulfonate or nitrilotriacetate. Reaction of N, N-diethyldithiocarbamate with ascorbate oxidase under aerobic conditions in solution gave the type-2 depleted enzyme.[97] Many experiments were carried out on T2D multi-copper oxidases in the past and to interpret these experiments it is important to know the actual occupation of the copper sites in the depleted enzyme. An X-ray structure analysis of the depleted enzyme of ascorbate oxidase[88] delivered new information on this point.

Crystals of native oxidized ascorbate oxidase were anaerobically dialyzed in microcells against 50 mM sodium phosphate buffer, pH 5.2, containing 25 % (v/v) methylpentanediol (MPD), 1 mM EDTA, 2 mM dimethyl-glyoxime (DMG) and 5 mM ferrocyanide for 7 and 14 hr. Thereafter, crystals were brought back to the aerobic 25 % MPD solution, buffered with 50 mM sodium phosphate, pH 5.5. This procedure is based on the method of Avigliano *et al.*'s[96] to prepare T2D ascorbate oxidase in solution and was modified by Merli *et al.*[98] for use with ascorbate oxidase crystals. The 2.5-Å-resolution X-ray structure analysis by difference-Fourier techniques and crystallographic refinement shows that about 1.3 copper ions per ascorbate oxidase monomer are removed. The copper is lost from all three copper sites of the trinuclear copper species, whereby the EPR-active type-2 copper is most depleted (see Fig. 10). Type-1 copper is not affected. The EPR spectra from polycrystalline samples of the respective native and T2D ascorbate oxidase were recorded. The native spectrum exhibits the type-1 and type-2 EPR signals in a ratio of about 1:1 as expected from the crystal structure. The T2D spectrum reveals the characteristic resonances of the type-1 copper center, also observed for T2D ascorbate oxidase in frozen solution, and the complete disappearance of the spectroscopic type-2 copper. This observation indicates preferential formation of a Cu-depleted form with the hole equally distributed over all three copper sites. Each of

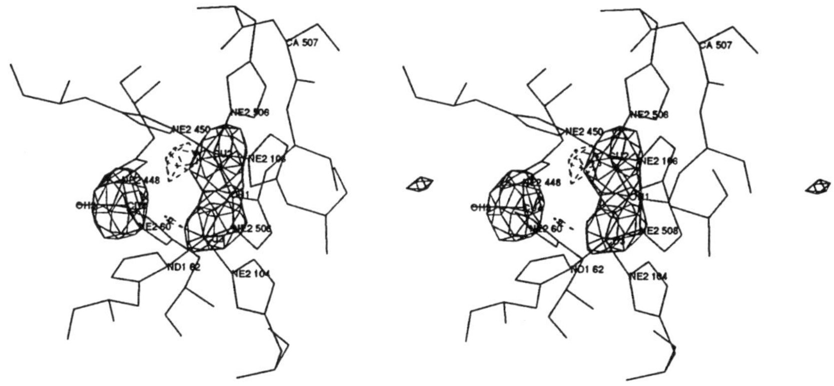

Fig. 10. Averaged FO_{T2D}-FC_{T2D}-difference electron density map plus atomic model around the trinuclear copper site. Contour levels: −18.0 solid line, 18.0 dashed line, magnitudes of the hole are less than −35.0.

these Cu-depleted species may represent an antiferromagnetically coupled copper pair that is EPR-silent and that could explain the disappearance of the type-2 EPR-signal.

8.3. *X-ray Structure of the Reduced Form of Ascorbate Oxidase*

Crystals of the reduced form (REDU) of ascorbate oxidase[89] had to be prepared and mounted in the glass capillary in a glove box that was flushed with argon gas and operated with a slight overpressure of argon. The degassed buffer solution was stored in the glove box. Dithionite was added to the buffer solution to a concentration of 10 mM. The siliconized X-ray capillary was washed with the buffer. The crystals were soaked in the reducing buffer for half an hour. After 15 min, the crystals had lost their blue colour. They were mounted in the X-ray capillary and carefully sealed with wax that had been degassed in the desiccator. Crystals mounted this way remained colourless and reduced over weeks.

The 2.2-Å resolution X-ray structure analysis by difference-Fourier techniques and crystallographic refinement delivered the following results.[89] The geometry at the type-1 copper remains much the same compared to the oxidized form. The mean copper-ligand bond lengths of both subunits increased by 0.04 Å in average, which is insignificant but may indicate a trend. Similar results have been obtained for the reduced forms of poplar plastocyanin at pH 7.8,[99] azurin from *Alcaligenes denitrificans*[100] and azurin from *Pseudomonas aeruginosa*.[101] In reduced poplar plastocyanin at pH 7.8 a lengthening of the two CU-N(His) bonds by about 0.1 Å is observed. In reduced azurin, pH 6.0, from *Alcaligenes denitrificans*, the distances from copper to the axial methionine and the carbonyl oxygen each increase by about 0.1 Å. The same shifts are found in the refined structures of reduced azurin from *Pseudomonas aeruginosa* determined at pH values of 5.5 and 9.0.[101] The estimated accuracy of the copper-ligand bond lengths in the high-resolution structures of above-mentioned small blue copper proteins is about 0.05 Å. The type-1 copper sites in the small blue copper proteins as well as in ascorbate oxidase require little reorganization in the redox process.

A schematic drawing of the reduced form of ascorbate oxidase is shown in Fig. 11. The structural changes are considerable at the trinuclear copper site. Thus on reduction the bridging oxygen ligand OH1 is released and

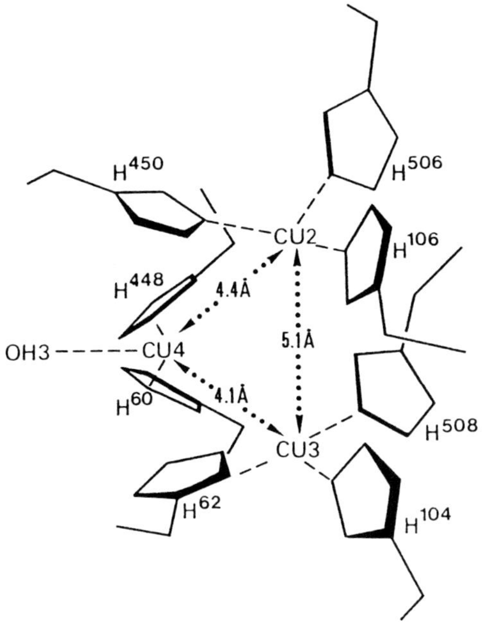

Fig. 11. Schematic drawing of the reduced form of ascorbate oxidase around the trinuclear copper site. The included copper–copper distances are the mean values between both subunits.

the two coppers, CU2 and CU3, move toward their respective histidines and become three coordinated, a preferred stereo-chemistry for Cu(I). The copper–copper distances increase from an average of 3.7 to 5.1 Å for CU2–CU3, 4.4 Å for CU2–CU4 and 4.1 Å for CU3–CU4. The mean values of the copper-ligand distances of the trinuclear copper site are comparable to those of native oxidized ascorbate oxidase and binuclear copper model compounds with nitrogen and copper ligands.[20,21] CU4 remains virtually unchanged between reduced and oxidized forms. Coordinatively unsaturated copper(I) complexes are reported in the literature. Linear two-coordinated[102] and T-shaped three-coordinated[103] copper(I) compounds have been reported. The copper nitrogen distances for both linearly arranged nitrogen ligands are about 1.9 Å, about 0.1 Å shorter than copper nitrogen bond lengths in copper(II) complexes. In T-shaped copper(I) complexes, the bond length of the third ligand is increased. Copper ion CU4 is in a T-shaped threefold coordination not unusual for copper(I) compounds. The structure of the fully reduced trinuclear copper site is quite different therefore from that of the fully oxidized resting form of the enzyme and its implications for the enzymatic mechanism will be discussed below.

8.4. *X-ray Structure of the Peroxide Form of Ascorbate Oxidase*

Native crystals of ascorbate oxidase were soaked in harvesting buffer solution (50 mM sodium phosphate, 25 % MPD, pH 5.5) containing different H_2O_2 concentrations.[89] Concentrations of H_2O_2 greater than 20 mM caused the crystals to crack, to become brownish and finally to decompose within several hours depending on the concentration of H_2O_2. At 10 mM H_2O_2, the crystals remained blue and did not crack for $1\frac{1}{2}$ to 2 days. A native crystal was therefore soaked for two hr in 10 mM H_2O_2 containing harvesting buffer solution, mounted in the X-ray capillary and immediately used for the X-ray intensity measurements.

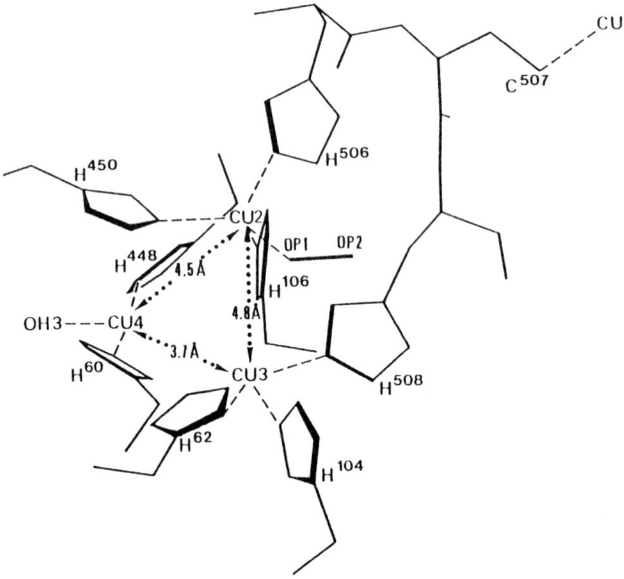

Fig. 12. Schematic drawing of the peroxide form of ascorbate oxidase around the trinuclear copper site. The included copper–copper distances are the mean values between both subunits.

The 2.6-Å resolution X-ray structure analysis by difference-Fourier techniques and crystallographic refinement reveal the data illustrated in Fig. 12.[89] The geometry at the type-1 copper site is not changed compared with that of the oxidized form. The copper-ligand bond distances averaged for both subunits show no significant deviations from those of the oxidized form. As in the reduced form, the structural changes are remarkable at the trinuclear copper site. The bridging oxygen ligand OH1 is absent, the peroxide binds terminally to the copper atom CU2 as hydroperoxide and the copper–copper distances increase from an average of 3.7 to 4.8 Å for CU2-CU3 and 4.5 Å for CU2-CU4. The distance CU3-CU4 remains at 3.7 Å. The mean values of the copper-ligand distances of the trinuclear copper site are again comparable to those of native oxidized ascorbate oxidase and corresponding copper model compounds.

Copper ion CU3 is threefold coordinated as in the reduced form but the coordination by the ligating N-atoms of the corresponding histidines is not exactly trigonal-planar and the CU3 atom is at the apex of a flat trigonal pyramid. The coordination sphere around CU4 is not affected and similar in all three forms. Copper atom CU2 is fourfold coordinated to the NE2 atoms of the three histidines, as in the oxidized form, and by one oxygen atom of the terminally bound peroxide molecule in a distorted tetrahedral geometry. Its distance to CU3 increases from 4.8 Å in the oxidized peroxide derivative to 5.1 Å in the fully reduced enzyme. The bound peroxide molecule is directly accessible from the solvent through a channel leading from the surface of the protein to the CU2–CU3 copper pair. This channel has already been described earlier and its possible role as dioxygen transfer channel has been discussed. An interesting feature is the close proximity of the imidazole ring of histidine 506 to the peroxide molecule. Histidine 506 is part of one possible electron transfer pathway from the type-1 copper to the trinuclear copper site and could indicate a direct electron pathway from CU1 to dioxygen. It may also help to stabilize important intermediate states in the reduction of dioxygen.

The strong positive peaks at CU2 in both FO_{NATI}-FO_{PEOX} and FO_{REDU}-FO_{PEOX} electron density maps could not be explained by a shift of CU2 alone. Occupancies of the copper atoms, the oxygen atom OH3 and the peroxide molecule were refined. Type-1 copper CU1 is nearly not affected. Copper atoms CU3 and CU4 are only partly removed but copper atom CU2 is about 50 % depleted. The oxygen ligands exhibit full occupancy. The treatment of crystals of ascorbate oxidase with hydrogen peroxide generates not only a well-defined peroxide binding but also a preferential depletion of copper atom position CU2. In the copper depleted molecules the ligating histidine 106 adopts an alternative side chain conformation detected in the 2FO-FC map, calculated with the final peroxide derivative model coordinates. This map shows that histidine 106 moves away when copper atom CU2 is removed and opens the trinuclear site even more. From the T2D crystal structure of ascorbate oxidase, it is apparent that copper from all three metal binding sites of the trinuclear copper species is removed in different amounts. The movement of the histidine 106 side chain could explain how this process is accomplished.

Copper depletion may also cause instability of the protein against hydrogen peroxide. Reaction of hydrogen peroxide with ascorbate oxidase in solution in excess leads to a rapid degradation of the enzyme.[104] This can be monitored in the UV/vis PEOX–NATI difference spectrum by a negative band at 610 nm and a positive band at 305 nm. Adding four equivalents hydrogen peroxide per monomer ascorbate oxidase does not lead to enzyme degradation and gives a positive peak at 305 nm indicative for the peroxide binding. Unfortunately, is was not possible to monitor a UV/vis spectrum of dissolved crystals after X-ray data collection because of the dissociation of the bound peroxide in solution.

The reaction of dioxygen with laccase or ascorbate oxidase has been reviewed in Sec. 7 and by Messerschmidt *et al.*,[9] where the possible binding modes of dioxygen to binuclear and trinuclear copper centers are discussed. A novel mode of dioxygen binding to a binuclear copper complex was found in a compound synthesized by Kitajimia *et al.*[105] The complex contains the peroxide in μ–$\eta^2 : \eta^2$ mode, i.e., side-on between the two copper(II) ions. Such a binding mode of dioxygen has been detected in the crystal structure of the oxidized form of *Limulus Polyphemus* subunit II hemocyanin.[106] However, the binding mode of dioxygen to the trinuclear copper site in the blue oxidases appears to be different, as can be seen from the X-ray structure of the peroxide derivative of ascorbate oxidase.

During its reaction with fully reduced laccase dioxygen binds to the trinuclear copper species and three electrons are very rapidly transferred to it, resulting in the formation of an "oxygen intermediate" with a characteristic optical absorption near 360 nm[73,107] and a broad low temperature EPR signal $g = 1.7$.[108,109] The type-1 copper is concomitantly reoxidized when the low-temperature EPR signal is formed. The oxygen intermediate decays very slowly ($t_{1/2} \sim 1$ to 15 sec), correlated with the appearance of the type-2 EPR signal.[110]

Solomon and coworkers[19,77,111] have identified and spectroscopically characterized an oxygen intermediate during the reaction of either fully reduced native tree laccase or T1Hg-laccase with dioxygen. They concluded from their spectroscopic data that the intermediate binds as 1,1-μ hydroperoxide between either CU2 and CU4 or CU3 and CU4. As it is unlikely that the dioxygen migrates or rearranges coordination during reduction, Messerschmidt *et al.*[89] proposed that the binding site and mode determined in the peroxide derivative of ascorbate oxidase is representative for all reaction intermediates of dioxygen, and by homology arguments, is in all blue oxidases. The relevance of this binding mode for the catalytic mechanism will be discussed later.

8.5. *X-ray Structure of the Azide Form of Ascorbate Oxidase*

The azide derivative was obtained by soaking the native crystals in the harvesting buffer solution containing 50 mM sodium azide for 24 hr.[89] Binding of azide was indicated by a change of colour of the crystal from blue to brownish. After X-ray intensity data collection, the crystal was dissolved in a solution containing 50 mM azide, 5 mM phosphate buffer, pH 5.5, and 8 % MPD and the UV/vis-spectrum was recorded at room temperature.

The results of the 2.3-Å resolution X-ray structure analysis by difference-Fourier techniques and crystallographic refinement are depicted in Fig. 13.[89] The geometry at the type-1 copper site is not changed compared with that of the native form. The copper-ligand bond distances averaged for both subunits show no significant deviations from those of the native form. Again, the structural changes are large at the trinuclear copper site. The bridging oxygen ligand OH1 and water molecule 145 have been removed, CU2 moves toward the ligating histidines and two azide molecules bind terminally to it. The copper–copper distances increase from an average of 3.7 to 5.1 Å for CU2-CU3 and 4.6 Å for CU2–CU4. The distance CU3–CU4 is decreased to 3.6 Å. The mean values of the copper–ligand distances of the trinuclear copper site are again comparable to those of native ascorbate oxidase and corresponding copper model compounds.

The coordination of CU3 resembles that in the peroxide form. The threefold coordination by histidines is a very flat trigonal pyramid. The

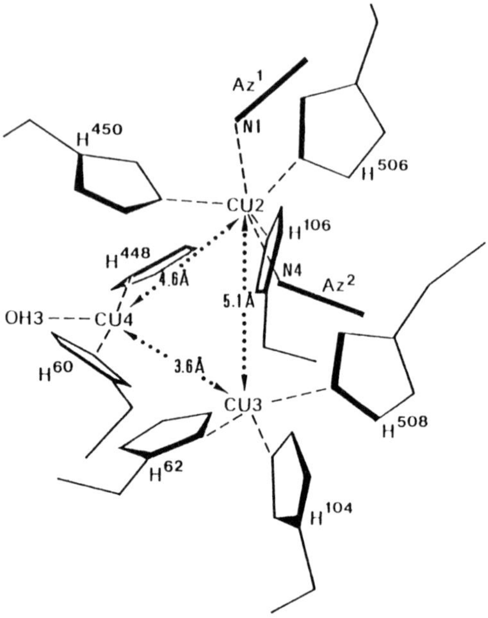

Fig. 13. Schematic drawing of the azide form of ascorbate oxidase around the trinuclear copper site. The included copper–copper distances are the mean values between both subunits.

coordination sphere around CU4 is not affected. CU2 is fivefold coordinated to the NE2 atoms of the three histidines, as in the reduced form, and to the two azide molecules. The two azide molecules are terminally bound at the apexes of a trigonal bipyramid. Both azide molecules bind to the copper atom CU2, which is well accessible from the broad channel leading from the surface of the protein to the CU2–CU3 copper pair. It is not unexpected that the second azide molecule (az^2 in Fig. 13) binds similarly to the peroxide molecule, as azide is regarded as a dioxygen analogue. There is no azide molecule bound bridging either CU2 with CU4 or CU3 with CU4.

The binding of azide to laccase as well as to ascorbate oxidase has been studied extensively by Solomon and co-workers,[19,112–114] and by Marchesini and associates[115,116] by spectroscopic techniques. The derived spectroscopic models involve the binding of two azide molecules to laccase and three azide molecules to ascorbate oxidase with different affinities. As the binding of the high-affinity azide seems to generate spectral features related to the type-2 and type-3 coppers, the spectroscopic data were interpreted as the binding of at least one azide as a 1,3-μ bridge between the type-3 copper ions and the type-2 copper ion.

After the X-ray data collection for the azide derivative, the crystal was dissolved in azide containing buffer and a UV/vis spectrum was recorded to check the spectral properties of the sample.[89] The spectrum was characterized by a broad increase of absorption in the 400- to 500-nm region and an intense absorption maximum at 425 nm, very similar to the results of Casella *et al.*[115]

There are many structural studies of copper coordination compounds with azide ligands, mainly of mononuclear and binuclear copper complexes but a few also of trinuclear copper complexes. A comprehensive review on copper coordination chemistry has been written by Hathaway.[117] Azide binds only terminally to mononuclear systems. Fivefold coordination of nitrogen ligands including azide to Cu^{2+} is frequently found to be arranged as a trigonal bipyramid. In binuclear systems azide may bind terminally as 1,1-μ or bridging as 1,3-μ. Similarly two azides may bind di-1,1-μ or di-1,3-μ. The interaction with all three copper ions of a trinuclear complex may be either terminally as 1,1,1-μ or bridging as 1,1,3-μ. In the X-ray crystal structure of ascorbate oxidase two azide molecules bind terminally to the type-3 CU2. Azide binding in ascorbate oxidase resembles therefore the binding of azide to an isolated copper ion. In fact there is little interaction of CU2 with CU3 and CU4, which are 5.1 and 4.6 Å away, respectively.

The coordination of copper ion CU4 in the native oxidized structure is of some interest. It has only three ligating atoms at close distances, forming a T-shape coordination that is known for Cu(I) complexes (see discussion of the reduced form). However, the ligand field is completed if we take into account the π-electron systems of the imidazole rings of histidines 62 and 450 (see Fig. 6). A ligand with tetragonal-pyramidal symmetry around CU4 is then formed. The shortest distances of CU4 are to CD2 450 with 3.4 Å and to CG 62 with 3.6 Å. These distances are too long for strong copper-π-electron interactions but the histidines will contribute to the CU4 ligand field.

9. The Catalytic Mechanism

Catalytic reaction schemes for laccase and ceruloplasmin have been formulated on the basis of the mechanistic studies and the state of characterization of the copper redox centers at this time. They are outlined in the reviews on laccase by Reinhammar[60] and on ceruloplasmin by Rydén.[2] The degree of correctness of these reaction schemes is rather limited due to the fact that structure and spatial arrangement of the copper centers were unknown at this time.

A tentative mechanism of ascorbate oxidase has been proposed based on the refined X-ray structure and on spectroscopic and mechanistic studies of ascorbate oxidase and the related laccase. The results of these studies have been discussed in detail.[9] The X-ray structure determinations of the fully reduced and peroxide derivatives define two important intermediate states during the catalytic cycle. A proposal for the catalytic mechanism

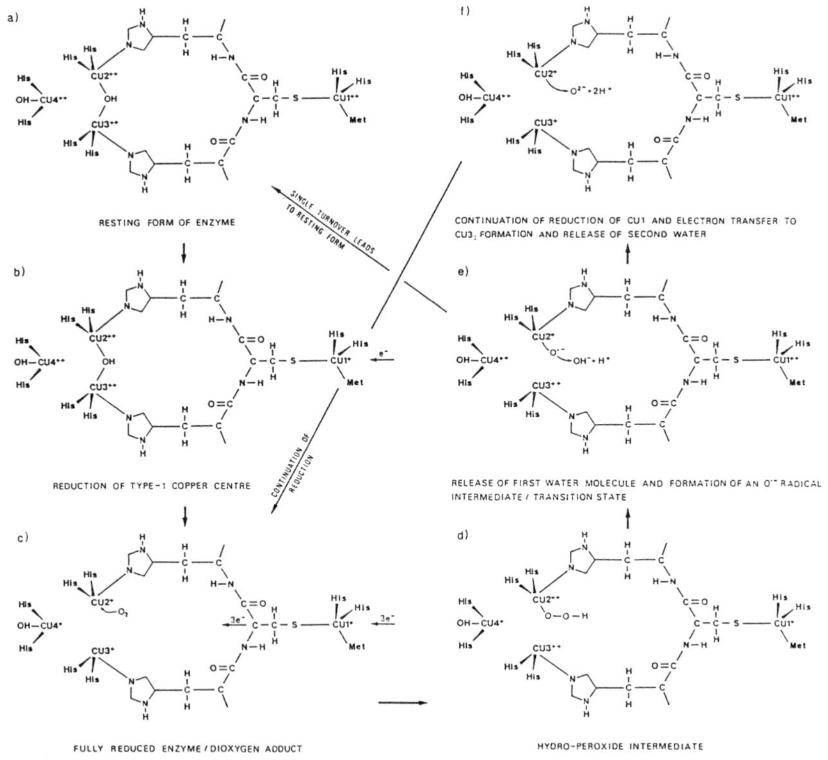

Fig. 14. Proposal for the catalytic mechanism of ascorbate oxidase.

incorporating this new information is given in Ref. 89 and presented in Fig. 14. This scheme should be valid in principle also for laccase due to the close relatedness between both blue oxidases.

The catalytic cycle starts from the resting form (Fig. 14a) in which all four copper ions are oxidized and CU2 and CU3 are bridged by an OH^- ligand. CU2 and CU3 are most likely the spin-coupled type-3 pair of copper and CU2 is the type-2 copper. The first step is the reduction of the type-1 copper CU1 by the reducing substrate in a one-electron transfer step (Fig. 14b). The electrons are transferred through the protein either to CU2 or CU3. Electron transfer may be through-bond, through-space or a combination of both. A branched through-bond pathway is available, leading to CU2 with 9 bonds (including a hydrogen bond) and to CU3 with 11 bonds, respectively. The fully reduced enzyme requires four electrons to be transferred (Fig. 14c). Its structure has been described in a previous section. The hydroxyl bridge between the copper pair has been released and the distance from copper atom CU2 to CU3 has been increased to about 5.1 Å. Considerable reorganization energy will be neccessary to reach this state from the resting form of enzyme. At this stage, dioxygen may bind to the enzyme at CU2, probably in the manner shown in the peroxide derivate described in the appropriate section. A transfer of two electrons from the copper pair to dioxygen leads to the formation of a hydroperoxide intermediate (Fig. 14d). A third electron may be transferred from CU4 to the hydroperoxide intermediate, and a fourth electron from type-1 copper to copper ion CU2. The O–O bond is broken at this stage and the first water molecule released (Fig. 14e). An oxygen radical has been detected in laccase by EPR. The EPR spectrum indicates that the type-1 copper has been reoxidized and the EPR signals of the oxygen radical intermediate and type-1 copper are present. The CU2 is in the reduced state, whereas the oxidized copper atoms CU3 and CU4 may be spin-coupled and EPR silent. The reduced CU2 may facilitate O–O bond breakage and release of water. The catalytic cycle is continued by a further reduction of the type-1 copper center by the reducing substrate. This electron may be transferred to CU3 of the copper pair via the 11 bond pathway. Now, the fourth electron may be transferred to the oxygen radical intermediate from copper atom CU2 and the second water molecule released (Fig. 14f). In the case of only four electron equivalents, the reaction may lead to the resting form and the second water may remain bound as the bridging ligand between CU2 and CU3, concomitant with a substantial rearrangement within the trinuclear copper site and its coordinating ligands. If turnover is continued, this will not occur and the trinuclear copper site may maintain a structure very close to that found in the fully reduced form. Only minor rearrangements will take place at the trinuclear copper site during the catalytic cycle, a prerequisite for facile electron transfer reactions. The four protons required for the formation of the two water molecules from dioxygen may be supplied from bulk water through the dioxygen channel via the water molecules bound in the vicinity of CU2 and CU3.

10. Electron Transfer Processes

10.1. *Electron Transfer to the Type-1 Copper Redox Center*

As previously mentioned, the electron transfer from one-electron-reduced nitroaromates ($ArNO_2^{\cdot -}$), $CO_2^{\cdot -}$, $MV^{\cdot +}$, lumiflavin or deazaflavin to the type-1 copper center (see Table 6) takes place in a bimolecular second-order reaction with rates compatible with or higher than the turnover number, with ascorbate as reducing substrate. The electron transfer from ascorbate to the type-1 copper center can be even faster and is completed within the dead time of the stopped-flow instrument. It is therefore not the rate limiting step in the overall reaction.

The factors governing electron transfer may be described within the framework of Marcus electron-transfer theory.[118] They are expressed in terms of driving force, distance of redox centers, reorganization energy etc.

The driving force, calculated from the difference in the redox potentials (+344 mV for the type-1 copper in ascorbate oxidase (see Table 5); +295 mV for the couple ascorbate/ascorbate-free radical,[119]) is 49 mV. In the proposed modeled encounter complex,[9] there is a short distance of about 7 Å between the two redox centers (distance CU1–O1 ASC = 6.8 Å; distance CU1–O2 ASC = 7.5 Å) and an effective parallel arrangement of the rings, with good overlap of the π-electron density systems facilitating a rapid electron transfer (see Fig. 15). It is well documented for small blue copper proteins, such as plastocyanin, that there are minimal structural changes upon reduction and reoxidation.[99] The reorganization energy is, therefore, probably small.

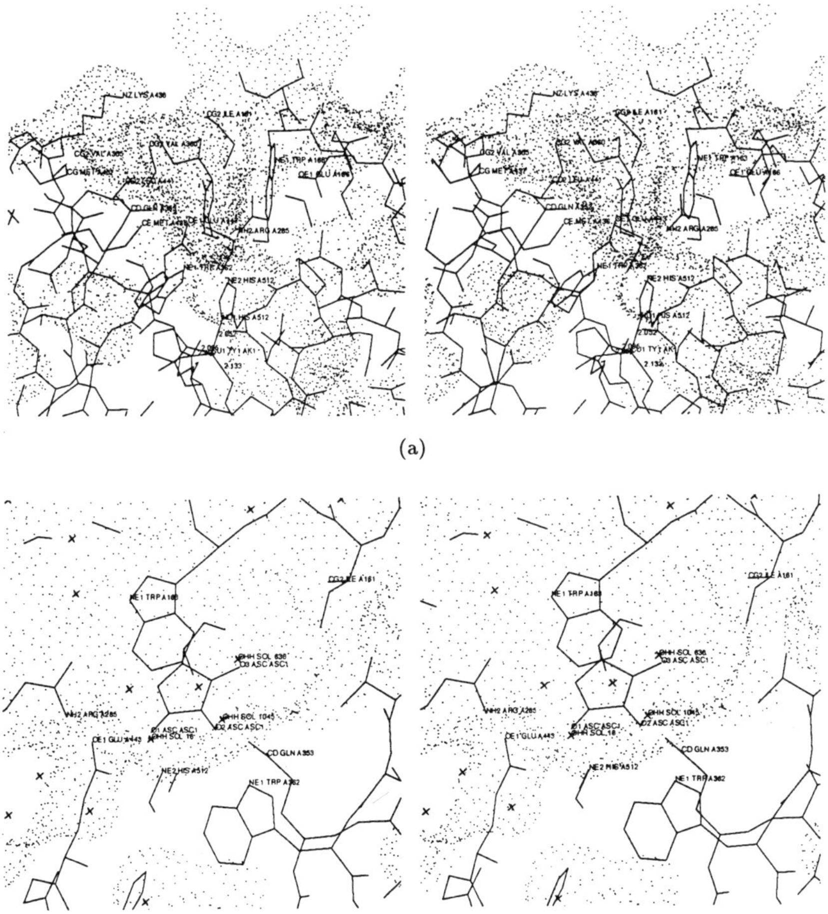

(a)

(b)

Fig. 15. Stereo drawing of the binding site near the type-1 copper plus docked, L-ascorbate. Atomic model plus Conolly dot surface. (a) Viewed perpendicular to the ND1 His512-CU1 bond. (b) Viewed parallel to the ND1 His512-CU1 bond.

10.2. *Intramolecular Electron Transfer from the Type-1 Copper Center to the Trinuclear Copper Center*

Long-distance intramolecular electron transfer can be described in the frame of the theory of Marcus.[118] In the formulation of Lieber *et al.*,[120] the intramolecular electron transfer rate constant k_{ET} can be written as

$$k_{ET} = \nu_n \Gamma \exp(-\Delta G^*/RT)\,, \tag{1}$$

where ν_n is the nuclear frequency factor, normally 10^{13} sec^{-1}, Γ is the electronic factor and ΔG^* is the activation free energy for ET. The electronic factor, Γ, is at unity when the donor and acceptor are strongly coupled but is much smaller at long donor–acceptor distances. In such cases, Γ is expected to fall off with distance, d,

$$\Gamma = \Gamma(d_0) \exp[-\beta(d - d_0)]\,, \tag{2}$$

where d_0 is the van der Waals contact distance, normally taken as 3 Å[118] and β is the electron coupling factor, which decreases with increasing distance and depends on the nature of the intervening medium. ΔG^* depends on the reaction free energy, ΔG^0, and the nuclear reorganization energy, λ,[118] according to the equation

$$\Delta G^* = (\Delta G^0 + \lambda)^2/4\lambda\,. \tag{3}$$

The ET rate is maximal when $-\Delta G^0 = \lambda$.

The shortest distances, d, between the type-1 copper and the coppers of the trinuclear site are 12.2 Å (CU1–CU2) and about 12.7 Å (CU1–CU3) (see Table 1). Figure 7 shows that His506–Cys507–His508 serve as bridging ligand between the two redox centers, providing a bifurcated pathway for ET from the type-1 copper center to the trinuclear copper species. The difference in redox potential of the type-1 copper center and the type-3 coppers, the driving force, measured at 10°C, is $-\Delta G^\circ = 30$ mV.[59] However, the binding of dioxygen to the partly reduced protein and the presence of reduction intermediates may affect this redox-potential (a very slow equilibration was found between type-1 and type-3 coppers in ascorbate oxidase in the absence of dioxygen[121]). For the reorganization energy, λ, and the electronic coupling factor, β, no estimates can be derived for ascorbate oxidase but reasonable values for proteins are $\lambda = 1$ eV and $\beta = 1.2$ Å^{-1}, according to Gray and Malmström.[122] These values inserted into Eq. (1) yield $k_{ET} \sim 10^5$ sec^{-1}. Changing β to 1.6 Å^{-1} gives $k_{ET} \sim 4 \times 10^3$ sec^{-1}, a value closer to the observed turnover number of 8×10^3 sec^{-1}. It has been suggested that the electron transfer in proteins may not be designed for very fast intramolecular ET, with the exception of light-induced ET in photosynthetic reaction centers.[123] They could even be designed to slow down these rapid rates, which might otherwise lead to biological "short circuits." Related to this point is the observation that maximal rates for intramolecular ET in organic donor–acceptor molecules with rigid spacers are significantly faster than those for Ru-labeled protein systems at similar distances.[124]

In the case of laccase and ascorbate oxidase, the observed ET rates for the reduction of the type-3 coppers (see Table 6) are lower than the observed turnover number. This can be explained only by the possibility that the enzymes are in the resting form under the experimental conditions. A considerable reorganization energy seems to be necessary to get to the reduced state of the type-3 coppers (release of the bridging OH^- and movement of the copper CU2 and CU3). From these data it cannot be decided what the rate-limiting step is in the catalytic cycle, either this intramolecular ET or the reaction of the dioxygen at the trinuclear copper site.

ET from the type-1 copper to the type-3 copper pair of the trinuclear copper site may be through-bond, through-space or a combination of both. A through-bond pathway is available for both branches each with 11 bonds (see Fig. 7). The alternative combined through-bond and through-space pathway from the type-1 copper CU1 to CU2 of the trinuclear center involves a transfer from the SG atom of Cys507 to the main-chain carbonyl of Cys507 and through the hydrogen bond of this carbonyl to the ND1 atom of the His506.

Kyritsis *et al.*[68] carried out a theoretical pathway analysis for ascorbate oxidase using an algorithm, which was developled by Beratan and Onuchic[125–127] to help identify the most favourable long-range electron transfer pathways in metalloproteins. According to this analysis, the most favoured route consists of four covalent bonds and a hydrogen-bond between main-chain carbonyl O of Cys507 and the ND1 atom of His506 (hydrogen-bond length 2.9 Å), (see Fig. 7), which gives an electronic coupling ε^2 value of 2.3×10^{-3}. The second most efficient pathway contains an extra covalent bond between the ND1 and the CG atoms of His506, and gives an ε^2 value of 8.3×10^{-4}. For the second imidazole ring the best pathway consists of seven covalent bonds between SG of Cys507 and CG of His508, with an ε^2 value 7.8×10^{-4}. Thus the hydrogen-bonded Cys507-His506 pathway gives approximately three times more efficient electronic coupling than the Cys507-His508 route. In the case of the type-1 blue copper protein plastocyanin, it is believed that a similar electron-transfer pathway, consisting of the copper ligand Cys84 and the adjacent highly conserved Tyr83 is relevant and made use in the reaction with cytochrome f.[128]

10.3. *Electron Transfer within the Trinuclear Copper Site*

Electron exchange within the trinuclear copper site is expected to be very fast due to the short distances between the copper atoms (from 3.7 Å to 5.2 Å in the reduced form), as is ET to the bound dioxygen. This fast electron exchange is necessary to include CU4 into the redox processes. CU4 is at a longer distance to the type-1 copper CU1 than CU2 and CU3.

Acknowledgments

The author thanks Professors R. Huber and R. Ladenstein and other colleagues involved in the X-ray structural work on ascorbate oxidase. These results would not have been possible without their expertise, cooperation and support. Furthermore, he is indebted to Professor E. Adman for supplying the coordinates of nitrite reductase prior to being deposited in the data bank.

References

1. H. F. Deutsch, C. D. Kasper and D. E. Walsh, *Arch. Biochem. Biophys.* **99**, 132 (1962).
2. L. Rydén, in *Copper Proteins and Copper Enzymes*, ed. R. Lontie (CRC Press, Boca Raton, FL, 1 1984) Vol. 3, pp. 37–100.
3. B. Magdoff-Fairchild, F. M. Lovell and B. W. Low, *J. Biol. Chem.* **244**, 3497 (1969).
4. V. N. Zaytsev, K. A. Moshkov, M. M. Shavlowski and S. A. Neifakh, *Bioorg. Khim.* **1**, 1521 (1975).
5. V. N. Zaytsev, A. A. Vagin, Yu. V. Nekrasov and K. A. Moshkov, *Kristallografiya* **31**, 937 (1986).
6. R. Ladenstein, A. Marchesini and S. Palmieri, *FEBS Lett.* **107**, 407 (1979).
7. M. Bolognesi, G. Gatti, A. Coda, L. Avigliano, G. Marcozzi and A. Finazzi-Agró, *J. Mol. Biol.* **169**, 351 (1983).
8. A. Messerschmidt, A. Rossi, R. Ladenstein, R. Huber, M. Bolognesi, G. Gatti, A. Marchesini, R. Petruzzelli, and A. Finazzi-Agró, *J. Mol. Biol.* **206**, 513 (1989).

8a. P. Kraulis, *J. Appl. Crystallogr.* **24**, 946 (1991).

8b. E. A. Merritt and M. E. P. Murphy, *Acta Crystallogr.* **D50**, 869 (1994).

8c. J. P. Priestle, *J. Appl. Crystallogr.* **21**, 572 (1988).

9. A. Messerschmidt, R. Ladenstein, R. Huber, M. Bolognesi, L. Avigliano, R. Petruzzelli, R. Rossi and A. Finazzi-Agró, *J. Mol. Biol.* **224**, 179 (1992).

10. W. Kabsch and S. Sander, *Biopolymers* **22**, 2577 (1983).
11. D. J. Barlow and J. M. Thornton, *J. Mol. Biol.* **168**, 867 (1983).
12. J. L. Crawford, W. N. Lipscomb and C. C. Schellmann, *Proc. Natl. Acad. Sci. USA.* **70**, 538 (1973).
13. J. S. Richardson, *Adv. Protein Chem.* **34**, 167 (1981).
14. R. Huber, and W. Steigemann, *FEBS Lett.* **48**, 235 (1974).
15. E. N. Baker and R. E. Hubbard, *Prog. Biophys. Mol. Biol.* **44**, 97 (1984).
16. B. K. Lee and F. M. Richards, *J. Mol. Biol.* **55**, 379 (1971).
17. J. M. Guss and H. C. Freeman, *J. Mol. Biol.* **169**, 521 (1983).
18. H. Nar, A. Messerschmidt, R. Huber, M. van de Kamp and G. W. Canters, *J. Mol. Biol.* **221**, 765 (1991).
19. J. L. Cole, P. A. Clark and E. I. Solomon, *J. Am. Chem. Soc.* **112**, 9534 (1990).
20. K. D. Karlin, J. C. Hayes, Y. Gultneh, R. W. Cruse, J. W. McKnown, J. P. Hutchinson and J. Zubieta, *J. Am. Chem. Soc.* **106**, 2121 (1984).
21. P. Chaudhuri, D. Ventor, K. Wieghardt, E. Peters and K. Peters, *Angew. Chemie* **97**, 55 (1985).
22. A. Volbeda and W. G. J. Hol, *J. Mol. Biol.* **209**, 249 (1989).
23. A. Messerschmidt and R. Huber, *Eur. J. Biochem.* **187**, 341 (1990).
24. G. A. Vehar, B. Keyt, D. Eaton, H. Rodriguez, D. P. O'Brien, F. Rotblat, H. Oppermann, R. Keck, W. I. Wood, R. N. Harkins, E. G. D. Tuddenham, R. M. Lawn and D. J. Capon, *Nature* **312**, 337 (1984).
25. L. D. Cripe, K. D. Moore and W. H. Kane, *Biochemistry* **31**, 3777 (1992).
26. Program Manual for the Wisconsin Package, Version 8, 1994, Genetics Computer Group, Madison, Wisconsin.
27. G. Karlsson, R. Aasa, B. G. Malmström and L. G. Lundberg, *FEBS Lett.* **253**, 99 (1989).
28. J. Deinum and T. Vänngard, *Biochim. Biophys. Acta* **310**, 321 (1973).
29. R. J. Carrica, B. G. Malmström and T. Vänngard, *Eur. J. Biochem.* **22**, 127 (1971).
30. T. L. Ortel, N. Takahashi and F. W. Putnam, *Proc. Natl. Acad. Sci. USA.* **81**, 4761 (1984).
31. K. A. Moshkov, A. A. Vagin and V. N. Zaytsev, *Mol. Biol.* **21**, 1124 (1987).
32. L. Rydén, *Proc. Natl. Acad. Sci. USA* **79**, 6767 (1982).
33. J. W. Godden, S. Turley, D. C. Teller, E. T. Adman, M. Y. Liu, W. J. Payne and J. LeGall, *Science* **258**, 438 (1991).
34. F. F. Fenderson, S. Kumar, E. T. Adman, M. -Y. Liu, W. J. Payne, and J. LeGall, *Biochemistry* **30**, 7180 (1991).
35. E. T. Adman, *Adv. Protein Chem.* **42**, 145 (1991).
36. M. Saloheimo, M. L. Niku-Paalova and J. K. C. Knowles, *J. Gen. Microbiol.* **137**, 1537 (1991).
37. Y. Kosima, Y. Tsukuda, Y. Kawai, A. Tsukamota, J. Sugiura, M. Sakaino and Y. Kita, *J. Biol. Chem.* **265**, 15, 224 (1990).

38. G. Sannia, *EMBL/GenBank/DDBJ databases* (1994) Accession Code: POLACCSA.
39. C. R. Perry, M. Smith, C. H. Britnell, D. A. Wood and C. F. Thurston, *J. Gen. Microbiol.* **139**, 1209 (1993).
40. P. M. Coll, C. Tabernero, R. Santamaria and P. Perez, *Appl. Environ. Microbiol.* **59**, 4129 (1993).
41. U. A. Germann, G. Müller, P. E. Hunziker and K. Lerch, *J. Biol. Chem.* **263**, 885 (1988).
42. G. H. Choi, T. G. Larson and D. L. Nuss, *Mol. Plant Microb. Interact.* **5**, 119 (1992).
43. R. Aramayo and W. E. Timberlake, *Nucleic Acids Res.* **18**, 3415 (1990).
44. P. R. Williamson, *J. Bacteriol.* **176**, 656 (1994).
45. P. R. LaFayette, K. -E. L. Eriksson and J. F. Dean, *Plant Physiol.* **107**, 667 (1995).
46. J. Ohkawa, N. Okada, A. Shinmyo and M. Takano, *Proc. Natl. Acad. Sci. USA* **86**, 1239 (1989).
47. M. Esaka, T. Hattori, K. Fujisawa, S. Sakajo and T. Asahi, *Eur. J. Biochem.* **191**, 537 (1990).
48. A. Rossi, R. Petruzzelli, A. Messerschmidt and A. Finazzi-Agró, private communication (1991).
49. G. A. Jones, *EMBL/GenBank/DDBJ databases* (1993). Accession Code: EMBL:U04283.
50. C. E. Barry, P. G. Nayar and T. P. Begley, *Biochemistry* **28**, 6323 (1989).
51. M. A. Mellano and D. A. Cooksey, *J. Bacteriol.* **170**, 2879 (1988).
52. Y. A. Lee, M. Hendson, N. J. Panopoulos and M. N. Schroth, *J. Bacteriol.* **176**, 173 (1994).
53. D. Albani, R. Sardana, L. S. Robert, I. Altossar, P. N. Arnison, and S. F. Fabijanski, *Plant J.* **2**, 331 (1992).
53a. N. Takahashi, T. L. Ortel and F. W. Putnam, *Proc. Natl. Acad. Sci. USA* **81**, 390 (1984).
54. C. Askwith, D. Eide, A. Van Ho, P. S. Bernard, L. Li, S. Davis-Kaplan, D. M. Sipe and J. Kaplan, *Cell* **76**, 403 (1994).
54a. G. J. Barton, *Protein Eng.* **6**, 37 (1993).
55. F. E. Dwulet and F. W. Putnam, *Proc. Natl. Acad. Sci. USA* **78**, 2805 (1981).
56. L. Rydén, in *Oxidases and Related Redox Systems*, eds. T. S. King, H. S. Mason and M. Morrison (Alan R. Liss, New York, 1988) pp. 349–366.
57. D. G. George, *Multiple Sequence Alignment Program ALNED*, NRBF, Washington, DC (1988).
57a. B. Reinhammar, *Biochim. Biophys. Acta* **275**, 254 (1972).
57b. B. Reinhammar and T. Vänngard, *Eur. J. Biochem.* **18**, 463 (1971).
58. A. G. Sykes, *Adv. Inorg. Chem.* **36**, 377 (1991).
59. P. M. H. Kroneck, F. A. Armstrong, H. Merkle and A. Marchesini, *Adv. Chem. Ser.* **200**, 223 (1982).

59a. K. Kawahara, S. Suzuki, T. Sakurai and A. Nakahara, *Arch. Biochem. Biophys.* **241**, 179 (1985).
60. B. Reinhammar, in *Copper Proteins and Copper Enzymes*, ed. R. Lontie (CRC Press, Boca Raton, FL, 1984) Vol. 3, pp. 1–35.
61. B. Mondovi and L. Avigliano, in *Copper Proteins and Copper Enzymes*, ed. R. Lontie (CRC Press, Boca Raton, FL 1984) Vol. 3, pp. 101–118.
62. L. C. Petersen, and H. Degn, *Biochim. Biophys. Acta* **526**, 85 (1978).
63. L. -E. Andreasson and B. Reinhammar, *Biochim. Biophys. Acta* **445**, 579 (1976).
64. L. -E. Andreasson, B. G. Malmström, C. Stromberg and T. Vänngard, *Eur. J. Biochem.* **34**, 434 (1973).
65. P. O'Neill, E. M. Fielden, L. Morpurgo and E. Agostinelli, *Biochem. J.* **222**, 71 (1984).
66. P. O'Neill, E. M. Fielden, A. Finazzi-Agró and L. Avigliano, *Biochem. J.* **209**, 167 (1983).
67. T. E. Meyer, A. Marchesini, M. A. Cusanovich and G. Tollin, *Biochemistry* **30**, 4619 (1991).
68. P. Kyritsis, A. Messerschmidt, R. Huber, G. A. Salmon and A. G. Sykes, *Chem. Soc. Dalton Trans.*, 73 (1992).
69. O. Farver and I. Pecht, *Proc. Natl. Acad. Sci. USA* **89**, 8283 (1992).
70. I. Yamazaki and L. H. Piette, *Biochim. Biophys. Acta* **50**, 62 (1961).
71. O. Farver and I. Pecht, *Mol. Cryst. Liq. Cryst.* **194**, 215 (1991).
72. L. -E. Andreasson, and B. Reinhammar, *Biochim. Biophys. Acta* **568**, 145 (1979).
73. L. -E. Andreasson, R. Bränden and B. Reinhammar, *Biochim. Biophys. Acta* **438**, 370 (1976).
74. T. Nakamura and Y. Ogawa, *J. Biochem.* (Tokyo) **64**, 267 (1968).
75. O. Farver, M. Goldberg and I. Pecht, *FEBS Lett.* **94**, 383 (1978).
76. R. E. Strothkamp and C. R. Dawson, *Biochem. Biophys. Res. Commun.* **85**, 655 (1978).
77. J. L. Cole, D. P. Ballou and E. I. Solomon, *J. Am. Chem. Soc.* **113**, 8544 (1991).
77a. P. A. Clark and E. I. Solomon, *J. Amer. Chem. Soc.* **113**, 8544 (1991).
78. G. Curson, *Biochem. J.* **100**, 295 (1966).
79. G. Curson and B. E. Speyer, *Biochem. J.* **105**, 243 (1967).
80. P. -O. Gunnarsson, U. Nylén and G. Petersson, *Eur. J. Biochem.* **27**, 572 (1972).
81. P. -O. Gunnarsson and G. Petersson, *Eur. J. Biochem.* **27**, 564 (1972).
82. R. Malkin, B. G. Malmström and T. Vänngard, *FEBS Lett.* **1**, 50 (1968).
83. R. Bränden, B. G. Malmström and T. Vänngard, *Eur. J. Biochem.* **36**, 195 (1973).
84. R. E. Strothkamp and C. R. Dawson, *Biochemistry* **16**, 1926 (1977).

85. C. T. Martin, R. H. Morse, R. M. Kanne, H. B. Gray, B. G. Malmström and S. I. Chan, *Biochemistry* **20**, 5147 (1981).
86. F. X. R. van Leewen, R. Wever, B. F. van Gelder, L. Aviliano and B. Mondovi, *Biochim. Biophys. Acta* **285**, 285 (1975).
87. M. M. Morie-Bebel, M. C. Morris, J. L. Menzie and D. R. McMillin, *J. Am. Chem. Soc.* **106**, 3677 (1984).
88. A. Messerschmidt, W. Steigemann, R. Huber, G. Lang, and P. M. H. Kroneck, *Eur. J. Biochem.* **209**, 597 (1992).
89. A. Messerschmidt, H. Luecke and R. Huber, *J. Mol. Biol.* **230**, 997 (1993).
90. R. Malkin, B. G. Malmström and T. Vänngard, *Eur. J. Biochem.* **7**, 253 (1969).
91. P. H. Hanna, D. R. McMillin, M. Pasenkiewicz-Gierula, W. E. Antholine and B. Reinhammar, *Biochemical J.* **253**, 561 (1988).
92. M. T. Graziani, L. Morpurgo, G. Rotilio and B. Mondovi, *FEBS Lett.* **70**, 87 (1976).
93. B. Reinhammar and Y. Oda, *J. Inorganic Biochem.* **11**, 115 (1979).
94. A. Schmidt-Klemens and D. R. McMillin, *J. Inorg. Biochem.* **38**, 107 (1990).
95. M. T. Graziani, P. Loreti, L. Morpurgo, I. Savini and L. Avigliano, *Inorg. Chim. Acta* **173**, 261 (1990).
96. L. Avigliano, A. Desideri, S. Urbanelli, B. Mondovi and A. Marchesini, *FEBS Lett.* **100**, 318 (1979).
97. L. Morpurgo, I. Savini, B. Mondovi and L. Avigliano, *J. Inorg. Biochem.* **29**, 25 (1987).
98. A. Merli, G. L. Rossi, M. Bolognesi, G. Gatti, L. Morpurgo and A. Finazzi-Agró, *FEBS Lett.* **231**, 89 (1988).
99. J. M. Guss, P. R. Harrowell, M. Murata, V. A. Norris and H. C. Freeman, *J. Mol. Biol.* **192**, 361 (1986).
100. W. E. B. Shephard, B. F. Anderson, D. H. Lewandowski, G. E. Norris and E. N. Baker, *J. Amer. Chem. Soc.* **112**, 7817 (1990).
101. H. Nar, Ph. D. Thesis, Technische Universität München (1992).
102. M. J. Schilstra, P. J. M. W. Birker, G. C. Verschoor and J. Reedijk, *J. Inorg. Chem.* **21**, 2631 (1982).
103. T. N. Sorell and M. R. Malachowski, *J. Inorg. Chem.* **22**, 1883 (1983).
104. A. Marchesini and P. M. H. Kroneck, *Eur. J. Biochem.* **101**, 65 (1979).
105. N. Kitajima, K. Fujisawa and Y. Moro-oka, *J. Am. Chem. Soc.* **111**, 8975 (1989).
106. K. A. Magnus, B. Hazes, H. Ton-That, C. Bonaventura, J. Bonaventura and W. G. J. Hol, *Proteins: Struct. Funct. Genetics* **19**, 302 (1994).
107. L. -E. Andreasson, R. Bränden, B. G. Malmström and T. Vänngard, *FEBS Lett.* **32**, 187 (1973).
108. R. Aasa, R. Bränden, J. Deinum, B. G. Malmström, B. Reinhammar and T. Vänngard *FEBS Lett.* **61**, 115 (1976).

109. R. Aasa, R. Bränden, J. Deinum, B. G. Malmström and T. Vänngard, *Biochem. Biophys. Res. Comm.* **70**, 1204 (1976).
110. R. Bränden and J. Deinum, *Biochim. Biophys. Acta* **524**, 297 (1978).
111. P.A. Clark and E. I. Solomon, *J. Am. Chem. Soc.* **114**, 1108 (1992).
112. M. D. Allendorf, D. J. Spira and E. I. Solomon, *Proc. Natl. Acad. Sci. USA* **82**, 3063 (1985).
113. D. J. Spira-Solomon, M. D. Allendorf and E. I. Solomon, *J. Am. Chem. Soc.* **108**, 5318 (1986).
114. J. L. Cole, L. Avigliano, L. Morpurgo and E. I. Solomon, *J. Am. Chem. Soc.* **113**, 9080 (1991).
115. L. Cassella, M. Gulotti, G. Pallanza, A. Pintar and A. Marchesini, *Biochem. J.* **251**, 441 (1988).
116. L. Casella, M. Gulotti, A, Pintar, G. Pallanza and A. Marchesini, *J. Inorg. Biochem.* **37**, 105 (1989).
117. B. Hathaway, *Compr. Coord. Chem.* **5**, 1 (1987).
118. R. A. Marcus and N. Sutin, *Biochim. Biophys. Acta* **811**, 265 (1985).
119. O. Farver, M. Goldberg, S. Wherland and I. Pecht, *Proc. Natl. Acad. Sci. USA* **75**, 5245 (1978).
120. C. M. Lieber, J. L. Karas, S. L. Mayo, M. Albin and H. B. Gray, in *Proc. of the Robert A. Welsh Conf. on Chemical Research, Design of Enzymes and Enzyme Model*, Robert A. Welsh Foundation (Houston, TX, 1987), pp. 9–33.
121. L. Avigliano, G. Rotilio, S. Urbanelli, B. Mondovi and A. Finazzi-Agró, *Arch. Biochem. Biophys.* **185**, 419 (1978).
122. H. B. Gray and B. G. Malmström, *Biochemistry* **28**, 7499 (1989).
123. G. McLendon, *Acc. Chem. Res.* **21**, 160 (1988).
124. S. L. Mayo, W. R. Ellis, Jr., R. J. Crutchley and H. B. Gray, *Science* **233**, 948 (1986).
125. J. N. Onuchic and D. N. Beratan, *J. Chem. Phys.* **92**, 722 (1990).
126. D. N. Beratan, J. N. Onuchic and H. B. Gray, in *Metal Ions in Biological Systems*, eds. H. Sigel and A. Sigel, (Dekker, New York, 1991), pp. 97–127.
127. D. N. Beratan, J. N. Onuchic, J. N. Betts, B. Bowler and H. B. Gray, *J. Am. Chem. Soc.* **112**, 7915 (1990).
128. A. G. Sykes, *Struct. Bonding* (Berlin) **75**, 175 (1991).

CHAPTER 3

THE STRUCTURE OF HUMAN CERULOPLASMIN AT 3.1 Å RESOLUTION

Peter F. Lindley
CCLRC Daresbury Laboratory, Warrington, Cheshire WA4 4AD, UK
E-mail: p.f.lindley@daresbury.ac.uk

and

Irina Zaitseva[‡*], Vjacheslav Zaitsev[‡*], Graeme Card[†*],
Kirill Moshkov[♯] and Benjamin Bax[†]
[‡]*Institute of Crystallography, Russian Academy of Science*
59 Leninsky prosp., Moscow 117333, Russia
[♯]*Institute of Experimental Medicine, Russian Academy of Medical Science*
ul. Acad. Pavlova 12, St. Petersburg 197376, Russia
[†]*Department of Crystallography, Birkbeck College, London WE1 7HX, UK*
[*]*CCLRC Daresbury Laboratory, Warrington WA4 4AD, UK*

ABSTRACT

X-ray crystallographic techniques have been used to elucidate the structure of human ceruloplasmin at a resolution of 3.1 Å. The structure reveals that the molecule is comprised of six plastocyanin-type domains arranged in a triangular array. There are six copper atoms. Three of these form a trinuclear cluster sited at the interface of domains 1 and 6 and are bound by eight histidine residues, four from each of the two domains. The three remaining copper atoms occupy mononuclear sites in domains 2, 4 and 6 and are coordinated by a cysteine and two histidine residues; the copper atoms in domains 4 and 6 also bind to a methionine residue, but in domain 2 the methionine is replaced by a leucine residue. The trinuclear centre and the mononuclear copper in domain 6 form a cluster essentially the same as that found in ascorbate oxidase, strongly suggesting that ceruloplasmin also has an oxidase activity. The relationship of ceruloplasmin to other "blue" copper oxidases is described and a putative functional role for ceruloplasmin in iron metabolism is discussed.

Contents

1. Introduction

1.1. *Ceruloplasmin (hCP) and Iron Metabolism*

Human ceruloplasmin (hCP) is a copper containing glycoprotein with a molecular weight of some 132 kDa; it was first isolated by Holmberg[1] in 1944. The molecule is comprised of a single polypeptide chain of 1046 amino acid residues and has a carbohydrate content of between 7 and 8 %.[2] A protein of the size and complexity of ceruloplasmin is likely to be multi-functional[3] and at least four functions have been attributed to ceruloplasmin as follows; copper transport, as a ferroxidase, as an amine oxidase, and as an antioxidant in the prevention of the formation of free radicals in serum. However, recent studies on yeast *S. cerevisiae*[4,5] have focussed particular interest on its role as a ferroxidase. These studies show that yeast cells accumulate iron by a process requiring a ferrireductase and a ferrous transporter. Sequence analysis of the corresponding gene reveals a protein, Fet3, with strong similarity to the family of blue multi-copper oxidoreductases. Consistent with the role of copper in iron transport, growth of the wild-type cells in copper-deficient media resulted in decreased ferrous iron transport, suggesting that the catalytic activity of the Fet3 protein is required for cellular iron accumulation. In contrast, mammalian cells have an iron transport system involving transferrin, whereby ferric iron is transported from sites of storage and absorption to sites of utilisation through a mechanism of receptor mediated endocytosis.[6] However, the incorporation of iron into transferrin from sites of storage in mammalian systems requires the release of iron from the iron storing cells and its oxidation to ferric iron. Although *in vitro* transferrin can readily oxidise iron (a standard method of loading transferrin with ferric iron is to add ferrous sulphate to the *apo*-protein), *in vivo* some control may be required to minimise the chances of ferrous ions causing free radical formation and subsequent tissue damage; ceruloplasmin may fulfil this custodial role. The role of hCP may therefore be to facilitate the release of iron from the storage cells and to assist in its incorporation into transferrin, prior to transport to sites of utilisation. In circumstances where there is a deficiency of free oxygen, hCP may play a dominant role in iron oxidation.

Further evidence for the involvement of ceruloplasmin in iron metabolism is provided by the identification of a mutation in the ceruloplasmin gene which is associated with systemic hemosiderosis in humans.[7,8] (see Sec. 3.5).

1.2. *Ceruloplasmin and Hereditary Diseases*

In addition to its link with iron metabolism, ceruloplasmin is associated with two hereditary disorders of copper transport[9]; Menkes "kinky hair" disease and Wilson's disease (hepatolenticular degeneration). In Menkes disease, gene mutations cause the defective export and/or intracellular transport of copper concomitant with a deficiency of copper-dependent enzymes such as hCP. Wilson's disease is a rare genetic disorder whereby gene mutations indicate defective biliary excretion of copper and a reduction in the rate of copper incorporation into hCP, leading to accumulation of copper in the liver and brain.

1.3. *Ceruloplasmin: A Member of the Family of "Blue" Copper Oxidases*

hCP belongs to a family of multi-nuclear "blue" copper oxidases which include ascorbate oxidase and laccase. In addition, sequence analysis indicates a domain structure with internal triplication and a high homology with the A-type domains of the blood clotting factors V and VIII[10,11]; there is a 34 % sequence identity between hCP and the three A-type domains of factor VIII. The X-ray structure of ascorbate oxidase (AO) has been reported.[12,13] AO crystallises as a dimer or tetramer, but the subunit containing 552 amino acid residues is comprised of three domains, each of which has a fold typical of the small blue copper proteins, azurin[14,15] and plastocyanin.[16] The subunit of AO contains a typical blue mononuclear type I copper centre in domain 3 and a trinuclear copper centre at the interface between domains 1 and 3. The trinuclear centre comprises two type III coppers bridged by an oxygen atom and a type II copper, which also binds an oxygen atom, (see Ref. 17 for classification of the copper atoms according to their visible and EPR spectra). The centre is held by eight histidine residues, four from each of domains 1 and 3. Using this structure and sequence comparison techniques Messerschmidt and Huber[18] were able to provide evidence that hCP has a 6-domain structure involving a three-fold repeat and that domains 2, 4 and 6 should bind mononuclear coppers with a possible trinuclear site between domains 1 and 6. The X-ray structure of nitrite reductase (NR) has also been reported[19] showing that it comprises three dimers, each of two domains, arranged in a triangular array. Copper atoms in NR are located in domain 1 and at the interface of domain 1 and domain 2′ of another dimer. The combination of sequence and structural identity between AO, NR and hCP enabled Fenderson *et al.*[20] to formulate a plausible model for the domain structure of hCP. In this model the six domains of hCP are arranged in a triangular trimeric structure based on a two-domain monomer.

The intention of this article is to describe the overall organisation of the human ceruloplasmin molecule, as determined by X-ray diffraction techniques at a resolution of 3.1 Å, the nature of the 6 copper binding sites, the relationship of the enzyme to other blue copper oxidases, and a putative functional role for the enzyme in iron metabolism.

2. Experimental

2.1. *Crystallisation and Preliminary X-ray Studies*

Preparations of hCP exhibit heterogeneity, particularly with respect to the carbohydrate moieties, and the protein is very susceptible to proteolytic cleavage, aggregation and loss of copper.[21] The hCP samples used in the present study were isolated and purified as described by Moshkov *et al.*[22] with the protease inhibitors, phenylmethyl-sulphonylfluoride and ε-aminocaproic acid, being used at all stages. Crystals for the X-ray studies were obtained from sitting drops, held at 277 K, and initially containing 28–30 mg/ml hCP, 3 % (w/v) PEG-20,000, 250 mM NaCl, and 100 mM sodium acetate buffer at pH 5.45; all buffer solutions contained 4 μM sodium merthiolate as a bacteriostatic agent. Crystals appeared in 5–7 days and grew typically over a month to dimensions of $0.5 \times 0.5 \times 0.25$ mm. Prior to diffraction measurements or heavy atom soaking experiments, crystals were stabilised in 6 % PEG-20,000, 250 mM NaCl and 100 mM sodium acetate buffer at pH 5.65. The crystals belong to the trigonal space group $P3_221$ with cell dimensions, a = 213.9(2) Å and c = 85.6(1) Å and one molecule of hCP in the asymmetric unit. Native crystals diffract to at least 3.1 Å using synchrotron radiation. Two heavy atom derivatives were prepared by soaking crystals in the stock solution with *p*-chloromercurybenzoate and $KAu(CN)_2$, the concentration of the reagents being increased stepwise from 0.1 mM to 1.0 mM for *p*-CMB and from 0.05 mM to 0.1 mM for the gold derivative over a five day period.

2.2. *X-ray Data Collection*

Diffraction data were collected from the native protein crystals and the two heavy atom derivatives at 276K by the oscillation method using station 9.6 at the SRS, CCLRC Daresbury Laboratory. Station 9.6 is equipped with a platinum coated fused quartz cylindrical mirror for vertical focussing and a bent triangular Si(111) monochromator for horizontal focussing and wavelength selection; the wavelength used was 0.882 Å with a band pass, $\Delta\lambda/\lambda = 0.0004$. During the data collection the synchrotron storage ring was operating at an energy of 2 Gev with an average circulating ring current of 200 mA. The detector was a 30 cm diameter Mar-Research image plate system. All the oscillation images were processed using the MOSFLM[23] suite of programs based on the Cambridge program MOSCO[24] and final scaling and data reduction was achieved with ROTAVATA and AGROVATA (CCP4 suite of programs[25]). For the native data, collected from 2 crystals, some 131,000 measurements yielded 38,793 unique reflections ($R_{\text{merge}} = 0.057$) with an overall completeness of 94.7 % and with 83.1 % of the data having $I > 3\sigma(I)$; for the highest resolution shell $R_{\text{merge}} = 0.247$, completeness = 82.3 %, and $I > 3\sigma(I) = 53.8$ %.

2.3. *Structure Analysis and Refinement*

The structure was solved using the conventional Multiple Isomorphous Replacement technique in conjunction with density modification, (program DM[26]). Protein phases were generated using the program MLPHARE,[27] and map interpretation was undertaken using the computer graphics O program package.[28] Initial refinement has been undertaken using a combination of XPLOR[29] and RESTRAIN.[30] Positional parameters have been refined for all atoms, but only main chain and side chain group isotropic thermal parameters for the individual residues due to the resolution limit of 3.1 Å. The current value of the overall R-factor is 21.5 % for all data in the resolution range 12.0–3.1 Å. The present model is well-defined, but contains main chain breaks at the interconnecting peptide between domains 2 and 3, residues 341–346, in an exposed surface loop in domain 3, residues 471–482, at the interconnecting region between domains 5 and 6, residues 885–890, and at the C-terminus, residues 1041–1046. In these regions the electron density is either very difficult to interpret or missing entirely. In addition, parts of the extended loop regions are poorly defined in the electron density and further fine-tuning of the model is in progress. Validation of the current structure with PROCHECK[31] shows that both the main chain and side chain geometries are within or better than the expected limits for equivalent structures at 3.1 Å resolution. Some 74.5 % of the residues have φ/ψ torsion angle values in the allowed regions of the Ramanchandran plot, 24.2 % in the additionally allowed regions, and 1.2 % in the generously allowed regions; all the residues in this last batch lie in poorly defined regions of electron density.

3. The Structure of Ceruloplasmin: A Putative Functional Role

3.1. *The Structure of hCP*

3.1.1. *Overall Molecular Organisation*

Figure 1a is a schematic view of the overall organisation of the molecule showing the 6 domains and the distribution of the copper atoms. Figure 1b shows a trace of the polypeptide α-carbon backbone viewed along the pseudo three-fold axis of the molecule. Each of the domains comprises an eight stranded β-barrel with the strands organised in a manner typical of the small blue copper proteins such as azurin and plastocyanin.[32] The odd numbered domains comprise approximately 200 residues, but the even domains are shorter with only about 150 residues. The barrels are oriented so that they are all the same way up, but the even numbered domains point inwards towards the pseudo three-fold axis reminiscent of the feet of a tripod, whereas the odd numbered domains point outwards; this arrangement has a pronounced effect on the separation of the mononuclear copper sites

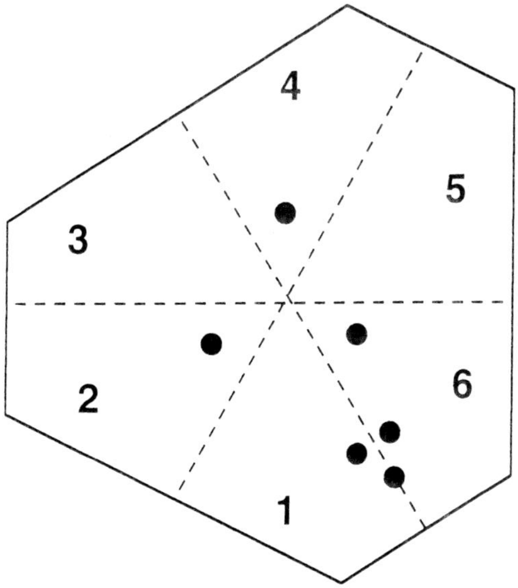

(a)

Fig. 1. The overall organisation of the hCP molecule. (a) A schematic representation of the overall organisation of the molecule of human ceruloplasmin. Domains 2, 4 and 6 contain mononuclear copper centres, whilst the trinuclear copper cluster is located at the interface of domains 1 and 6. (b) An α-carbon ribbon diagram of the human ceruloplasmin molecule viewed along the pseudo three-fold axis. Domains 1, 3 and 5 are depicted by striped motifs, whereas domains 2, 4 and 6 are dark shaded. The copper atoms are depicted as black circles and the amino acid side chains involved in copper binding are also drawn.

Residues 341–346, 885–890, and 1041–1046 cannot be located unequivocally in the present electron density syntheses. There is also a chain break in domain 3 between residues 476–482 (on the left-hand side of the figure). Interestingly this last region of ill-defined density corresponds to the cleavage site between the 67 and 50 kDa fragments.

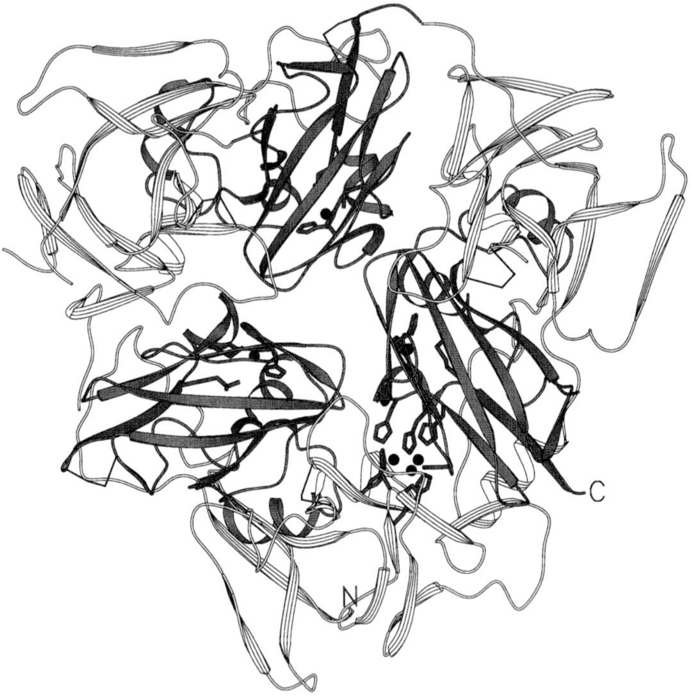

(b)

Fig. 1. *(Continued)*

(see Sec. 3.2.2). Figure 2 shows the fold of domains 1 and 2 of hCP respectively and key features are the extended loops between the first and second strands; these strands effectively cover the top of the molecule and may function to restrict access to the mononuclear copper binding sites. It is pleasing to note that the overall organisation of the molecule is consistent with that derived from electron microscopy.[33]

3.1.2. *Structural Similarity of the Domains within hCP*

There is considerable similarity between domains 1, 3 and 5 with excellent matching of the β-strands; 171 α-carbon atoms of domain 3 and 150 atoms of domain 5 can be superimposed on domain 1 with rms fits of only 1.0 Å and 1.1 Å respectively. Major deviations occur in the loops between

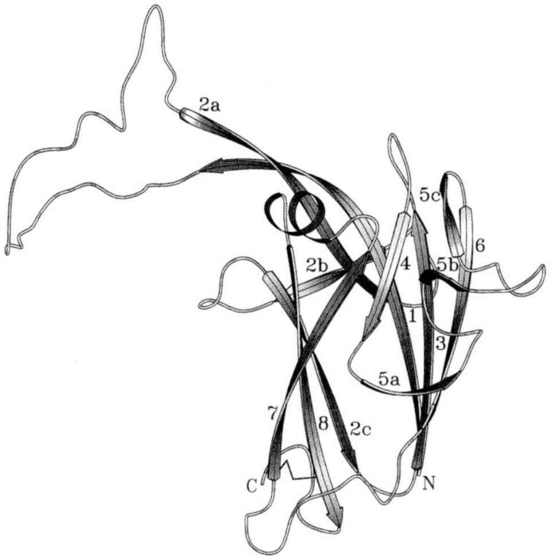

(a)

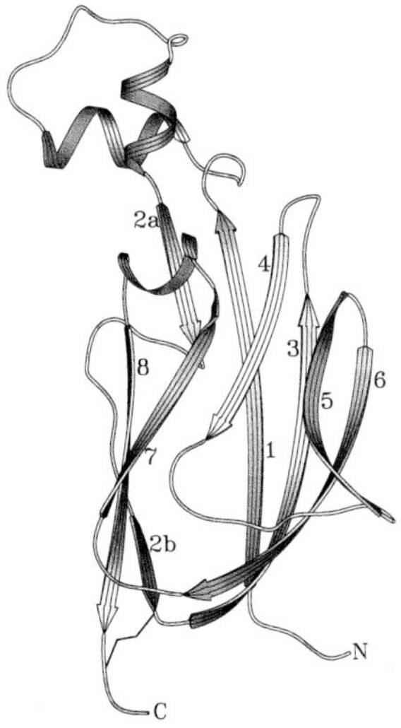

(b)

Fig. 2. The domain folds. The α-carbon ribbon diagrams of domains 1 and 2 in hCP, (shaded by 1 and 2 stripes respectively). Each domain has a plastocyanin-like fold. Major differences between the odd domains, which do not have mononuclear copper binding sites and the smaller even domains which bind type II coppers, lie in the loop regions between strands 1 and 2, and, 5 and 6.

Table 1. Superposition of α-carbon atoms for domains in hCP.

(a) Domains resembling the odd domains in hCP [rms fit (Å) and number of atoms].

	DM1	DM3	DM5	NR1	AO1
DM3	1.05				
	(171)				
DM5	1.08	0.99	—		
	(150)	(155)			
NR1	1.16	1.25	1.69		
	(103)	(104)	(111)		
AO1	1.29	1.31	1.37	1.34	—
	(110)	(110)	(106)	(102)	

(b) Domains resembling the even domains in hCP [rms fit (Å) and number of atoms].

	DM2	DM4	DM6	NR2	AO2	AO3
DM4	0.90	—				
	(146)					
DM6	0.92	0.69	—			
	(141)	(144)				
NR2	1.71	1.42	1.45			
	(100)	(99)	(100)			
AO2	1.79	1.70	1.64	1.64	—	
	(97)	(98)	(101)	(110)		
AO3	1.57	1.41	1.33	1.53	1.42	
	(100)	(103)	(101)	(110)	(98)	

Other important matches include;

DM2 to DM1	1.75 (91)
NR1 to DM2	1.52 (94)
NR2 to DM1	1.60 (87)

All superpositions were achieved with the "lsq_improve" option in the O program suite.[28]

strands 1 and 2, and in the 5b to 5c region in Fig. 2a. The similarity of the smaller mononuclear copper binding domains 2, 4 and 6, is even more pronounced with 146 atoms from domain 4 and 141 from domain 6 fitting domain 2 with rms deviations of only 0.9 Å in each case. However, although all 6 domains are based on a plastocyanin type fold, the various loop regions deteriorate the match between an even and an odd domain with a typical fit of 1.8 Å for only 91 atoms when domain 2 is superposed onto domain 1. The superposition results are summarised as part of Table 1.

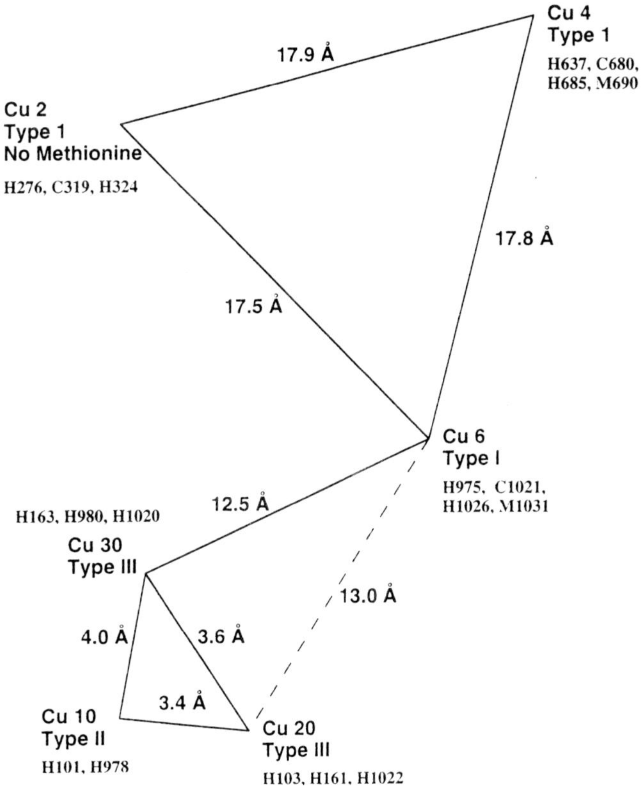

Fig. 3. Interatomic copper distances and copper binding residues. A schematic diagram describing the interatomic separations of the copper centres and their coordinating residues (see also Table 2).

3.2. *The Copper Binding Sites*

3.2.1. *Inter-Atomic Distances and Sequence Alignment*

The human ceruloplasmin structure contains six copper atoms; three of these occupy mononuclear centres in domains 2, 4 and 6, whereas the remaining three form a trinuclear cluster sited at the interface between domains 1 and 6. The disposition of the trinuclear cluster and the nearest mononuclear copper, that in domain 6, is closely similar to that found in the subunit of ascorbate oxidase,[12,13] strongly suggesting that hCP may also have an oxidase function. Figure 3 shows the distance between the copper atoms, the EPR and visible spectra classification,[17] and the binding ligands. All of the copper binding residues are sited either on strand 4 of the β-barrel or on the loop region connecting strands 7 and 8; this is shown in Table 2, together with the corresponding sequences for nitrite reductase,[19] ascorbate oxidase[12,13] and plastocyanin.[16]

Table 2. Sequence alignment for copper binding sites.

(a) β-Strand 4.

(i) Domains which do NOT bind mononuclear copper.

					↓		↓				
DM1	98	Y	T	F	**H**	S	**H**	G	I	T	106
DM3	454	L	S	I	E	P	I	G	V	R	462
DM5	813	Y	S	I	H	A	H	G	V	Q	821
NR2	A252	T	R	P	**H**	L	I	G	G	H	A260
AO1	A57	V	V	I	**H**	W	**H**	G	I	L	A65
AO2	A226	A	L	N	F	A	I	G	N	H	A234

(ii) Domains which contain mononuclear copper sites.

		⇓			↓		↓				
DM2	276	**H**	A	A	F	F	H	G	Q	A	284
DM4	637	**H**	G	I	Y	F	S	G	N	T	645
DM6	975	**H**	T	V	**H**	F	**H**	G	H	S	983
NR1	A95	**H**	N	I	D	F	**H**	A	A	T	A103
AO3	A445	**H**	P	W	**H**	L	**H**	G	H	D	A453
PCY	37	**H**	N	I	V	F	D	.	.	.	42

(b) β-Strands 7 through to 8.

(i) Domains which do NOT bind mononuclear copper.

										↓		↓																						
DM1	155	C	–	–	V	T	R	I	Y	H	S	H	I	–	–	D	A	P	K	D	I	A	S	–	G	L	I	G	P	L	I	I	C	181
DM3	515	C	–	–	L	A	K	M	Y	Y	S	A	V	–	–	D	P	T	K	D	I	F	T	–	G	L	I	G	P	M	K	I	C	541
DM5	855	C	–	–	I	P	W	A	Y	Y	S	T	V	–	–	D	Q	V	K	D	L	Y	S	–	G	L	I	G	P	L	I	V	C	881
NR2	A298	P	–	–	G	V	Y	A	Y	V	N	H	N	–	–	–	–	L	I	E	A	F	E	L	G	A	A	A	H	F	K	V	T	A323
AO1	A99	·	·	·	G	T	F	F	Y	H	G	H	L	–	–	G	–	–	M	Q	R	S	A	–	G	L	Y	G	S	L	I	V	D	A122
AO2	A271	N	P	S	E	N	Y	W	V	S	V	G	T	–	–	·	·	·	·	·	·	·	·	–	·	G	L	T	L	L	N	Y	L	A299

(ii) Domains which contain mononuclear copper sites.

										↓	⇓	↓								⇓						⇓								
DM2	312	P	–	–	G	A	W	M	L	A	C	Q	N	–	–	–	–	L	N	H	L	K	A	–	G	L	Q	A	F	F	Q	V	Q	336
DM4	673	E	–	–	G	T	F	N	V	E	C	L	T	–	–	–	–	T	T	H	Y	T	G	–	G	M	K	Q	K	Y	T	V	N	697
DM6	1014	P	–	–	G	I	W	L	L	H	C	H	V	–	–	–	–	T	D	H	I	H	A	–	G	M	E	T	T	Y	T	V	L	1038
NR1	A130	·	·	·	G	V	F	V	Y	H	C	A	P	P	G	M	V	P	W	H	V	T	S	–	G	M	N	G	A	I	M	V	L	A157
AO3	A500	P	–	–	G	V	W	A	F	H	C	H	I	–	–	–	–	E	P	H	L	H	M	–	G	M	G	V	V	F	A	E	G	A524
PCY	77	P	–	–	G	T	Y	G	V	Y	C	D	P	–	–	–	–	–	–	H	S	C	A	–	G	M	K	M	T	I	T	V	Q	99

A dash indicates a sequence deletion, a dot indicates a structural mismatch.

DM1-6, NR1-2, AO1-3 refer to domain 1, 2, 3, etc in human ceruloplasmin, nitrite reductase,[19] and ascorbate oxidase[12,13] respectively. PCY refers to the plastocyanin from green algae (*Enteromorpha prolifera*).[16]

The sequence matches were obtained using the "lsq_improve" option in the O program suite.[28]

A single arrow over a column indicates residues binding to the trinuclear centres, a double arrow indicates residues binding to the type I copper sites.

3.2.2. *The Mononuclear Copper Centres*

The mononuclear copper atoms in domains 4 and 6 have typical type I blue copper environments with two histidine ligands, a cysteine and a methionine. The Cu-N and Cu-S(Cys) distances are around 2.1(1) Å with the Cu-S(Met) at a distance of approximately 3.0(1) Å. In the case of the domain 6 copper the main chain carbonyl O of residue L974 appears to point towards the Cu atom as found in the oxidised form of azurin from *Alcaligenes denitrificans*.[14,15] However, the precise orientation of this peptide group must await a higher resolution structure. In EPR terms the type-I coppers in domains 4 and 6 are likely to be "fast" coppers, because of their resemblance to those in azurin, plastocyanin and ascorbate oxidase. The mononuclear copper in domain 2 is structurally different from those in domains 4 and 6 in that it lacks the methionine residue; a non-polar leucine residue replaces the methionine and does not appear to have any significant interaction with the copper atom.

The disposition of the mononuclear coppers with interatomic separations of around 18 Å is intriguing; had nature organised the molecule so that the copper atoms were bound in the odd numbered domains these inter-mononuclear copper distances would increase substantially since these domains point outwards from the pseudo three-fold axis. Langen *et al.*[34] have recently studied the networks for coupling the internal redox centres in azurin and cytochrome c with the surface of these proteins. A distance of 18 Å appears within the range for efficient coupling and it is interesting to speculate that the coppers in domain 2 and 4 in hCP increase the possibility of electron capture, prior to the transfer of the electron *via* the domain 6 copper to the oxidase centre involving the trinuclear cluster. Furthermore, the different environment of the domain 2 copper, where a leucine residue replaces the methionine found in the other two mononuclear sites, may have an effect on the speed of electron transfer.

3.2.3. *The Trinuclear Copper Cluster*

The trinuclear cluster is surrounded by four pairs of histidine residues with Cu-N distances of approximately 2.1(1) Å, Fig. 4. Domain 1 donates two pairs, H101, H103 and H161, H163 with a serine residue separating the components of each pair and domain 6 the remaining pairs H978, H980 and H1020, H1022, separated by phenylalanine and cysteine residues respectively. The cysteine residue between H1020 and H1022 coordinates to the mononuclear centre in domain 6. Two of the copper atoms (type III coppers) in the trinuclear centre are bound by three histidines, (H103, H161 and H1022; H163, H980 and H1020), whilst the copper furthest away from the domain 6 copper atom is bound by only two histidine residues (H101 and H978) and is therefore designated type II; in all cases the pairs of histidine residues bridge two copper atoms. However, the electron density at the trinuclear centre is consistent with the model of the trinuclear centre found in AO whereby there is one oxygen atom bridging the type III coppers and a second attached to the type II copper. The presence of a third oxygen atom in the middle of the trinuclear cluster can neither be discounted nor substantiated with the present resolution and quality of data. The oxygen atom attached to the type II copper is well-defined and hydrogen bonds to

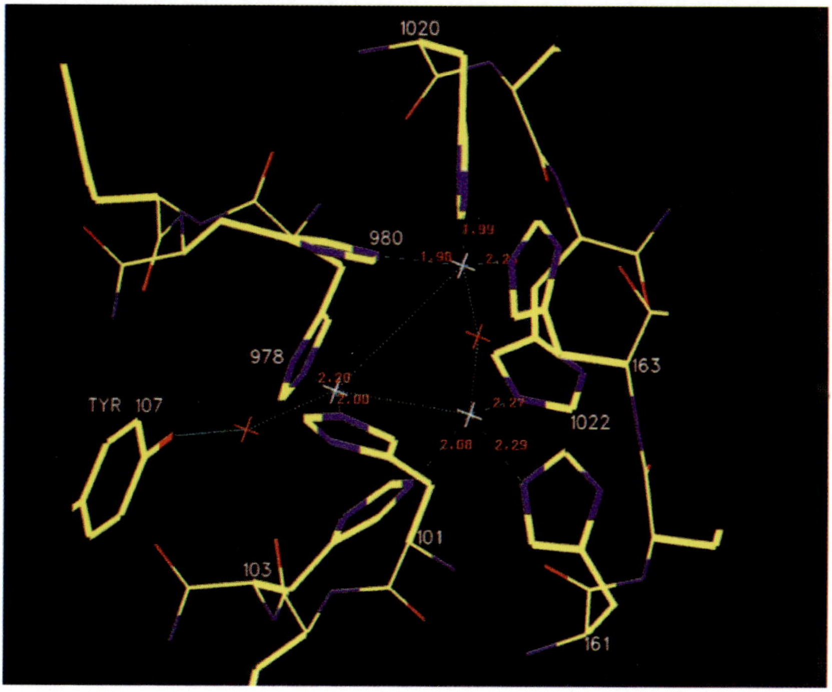

Fig. 4. The trinuclear copper centre. The trinuclear copper centre showing the 8 coordinating histidine residues, 4 each from domains 1 and 6. Oxygen atoms have been placed adjacent to the type II copper, Cu 10, and between the two type III coppers in accordance with the structure of AO. This does NOT discount the presence of a further oxygen atom in between the copper atoms; higher resolution data is required to resolve this detail.

the hydroxyl group of Y107 and the main chain carbonyl oxygen of S102; in AO the OH group of the tyrosine is replaced by a water molecule which hydrogen bonds to the side chain of Q66.

3.3. *Disulphide Bridges and Cleavage Sites*

hCP possesses five disulphide bridges and one free cysteine residue. The disulphide bridges are located in domains 1 to 5 and serve to anchor the last strand of each domain; the C-terminal strand in domain 6 is not anchored to domain 6 through a disulphide bridge. In domains 1, 3 and 5 the disulphide bridges link strands 7 and 8 with the cysteines being separated along the chain by 25 residues; for domains 2 and 4, the disulphide bridges link strands 2b and 8 with 80 residues separating the cysteines. The free cysteine residue, C221, is located in domain 2, on the extended loop between the first and second strands and appears to be readily accessible.

Ceruloplasmin is readily cleaved into three fragments with molecular weights of 67, 50 and 19 kDa respectively.[35] The 19 kDa fragment corresponds almost exactly to domain 6, so that the cleavage site is inter-domain in nature. However, the cleavage site for the larger fragments occurs at residue 479, a residue in domain 3 on an exposed loop prior to strand 6 (see Fig. 2a).

3.4. *The "Blue" Copper Oxidase Family*

Ceruloplasmin is a member of the family of blue copper oxidases which also contains laccase and ascorbate oxidase. The relationship between hCP and AO is quite marked and extends beyond the presence of a trinuclear copper centre sited some 12–13 Å away from a mononuclear copper atom. Superposition of the three domains of AO onto the hCP structure shows that domains 1 and 3 of AO correspond to domains 1 and 6 of hCP, so that AO domain 3 and hCP domain 6 both contain a type I copper and that the trinuclear copper cluster lies at the domain interface. Furthermore, AO domain 2, although it does not bind copper, is spatially equivalent to domain 2 of hCP; that is, the subunit of AO is very similar to half the hCP pseudo hexamer. Details of these superpositions are given in Table 1. A superposition of the NR trimer and hCP further shows that the six respective domains have a similar spatial arrangement, with one of the interdomain mononuclear copper atoms of NR occupying a position equivalent to the trinuclear cluster in hCP. However, the intradomain coppers are bound in the three domains 1 and not domains 2.

The data in Table 1 appears to indicate that domain 1 of nitrite reductase is structurally closer to domain 1 of hCP rather than to domain 2; in a similar manner domain 2 of NR appears structurally more similar to domain 2 of hCP than domain 1, despite the reversal in copper binding capability. It therefore seems likely that AO, NR and hCP have all evolved from a two domain ancestral protein (gene duplication of a cupredoxin) which had type I coppers in both domains. Various evolutionary pathways can be considered from the formation of this precursor. For example, loss of copper binding capability in domain 1, followed by the insertion of a spacer domain, domain 2, would enable the formation of a three-domain subunit as found in AO; the formation of a trinuclear binding site at the interface of domains 1 and 3 would be the last stage in the evolutionary pathway. In the case of hCP gene triplication would lead to a six-domain protein, followed by the formation of a trinuclear binding site between domains 1 and 6 (equivalent to AO domains 1 and 3). On the other hand, the loss of copper binding capability from domain 2 of the two-domain precursor, followed by trimerisation and the formation of a mononuclear binding site between domain 1 and domain 2′ from an adjacent element of the trimer would lead to the nitrite reductase structure.[19] The relationships between hCP, AO and NR are shown schematically in Fig. 5.

Figure 5 also indicates a putative model for the blood clotting factor VIII[10,11] in which the A_1, A_2 and A_3 domains can be considered

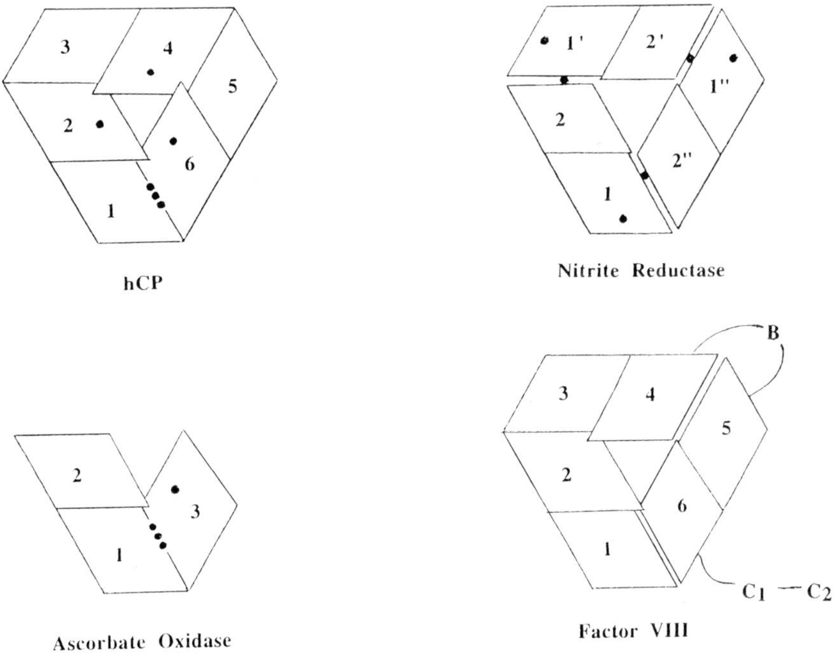

Fig. 5. The family of "blue" copper oxidases. A schematic diagram indicating the putative structural relationships between hCP, AO, NR and factor VIII.

equivalent to hCP domains 1 and 2, 3 and 4, 5 and 6 respectively; i.e. each A domain consists of 2 hCP domains. The C_1 and C_2 domains of factor VIII may form a C-terminal tail and the B domain, which is excised during factor VIII activation, could form a protective cover over the bottom of the molecule. This model and further details of the structural relationships in the blue copper oxidase family of proteins will be pursued elsewhere.

3.5. *A Putative Role for Ceruloplasmin in Iron Metabolism*

As indicated in Sec. 1.1 the view that ceruloplasmin is intricately linked with iron metabolism has recently received renewed attention. Recent papers[7,8] associating a hereditary ceruloplasmin deficiency with aceruloplasminemia and severe iron deposition in visceral organ and brain tissues have focused attention on the putative role of ceruloplasmin as a ferroxidase. In certain cases, systemic hemosiderosis appears to be caused by incomplete expression of the hCP molecule, with the polypeptide chain being terminated at residue 991. This observation gives credence to the proposition that the enzyme may be involved in facilitating the release of intracellular iron and its oxidation prior to uptake and transport by transferrin. Further, the X-ray structure provides structural explanations as to why the truncated molecule may have its oxidase function severely impaired. Thus, Fig. 6 shows the trace of the α-carbon polypeptide backbone for domains 1 and 6 of human ceruloplasmin only; domain 1 is coloured red on the left-hand side of the figure. The magenta and blue structures on the right-hand side of the figure represent domain 6 and, in particular, the blue trace is the portion of the polypeptide chain, residues 992 to 1046 inclusive, which would be missing if incorrect expression terminated the chain at 991. The copper binding ligands that would be concomitantly missing are shown in yellow and these comprise three of the ligands that bind to the domain 6 mononuclear copper (C1021, H1026 and M1031). In the absence of these ligands, it can be concluded that domain 6 is very unlikely to bind copper and this may have a profound effect on the oxidase efficiency of the enzyme. In addition, two of the histidine residues that bind to the trinuclear cluster (H1020 and H1022) are also absent which gives rise to the possibilities that this cluster may not bind in the same way as in the intact enzyme or that it may not bind at all; in either case this will affect the oxidase activity of the enzyme. Furthermore, the amino acid residues on the missing peptide

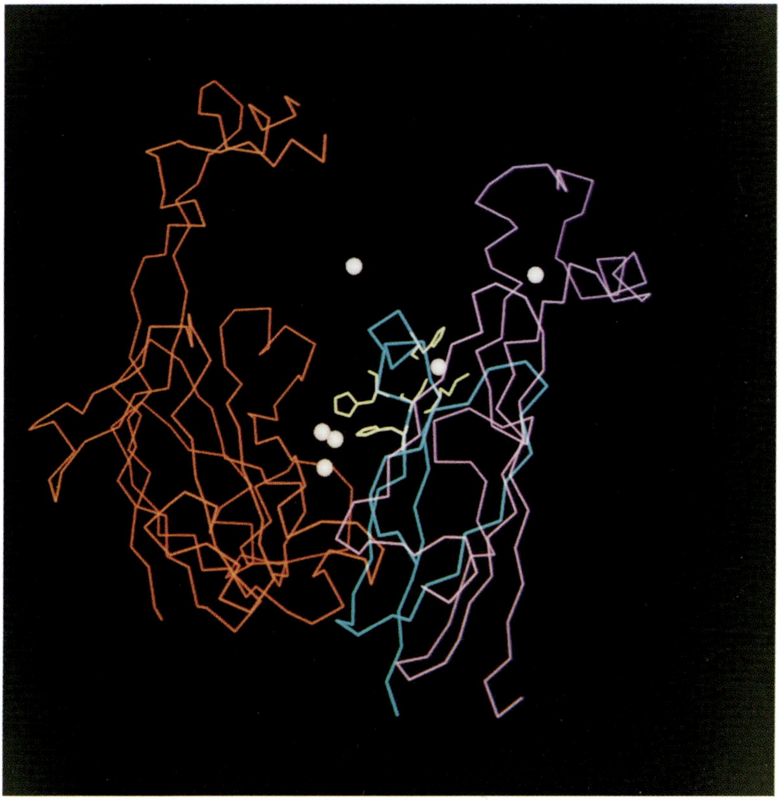

Fig. 6. Ceruloplasmin and systemic hemosiderosis. Traces of the α-carbon polypeptide backbone of domains 1 and 6 in the hCP structure. Domain 1 is shown on the left-hand side in red; this domain contributes 4 histidine residues (not shown) to the trinuclear cluster; copper atoms are depicted as white spheres. Domain 6 is on the right-hand side of the figure and also contributes 4 histidine residues to the cluster. The portion of the polypeptide chain coloured blue is that which is missing in the truncated enzyme. This polypeptide, residues 992 to 1046 inclusive, includes 2 histidine residues bound to the trinuclear copper centre and 3 residues bound to the mononuclear copper in domain 6; these residues are depicted in yellow. The absence of the C-terminal polypeptide would also remove over 50 % of the interdomain hydrogen-bond and iron pair interactions observed in the intact enzyme.

(coloured blue in Fig. 6) contribute over 50 % of the hydrogen-bond and ion pair interactions between domains 1 and 6 in the intact enzyme. It is therefore highly probable that the incomplete enzyme will adopt a different organisation of the tertiary structure and may well exist in an open configuration. Such a configuration will be even more susceptible to proteolytic cleavage than the correctly folded enzyme.

The net effect of the missing C-terminal portion of the polypeptide chain is that the enzyme may not fold into the organisational tertiary structure depicted schematically in Fig. 1 and that its ability to act as an oxidase will be severely impaired with respect to the intact enzyme. Incomplete folding and increased susceptibility to proteolytic cleavage would be consistent with the observed aceruloplasminemia such as the failure to find ceruloplasmin in liver extracts by the Western blot and ELISA methods. An impairment of the ferroxidase activity would be consistent with a gradual accumulation of intracellular iron leading to systemic hemosiderosis; the truncated enzyme would not mediate the release of iron from cells nor its oxidation prior to uptake by transferrin.

Acknowledgments

We gratefully acknowledge the Medical Research Council (UK) for financial support (IZ & VNZ), the Engineering and Physical Sciences Research Council (formerly SERC) for a research studentship (GC), the support of our colleagues at CCLRC Daresbury Laboratory (the Protein Crystallography Project Team), and Professor B. Vainshtein (Institute of Crystallography, Moscow) who enabled and encouraged the collaboration between the UK and Russian research teams. Figures 1b, 2, and 4 were drawn using the SETOR[38] suite of programmes.

References

1. C. G. Holmberg, *Acta Physiol. Scand.* **8**, 227 (1944).
2. N. Takahashi, T. L. Ortel and F. W. Putnam, *Proc. Natl. Acad. Sci. (USA)* **81**, 390 (1984).
3. E. Frieden, *Metal Ions in Biological Systems* **13**, ed. H. Sigel (Marcel Dekker, New York, 1981), p. 110.
4. C. Askwith, D. Eide, A. Van Ho, P. S. Bernard, L. Li., D. Davis-Kaplan, D. M. Sipe and J. Kaplan, *Cell* **76**, 403 (1994).
5. A. Dancis, D. S. Yuan, D. Haile, C. Askwith, D. Eide, C. Moehle, J. Kaplan and R. D. Klausner, *Cell* **76**, 393 (1994).
6. A. Dautry-Varsat and H. F. Lodish, *Sci. Am.* **250**, 52 (1984).
7. H. Morita, S. Ikeda, K. Yamamoto, S. Morita, K. Yoshida, S. Nomoto, M. Kato and N. Yanagisawa, *Ann. of Nerology* **37**, 646 (1995).
8. K. Yoshida, K. Furihata, S. Takeda, A. Nakamura, K. Yamamoto, H. Morita,

S. Hiyamuta, S. Ikeda, N. Shimizu and N. Yanagisawa, *Nature Genetics* **9**, 267 (1995).
9. D. M. Danks, *The Metabolic Basis of Inherited Diseases*, eds. C. R. Scriver, A. L. Beaudet, W. S. Sly and D. Valle (McGraw-Hill, New York, 1989), pp. 1411–1431.
10. W. R. Church, R. L. Jernigan, J. Toole, R. M. Hewick, J. Knopf, G. J. Knutson, M. E. Nesheim, K. G. Mann and D. N. Fass, *Proc. Natl. Acad. Sci. (USA)* **81**, 6935 (1984).
11. Tuddenham, E. G. D. *Bailliâte's Clin. Haematology* **2**, 849 (1989).
12. A. Messerschmidt, A. Rossi, R. Ladenstein, R. Huber, M. Bolognesi, G. Gatti, A. Marchesini, R. Petruzzelli and A. Finazzi-Agró, *J. Mol. Biol.* **206**, 513 (1989).
13. A. Messerschmidt, R. Ladenstein, R. Huber, M. Bolognesi, L. Avigliano, R. Petruzzelli, A. Rossi and A. Finazzi-Agró, *J. Mol. Biol.* **224**, 179 (1992).
14. E. N. Baker, *J. Mol. Biol.* **203**, 1071 (1988).
15. H. Nar, A. Messerschmidt, R. Huber, M. van der Kamp and G. W. Canters, *J. Mol. Biol.* **221**, 765 (1991).
16. C. A. Collyer, J. M. Guss, Y. Sugimura, F. Yoshizaki and H. C. Freeman, *J. Mol. Biol.* **211**, 617 (1990).
17. B. G. Malmström, *Ann. Rev. Biochem.* **51**, 21 (1982).
18. A. Messerschmidt and R. Huber, *Eur. J. Biochem.* **187**, 341 (1990).
19. J. W. Godden, S. Turley, D. C. Teller, E. T. Adman, M. -Y. Liu, W. J. Payne and J. LeGall, *Science* **253**, 438 (1991).
20. F. F. Fenderson, S. Kumar, E. T. Adman, M. -Y. Liu, W. J. Payne and J. LeGall, *Biochem.* **30**, 7180 (1991).
21. L. Ryden, *Copper Proteins and Copper Enzymes* **3**, ed. R. Lontie (CRC Press, Boca Raton, Florida, 1984), pp. 37–99.
22. K. A. Moshkov, S. Lakatos, J. Hajdu, P. Zavodsky and S. A. Neifakh, *Eur. J. Biochem.* **94**, 127 (1979).
23. A. G. W. Leslie, *Joint CCP4 and ESF-EACMB Newsletter on Protein Crystallography* **26** (CCLRC Daresbury Laboratory, Warrington WA4 4AD, UK, 1992).
24. J. Nyborg and A. J. Wonacott, *The Rotation Method in Crystallography*, eds. U. W. Arndt and A. J. Wonacott (Amsterdam, North-Holland, 1977), pp. 139–152.
25. CCP4 Suite: Programs for Protein Crystallography, *Acta Cryst.* **D50**, 760 (1994).
26. K. Cowtan, *Joint CCP4 and ESF-EACBM Newsletter in Protein Crystallography* **31**, eds. S. Bailey and K. Wilson (CCLRC Daresbury Laboratory, Warrington WA4 4AD, UK, 1994), pp. 34–38.
27. Z. Otwinowski, *Isomorphous Replacement and Anomalous Scattering; Proceedings of the CCP4 Study Weekend*, eds. W. Wolf, P. R. Evans and A. W. G. Leslie (CCLRC Daresbury Laboratory, Warrington WA4 4AD, UK; DL/SCI/R32, 1993), pp. 80–86.

28. T. A. Jones, J. -Y. Zou, S. W. Cowan and M. Kjeldgaard, *Acta Cryst.* **A47**, 110 (1991).
29. A. T. Brünger, J. Kuriyan and M. Karplus, *Science* **235**, 458 (1987).
30. H. Driessen, M. I. J. Haneef, G. W. Harris, B. Howlin, G. Khan and D. S. Moss, *J. Appl. Cryst.* **22**, 510 (1989).
31. R. A. Laskowski, M. W. MacArthur, D. S. Moss and J. M. Thornton, *J. Appl. Cryst.* **26**, 283 (1993).
32. E. T. Adman, *Advances in Prot. Chem.* **32**, 145 (1991).
33. T. G. Samsonidze, K. A. Moshkov, N. A. Kiselev and S. A. Neifakh, *Int. J. Peptide Protein Res.* **14**, 161 (1979).
34. R. Langen, I. -Jy. Chang, J. P. Germanas, J. H. Richards, J. R. Winkler and H. B. Gray, *Science* **268**, 1733 (1995).
35. T. L. Ortel, N. Takahashi, R. A. Bauman and F. W. Putnam, *Protides Biol. Fluids* **31**, 243 (1983).
36. K. Yamashita, C. J. Liang, S. Funakoski and A. Kobata, *J. Biol. Chem.* **256**, 1283 (1981).
37. M. Endo, K. Susuki, K. Schmidt, B. Fournet, Y. Karamanos, J. Montrueil, L. Dorland, H. Van Halbeek and J. F. G. Vliegenthart, *J. Biol. Chem.* **257**, 8755 (1982).
38. S. V. Evans, SETOR: Hardware lighted three-dimensional solid model representations of macromolecules, *J. Mol. Graphics* **11**, 134 (1993).

CHAPTER 4

SPECTROSCOPY OF MULTI-COPPER OXIDASES

Edward I. Solomon*, Timothy E. Machonkin and Uma M. Sundaram
Department of Chemistry, Stanford University, Stanford, CA 94305, USA
E-mail: fb.eis@forsythe.stanford.edu

ABSTRACT

Spectroscopy has played a major role in developing an understanding of the copper centers in the multi-copper oxidases and their interactions with O_2. The four key spectroscopic methods, Electron Paramagnetic Resonance (EPR), Electronic Absorption (Abs), Magnetic Circular Dichroism (MCD), and X-ray Absorption Spectroscopies (XAS), are briefly reviewed and then applied to native laccase, exogenous ligand-binding to the native enzyme, and the T2D and T1Hg derivatives. The results of spectroscopic studies are correlated with those from crystallography to obtain geometric and structural insight into function and to define questions which still exist for future research.

Contents

*To whom all correspondences should be addressed.

1. Introduction

Spectroscopy has played a major role in developing our understanding of the active sites in the multi-copper oxidases and their reactivity with dioxygen.[1–4] In the early days of this field, the copper centers in the multi-copper oxidases were characterized by their different and often unique spectral features: Type 1, or blue copper, centers have an extremely intense absorption band at 600 nm ($\varepsilon \sim 5000\ M^{-1}\ cm^{-1}$) and an unusually small parallel hyperfine splitting ($A_{||} < 100 \times 10^{-4}\ cm^{-1}$) in the EPR spectrum; Type 2 copper centers have a "normal"-looking EPR signal with a large $A_{||} (> 150 \times 10^{-4}\ cm^{-1})$ and no obvious absorption bands in the visible/UV spectral region; Type 3 or coupled binuclear copper centers exhibit no EPR signal but have an absorption band at 330 nm ($\varepsilon \sim 3000\ M^{-1}\ cm^{-1}$).[5] As will be described later in this review, these spectral features are now well-understood and provide important insight into the geometric and electronic structure of these centers and, in turn, into their function.

Laccase is the simplest multi-copper oxidase, with one of each type of copper center for a total stoichiometry of four metal ions per protein molecule. Ascorbate oxidase can be viewed as a dimer of laccase sites with a Cu connecting the two subunits for a total of nine copper centers per protein molecule.[6–8] Ceruloplasmin is more complex. This human enzyme clearly has two Type 1 and one Type 2 centers detectable by EPR and a Type 3 site. There also appears to be an additional Type 1 copper from X-ray crystallography[9] which does not appear in the EPR spectrum.[10] At this point it is important to emphasize that EPR and UV/vis absorption spectroscopy only probe the oxidized, cupric state. Much confusion has existed in the multicopper oxidase literature due to the possible presence of reduced cuprous sites in these enzymes, since reduced copper cannot be detected by EPR, optical absorption or other open-shell spectroscopic methods. X-ray absorption edge spectroscopy has thus been developed as a direct probe of reduced copper centers in the multicopper oxidases[11] and this method will be described in Sec. 2.4.

While laccase has the simplest stoichiometry, it is still difficult to study since the blue copper site dominates the spectral features, obscuring those of the other copper centers, including the Type 2 site whose role in dioxygen reduction has been unclear. Thus, two derivatives of laccase have played an important role in understanding these sites and their reactions with dioxygen. These are the Type 2 depleted (T2D) derivative, where the Type 2 center is reversibly removed,[12,13] and a Type 1 mercury-substituted (T1Hg) derivative,[14] where the Type 1 site is replaced by a spectroscopically-silent and redox-innocent mercuric ion.

In the next section, we briefly review the key spectroscopic methods that have provided significant insight into the structure and reactivity of the multi-copper oxidases. In Sec. 3, we apply a combination of these spectroscopic methods to native laccase (including the assignments of the characteristic Type 1, Type 2 and Type 3 spectral features listed above), the Type 2 depleted derivative, exogenous ligand-bound forms of the native enzyme, and the Type 1 mercury-substituted derivative. Finally in Sec. 4, we correlate the spectroscopic results to those from protein crystallography. Significant insight is obtained from this combination of methods, although specific problems exist which still need to be resolved.

2. Spectroscopic Methods

2.1. *Electron Paramagnetic Resonance* (*EPR*)

Placing a cupric ion with its nine d-electrons in an octahedral ligand field produces a ^2E_g ground state (Fig. 1B). This ground state is unstable to a Jahn-Teller distortion that lowers the symmetry and energy of the complex. The Jahn-Teller distortion normally observed is a tetragonal elongation along the z-axis and contraction in the equatorial x, y-plane that can ultimately result in a square-planar ligand environment, as in the "normal" copper complex D_{4h}-$CuCl_4^{2-}$ (Fig. 1A). A key feature of ligand field theory[15–19] is that the d-orbital splitting is very sensitive to the environment of the ligands around the metal center. The d-orbital splitting experimentally determined using optical spectroscopy[20] for D_{4h}-$CuCl_4^{2-}$ is given in Fig. 1B. The half-occupied $d_{x^2-y^2}$ orbital is at highest energy as it has the largest repulsive interaction with the ligands in the equatorial plane. A more complete description of this half-occupied ground state is provided by molecular orbital (MO) theory. In particular, self-consistent field-Xα-scattered wave (SCF-Xα-SW) calculations,[21–25] adjusted to ground-state parameters[26–29] are in good agreement with spectral data over many orders of magnitude in photon energy. These calculations[26,27] generate a description of the ground state of D_{4h}-$CuCl_4^{2-}$ that has 61 % Cu $d_{x^2-y^2}$ character with the remaining part of the wavefunction being delocalized equivalently into the four p_σ orbitals of the chloride ligands that are involved

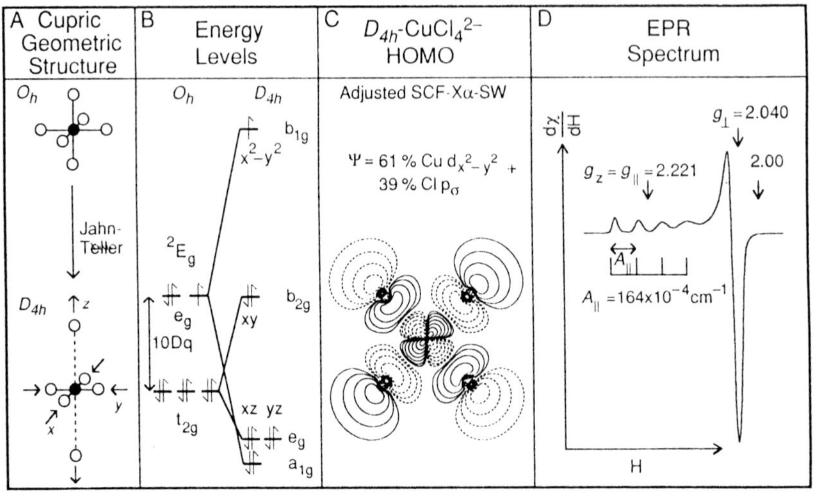

Fig. 1. Normal copper complexes — ground state spectral features. (A) Jahn–Teller tetragonal elongation of an octahedral CuL_6 complex to the square planar limit. (B) Energy level correlation diagram for the Jahn–Teller distortion depicted in A. (C) SCF-X_α-SW wave function contour and charge decomposition for the HOMO of D_{4h}-$CuCl_4^{2-}$ (Refs. 26 and 27). (D) X-band EPR spectrum for tetragonal Cu(II) with D_{4h}-$CuCl_4^{2-}$ parameters.

in antibonding interactions with the metal ion (Fig. 1C). The unpaired electron in this wavefunction produces the EPR spectrum shown in Fig. 1D, where $g_{||}$ (corresponding to the magnetic field oriented along the z-axis of the complex) $> g_\perp > 2.00$, that is characteristic of this $d_{x^2-y^2}$ ground state. Additionally, copper has a nuclear spin ($I_N = 3/2$) that couples to the electron spin to produce a four-line hyperfine splitting (A) of the EPR spectrum. Normal tetragonal cupric complexes generally have a large hyperfine splitting in the $g_{||}$-region ($A_{||} > 150 \times 10^{-4}$ cm^{-1}); that of D_{4h}-$CuCl_4^{2-}$ is 164×10^{-4} cm^{-1} (Ref. 30) while that of the Type 2 center in laccase is 206×10^{-4} cm^{-1} (Ref. 31).

Upon going to a binuclear cupric site with a bridging ligand, the ground state spectroscopic properties can dramatically change. The bridging ligand orbitals provide a superexchange pathway for antiferromagnetic coupling between the $S = 1/2$ ground state on each of the two copper (II)'s leading to an $S_{\text{total}} = 0$ ground state and a $S_{\text{total}} = 1$ excited state at $2J$ above the ground state where J is the exchange parameter associated with antiferromagnetic coupling in the dimer ($\mathcal{H} = -2J\ S_1.S_2$). The $S_{\text{total}} = 0$ ground state is EPR-silent and diamagnetic. If the antiferromagnetic coupling ($2J$) is large enough, a binuclear cupric site can appear diamagnetic even to the most sensitive (i.e. SQUID) magnetometer. This is the case for the Type 3 centers in the multi-copper oxidases. Thus, the lack of a paramagnetic signal in the EPR spectrum and SQUID magnetic susceptibility in a copper protein cannot distinguish between the two possible explanations: reduced copper or antiferromagnetic coupling in a binuclear cupric T3 state.

2.2. *Electronic Absorption* (*Abs*)

With respect to excited states, optical excitation of electrons from the filled Cu d-orbitals to the half-filled $d_{x^2\text{-}y^2}$ orbital (Fig. 2A) produces Laporté-forbidden ligand-field transitions.[5] These transitions are weak in the absorption spectrum, with molar extinction coefficients (ε) of 30–50 M^{-1} cm^{-1}

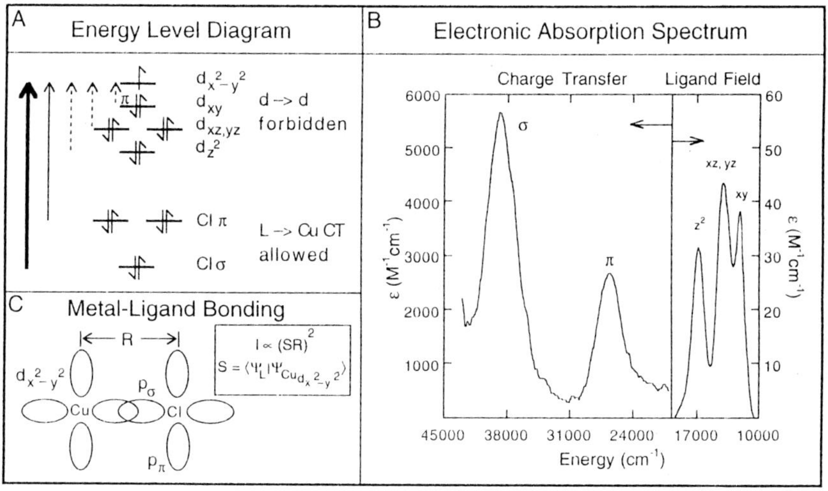

Fig. 2. Excited state spectral features of D_{4h}-$CuCl_4{}^{2-}$. (A) Energy level diagram showing the ligand-field ($d \rightarrow d$) and charge transfer (CT) optical transitions. The intensity of the transitions is approximated by the thickness of the arrow with the very weak ligand-field transitions represented as dotted arrows. (B) Electronic absorption spectrum for D_{4h}-$CuCl_4{}^{2-}$ (Refs. 20 and 32). (C) Schematic of the σ and π bonding interactions between the Cu $3d_{x^2\text{-}y^2}$ and Cl $3p$ orbitals and their correlations to CT intensity.

in the 12,000–16,000 cm^{-1} region[20] (Fig. 2B, right). To higher energy in the absorption spectrum are the Laporté-allowed ligand-to-metal charge-transfer (CT) transitions that are at least two orders of magnitude more intense than the ligand-field transitions[32] (Fig. 2B, left). The energies and intensities of these CT transitions allow one to probe the specific bonding interactions of the ligand and the metal center.[32] In the square-planar $CuCl_4{}^{2-}$ example, chloride has three valence $3p$-orbitals which split into two sets (σ and π) on binding to copper (Fig. 2C). The p_σ-orbital is oriented along the Cl–Cu bond and is stabilized to higher binding energy due to strong overlap with the Cu^{2+} ion. The two chloride p_π-orbitals are perpendicular to the Cl–Cu bond and hence are more weakly interacting with the metal and are at lower binding energy.

The intensity associated with the charge-transfer excitation of an electron from these filled ligand orbitals into the half-occupied Cu $d_{x^2\text{-}y^2}$ orbital also reflects metal-ligand bonding. Charge-transfer intensity is proportional to $(RS)^2$, where S is the overlap of the donor and acceptor orbitals involved in the charge-transfer transition and R is the metal-ligand bond length.[33] Thus the higher energy Cl $p_\sigma \rightarrow$ Cu $d_{x^2\text{-}y^2}$ charge-transfer transition is intense due to strong overlap, while the lower energy Cl $p_\pi \rightarrow d_{x^2\text{-}y^2}$ charge-transfer transition is weak due to poor overlap (Fig. 2C). The key points to be emphasized here are that the charge-transfer transitions sensitively probe the ligand-metal bond, and that for "normal" complexes one should observe a lower energy weak π and higher energy intense σ charge-transfer transition as is observed experimentally for D_{4h}-$CuCl_4{}^{2-}$ in Fig. 2B, left.

2.3. *Magnetic Circular Dichroism (MCD)*

Low temperature MCD spectroscopy provides a powerful means of distinguishing excited states associated with paramagnetic ground states from those associated with diamagnetic ground states. Thus, one can resolve excited-state features of the Type 1 and Type 2 with $S = 1/2$ ground-state centers from those of the antiferromagnetically coupled $S_{\text{total}} = 0$ Type 3 center.[34,35] As illustrated at the top of Fig. 3A, application of a magnetic field to the Type 2 or Type 1 (S=1/2) center splits both the ground and the excited state by $g\beta H$. The selection rules for MCD transitions are $\Delta m = +1$ for left circularly polarized (LCP) light and $\Delta m = -1$ for right circularly polarized (RCP) light. Thus, one predicts for the Type 2 or Type 1 center two transitions with equal intensity and opposite sign. Since

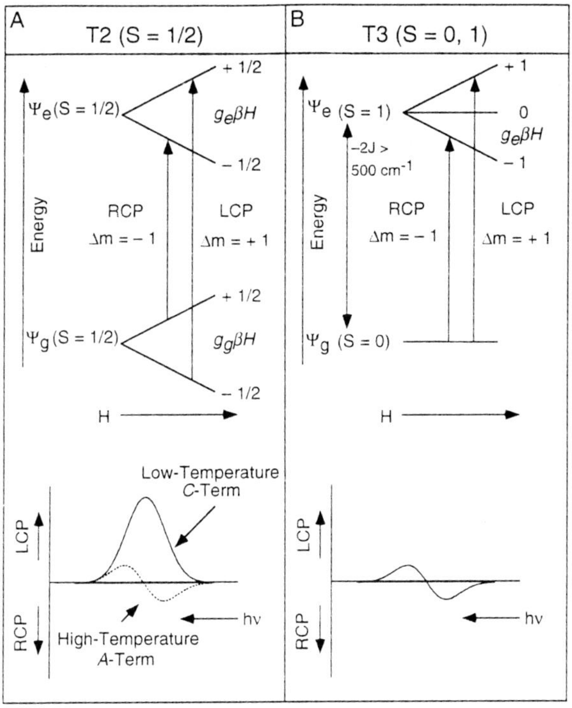

Fig. 3. Model for the temperature dependence of the MCD bands of native laccase. (A) Transitions and band profiles associated with type 2 copper. (B) Transitions and band profiles associated with type 3 copper. Note the difference in temperature dependence of the MCD signal as described in the text.

the Zeeman splitting is on the order of 10 cm^{-1} and absorption bands are several thousand cm^{-1} wide, the transitions will mostly cancel except for a broad, weak derivative-shaped signal at high temperatures (an A term). However, as the temperature is lowered, the Boltzmann population of the $m_s = +1/2$ level is reduced relative to that of the $m_s = -1/2$ level. This leads to a large increase in absorption of LCP light and results in a temperature-dependent C term (Fig. 3A, bottom). At very low temperatures (4 K), these C terms will be two to three orders of magnitude more intense than the A terms. In contrast, the strong antiferromagnetic coupling in the Type 3 center leads to an $S = 0$ ground state and a thermally inaccessible $S = 1$ excited state ($-2J > 550$ cm^{-1}, Fig. 3B, top). Since the $S = 0$ level is nondegenerate, it cannot be split by a magnetic field and there is no change in ground state population with temperature. Thus, there is no C term associated with the Type 3 center (Fig. 3B, bottom) and the low temperature MCD spectrum of native laccase should be dominated by the contributions from the paramagnetic Type 2 and Type 1 centers.

2.4. *X-Ray Absorption Spectroscopy* (*XAS*)

In contradistinction to EPR and Abs spectroscopy, K-edge XAS provides a direct and sensitive probe to the oxidation state of a copper site even in an antiferromagnetically coupled binuclear cupric center. A systematic study of a series of twenty Cu(I) and forty Cu(II) model complexes, in collaboration with Professor Keith O. Hodgson, was used to define and quantitate Cu K-edge XAS features.[11,36] The Cu K-edge band energies and shapes depend

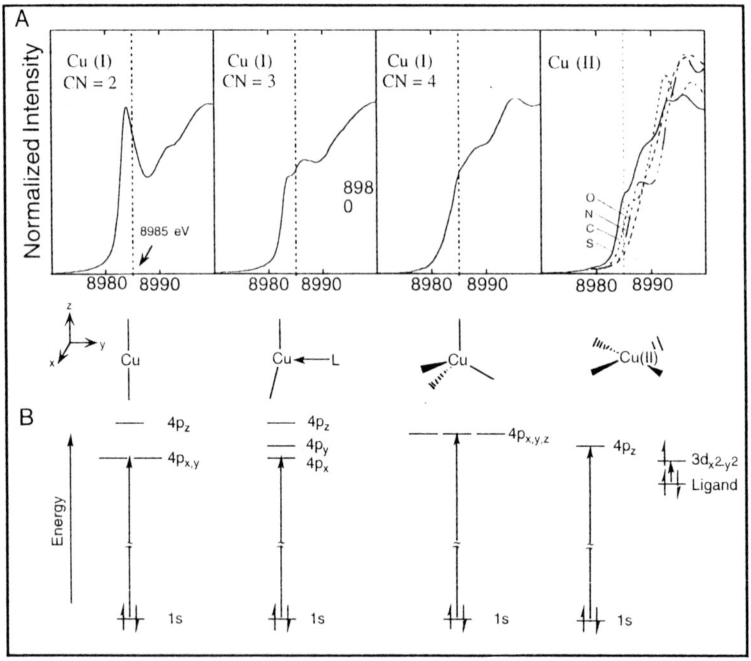

Fig. 4. Representative Cu K-edge X-ray absorption data for Cu(I) and Cu(II) compounds. (A) Edge spectra for a variety of coordination environments. (B) Corresponding ligand field splitting diagrams for the Cu $1s \rightarrow 4p$ transitions.

on oxidation state, ligation and coordination geometry. Figure 4A shows reprepresentative edges for two-, three- and four-coordinate Cu(I) and tetragonal Cu(II) complexes with a variety of ligand sets. Two- or three-coordinate Cu(I) complexes always exhibit a peak at energies below 8985 eV. In addition, the three-coordinate complexes display a double peak with lower intensity. No Cu(II) complex studied shows a peak below 8986 eV. There is, however, a low-energy tail in the 8984 eV region which is associated with a peak at higher energy.

These spectroscopic differences can be understood through ligand field theory.[11,36] For two-coordinate Cu(I) with an approximately linear geometry, the electric dipole allowed $1s \rightarrow 4p$ transition is predicted to split into $1s \rightarrow 4p_z$ and $1s \rightarrow 4p_{x,y}$ components. Furthermore, the $1s \rightarrow 4p_z$ transition is expected to be highest in energy due to repulsive interactions with the axial ligands. Polarized single-crystal X-ray edge spectroscopy[37] confirms the assignment of the lowest energy peak as $1s \rightarrow 4p_{x,y}$. As shown in Fig. 4B, increasing the ligand field strength by adding a third ligand along the y-axis leads to an additional repulsive interaction which raises the energy of the $4p_y$ level relative to the $4p_x$. This accounts for the experimentally observed edge splitting in the three-coordinate Cu(I) complexes. For a tetrahedral geometry, all p-orbitals are equally destabilized by the ligand field and the $1s \rightarrow 4p$ transitions are all above 8985 eV.

For the tetragonal Cu(II) complexes, both the low energy tail and its associated higher energy peak shift to lower energy as the covalent interaction with the equatorial ligand set increases. Since this band is z-polarized[37] with an energy shift dependent on equatorial ligand ionization energy, this transition can be assigned as the $1s \rightarrow 4p_z$ combined with a ligand $\rightarrow$ Cu(II) charge-transfer shake-up which is also supported by final-state calculations.[26,38–40]

X-ray absorption spectroscopy has played a key role in clearing up the confusion concerning the nature of the Type 2 depleted derivative and the reactions of the different derivatives of laccase with dioxygen as described in the following sections.

3. Spectroscopic Properties Of Laccase Derivatives

3.1. *The Native Enzyme*

The EPR and absorption spectra of native laccase are shown in Figs. 5A and 5B. The EPR spectrum exhibits a superposition of contributions from the

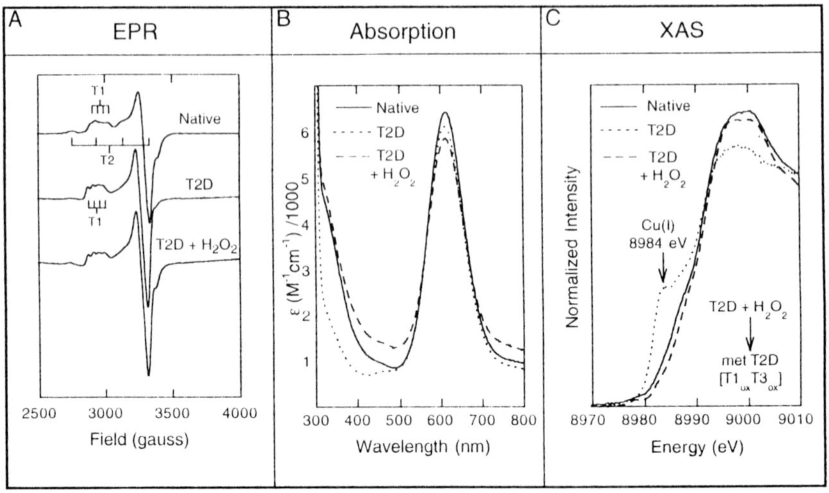

Fig. 5. Spectral features of native and T2D laccase. (A) EPR, (B) visible absorption, and (C) X-ray absorption spectra of native, T2D, and T2D laccase following reaction with hydrogen peroxide.

Type 1 and Type 2 centers. The Type 3 center is strongly antiferromagnetically coupled through a bridging hydroxide ($-2J > 550$ cm^{-1} from SQUID magnetic susceptibility studies) and thus does not contribute to the EPR spectrum. The Type 2 site has tetragonal Cu(II) spin Hamiltonian parameters with a large parallel hyperfine splitting of 206×10^{-4} cm^{-1} (Ref. 31). The Type 1 center exhibits a small $A_{||}$ value of 43×10^{-4} cm^{-1} (Refs. 31 and 41). Originally this was ascribed to a geometry effect, wherein the distorted tetrahedral structure of the blue copper center (see structure[42] in Fig. 6) causes the $4p_z$-orbital to mix with the $d_{x^2\text{-}y^2}$ ground state, thus reducing the $A_{||}$ value.[43–45] However, we have shown through polarized single-crystal XAS studies that blue copper sites in fact have $4p_{x,y}$-mixing, which would increase, not decrease, the $A_{||}$ value.[46,47] This allowed an alternative possible explanation for the small $A_{||}$ value as being due to high covalency wherein the unpaired electron of Cu is delocalized onto the ligands reducing its interaction with the nuclear spin on the copper.[48,49] This has been confirmed by Cu L-edge XAS studies.[50] In addition, S K-edge XAS studies[47] have demonstrated that the strong covalency involves the cysteine sulfur π-copper bond in Fig. 6 and is significant with respect to electron-transfer reactivity.[51]

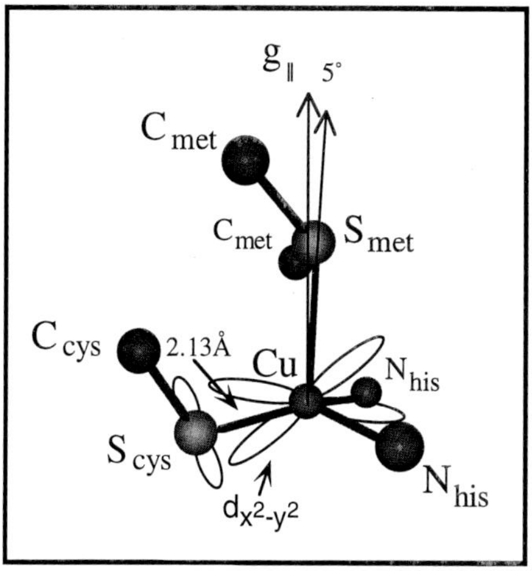

Fig. 6. X-ray structure of poplar plastocyanin[42] with orientation of the $g_{||}$ direction and the $d_{x^2-y^2}$ and Cys S_π orbital superimposed.[70]

The absorption spectrum of the multi-copper oxidases (Fig. 5B) has two dominant features. The intense band at 600 nm is associated with the blue copper center[3,52] and obscures the ligand field transitions associated with the Type 2 and Type 3 centers, which also contribute in this energy region but have low intensity. This band has been assigned as a cysteine sulfur $\pi \rightarrow$ Cu $d_{x^2-y^2}$ charge-transfer transition.[48] It is important to note that, as described in Sec. 2.2, normal ligand-to-metal charge-transfer involves an intense σ charge-transfer transition. The fact that the π charge-transfer is dominant for the Type 1 center indicates that the $d_{x^2-y^2}$ orbital is rotated such that its lobes bisect the cysteine S-Cu bond (see Fig. 6). This is due to a strong π antibonding interaction between the cysteine and the copper, which, in turn, is due to the short Cys S-Cu bond length of 2.1 Å. The 330 nm band in Fig. 5B reduces with two electrons at the same potential and had been associated with the oxidized Type 3 center (although from MCD there is also some Type 2 charge-transfer contribution in this region).[35] Based on energy and behavior with one-electron reduction in the Type 2 depleted derivative (*vide infra*) this feature can be assigned as involving charge-transfer transitions from the histidine and hydroxide ligands to the Type 3 coppers.[53] It is important to note that charge-transfer transitions of bridging ligands (i.e. the OH^- at the Type 3 center *vide infra*) are shifted to lower energy due to excited state antiferromagnetic coupling in metal dimers.[54]

3.2. *Type 2 Depleted Derivative*

Early data on the Type 2 depleted derivative[55–60] initially complicated the above spectroscopic assignment of native laccase. From Fig. 5A, the Type 2 contribution is clearly eliminated from the EPR spectrum of T2D and the Type 1 contribution to both the EPR and absorption spectrum remains. However, upon removal of the Type 2 copper the 330 nm absorption is eliminated, raising concerns with respect to the involvement of the Type 2 copper in this spectral feature. When T2D laccase is reacted with peroxide the 330 nm band reappears (Fig. 5B). This observation suggested the possibility that the Type 3 site might be reduced in the T2D derivative but could be oxidized by H_2O_2.[61] Alternatively, other researchers assigned the 330 nm band as a $O_2^{2-} \rightarrow Cu(II)$ charge transfer transition in analogy to oxyhemocyanin and therefore associated this spectral feature with peroxide binding to the Type 3 site.[57–60] An absorption feature at 330 nm, however, cannot directly distinguish between these possibilities. Also, the EPR spectrum is not diagnostic of the Type 3 oxidation state since both a reduced site and an antiferromagnetically coupled oxidized binuclear cupric site would be EPR-nondetectable. A direct probe of oxidation state is, however, available through X-ray absorption edge spectroscopy. Importantly the T2D form exhibits a peak in XAS at 8984 eV which is eliminated by reaction with peroxide (Fig. 5C).

This energy is characteristic of Cu(I) while the shape of the peak suggests that it is due to three-coordinate copper. The amount of reduced copper present can then be quantitated from the normalized edge intensities of copper model complexes with the appropriate coordination number. For T2D laccase, the 8984 eV peak indicates that 98 ± 22 % of the Type 3 sites are reduced.[11] Peroxide-reacted T2D laccase shows no 8984 eV feature and corresponds quantitatively to a fully oxidized Type 3 site.

The possibility of the 330 nm absorption band being a $O_2^{2-} \rightarrow Cu(II)$ charge transfer transition was investigated through a correlation of the 330 nm absorption intensity increase with the loss of the 8984 eV Cu(I) peak as T2D was titrated with H_2O_2. The linear correlation in Fig. 7 demonstrates that all change at 330 nm is associated with oxidation of the

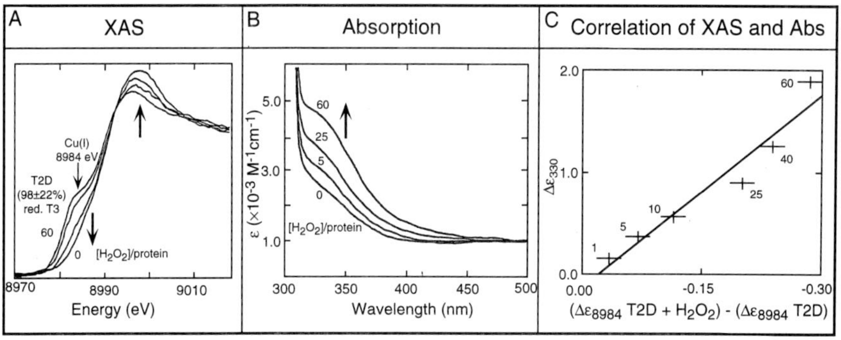

Fig. 7. Correlation of X-ray edge data and optical absorption for the reaction of T2D laccase with peroxide. (A) Copper K-edge XAS. (B) Electronic absorption spectra. (C) Plot of changes in intensity of the 330 nm band vs. changes in intensity of the 8984 eV feature upon addition of H_2O_2.

Type 3 site with no evidence for additional peroxide-binding to the oxidized Cu(II). Thus, as indicated in Scheme I, the Type 3 site in T2D laccase is in a deoxy[Cu(I)Cu(I)] form which, in strong contrast to deoxyhemocyanin and deoxytyrosinase, does **not** react with dioxygen. It is, however, oxidized by peroxide to give the binuclear cupric met derivative, [Cu(II)Cu(II)].[62]

Scheme I

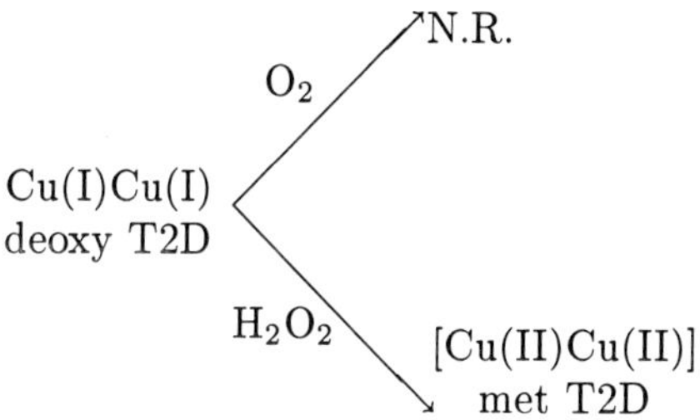

In the above studies the T1 center in T2D is oxidized, T2D [$T1_{ox}$–$T3_{red}$]. We next looked at the reactivity of the fully reduced T2D [$T1_{red}$ $T3_{red}$] derivative with O_2. This had been generally viewed as the combination of copper centers in laccase required for dioxygen reactivity in the mechanistic proposals in the literature.[55,63] From Fig. 8 it is clear that the 8984 eV Cu

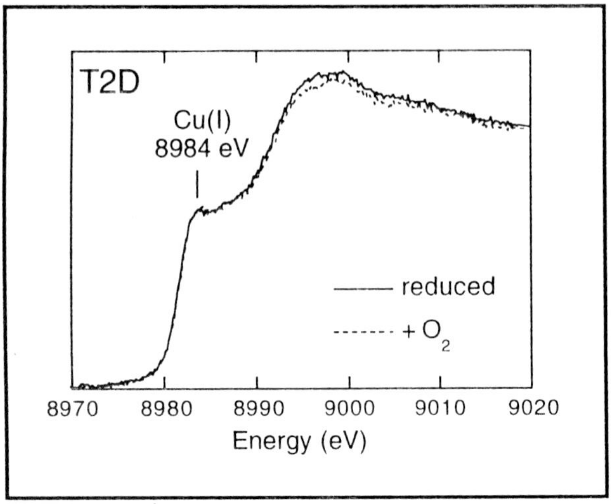

Fig. 8. XAS of fully reduced T2D laccase and fully reduced T2D laccase following exposure to dioxygen.

K-edge peak associated with reduced Cu does not change on exposure of fully reduced T2D laccase to O_2. This indicates that the Type 2 center is, in fact, required for dioxygen reactivity.

3.3. *Exogenous Ligand-Binding To Native Laccase*

In order to determine the role played by the Type 2 and Type 3 centers in the oxygen reactivity of the native enzyme, exogenous ligand-binding to native laccase was studied. To this end, low-temperature MCD spectroscopy was employed, since the optical features associated with the paramagnetic Type 2 site can be distinguished from those associated with the antiferromagnetically coupled Type 3 center.[34,35] There are no C terms associated with the Type 3 center and the low-temperature MCD spectrum of native laccase should be dominated by the contributions from the paramagnetic Type 2 and Type 1 centers, where the Type 1 center is believed to be incapable of binding exogenous ligands.[64]

The binding of azide to the Type 3 and Type 2 sites in native laccase has been probed using absorption and low-temperature MCD spectroscopic studies of the $N_3^- \rightarrow Cu(II)$ charge-transfer transitions in the 400–500 nm region (Figs. 9A and 9B). The low-temperature MCD and Abs behavior in

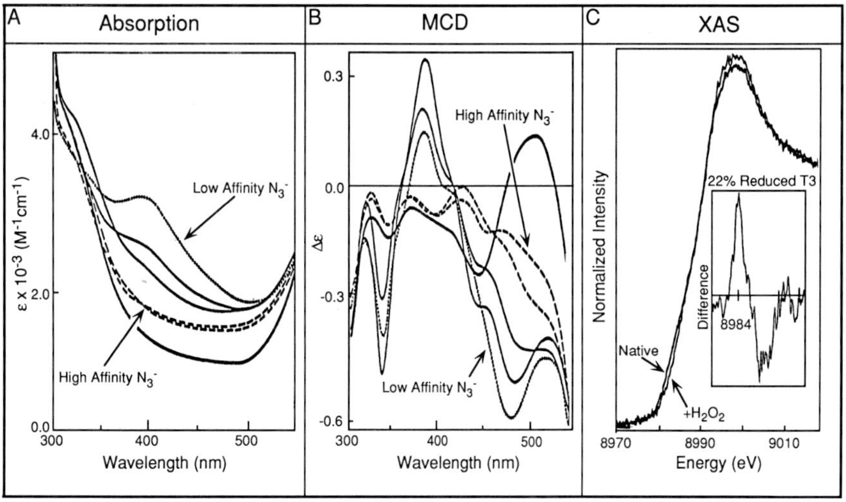

Fig. 9. (A) Absorption spectra of azide binding to native laccase. (B) Low-temperature MCD spectra of azide binding to native laccase. (C) X-ray absorption edge data for reaction of native laccase with peroxide (insert presents normalized difference edge).

this region is quite complex and consists of two contributions. The spectroscopic changes associated with high-affinity (HA) binding are completed with the addition of less than one equivalent of azide ($K_{\mathrm{HA}} \geq 10^4\mathrm{M}^{-1}$). Prior addition of H_2O_2 eliminates these features, suggesting that they may be associated with ligand-binding to a fraction of native laccase molecules in which the Type 3 site is reduced. This interpretation has been confirmed through X-ray absorption edge studies, shown in Fig. 9C in which the edge features of H_2O_2-treated laccase have been subtracted from those of native enzyme.[11,36] A peak is present at 8984 eV with an intensity that corresponds to $\sim$ 22 % reduced Type 3 sites in the native enzyme.

The features associated with low-affinity (LA) azide binding to fully oxidized laccase sites can then be obtained by subtraction of the high-affinity azide spectral contributions. The low-affinity Abs and low-temperature MCD spectra are shown in Figs. 10A and 10B for increasing azide concentration. Two $N_3^- \rightarrow Cu(II)$ charge-transfer bands are observed in absorption, one at 500 nm and a more intense band at 400 nm. Both increase with azide concentration; the 400 nm absorption intensity is plotted as a function of $[N_3^-]$ in Fig. 10C. In the low-temperature MCD spectrum a negative peak

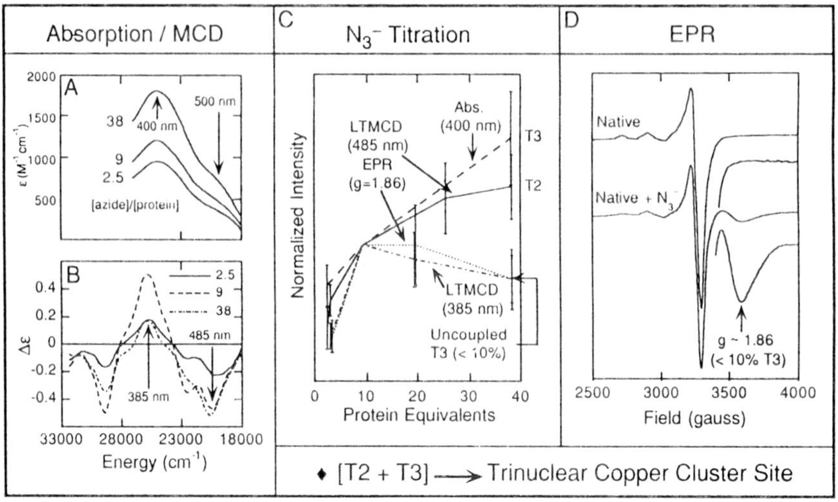

Fig. 10. Azide-binding to fully oxidized native laccase. (A) Electronic absorption spectra at 298 K with HA N_3^- form subtracted. (B) Difference MCD spectra at 4.9 K (5 T). (C) Changes in EPR, LT MCD, and absorption intensities plotted as a function of increasing azide concentration. (D) EPR at 8 K of laccase titrated with azide. Arrow indicates the new signal present at $g = 1.86$.

is observed at 485 nm which increases in magnitude with increasing azide concentration. It is thus clear that the corresponding 500 nm absorption band is associated with a paramagnetic ground state and must correspond to azide binding to the Type 2 center. In the region of the intense 400 nm absorption, a positive feature at 385 nm is observed in the low-temperature MCD spectrum. The intensity of this feature first increases, then decreases with increasing azide concentration. As shown in Fig. 10C, the intensity of this low-temperature MCD feature at 385 nm does not correlate with the 400 nm absorption intensity. Thus, no low-temperature MCD intensity is associated with the intense 400 nm absorption band and it must be associated with azide-binding of the antiferromagnetically coupled Type 3 site. The 385 nm low temperature MCD feature however parallels the appearance of a $g = 1.86$ EPR signal which is associated with a small fraction (< 10 %) of Type 3 sites which are protonatively uncoupled in the presence of azide (Fig. 10D).

From the low-temperature MCD data it is thus found that azide binds to both the Type 2 and Type 3 centers with similar binding constants. The stoichiometry of this binding has been determined through a series of ligand competition experiments, one of which is summarized in Fig. 11. From Fig. 11A, addition of F^- to native laccase containing 9-fold excess $N_3{}^-$ (corresponding to the presence of both HA and LA forms) produces a superhyperfine splitting ($I_N = 1/2$ for ^{19}F) of the Type 2 hyperfine line, indicating that one F^- binds to the Type 2 center. In the absorption and low-temperature MCD spectra (Figs. 11B and 11C), the low-affinity $N_3{}^- \rightarrow$ Cu(II) charge-transfer transitions associated with the Type 2 (485 nm MCD) and the Type 3 (400 nm absorption) centers are eliminated with the addition of increasing concentrations of fluoride. The quantitative loss of these features parallel each other as shown by the solid lines in Fig. 11D. Fitting these data to models[35] for competitive binding of one or two fluorides clearly demonstrates that one fluoride binds to the Type 2 center and displaces azide from both the Type 2 and Type 3 centers. Therefore, one

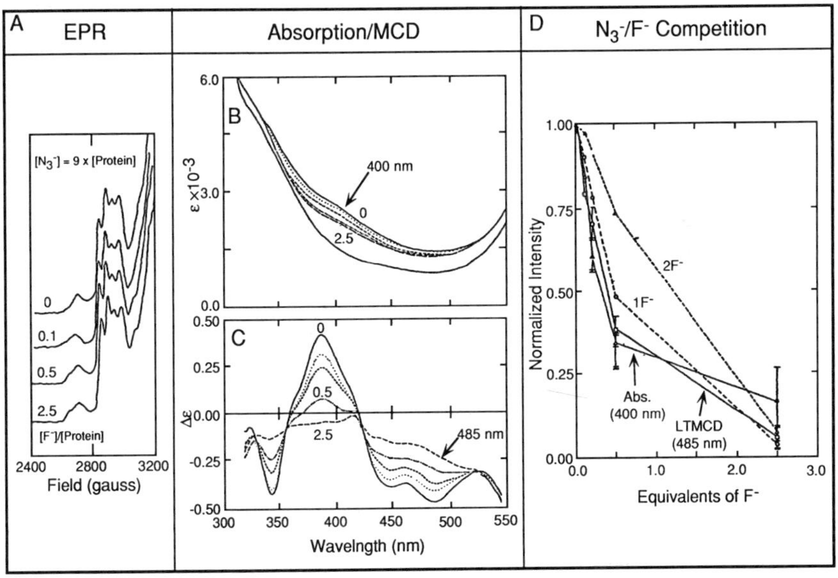

Fig. 11. Summary of N_3^-/F^- binding competition experiments for native laccase. (A) EPR spectra. (B) Electronic absorption spectra. (C) Low-temperature MCD spectra. (D) Plot of normalized intensity of spectral features indicated as a function of equivalents of F^-. Dashed lines give simulated titration results for models assuming 1 and 2 F^- binding.

exogenous azide ligand appears to bridge between the Type 2 and Type 3 centers in native laccase. This combined with a series of other MCD and XAS studies first defined the presence of a trinuclear copper cluster active site in the multi-copper oxidases.[34,35]

3.4. *The Type 1 Hg^{2+}-Substituted Derivative: Reactions With O_2*

Having demonstrated that the Type 2 center is required for dioxygen reactivity and that the Type 3 and Type 2 centers must be viewed together as part of a trinuclear copper cluster, we then investigated the O_2 reactivity of the T1Hg derivative, as it contains a valid Type 2/Type 3 trinuclear cluster. These studies used XAS as a sensitive probe of the presence of reduced copper. From Fig. 12, the fully reduced trinuclear copper cluster site rapidly reacts with O_2 eliminating the 8984 eV peak. This corroborates any previous studies on the T2D derivative, indicating that the trinuclear copper cluster is the minimum structural unit required for O_2-reduction.[65]

Since the mercuric ion in T1Hg laccase is redox-inactive, this derivative has one less electron equivalent available for O_2-reduction than native laccase. This property enabled us to stabilize an oxygen intermediate in

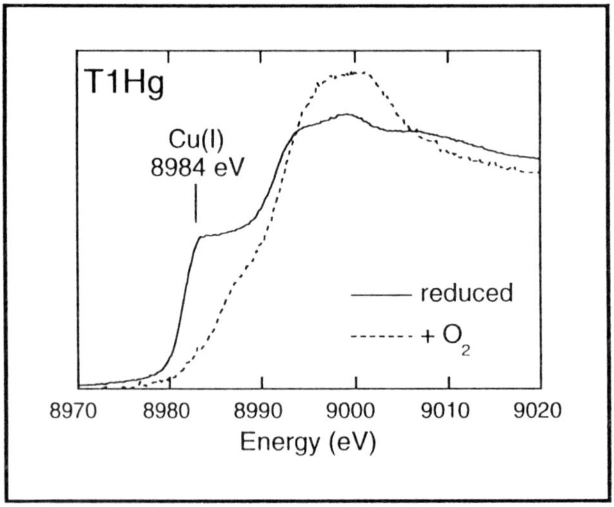

Fig. 12. XAS of reduced T1Hg laccase and reduced T1Hg laccase following exposure to dioxygen.

T1Hg laccase. A combination of spectroscopic methods has demonstrated that two coppers of the trinuclear cluster are oxidized in this intermediate.[66] Thus, two electrons have been transferred to dioxygen and this species corresponds to a peroxide-level intermediate that can be compared to the other peroxo-binuclear cupric species in biology, the side-on bridged peroxo-binuclear cupric sites in oxyhemocyanin and oxytyrosinase. As is clear from Fig. 13A, the peroxide intermediate in laccase has a strikingly different charge transfer spectrum from that of oxyhemocyanin and oxytyrosinase. This requires a different geometric and electronic structure for this peroxo-trinuclear copper cluster site.[66] Detailed spectral studies on this intermediate are presently underway. However, at this point our data

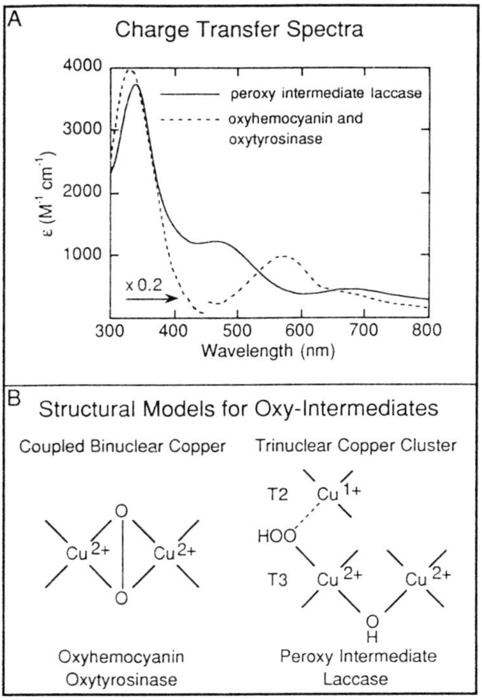

Fig. 13. Comparison — oxygen intermediates. (A) Electronic absorption spectra of the peroxy-intermediate in laccase versus oxyhemocyanin and oxytyrosinase. (B) Proposed structural differences between peroxide binding in oxyhemocyanin and oxytyrosinase relative to the end-on bound hydroperoxide intermediate at the trinuclear copper cluster in laccase.

indicate that it corresponds to a hydroperoxide bound end-on to one of the coppers of an oxidized Type 3 center and also likely bridging to a reduced Type 2 copper center (Fig. 13B). This difference in peroxide binding relative to hemocyanin and tyrosinase appears to play a key role in promoting its irreversible, further reduction to water at the trinuclear copper cluster site.

4. Correlation between Spectroscopy and Crystallography

Generally, the agreement between the spectroscopy and crystallography of the multi-copper oxidases is quite good and the combination provides further insight into function. This is clearly the case with respect to understanding electron transfer pathways from the Type 1 to the trinuclear copper cluster. Our detailed spectroscopic and theoretical studies of the blue copper center[67–69] have provided an accurate description of the electronic structure of its ground state, which can be correlated with that of Messerschmidt *et al.*'s X-ray structure of ascorbate oxidase.[8] The X-ray structure shows that the cysteine ligand of the blue copper site in this multicopper oxidase is flanked on either side in the sequence by histidines that are ligands to the two Type 3 coppers in the trinuclear copper cluster site. This blue copper site transfers an electron rapidly over ~13 Å in the reduction of O_2 at the trinuclear copper center.[66] As can be seen from the Xα calculated wavefunction contour that we superimposed on the crystal structure of the blue copper center in ascorbate oxidase (Fig. 14), the ground-state wavefunction provides a highly anisotropic covalent pathway involving the cysteine sulfur. The covalency activates this residue for directional electron transfer. In addition, the low-energy, intense Cys $\pi \rightarrow$ Cu $d_{x^2\text{-}y^2}$ charge-transfer transition in the blue copper absorption spectrum (i.e. the 600 nm band in Fig. 5B) provides an efficient hole superexchange pathway for rapid electron-transfer between the blue and trinuclear copper cluster sites.[51,68,70] Clearly, as shown in Fig. 14, the unique electronic structure of the blue copper center reflects a ground-state wavefunction which plays a critical role in its functioning of rapid long-range electron-transfer to the Type 2/Type 3 trinuclear cluster site.

MCD and XAS also predicted the presence of a trinuclear copper cluster with a Type 3 pair bridged by a protonatable residue, RO^-, which from the crystallography is hydroxide and with a non-bridged Type 2 center within about 4 Å of the Type 3 site (see Fig. 15). Both crystallography and spectroscopy also show that all three coppers are inequivalent. A limited

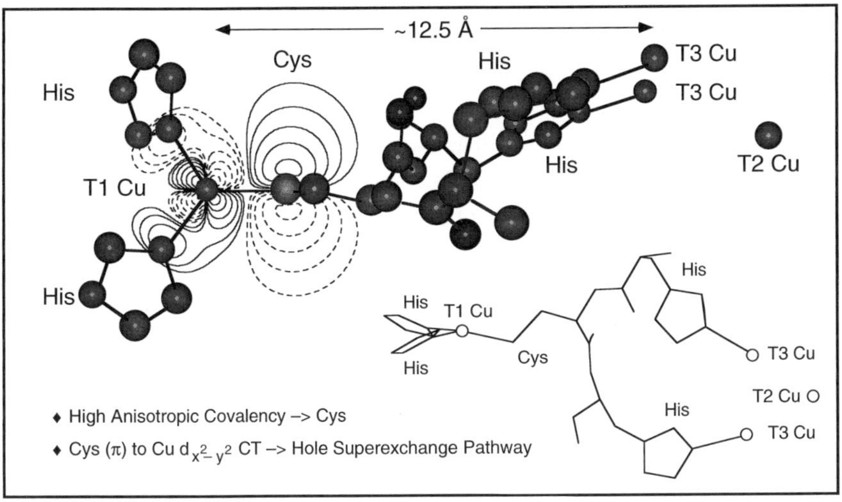

Fig. 14. Proposed electron transfer pathway in blue copper proteins. The plastocyanin wave function contours have been superimposed on the blue copper (type 1) site in ascorbate oxidase.[8] The contour shows the substantial electron delocalization onto the cysteine Sp_π orbital that activates electron transfer to the trinuclear copper cluster at $\sim$ 12.5 Å from the blue copper site. This low-energy, intense Cys $S_\pi \rightarrow$ Cu charge transfer transition provides an effective hole superexchange mechanism for rapid long-range electron transfer between these sites.[51,68,70]

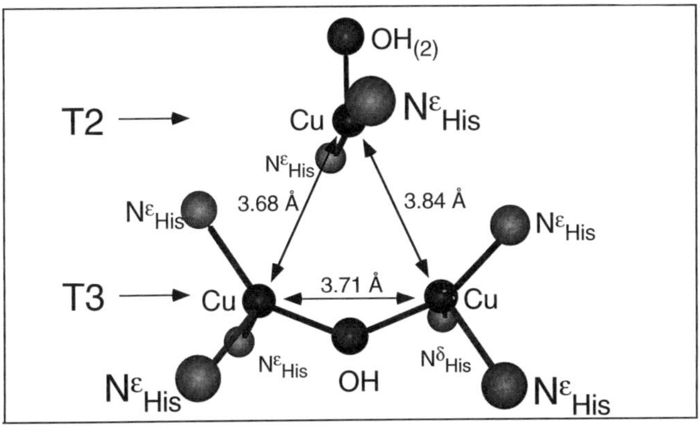

Fig. 15. The trinuclear Cu cluster in the multi-copper oxidase, ascorbate oxidase.[8] The T1 Cu site (not pictured) is $\sim$ 12.5 Å from the T3 Cu atoms, opposite the T2 center.

difference is that the crystallography indicates that the two Type 3 centers have close to tetrahedral geometries, while CD/MCD data show no bands at lower energy than 9,000 cm^{-1} which would imply a more tetragonal geometry for these copper centers.[71]

The fundamental disagreement between the crystallography and spectroscopy is in how exogenous ligands and oxygen bind to the trinuclear cluster site. Spectroscopy has and continues to indicate that exogenous ligands and oxygen intermediates bridge the Type 2 copper and one of the Type 3 coppers of the trinuclear cluster in laccase,[34,35,66,72] while the X-ray structures of azide- and peroxide-adducts of ascorbate oxidase[73] indicate that these ligands bind to only one copper and that all coppers are greater than 4 Å apart with no bridging ligands. This is an important difference, particularly with respect to oxygen-binding. This may relate to protein differences (laccase vs. ascorbate oxidase), differences in conditions (peroxide-adduct vs. oxygen intermediate) or differences in reactivity in solution vs. in the crystalline state. Resolving this difference and in particular defining the geometric and the electronic structure of oxygen intermediates at the trinuclear copper cluster site are major goals of future spectroscopic studies in this field. These studies should provide atomic-level information into the catalytic mechanism of the four-electron reduction of dioxygen to water.

Acknowledgments

This research was supported by the NIH (DK 31450). E.I.S. wishes to express his sincere appreciation to all his students and collaborators who are listed as co-authors in the References for their commitment and contributions to this research.

References

1. J. A. Fee, B. G. Malmström and T. Vänngård, in *Biochem. Sauerst., Colloq. Ges. Biol. Chem., 19th*, ed. B. Hess (Springer-Verlag, Berlin, 1968).
2. B. G. Malmström, B. Reinhammer and T. Vänngård, *Biochim. Biophys. Acta* **156**, 67 (1968).
3. E. I. Solomon, K. W. Penfield and D. E. Wilcox, *Structure and Bonding* **53**, 1 (1983).
4. E. I. Solomon, B. L. Hemming and D. E. Root, in *Bioinorganic Chemistry of Copper*, ed. K. D. K. a. Z. Tyeklar (Chapman and Hall, New York, 1992).

5. E. I. Solomon, M. D. Lowery, L. B. LaCroix and D. E. Root, in *Methods in Enzynology*, ed. J. F. Riordan and B. L. Vallee, 1993, Vol. 226; pp. 1–33.
6. B. Mondoví and L. Avigliano, in *Copper Proteins and Copper Enzymes*, Vol lll, ed. R. Lontie (CRC Press, Inc., Boca Raton, Florida, 1984), pp. 101–118.
7. A. Messerschmidt, A. Rossi, R. Ladenstein, R. Huber, M. Bolognesi, G. Gatti, A. Marchesini, R. Petruzzelli and A. Finazzi-Agro, *J. Mol. Biol.* **206**, 513 (1989).
8. A. Messerschmidt, R. Ladenstein, R. Huber, M. Bolognesi, L. Avigliano, R. Petruzzelli, A. Rossi and A. Finazzi-Agro, *J. Mol. Biol.* **224**, 179 (1992).
9. V. Zaitsev, I. Zaitseva, G. Card, A. Ralph, B. Bax and P. Lindley, *J. Inorg. Biochem.* **59**, 719 (1995).
10. L. Ryden, in *Copper Proteins and Copper Enzymes*, Vol III, ed. R. Lontie (CRC Press, Boca Raton, Florida, 1984), pp. 37–100.
11. L. S. Kau, D. J. Spria, J. E. Penner-Hahn, K. O. Hodgson and E. I. Solomon, *J. Am. Chem. Soc.* **109**, 6433 (1987).
12. M. T. Graziani, L. Morpurgo, G. Rotilio and B. Mondoví, *FEBS Lett.* **70**, 87 (1976).
13. M. T. Graziani, P. Loreti, L. Morpurgo, I. Savini and L. Avigliano, *Inorg. Chim. Acta* **173**, 261 (1990).
14. M. M. Morie-Bebel, M. C. Morris, J. L. Menzie and D. R. McMillin, *J. Am. Chem. Soc.* **106**, 3677 (1984).
15. C. J. Ballhausen, *Introduction to Ligand Field Theory* (McGraw-Hill, New York, 1962).
16. D. S. McClure, *Electronic Spectra of Molecules and Ions in Crystals* (Academic Press, New York, 1959).
17. J. S. Griffith, *The Theory of Transition Metal Ions* (Cambridge Univ. Press, London, 1964).
18. S. Sugano, Y. Tanabe and H. Kamimura, *Multiplets of Transition Metal Ions in Crystals* (Academic Press, New York, 1970).
19. B. N. Figgis, *Introduction to Ligand Fields* (Interscience, New York, 1967).
20. M. A. Hitchman and P. J. Cassidy, *Inorganic Chemistry* **18**, 1745 (1979).
21. K. H. Johnson, *Advances in Quantum Chemistry* **7**, 143 (1973).
22. K. H. Johnson, J. G. Norman Jr. and J. W. D. Connolly, in *Computational Methods for Large Molecules and Localized States in Solids*, eds. F. Herman, A. D. McLean and R. K. Nesbet (Plenum, New York, 1973), pp. 161–201.
23. J. W. D. Connolly, in *Semiempirical Methods of Electronic Structure Calculations, Part A: Techniques*, ed. G. A. Segal (Plenum, New York, 1977).
24. N. Rosch, in *Electrons in Finite and Infinite Structures*, eds. P. Phariseu and L. Scheire (Wiley, New York, 1977).
25. J. C. Slater, *The Calculation of Molecular Orbitals* (John Wiley & Sons, New York, 1979), p. 104.
26. A. A. Gewirth, S. L. Cohen, H. J. Schugar and E. I. Solomon, *Inorganic Chemistry* **26**, 1133 (1987).

27. E. I. Solomon, A. A. Gewirth and S. L. Cohen, in *Understanding Molecular Properties*, eds. J. Avery, J. P. Dahl and A. E. Hansen (D. Reidel, Dordrecht, 1987), pp. 27–68.
28. S. V. Didziulis, S. L. Cohen, A. A. Gewirth and E. I. Solomon, *J. Am. Chem. Soc.* **110**, 250 (1988).
29. A. Bencini and D. Gatteschi, *J. Am. Chem. Soc.* **105**, 5535 (1983).
30. C. Chow, K. Chang and R. D. Willet, *J. Chem. Phys.* **59**, 2629 (1973).
31. B. G. Malmström, B. Reinhammer and T. Vänngård, *Biochim. Biophys. Acta* **205**, 48 (1970).
32. S. R. Desjardins, K. W. Penfield, S. L. Cohen, R. L. Musselman and E. I. Solomon, *J. Am. Chem. Soc.* **105**, 4590 (1983).
33. R. S. Mulliken, *J. Chem. Phys.* **7**, 20 (1939).
34. M. D. Allendorf, D. J. Spira and E. I. Solomon, *Proc. Natl. Aca. Sci. USA* **82**, 3063 (1985).
35. D. J. Spira, M. D. Allendorf and E. I. Solomon, *J. Am. Chem. Soc.* **108**, 5318 (1986).
36. L. S. Kau, E. I. Solomon and K. O. Hodgson, *J. Physique* **47**, 289 (1986).
37. T. A. G. Smith, Ph. D. Thesis, Stanford Univ., 1985.
38. R. A. Blair and W. A. Goddard, *Phys. Rev.* **B22**, 2767 (1980).
39. N. Kosugi, T. Yokoyama, K. Asakuna and H. Kuroda, *Springer Proc. in Phys.* **2**, 45 (1984).
40. N. Kosugi, T. Yokoyama, K. Asakuna and H. Kuroda, *Chem. Phys.* **91**, 249 (1984).
41. B. Reinhammer, *J. Inorg. Biochem.* **15**, 27 (1981).
42. J. M. Guss and H. C. Freeman, *J. Mol. Biol.* **169**, 521 (1983).
43. C. A. Bates, W. S. Moore, K. J. Standley and K.W. H. Stevens, *Proc. Phys. Soc.* **79**, 73 (1962).
44. M. Sharnoff, *J. Chem. Phys.* **42**, 3383 (1965).
45. J. E. Roberts, T. G. Brown, B. M. Hoffman and J. Peisach, *J. Am. Chem. Soc.* **102**, 825 (1980).
46. R. A. Scott, J. E. Hahn, S. Doniach, H. C. Freeman and K. O. Hodgson, *J. Am. Chem. Soc.* **104**, 5364 (1982).
47. S. E. Shadle, J. E. Penner-Hahn, H. J. Schugar, B. Hedman, K. O. Hodgson and E. I. Solomon, *J. Am. Chem. Soc.* **115**, 767 (1993).
48. A. A. Gewirth and E. I. Solomon, *J. Am. Chem. Soc.* **110**, 3811 (1988).
49. K. W. Penfield, A. A. Gewirth and E. I. Solomon, *J. Am. Chem. Soc.* **107**, 4519 (1985).
50. S. J. George, M. D. Lowery, S. P. Cramer and E. I. Solomon, *J. Am. Chem. Soc.* **115**, 2968 (1993).
51. M. D. Lowery, J. A. Guckert, M. S. Gebhard and E. I. Solomon, *J. Am. Chem. Soc.* **115**, 3012 (1993).
52. E. I. Solomon, J. W. Hare and H. B. Gray, *Proc. Natl. Acad. Sci. USA* **73**, 1389 (1976).
53. D. J. Spira-Solomon and E. I. Solomon, *J. Am. Chem. Soc.* **109**, 6421 (1987).

54. F. Tuczek and E. I. Solomon, *J. Am. Chem. Soc.* **116**, 6916 (1994).
55. B. Reinhammar and Y. Oda, *J. Inorg. Biochem.* **11**, 115 (1979).
56. L. Morpurgo, M. T. Graziani, A. Finazzi-Argó, G. Rotilio and B. Mondovì, *Biochem. J.* **187**, 361 (1980).
57. O. Farver, P. Frank and I. Pecht, *Biochem. and Biophys. Research Comm.* **108**, 273 (1982).
58. P. Frank, O. Farver and I. Pecht, *J. Biol. Chem.* **258**, 11112 (1983).
59. P. Frank and I. Pecht, *Biochem. Biophys. Research Comm.* **114**, 57 (1983).
60. P. Frank, O. Farver and I. Pecht, *Inorganica Chimica Acta* **91**, 81 (1984).
61. C. D. LuBein, M. E. Winkler, T. J. Thamann, R. A. Scott, M. S. Co, K. O. Hodgson and E. I. Solomon, *J. Am. Chem. Soc.* **103**, 7014 (1981).
62. D. J. Spira and E. I. Solomon, *J. Am. Chem. Soc.* **106**, 6421 (1987).
63. O. Farver, M. Goldberg and I. Pecht, *Eur. J. Biochem.* **104**, 71 (1980).
64. D. J. Spira, M. S. Co, E. I. Solomon and K. O. Hodgson, *Biochem. and Biophys. Research Comm.* **112**, 746 (1983).
65. J. L. Cole, G. O. Tan, E. K. Yang, K. O. Hodgson and E. I. Solomon, *J. Am. Chem. Soc.* **112**, 2243 (1990).
66. J. L. Cole, D. P. Ballou and E. I. Solomon, *J. Am. Chem. Soc.* **113**, 8544 (1991).
67. E. I. Solomon, A. A. Gewirth and S. L. Cohen, in *Excited States and Reactive Intermediates: Photochemistry, Photophysics, and Electrochemistry*, ed. A. B. P. Lever (York Univ. Press, York, 1986), Vol. 307, pp. 236–266.
68. E. I. Solomon, M. J. Baldwin and M. D. Lowery, *Chem. Rev.* **92**, 521 (1992).
69. E. I. Solomon and M. D. Lowery, in *The Chemistry of the Copper and Zinc Triads*, eds. A. J. Welch and S. K. Chapman (The Royal Society of Chemistry, Cambridge, 1993), pp. 12–29.
70. E. I. Solomon and M. D. Lowery, *Science* **259**, 1575 (1993).
71. J. L. Cole, P. A. Clark and E. I. Solomon, *J. Am. Chem. Soc.* **112**, 9534 (1990).
72. P. A. Clark and E. I. Solomon, *J. Am. Chem. Soc.* **114**, 1108 (1992).
73. A. Messerschmidt, H. Luecke and R. Huber, *J. Mol. Biol.* **230**, 997 (1993).

CHAPTER 5

BIOINORGANIC CHEMISTRY OF LACCASE

David R. McMillin and Michèle K. Eggleston
Chemistry Department, Purdue University, 1393 Brown Building
West Lafayette, Indiana 47907-1393, USA
E-mail: McMillin@vm.cc.purdue.edu

ABSTRACT

This chapter describes a series of studies involving site-specific metal replacement or site-specific metal depletion experiments with a blue copper oxidase known as laccase. The most important derivatives are T1Hg laccase, in which mercury(II) replaces copper(II) in the type 1 (or blue) copper site, and T2D (type 2-depleted) laccase, in which the type 2 site is empty. Physico-chemical studies of the various derivatives have provided new and unique structural and mechanistic insights into this enzyme system. In particular, the results reveal the flexible nature of the dioxygen-reduction site that includes the three coppers traditionally known as the type 2 copper and the type 3 pair. Within the cluster there is, generally, one EPR-active metal ion, but the particular ion that is EPR active varies with conditions. The results of several experiments have revealed key facets of the mechanism for the removal of the type 2 copper. There are, in fact, several routes to T2D laccase available, but each probably involves the release of copper(I) from the enzyme.

Contents

1. Introduction to Laccase

1.1. *Function*

Laccases (benzenediol: oxygen oxidoreductase, EC 1.10.3.2) are members of a class of proteins known as the blue oxidases that have the unusual ability to orchestrate the four-electron reduction of dioxygen to water. In the context of living systems this is a useful reaction because it is a source of free energy and because it avoids the release of reactive, partially reduced forms of the dioxygen molecule such as hydrogen peroxide or hydroxyl radical which would pose a threat to cells. Other members of this class of proteins include the plant protein ascorbate oxidase and the plasma protein ceruloplasmin. A number of review articles discuss these systems in detail.[1–4] A key, membrane-bound enzyme known as cytochrome *c* oxidase carries out the same reduction in many aerobes and appears at the end of the electron transport chain. Unlike cytochrome *c* oxidase, however, the blue oxidases are water-soluble enzymes generally consisting of a single polypeptide chain. Accordingly, they are much more tractable enzymes, and laccase is the simplest of the blue oxidases.

Various fungi and certain types of plants, notably the Oriental lacquer tree, produce laccases. Independent of source, laccases have a chain length of about 500 amino acids; however, the molecular weight of the tree enzyme (125 ± 15 kD) is approximately twice that of the fungal enzyme due to a more extensive carbohydrate coat.[1,5] Tree laccase catalyzes the hardening of the sap of the lacquer tree and is part of the wound-healing apparatus. The hardening reaction involves the oxidative polymerization of urushiols, various catechol derivatives found in the latex.[5] Recent results suggest that the tree enzyme may also have a role in lignification.[6,7] In contrast, fungal laccase is an extracellular enzyme that probably plays some role in lignin degradation.[1,5]

1.2. *Metal Centers*

Well characterized laccases all contain four copper atoms per mole of protein. The distinctive spectroscopic and physical properties of the various copper centers within the protein allow for classification into three different kinds of copper sites: type 1, type 2 and type 3.[8] Now that the crystal structure of the related enzyme, ascorbate oxidase, is available,[9,10,11] it is clear that the type 2 and type 3 sites lie very close to one another, so close that many investigators now regard them as components of a flexible trinuclear cluster, vide infra. Because of the similar spectroscopic properties and the homologies within the sequences of the blue oxidases,[11,12] the copper sites of ascorbate oxidase are useful structural models for those of laccase.

In fact, the type 1 coppers in laccase and ascorbate oxidase are similar to type 1 copper centers found in a number of smaller proteins such as azurin and plastocyanin that function as electron transferases. Adman has termed these proteins cupredoxins in recognition of the fact that they are copper-containing redox proteins.[2] A trigonal arrangement of a cysteine sulfur and two nitrogen atoms from histidine sidechains serve as the primary donors to the copper in the type 1 site. Weaker axial interactions with a methionine sulfur and/or a peptide oxygen may also occur. Such a low primary coordination number is very unusual for copper(II), and as a consequence, the copper-to-cysteine bond in the type 1 site is quite short and extremely covalent. In fact, a substantial delocalization of spin density onto the sulfur center accounts for the unusually small hyperfine coupling constant of copper in the EPR spectrum.[13] The copper-thiolate interaction is also responsible for the other important spectroscopic signature of the oxidized type 1 site, an intense visible absorption ($\varepsilon \approx 5000\ M^{-1}\ cm^{-1}$) that occurs at ca. 610 nm. The absorption corresponds to a charge-transfer transition between the thiolate sulfur and the copper(II) center.[14,15] The ligand field, or *d-d*, transitions occur at lower energies, in the near infrared region of the spectrum.[3]

The spectral properties of the type 2 copper, the other EPR-active center in resting laccase, are more typical of copper(II) systems. Thus, the hyperfine coupling constant for this copper center is representative of a complex with four ordinary oxygen- and/or nitrogen-donating ligands interacting with an unpaired electron in the $d(x^2\text{-}y^2)$ orbital.[8] The *d-d* transitions of the type 2 copper are so weak that no one has yet identified them in the visible absorption spectrum. Using MCD and CD spectroscopies in conjunction with chemical modifications, Solomon and co-workers have, however, identified several transitions that they regard as *d-d* bands.[16]

Previous work has shown that the other two copper centers in laccase, the so-called type 3 coppers, are EPR silent in the oxidized form of the protein due to a strong antiferromagnetic coupling interaction.[8] On the other hand, the type 3 coppers are responsible for a shoulder at around 330 nm in the absorption spectrum of oxidized laccase, and titrations with a reducing agent show that this chromophore behaves as a two-electron acceptor.[17] These observations formed the basis of the original formulation of the type 3 site as a binuclear copper-binding site.[1,8] Much later, the results of inhibitor-binding studies raised the possibility that the type 2 and type 3 coppers lie close enough to each other to share a bridging azide ligand.[18,19] Finally, the crystal structure of ascorbate oxidase established the proximity of the type 2 and type 3 coppers and solidified the view that the three copper ions form a trinuclear cluster.[9]

Within each subunit of ascorbate oxidase the structure shows that the trinuclear cluster lies in the interface between two barrel-shaped domains of the folded polypeptide chain.[9,10] Figure 1 is a representation of the essential elements of the type 2/type 3 cluster. The copper–copper distances vary within the cluster, but they are all very similar (3.70 ± 0.05 Å). The distance between the type 1 copper atom (labeled CU1 in the published structure) and CU2 or CU3 in Fig. 1 is about 12.5 Å, and a run of three amino acids links these coppers. Thus, His 506 is a ligand of CU2 while Cys 507 and His 508 bind to CU1 and CU3, respectively.[10] Because of the presence of the bridging hydroxide or oxide ligand (OH1), Messerschmidt *et al.* assigned CU2 and CU3 as the antiferromagnetically coupled coppers. Donor nitrogens from six different histidine sidechains complete the coordination spheres of this pair. The disposition of these six nitrogens is roughly trigonal prismatic with the coppers sandwiched between the opposite triangular faces.

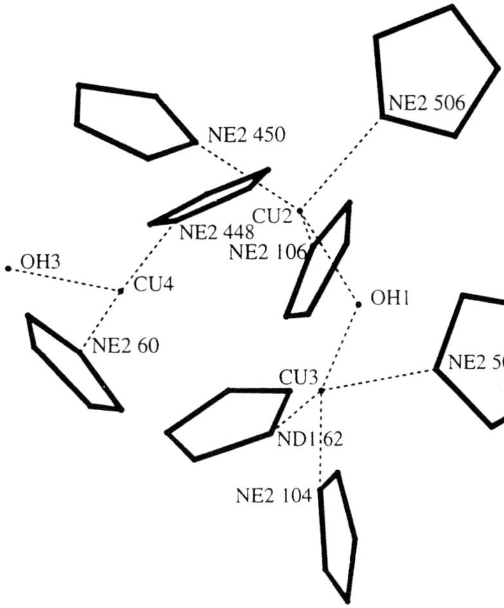

Fig. 1. A view of the type 2/type 3 cluster in ascorbate oxidase. The OH1 bridging oxygen ligand suggests that CU2 and CU3 comprise the type 3 pair. In this form the type 2 copper is three coordinate. (Figure produced by A. Messerschmidt)

The crystal structure of ascorbate oxidase provides an enormous amount of useful information; however, there are many questions that remain. One concerns the coordination environment of the type 2 copper. According to the structure, CU4 has only three ligands arranged in a T-shaped fashion: a pair of imidazole groups and a water molecule or a hydroxide group (OH3 in Fig. 1). Although there is room for another water ligand opposite OH3, nothing shows up in the electron density map.[10] As pointed out earlier,[4] it is possible that the published structure relates to a partially reduced cluster. Several related observations have motivated this suggestion. One is the fact that three-coordinate copper(II) is quite rare, and there are no short metal-ligand bonds to the copper center to compensate for the paucity of ligands (CU4-NE2 60: 2.00 Å, CU4-NE2 448: 2.09 Å, and CU4-OH3: 2.02 Å).[10] Besides, the type 2 copper has a very ordinary EPR spectrum; a three-coordinate system would undoubtedly give a much more distinctive signal. Messerschmidt and co-workers made another observation recently that could have an important bearing on the type 2 copper, when they solved the structures of the diazide and hydroperoxide derivatives of ascorbate oxidase.[20] In each case binding occurs at what corresponds to CU2 in Fig. 1. Since previous work has shown that azide binds at an EPR-detectable copper, this reopens the question of the actual location of the type 2 copper within the cluster. To add to the mystery, when Li *et al.* studied the temperature dependence of the EPR spectrum of a mixed-isotope derivative of laccase, they found evidence that different copper centers within the cluster were EPR active at different temperatures.[21] It is no surprise that such a complex protein system can exist in different redox states or in more than one conformation. The kinetics studies mentioned below illustrate some of the possible states.

1.3. *Mechanistic Issues*

Kinetics studies indicate that under anaerobic conditions the reduction of the type 1 and type 2 coppers of laccase occur in sequence and that these centers then simultaneously transfer their electrons to the type 3 coppers by an intramolecular process.[1,22,23] At least in the case of ascorbate oxidase, this model squares with the crystal structure because it shows that the natural substrate, ascorbate, has a binding site near the type 1 copper, almost certainly the main port of entry for the electrons into the protein.[10] This is consistent with the normal electron transferase function of type 1 copper. Further support for this model comes from the observation that the reduction of the type 3 copper in laccase occurs very slowly when a redox-inactive ion like Hg(II) replaces copper in the type 1 site.[24] With regard to the kinetics of oxidation most studies have focused on the reaction of fully reduced laccase with dioxygen. Using this approach, Reinhammar and co-workers have identified a transient intermediate with maximum absorption at 360 nm.[25] When they used a rapid-freeze method to quench the reaction, they also detected a low-temperature EPR signal ($g < 2$) from an intermediate that they took to be a three-electron reduced form of dioxygen, formally a bound form of O^-.[26] Recently, Solomon and co-workers studied the reoxidation of the reduced form of a derivative of laccase with Hg(II) residing in the type 1 site. They also detected the intermediate that absorbs at 360 nm, and they assigned the absorbance to a hydroperoxide-containing form of the enzyme.[27] Since they did not detect the $g < 2$ EPR signal, they concluded that it relates to a different state of the enzyme system.[28]

In view of the number of redox-active metal centers in laccase and the number of partially reduced forms of dioxygen that exist, there are obviously a host of states available in the catalytic cycle. Studies involving substrate analogues and/or inhibitors can therefore be informative. Moreover, the replacement or removal of a specific metal center can simplify the spectral analysis. After a brief discussion of the strategies and principles involved, the rest of this chapter focuses on our laboratory's recent efforts in these two areas.

1.4. *Principles and Considerations for Specific Modifications*

As a consequence of the replacement or removal of a specific metal center, an enzyme may lose all activity. However, this does not necessarily compromise the strategy. Remember that the blockage of just one step in a complex pathway is enough to destroy the activity even if the rest of the machinery is in perfect working order. Another possible concern is that structural changes may occur with metal removal or metal replacement, but the evidence to date suggests that the changes are minimal when the metal is an integral part of the active site.[29]

Thus, structural data available for a variety of derivatives of type 1 copper-containing proteins reveal only minimal structural changes. Consider, for example, plastocyanin. Freeman and co-workers solved the structure of Hg(II)Pl, i.e., the mercury(II) derivative of plastocyanin, and found that there is a remarkable correspondence with the structure of the native protein, Cu(II)Pl.[30] In both structures, the metal resides in a pseudotetrahedral coordination environment that exhibits a pronounced trigonal distortion. Sidechain functions of the His 37, Cys 84 and His 87 residues serve as the primary ligands, and a rather long bond to the sulfur of Met 92 completes the coordination sphere of the metal. With the insertion of mercury(II), the metal-ligand bond lengths increase by about 0.28 Å on average in accordance with the larger ionic radius. The flexibility that permits the radial movement of the ligands is undoubtedly an intrinsic property of the metal-binding site; comparable to inward displacements of the donor atoms which occur upon removal of the metal center.[31] The type 1 copper binds to the same primary donor set in the azurin, (Cu(II)Az) from *Pseudomonas aeruginosa*, but the copper-to-methionine bond is even longer than it is in Cu(II)Pl.[32] There is also another long range, axial interaction to a glycine oxygen, opposite the methionine. The structures of the Zn(II)Az[33] and Ni(II)Az[34] derivatives reveal only minor differences. Thus, the same three sidechains define the primary coordination shell; however, small atomic displacements occur such that the Zn(II) and Ni(II) ions reside on the Gly 45 oxygen side of the plane defined by the primary donor atoms. At the same time, the Gly 45 oxygen moves toward the metal ion to give a more nearly tetrahedral donor environment. Cadmium(II) binds similarly in the azurin from *Alcaligenes denitrificans*.[35] In all of these derivatives the structural reorganization is minimal, almost certainly because the protein essentially controls the ligand disposition. In fact, the structures of the non-copper derivatives probably reveal the geometry that the protein favors. The equilibrium geometry differs somewhat when copper(II) is present because the Jahn-Teller effect comes into play when the metal ion has a d^9 electronic configuration.[36]

By comparison with the type 1 copper site, the prospects for structural variability are probably greater for the type 2/type 3 cluster.Flexibility is inherent in this site because it must bind the various diatomic and monatomic intermediates and reaction products that appear during the reduction of dioxygen to water. In other words, the protein structure must accommodate several different energy minima within the multidimensional configuration space that describes the state of the type 2/type 3 cluster. In any given redox state, a chemically modified form and the native form may occupy different minima.

Metal replacement and metal removal are the two main modification techniques we have employed in our studies of laccase. In a multi-copper protein like this, the characterization of a new derivative begins with the identification of the site of modification. To achieve specificity, we try to exploit thermodynamic and kinetic factors as much as possible. As a prelude to the discussion of results, we will briefly describe what we know about the complete removal and replacement of the copper in laccase.

2. Preparation and Remetalation of Apolaccase

About a quarter of a century ago, Ando reported a reliable method for removing the copper from laccase. The procedure employed cyanide as the metal-scavenging agent,[37] although Malmström *et al.* have pointed out that the "apoprotein" retained a significant amount of copper, enough to permit an individual site to remain essentially intact.[38] Nevertheless, the copper-depleted protein did not exhibit the characteristic visible absorbance at 614 nm, the EPR signals or the activity of the native enzyme. Ando found that reconstitution was possible to some extent with the addition of either copper(II) or copper(I), but that the addition of copper(I) gave better results. He also found it necessary to control the pH during the cyanide treatment. Thus, the yield of reconstituted protein was much higher if the exposure to cyanide occurred at pH 7.2 as compared to pH 8. Hauenstein and McMillin subsequently demonstrated that efficient dialysis against a solution with a total cyanide concentration of 50 mM at around pH 7.2 and at 5°C for a few hours removed essentially all of the copper from tree laccase.[39] Moreover, they showed that they could reconstitute the native enzyme by treatment with a slight excess of copper(I) under a nitrogen atmosphere.

In contrast to tree laccase, fungal laccase does not release its copper very readily. For example, in the case of the enzyme from *Polyporus versicolor*, Hanna *et al.* found that copper removal was incomplete even after dialysis against 0.2 M potassium cyanide at pH 9.5 for 48 h.[40] They succeeded in removing the copper by means of a 48 h dialysis at pH 10.3, but they failed to reconstitute the apoprotein. The apoprotein took up copper(I), but neither the 610 nm absorbance nor the EPR signal of type 1 copper reappeared.

3. Derivatives of Tree Laccase that Include a Non-Native Metal Ion

3.1. *T1Hg Laccase*

Since 1978, when the structure of plastocyanin appeared,[41] the unique architecture of the type 1 copper site has been apparent. To repeat, the most distinctive properties include the fact that there are strong bonds to only three ligands, one of which is a coordinate-covalent bond to the cysteine sulfur. If only because of its namesake functionality, the mercaptide ligand, the mercury ion was an obvious candidate to replace the type 1 copper. And, in fact, our group reported the successful incorporation of mercury into laccase in 1984.[42] The preparation involves the preincubation of a solution of apolaccase with a slight excess of Hg(II) before incubation with excess Cu(I). After exposure to air and purification by ion-exchange chromatography, the product, now called TlHg laccase, analyzed for one mercury and three copper atoms per mole of protein. The UV-visible spectrum of the T1Hg derivative exhibited a shoulder at 330 nm, but there was no maximum at 614 nm. Moreover, the EPR spectrum showed the presence of the type 2 but not the type 1 copper. These data were consistent with the presence of the type 2 and type 3 coppers and stoichiometric binding of mercury in the type 1 site. To define the environment of the bound mercury more explicitly, we measured the mercury L-edge X-ray absorption spectrum in conjunction with Penner-Hahn and co-workers.[43] For comparison, we also measured the corresponding spectra of Hg(II)Pl, Hg(II)Az and Hg(II)St, where Hg(II)St denotes the mercury derivative of stellacyanin. The EXAFS spectra of all four proteins were very similar. Since X-ray crystallographic studies have shown that mercury binds at the type 1 site of plastocyanin,[30] the results confirmed that the mercury ion occupies the type 1 copper site of laccase. As a bonus, the EXAFS analysis provided the first direct structural information about the type 1 site in laccase. The analysis revealed that the metal ion forms bonds to two nitrogen atoms from imidazole rings and a short bond to a sulfur center and is thus very analogous to the type 1 copper site in the smaller proteins. Table 1 contains the results derived from the best fits of the first-shell data. We describe other spectroscopic studies of T1Hg laccase below.

Table 1. Best fit for the first shell of donor atoms.[43]

Protein	Number of Atoms and Bond Distance			
	Nitrogens	R(Å)	Sulfurs	R(Å)
TlHg laccase	2	2.24	1	2.32
Hg(II)Pl	2	2.25	1	2.32
Hg(II)Az	2	2.24	1	2.38
Hg(II)St	2	2.24	1	2.37

In the case at hand, thermodynamic considerations are almost certainly responsible for the observed site-specific incorporation of the replacement metal ion. The low primary coordination number and the presence of the cysteine thiolate ligand predispose the type 1 site for a strong interaction with the divalent mercury ion. In particular, the relativistic effect that strongly stabilizes the 6*s* valence orbitals of mercury relative to the 6*p* valence orbitals, favors the formation of complexes with low coordination numbers.[44] In addition, hybridization of the 6*s* orbital with the filled $5d(z^2)$ orbital has an influence on the preferred coordination geometry. In line with this reasoning, studies of the thermal denaturation of various derivatives of azurin, including Hg(II)Az, have demonstrated that, compared with other divalent metal ions, Hg(II) has a very favorable enthalpy of adduct formation for the type 1 binding site.[45]

3.2. *Cobalt Derivatives*

Cobalt(II) is often useful in metal-replacement studies of metalloproteins because it is compatible with a wide variety of coordination geometries and because the spectral properties of the cobalt center provide useful information about the ligand environment. Actually, studies of cobalt binding to apolaccase preceded those of mercury binding, but the introduction of Hg(II) has proved useful in the interpretation of the cobalt-binding studies.

In 1979, Larrabee and Spiro reported that they had successfully introduced cobalt(II) into the type 1 site of tree laccase.[46] The supporting evidence included the charge-transfer absorption bands that they identified in the UV spectrum and assigned to S(Cys)→Co(II) transitions by analogy with other studies of cobalt(II)-substituted type 1 centers. However, Sakurai *et al.* subsequently reinvestigated this system, and they did not observe any CT transitions in the anticipated region of the spectrum,[47] although they did find a series of *d-d* transitions maximizing in the visible region at 528, 558 and 589 nm. The spread in the wavelengths is narrower than expected for Co(II) in a strongly distorted tetrahedral binding site, such as the type 1 site, and a plot of the absorbance intensity versus the cobalt-to-protein ratio revealed that the appearance of the absorption bands coincided with the uptake of not one, but two cobalt ions. Hence, Sakurai *et al.* assigned the transitions to a pair of cobalt centers bound in pseudo-tetrahedral coordination environments within the type 3 site.[47] To account for the surprising absence of cobalt in the type 1 site, they suggested that the type 1 copper site in laccase differed in some way from the sites found in smaller blue copper proteins.

Later, Li and McMillin reexamined this system and showed that Co(II) did bind in the type 1 site but with low affinity compared to smaller proteins.[48] One of the complications investigators faced was the number of accessible binding sites in laccase. The other problem was the congestion in the UV region of the spectrum because laccase has so many aromatic residues. Fortunately, when Hg(II) is present there is no absorption from the type 1 site down to about 280 nm. Thus, Li and McMillin isolated the CT bands of interest by means of a difference technique. More specifically, they placed a solution of apolaccase containing excess Co(II) in the sample beam and a solution of apolaccase incubated first with one eq. of Hg(II) and then excess Co(II) in the reference beam. The difference spectrum exhibited a CT maximum at ca. 303 nm and a shoulder at ca. 360 nm. The CT maximum occurred at relatively high energy (short wavelength) in that other Co(II) derivatives of blue copper proteins exhibit a CT maximum at around 330 nm. Stellacyanin, which has its CT maximum at 310 nm, is an exception. Stellacyanin is also unusual in that it contains no methionine residues, and in this case it is possible that the CT band occurs at a relatively high energy due to the formation of a stronger-than-normal bond to the fourth ligand, possibly a glutamine oxygen.[49]

Li and McMillin suggested that the unusually low affinity for the binding of Co(II) could be a reflection of the conformational flexibility of the protein.[48] In the case of plastocyanin the natural fold of the protein shapes the copper binding site, even in the absence of the metal ion. However, this may not be the case with laccase, and any expenditure of free energy for the reorganization of the donor framework during metal uptake would result in a reduction in the binding constant. Another possibility is that the incorporation of Co(II) into the type 2/type 3 cluster alters the structure at the type 1 site. This is quite feasible because there are direct covalent bonds that link the cysteine ligand of the type 1 site and two histidine residues that act as ligands for the metal ions bound in the type 2/type 3 cluster. Thus, if these histidines move to accommodate the binding of Co(II), the cysteine ligand may have to move as well, and this would alter the binding constant for the metal ion in the type 1 site. Another interesting observation is that the visible spectrum of Co(II) in the type 1 site of laccase is pH dependent.[48] This could reflect a pH-dependent structural transition within the type 1 site itself, or it could arise because of a transition that occurs elsewhere, such as in the type 2/type 3 cluster.

4. Studies Utilizing T1Hg Laccase

4.1. *Physical Studies*

The EPR spectra of native and TlHg laccase EPR spectra of native and TlHg laccase appear in Fig. 2. Table 2 contains EPR parameters derived

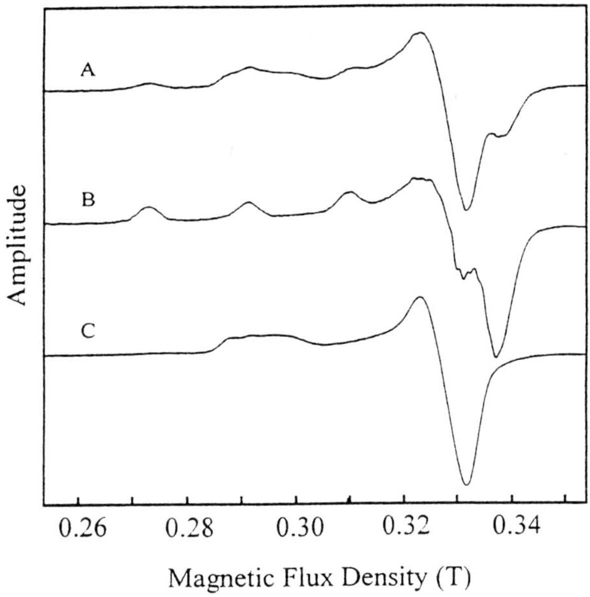

Fig. 2. The X-band EPR spectra of three forms of tree laccase. In each case the temperature is −150°C, and the buffer is 0.2 M pH 6 potassium phosphate. A: Native laccase; B: T1Hg laccase; C: T2D laccase.

Table 2. EPR parameters for the T2 copper of tree laccase.

$g_{\|}$	$10^4 \times A_{\|}$ (cm^{-1})	buffer	pH (77 K)	Ref.
	^{65}Cu-enriched T1Hg Laccase			
2.25	201	sodium phosphate	3–4	50
2.24	204	sodium phosphate	3–4	51
2.24	209	potassium phosphate	~ 6	51
	Natural Abundance Native Laccase			
2.237	200	sodium phosphate	?	52

from simulations of the spectrum of TlHg laccase in two different buffers along with those of the native protein. All of the spectra relate to frozen solutions, and Table 2 also lists estimates of the pH when available. The reason is that, at reduced temperatures, the precipitation of a disodium phosphate salt drastically alters the pH of a sodium phosphate buffer.[53,54]

The correspondence between the spectral data for the type 2 copper in T1Hg laccase and in native laccase proves that the EPR-active copper has the same structure in both forms. Morie-Bebel *et al.* also used EPR methods to demonstrate that fluoride binds to the EPR-active copper in T1Hg laccase in the same way as with the native enzyme.[50] Thus, the type 2 copper has the same reactivity whether Hg(II) or Cu(II) is present in the type 1 site.

In connection with inhibitor binding studies involving T1Hg laccase, Solomon and co-workers have carried out extensive spectral studies of T1Hg laccase with and without the presence of azide.[16] Their analysis predicts that the oxidized type 2 and type 3 coppers are five or six coordinate with tetragonally elongated coordination environments. On the basis of absorbance, CD, MCD and EPR results, they also concluded that azide acts as bridging ligand between the type 2 copper and one of the type 3 coppers.[3,16] Moreover they assigned transitions occurring at 340 nm and 500 nm as azide-to-type 3 copper and azide-to-type 2 copper CT bands, respectively. Although the crystal structure of ascorbate oxidase leaves little doubt but that the type 2 and the type 3 coppers lie very near each other, the MCD results provided some of the first experimental evidence for the proximity of the type 2 and the type 3 coppers. However, the argument is complex because it requires the assumption of at least three types of azide adducts, including two monoazide derivatives. One of the latter is a minor species with EPR-active type 3 coppers that give a $g < 2$ signal. In the presence of sufficient inhibitor, both monoazide adducts apparently convert to the same diazide adduct.[16]

Since there are three coppers so near each other in the protein, there is still some doubt about the identity of the type 2 copper. As pointed out earlier, Messerschmidt and co-workers have identified atom CU4 in ascorbate oxidase as the type 2 copper because of the bridging ligand between atoms CU2 and CU3. This implies that there are two imidazole ligands bound to the type 2 copper, and pulsed EPR data for tree laccase seem to be in accord with this model.[55] However, we have pointed out that the $g_{\parallel}$ value is more consistent with a coordination sphere containing three nitrogen ligands.[50] Thus, in small molecule complexes of copper(II), systems with three nitrogen ligands tend to exhibit a $g_{\parallel}$ value in the range of 2.23–2.25 whereas systems with two nitrogen ligands have a $g_{\parallel}$ in the range of 2.27–2.29.[56,57] In principle, the hyperfine structure in the $g_{\perp}$ region of the X-band EPR spectrum of the T1Hg derivative (Fig. 2) could answer the question because it reflects coupling interactions with the ligand nitrogens. However, this region of the spectrum is quite difficult to analyze because there are so many contributing effects, e.g., the hyperfine interaction with the copper nucleus, the possibility that $g_x \neq g_y$, etc.

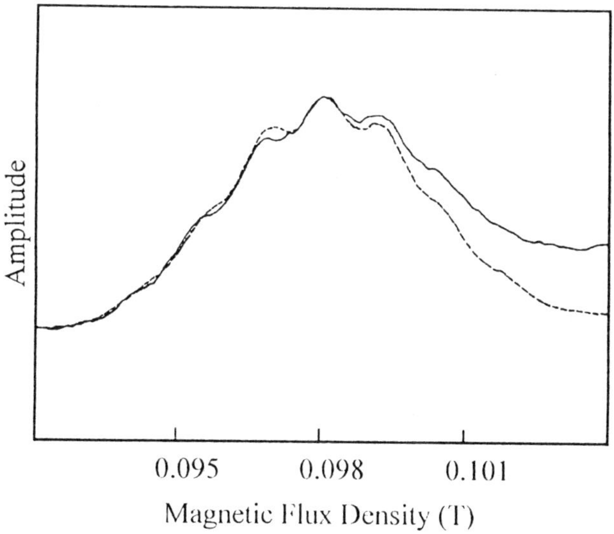

Fig. 3. The $M_{Cu} = -\frac{1}{2}$ transition (—) in the parallel region of the S-band EPR spectrum of T1Hg laccase at $-150°C$. The buffer was sodium phosphate, and the room-temperature pH was 6. The simulation (– – – –) assumes that there is splitting from three equivalent nitrogens. (Reprinted with permission from Morie-Bebel *et al.*[50])

The prospects of resolving the ligand hyperfine structure are better at lower frequencies, e.g., *S*-band, because the influence of *g*-strain broadening is less important.[58] This type of heterogeneous broadening arises with metalloproteins because of the conformational flexibility. In effect, the EPR spectrum reflects a distribution of copper centers with varying local environments and ligand fields, and the spread in the g values and A values results in a broadening of the spectrum. At *S*-band, the best resolution of ligand hyperfine structure typically occurs on the $M_{Cu} = -\frac{1}{2}$ transition in the $g_{\|}$ region of the spectrum.[58] Figure 3 shows the structure that appears on this transition in the spectrum of T1Hg laccase. Five lines are evident in the spectrum, and this is consistent with the presence of two approximately equivalent nitrogen ligands. However, the wings of the signal are quite prominent and could reflect additional lines. We should note, in this regard, that the most severe broadening due to chemical nonequivalence of the ligand nuclei occurs in the outermost lines. Indeed, the relative intensities of the resolved lines are more consistent with the intensity ratios 3 6 7 6 3 expected for the center lines of a seven-line pattern (from three nitrogens) than they are with the 1 2 3 2 1 pattern of a five-line multiplet. See below for additional observations that support the idea that the EPR-active copper of the trinuclear cluster actually has three imidazole ligands in its coordination sphere.

We have also studied the pH dependence of the spectrum of T1Hg laccase at $-150°C$ where we estimated the pH (± 0.5 units) from the color of a suitable indicator dye.[54] The data showed that within experimental error, $g_{\|}$ and $A_{\|}$ remained constant in the pH range 5 to 7, but there was a decrease in $g_{\|}$ by around pH 4. We concluded that the pK was too low for the deprotonation of a terminal water ligand, but it was plausible for a bridging water connecting two copper ions. This is an intriguing possibility in view of the pH dependence of inhibitor binding described later.

4.2. *Inhibitor Binding Studies*

4.2.1. *Motivation*

Inhibitor-binding studies involving an ion like fluoride or azide are of interest because of the insight they give into structure as well as the enzyme mechanism.[3,8,54,59] Moreover, the adducts that form may mimic physiologically relevant intermediates that involve bound forms of partially reduced dioxygen, e.g., the hydroperoxide ion. Inhibitor-binding results also provide important contrasts within this class of oxidases, e.g., differences between fluoride binding with tree laccase and fungal laccase. Although fluoride binds tenaciously to the fungal enzyme,[8] EPR data show that with tree laccase even the formation of the monofluoride adduct of the type 2 copper requires the addition of excess fluoride. Moreover, quantitative formation of the difluoride adduct requires a huge excess of fluoride.[60] Finally, results pertaining to the azide adduct have provided useful information about structure. Because the dioxygen reduction site contains the only EPR-active copper in T1Hg laccase, it is an ideal vehicle for EPR studies of inhibitor adducts. In the pursuit of this work, we discovered an incubation method for forming adducts that required much less anion in solution.

4.2.2. *Incubation Method*

John Severns originally developed the new method of adduct formation.[61] In his initial studies he used a T1Hg laccase concentration of about 0.6 mM, added an excess of the inhibitor ion, mixed the solution and then immersed the sample tube in a cryogenic solution in order to freeze the contents. With a six-fold excess of azide there was no hint of adduct formation in the EPR spectrum right after freezing. However, after he incubated the still nominally frozen sample at $-20°C$ for a couple of hours, the formation of the azide adduct was virtually complete. The spectral evolution is dramatic in a sodium phosphate buffer; however, a similar phenomenon occurs in other buffers as well.[54] This method of promoting adduct formation is not specific to laccase, but works for other systems, too. Independently, Yang and Brill have devised a similar method to stimulate fluoride uptake by methemoglobin.[62]

Yang and Brill correctly interpreted the effect in terms of a diffusive encounter of the substrate and the protein,[62] even though the incubation takes place at subzero temperatures. As the work of Luyet shows,[63] with the onset of freezing, an aqueous solution of an electrolyte rapidly evolves into a two phase system. As the ice phase develops, the excluded solutes accumulate in fluid domains that remain liquid at temperatures at least as low as $-20°C$ due to the freezing point lowering phenomenon. At a low enough temperature the sample effectively fully freezes and becomes rigid; however, by then the speciation has changed if only because mass action drives adduct formation as the solutes concentrate in the liquid regions. On the other hand, because of the viscosity and/or because the rate of freezing is rapid, the system frequently does not achieve equilibrium in the new concentration regime before rigidity sets in. When this is the case, regenerating the liquid phase by incubating the sample at $-20°C$ promotes further adduct formation. Figure 4 is a schematic representation of the sample composition when two phases are present. It shows how segregation of the protein from the inhibitor ion is also possible. Of course, no amount of incubation will permit adduct formation to occur for the isolated protein molecules.

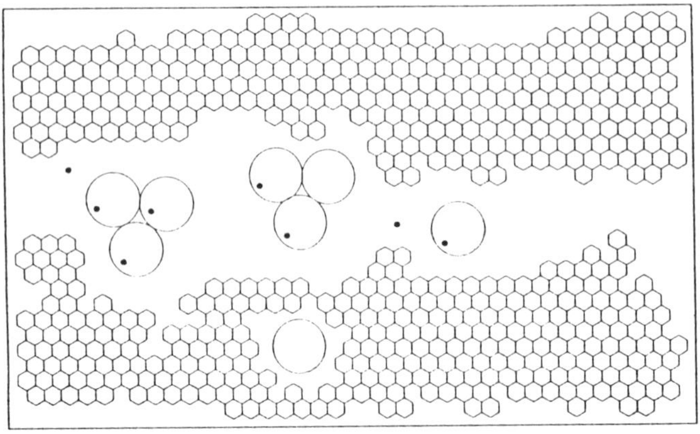

Fig. 4. A two-dimensional representation of the phases present in an aqueous buffer at −20°C. The hexagonal lattice represents pure ice while the circles and the dots represent protein molecules and inhibitor ions, respectively. For clarity we have omitted the buffer ions which also reside in the liquid domain. The sketch also indicates that aggregation effects and/or sporadic entrapment of protein may complicate the picture.

The spectral evolution may be quite different when glycerol is present in the solution. First, as a glassing agent, it inhibits the formation of long range ice structure and, hence, discourages solute pooling. Second, because the presence of glycerol increases the viscosity, diffusion is a slower process. Thus, the system equilibrates more slowly, and aggregation processes are less likely to occur. As a dramatic illustration of this point, the addition of glycerol suppresses the precipitation of disodium hydrogen phosphate and, therefore, the pH drop that ordinarily occurs with the freezing of an aqueous sodium phosphate buffer.[64]

4.2.3. *More About the Influence of pH*

The pH of the medium is an important factor influencing adduct formation by tree laccase because, as Morpurgo *et al.* have noted, the formation constant is much larger around pH 4 than it is at pH 6.[65] This effect explains, in part, some inconsistencies reported in the literature. For example, when Severns and McMillin combined T1Hg laccase with three eq. of fluoride in a sodium phosphate buffer, they found evidence of the monofluoride adduct and unreacted type 2 copper right after freezing the sample.[61] Upon incubation at −20°C, however, the system evolved into the difluoride adduct.

In contrast, when Miller and McMillin studied the same system in a pH 5 acetate buffer, they found that the sample contained approximately equal amounts of the mono- and difluoride adducts right after freezing and that incubation had little effect on the speciation.[54] The differences in the results are undoubtedly largely a reflection of changes in pH. Although the sodium phosphate buffer was pH 6 at room-temperature, it dropped to ca. pH 4 during the freezing process.[54] The pH drop stimulated adduct formation, but due to slow kinetics, an incubation step was necessary. The fact that the pH changed during the freezing process also exacerbated the kinetics problem in the sodium phosphate buffer. However, the situation is much different in an acetate buffer since there is no significant change in pH upon freezing. As discussed in more detail below, the uptake of anions may involve the cleavage of a bond to a bridging hydroxide ligand, and this may explain why a more acidic pH favors adduct formation.

4.2.4. *Azide Adduct*

Table 3 summarizes the EPR parameters obtained for the azide adduct, where each entry is appropriate for ^{65}Cu in the type 2 site. The $A_{\|} = 155 \times 10^{-4}$ cm^{-1} species, assigned as the monoazide adduct, forms upon incubation with six eq. of azide in a sodium phosphate buffer. At higher levels of azide another species with $A_{\|} = 146 \times 10^{-4}$ cm^{-1} begins to grow in, and it probably corresponds to the diazide adduct.[61] A parallel study with [^{15}N]azide in a pH 4.5 sodium acetate buffer gave much better resolved hyperfine structure (Fig. 5). The absence of a central peak in the ligand hyperfine pattern is consistent with an odd number of $^{15}N(I = \frac{1}{2})$ nuclei, and a spectral simulation indicates that there is splitting from only one ^{15}N nucleus and three approximately equivalent ^{14}N nitrogens.[51] According to Fig. 1, this must mean that the azide binds to an EPR-active form of the CU2 or the CU3 copper ion.

This result is consistent with the findings of Messerschmidt *et al.* in their studies of the hydroperoxide adduct of ascorbate oxidase.[20] In particular, the X-ray structure showed that the hydroperoxide anion binds to CU2 and that there is no longer a bridging ligand between CU2 and CU3. As a result, the distance from CU2 to CU3 expands from 3.7 Å to 4.8 Å. In contrast, CU3 remains close to CU4, although there is still no indication of a bridge between these two copper centers in the structure. If the structure of the azide adduct is analogous, autoreduction of CU2 and CU3 or a

Table 3. EPR parameters for inhibitor adducts of TlHg tree laccase.

		^{65}Cu	^{14}N	[X]	
Adduct	$g_{\|}$	$10^4 \times A_{\|}$ (cm^{-1})	$10^4 \times A$ (cm^{-1})	$10^4 \times A$ (cm^{-1})	Ref.
Control	2.24	209			51
Monoazide	2.24	155	11	11[^{15}N]	51
Diazide	2.29	146			61
Dicyanide	2.16	215	19	45[^{13}C]	51

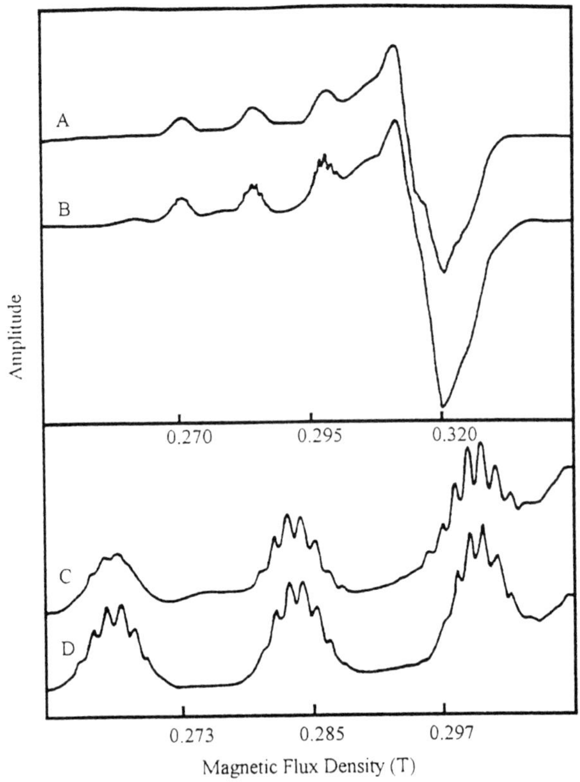

Fig. 5. X-band EPR spectra of monoazide adducts of T1Hg laccase at −150°C. A: After incubation with 6 eq. azide in 0.1 M sodium phosphate. B: After incubation with 6 eq. [^{15}N]azide in 0.1 M, pH 4.5, sodium acetate. C: Blow up of the parallel region of B. D: Simulation of C; see Table 3 for parameters. (Reprinted with permission from Peyratout *et al.*[51])

through-space antiferromagnetic coupling is a possible explanation for the fact that the two coppers are EPR silent. The structure of the hydroperoxide adduct is probably relevant because Messerschmidt and co-workers have also solved the structure of a diazide adduct of ascorbate oxidase, and it shows that both azide ions bind to CU2.[20] In this case, the copper center adopts a trigonal bipyramidal coordination geometry with two apical azide ligands and three histidine ligands in the equatorial plane. Uptake of azide at CU2 is also consistent with the observed pH dependence of inhibitor binding because a reduction in the pH would facilitate the loss of the hydroxide bridge between CU2 and CU3.

4.2.5. *Cyanide Adduct*

Although Malkin *et al.* characterized a dicyanide adduct of fungal laccase many years ago,[66] the corresponding adduct of tree laccase escaped detection until recently.[51] In view of the basicity of the cyanide ion, we originally attempted to form the adduct in a pH 7 potassium phosphate buffer. However, under these conditions we observed a significant loss of EPR signal intensity, probably as a result of reduction to copper(I) by cyanide. In addition, at this pH cyanide may remove copper from the protein. In view of the influence the pH had on azide binding, we then switched to a pH 5 ammonium acetate buffer and obtained better results. Figure 6A is the spectrum obtained immediately after freezing a sample containing 20 eq. of cyanide and is identical to that obtained in the absence of cyanide. However, after overnight incubation at –20°C, about 30–40 % of the protein converted into the cyanide adduct (Fig. 6B). The ligand hyperfine structure that appears in the spectrum is diagnostic of splitting from two equivalent $I = 1$ nuclei, almost certainly nitrogen-donating ligands. They are in fact protein nitrogens since the addition of [^{15}N]cyanide gave exactly the same spectrum. On the other hand, new hyperfine structure appeared with the addition of [^{13}C]cyanide (Fig. 6C). Analysis showed that two equivalent ^{13}C nuclei were responsible for the splitting.[51] See Table 3 for a summary of the derived coupling constants.

The EPR data indicate that the dicyanide adduct has a quite different structure than the monoazide adduct. In the first place, the relatively large hyperfine coupling constants for the protein nitrogens and $A_{\|}$(Cu) suggest that the cyanide adduct has a more tetragonal coordination geometry.[67,68] Spectral simulations indicate that the four in-plane donors in

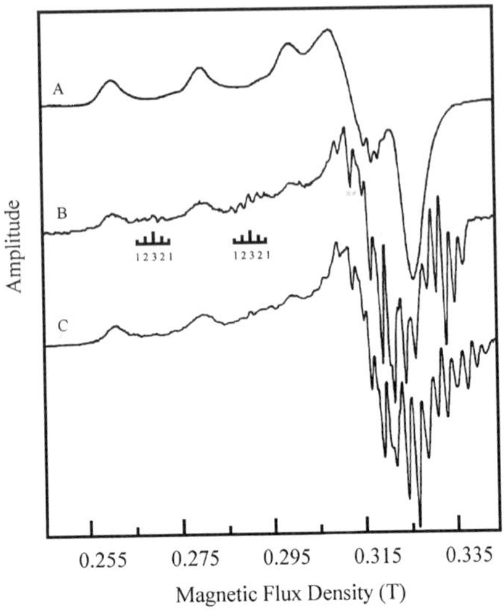

Fig. 6. X-band EPR spectra of T1Hg laccase in 0.1 M, pH 5, ammonium acetate buffer at −150°C. A: Sample with 20 eq. cyanide immediately after freezing. B: Sample in A after overnight incubation at −20°C. C: Same as in B except with [^{13}C]cyanide. Note the presence of unreacted protein in all three spectra. (Reprinted with permission from Peyratout *et al.*[51])

the tetragonal binding site involve two carbon atoms from the cyanide ions and two protein nitrogens. Clearly the CU4 copper in the type 2/type 3 cluster of ascorbate oxidase, i.e., the center Messerschmidt and co-workers assigned as the type 2 copper, is compatible with the proposed structure because it entails only two imidazole ligands. However, the pH has the same effect on the formation of the cyanide adduct or the monoazide adduct, and this suggests that both inhibitors compete for the same copper center. Furthermore, in the only structurally characterized adduct involving two inhibitor anions, both azides bind at the CU2 atom of ascorbate oxidase.[20] Instead of adopting the trigonal bipyramidal coordination geometry of the diazide adduct, the CU2 copper may assume a square pyramidal geometry with the strong-field cyanide ligands and two of the imidazole groups defining the basal plane. This would relegate the other imidazole group to the axial coordination position where it would experience at best a weak hyperfine interaction with the copper center. Such a structural reorganization seems feasible because five-coordinate square pyramidal and trigonal bipyramidal geometries normally have similar energies in copper(II) systems. The need for structural reorganization may also account for the absence of a monocyanide adduct. If the structural change occurs with the uptake of the first cyanide, the binding constant for the second cyanide could be larger than the first.

4.3. *State of Copper in T1Hg Laccase*

Simply exposing the reduced form of T1Hg laccase to air-saturated buffer results in the oxidation of the type 2/type3 cluster. Thus, Tamilarasan and McMillin[24] used a colorimetric assay[69] to show that in the resting form of T1Hg laccase upwards of 50 % of the copper is in the copper(II) oxidation state. And, on the basis of the pre-edge X-ray absorption spectrum (XAS), Cole *et al.* estimated that < 10 % of the copper in their preparation was in the reduced form.[28] The results parallel those obtained for native *Rhus* laccase where the resting enzyme has about 25 % copper(I) in the type 3 site.[28,70]

In contrast, type 2 depleted (T2D) laccase, which we discuss in detail below, tends to contain a considerable amount of reduced copper. The first definitive evidence came from XAS data which revealed almost complete reduction of its type 3 copper.[70] The spectrophotometric assay developed by Hanna *et al.* yielded similar results[69] but others have argued that their T2D preparations contain virtually no copper(I).[71] Although some doubt may remain, the preponderance of the evidence suggests that the protein must contain both the type 2 and the type 3 copper in order to react with molecular oxygen efficiently.

4.4. *Kinetics of T1Hg Laccase*

Although there have been relatively few studies of the kinetics of oxidation or reduction of T1Hg laccase, there are some interesting contrasts with the native enzyme. In terms of turnover activity with ordinary substrates, T1Hg laccase typically has no more than about 5 % of the activity of the native enzyme, and traces of holoprotein are probably mainly responsible for the residual activity. Tamilarasan and McMillin first reported that the reduction of the type 3 copper site is a very slow process when ascorbate is the reducing agent.[24] However, by the time they measured the room-temperature EPR spectrum (within minutes), the reduction of the type 2 copper in the same sample appeared to be complete. Cole *et al.* also reported that reduction by ascorbate was slow, but they found that reduction proceeded much more rapidly with dithionite.[28]

Solomon and co-workers' studies of the oxidation of reduced T1Hg laccase proved very interesting.[27] They found that the reduced form underwent a reaction with dioxygen with a second-order rate constant of 2.2×10^6 $M^{-1}s^{-1}$ in a pH 7.4 potassium phosphate buffer at 3°C. The initial product decayed with a halflife of ca. 1 h, and it exhibited absorption maxima at 340 nm ($\varepsilon = 3700\ M^{-1}cm^{-1}$) and at 470 nm. On the basis of comparisons with model compounds, Cole *et al.* assigned the intermediate as a hydroperoxide adduct. The corresponding reaction with reduced native laccase also formed an intermediate by means of a second-order pathway with a similar rate constant.[25,27] When Reinhammar and co-workers studied this same reaction by EPR methods, they also detected a transient species that gave a $g < 2$ signal.[26] Since Solomon and co-workers did not observe this signal in their studies of T1Hg laccase, they concluded that its formation required reduction beyond the level of hydroperoxide.[28]

Understanding the nature of the optically detected intermediate is an important aim. Solomon and co-workers have argued that the adduct is unlikely to involve a μ-1,2 peroxo bridge between the two type 3 coppers because that would entail a much larger molar absorptivity.[3,27] Since the CT energies are too high for a terminal peroxide ligand, they proposed that there is a μ-1,1 linkage of hydroperoxide between a type 3 copper in the copper(II) form and the type 2 copper in the copper(I) form.[3] They also concluded that MCD data were consistent with the proposed oxidation state of the type 2 copper. They also inferred that the corresponding intermediate does not persist in the case of the native enzyme because of further reduction due to rapid electron transfer from the type 1 (and perhaps the type 2) copper. The puzzle is why electron transfer from the type 2 copper would be so slow in the case of T1Hg laccase since peroxides typically react via inner-sphere redox mechanisms.[72] More recently, Messerschmidt and co-workers have structurally characterized a hydroperoxide adduct of ascorbate oxidase, and it turns out to be a terminal ligand bound to one of the copper centers originally designated as part of the type 3 pair.[20] Figure 7 portrays the structure. From this perspective, it is not clear why molecular oxygen does not rapidly oxidize the type 3 pair of T2D laccase.

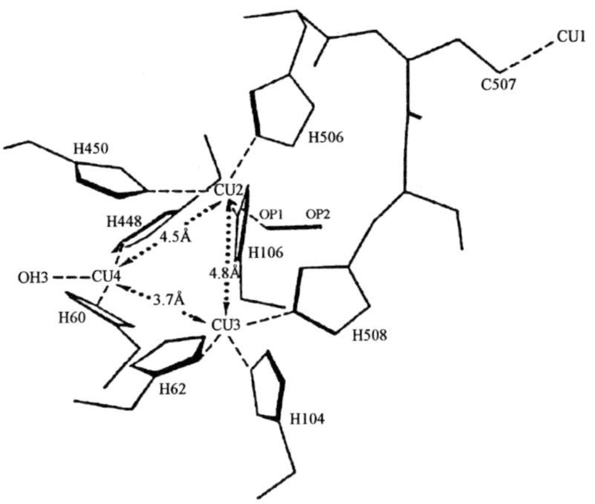

Fig. 7. Schematic view of the type 2/type 3 cluster in the hydroperoxide adduct of ascorbate oxidase. The distances listed are the mean values determined for the two subunits. The atoms OP1 and OP2 are part of the hydroperoxide ligand. (Adapted with permission from Messerschmidt *et al.*[20])

5. Type 2-Depleted Laccase

5.1. *Preparation and Characterization*

In 1969 Malkin *et al.* first reported the preparation of a type 2 depleted (T2D) derivative of fungal laccase from *Polyporus versicolor.* Since the procedure employed reducing conditions and a chelator specific for copper(I), the protein undoubtedly released the metal as copper(I).[73] Several years later Graziani *et al.* reported a procedure for the removal of the type 2 copper from tree laccase.[74] Their method called for anaerobic conditions and ferrocyanide as a reducing agent, but it employed EDTA and a lipophilic reagent such as dimethylglyoxime (DMG) as copper scavengers.[74] The fact that laccase readily releases a single copper atom argues that one of the copper binding sites is uniquely labile, and by the process of elimination, it appears to be the type 2 copper. One reason is that the EPR signal of the type 2 copper is almost completely absent in the spectrum of the product, whereas the EPR signal of the type 1 copper is still present. In addition, the oxidized T2D protein also exhibits the 330 nm absorption that is the hallmark of the type 3 pair. Finally, in studies involving ascorbate oxidase, crystallographic results show that the procedure preferentially removes what Messerschmidt has assigned as the type 2 copper, although there is some copper loss from the other two copper sites within the trinuclear cluster as well.[75]

While most workers agree that the type 3 coppers remain bound in the T2D form of the enzyme, there is no general agreement about their valence state. Graziani and co-workers originally reported that the absorbance at 330 nm varied somewhat from preparation to preparation.[74] But subsequent spectrophotometric titrations were consistent with an oxidized type 3 site since the complete reduction of the T2D protein required three electrons.[71,76,77] Moreover, Reinhammar and Oda reported that the molar absorbance of the 330 nm chromophore did not change upon removal of the type 2 copper.[78] In order to reconcile these results with other conflicting findings, Pecht and co-workers have proposed that there are as many as three different forms (isotypes) of T2D laccase.[79]

The conflicting findings come from two groups who have developed alternative means of probing the valence state of the T2D derivative. Solomon and co-workers first conducted an X-ray absorption (XAS) edge study on T2D tree laccase in 1980, and concluded that the type 3 coppers were in the reduced state.[80] Later, using an XAS difference technique, Kau *et al.* found that the extent of the reduction of type 3 site was 98 ± 22 % even after exposure to air or excess ferricyanide.[70] Upon addition of peroxide, however, the absorbance at 330 nm increased in conjunction with the oxidation of the binuclear type 3 site.[70] In another study, Hanna *et al.* developed a colorimetric assay specific for residual copper(I) in laccase.[69] This assay determined that 85 % of the type 3 coppers in T2D laccase are in the reduced form. Even after oxidation with hydrogen peroxide, about 10 % of the type 3 copper sites remain in the half-reduced state according to EPR integrations.[81] Although it is possible that the oxidation state of the type 3 site varies in different preparations, both groups who have studied the copper centers directly have found that the type 3 site remains reduced in the T2D derivative.

Figure 8 presents the EPR spectrum of T2D tree laccase. Aside from the type 1 signal, a signal from another center is evident at the low flux density end of the spectrum, near 0.265 T. Morpurgo *et al.* originally attributed this to a modified type 2 copper center.[82] But, on the basis of spectral data compiled by Reinhammar,[83] this signal probably corresponds to a half-reduced type 3 copper site, i.e. a [Cu(I)-Cu(II)] form. In

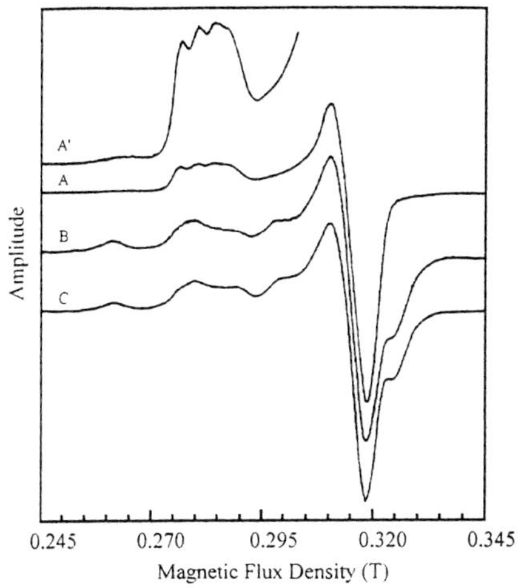

Fig. 8. X-band EPR spectra of tree laccase samples at −150°C. A: T2D laccase. A′: Same as A at higher amplification. B: Native laccase. C: Reconstituted T2D laccase. *A′* samples are in pH 6 sodium phosphate. (Reprinted with permission from Klemens *et al.*[81])

particular, the signal shows strong rhombic character like signals found in other proteins containing bimetallic centers including superoxide dismutase, half-met hemocyanin, half-met tyrosinase and cytochrome c oxidase.[84] Solomon and co-workers have found that the copper(II) side of the half-met type 3 site is capable of binding an anion as a terminal ligand.[19,85] This behavior contrasts that seen for anion binding to half-met hemocyanin and tyrosinase where exogenous ligands bridge the binuclear site. This difference in substrate binding carries over to the way in which the proteins bind dioxygen at the peroxide level of reduction.[3]

In the case of the T2D form of fungal laccase, the EPR spectrum shows well-resolved hyperfine structure in the $g_\perp$ region.[40] At a frequency of 2.38 GHz there is also well-resolved ligand hyperfine splitting on the $M_{\mathrm{Cu}} = -\frac{1}{2}$ transition in the parallel region (Fig. 9). Spectral simulations suggest that the pattern arises from splitting by two nitrogen and two hydrogen nuclei. The coupling constants for the hydrogens are large (25–26 MHz), consistent with ENDOR data for the type 1 site in fungal laccase.[86] We have also observed similar coupling constants in the spectrum of azurin.[87]

One other well studied property of T2D laccase is its ability to undergo reconstitution to native laccase. The reconstituted protein exhibits EPR

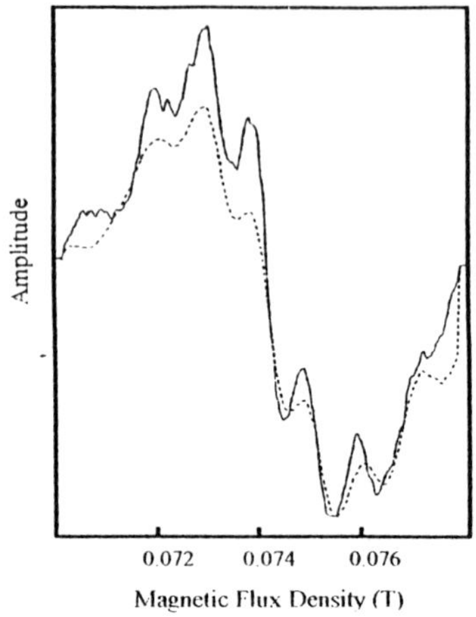

Fig. 9. $M_{\mathrm{Cu}} = -\frac{1}{2}$ transition in the S-band EPR spectrum of the T2D form of fungal laccase at −150°C. The microwave frequency was 2.38 GHz, and the buffer was 0.1 M sodium phosphate. Experimental spectrum (—). Simulation (− − − −). (Reprinted with permission from Hanna *et al.*[40])

signals from two EPR active copper centers, the characteristic 330 nm absorbance of the type 3 site, and full activity. To reconstitute the protein most workers have found reducing conditions give the best results.[40,73,81,82] Morpurgo *et al.* have, however, reported that the T2D protein slowly takes up Cu(II).[88]

5.2. *Mechanism of the Removal of the Type 2 Copper*

In their original report on the preparation of T2D laccase, Graziani *et al.* indicated that dialysis under a nitrogen atmosphere against a solution of DMG caused a significant modification of the type 2 copper in tree laccase but did not entirely remove the copper from the protein solution.[74] On the other hand, when the dialysis medium contained a mixture of DMG and EDTA, the removal of the type 2 copper went smoothly. Without the presence of DMG, however, EDTA had no effect on the type 2 copper. They reasoned that the smaller, more lipophilic reagent DMG attacked the copper in the type 2 site more effectively but that EDTA was superior at solubilizing copper(II) in the dialysis medium. At that time, they did not postulate a role for ferrocyanide. Later, the same group extended the method and showed that DDC (N,N-diethyldithiocarbamate) would replace DMG.[82] At this juncture, they suggested that the reduction of the type 3 pair by ferrocyanide might facilitate the removal of the type 2 copper.

In 1990, work by Klemens and McMillin offered a new perspective on the mechanism of removal of the type 2 copper.[81] They found it was possible to omit sparingly soluble lipophilic agents like DMG, if they dialyzed the protein under a nitrogen atmosphere against EDTA and a redox buffer comprised of a mixture of ferricyanide and ferrocyanide. The addition of a chromatographic elution step in the procedure provided a rational means of obtaining a more homogeneous preparation. EPR integrations showed that the T2D product contained on average 1.1 EPR active coppers. Moreover almost all of the coppers in the type 3 sites are in the reduced, i.e., copper(I) form. Since there is a redox buffer present, it is possible that the type 2 copper dissociates as copper(I) or as copper(II). As the authors pointed out, if it dissociates as copper(I), the ferricyanide present in solution can oxidize the metal to copper(II) for subsequent scavenging by EDTA.

Li and McMillin later reinvestigated the reaction in order to establish the role of the redox buffer.[89] They found that when they employed a soluble ferrocene system as the redox buffer, the procedure was ineffective, unless they also included a cyanometalate cofactor such as $Co(CN)_6{}^{3-}$ or $Ru(CN)_6{}^{4-}$. Three potential roles for the cofactor were apparent: a source of free cyanide, a ligand that attacks the type 2 copper directly, or a binding agent that induces an allosteric transition in the protein. The thinking was that a conformational transition could enhance the accessibility of the type 2 copper.

Li and McMillin also provided evidence that suggested that the type 2 copper dissociates as copper(I).[89] In particular, they showed that they could remove the type 2 copper in the presence of ascorbate, i.e., under strictly reducing conditions, provided that they incorporated a ligand suitable for scavenging copper(I) in place of EDTA. Significantly, the ascorbate method also required the presence of a cyanometalate cofactor.

With regard to the procedure that employs a redox buffer (the EDTA method), there are two obvious mechanistic possibilities.[89] According to the first mechanism, the type 2 copper dissociates from reduced laccase (with some type of assistance from the cyanometalate cofactor), and then the redox buffer provides for oxidation to copper(II) prior to scavenging by EDTA. The second possibility is that the type 2 copper dissociates as copper(II) from a mixed-valence form of laccase. In this view, one of the major roles of the redox buffer is to poise the potential so as to generate the mixed-valence form. Morpurgo *et al.* favored the latter mechanism because they found that the size and the lipophilicity of the copper(II) chelator strongly influenced the efficiency of the process.[90] The order of effectiveness was nitrilotriacetate (NTA) $>$ DMG $\approx$ DDC $\gg$ EDTA. They also reported that the solution potential necessary for reaction varied with the conditions. Thus, at pH 5.2 the procedure that used DMG required the presence of ferricyanide and ferrocyanide. But at pH 7 where DMG has a higher affinity for copper(II), the addition of ferricyanide was unnecessary.

In order to elucidate the role of the cyanometalate cofactor, we carried out some additional experiments that utilized the ascorbate method. Because of the low reactivity with regard to oxidation/reduction chemistry as well as ligand-substitution chemistry, we chose the tris(oxalato)-chromate(III) ion, $Cr(ox)_3{}^{3-}$, as a replacement ion for $Co(CN)_6{}^{3-}$.[91] The oxalate complex is relevant because it has the same charge and is about the same size as a typical cyanometalate cofactor. However, with the oxalate complex present in the reaction chamber, little or no depletion of the type 2 copper occurred. This result casts grave doubt on the notion that the cofactor is an allosteric agent. At the same time, we showed that it was possible to omit the cyanometalate cofactor provided that we continuously injected low levels of cyanide into the reaction chamber.[91] This strongly suggests that it is, in fact, free cyanide that attacks the type 2 copper and that the key role of the cofactor is to release cyanide into the solution. A continuous influx of cyanide is probably necessary because the reaction occurs under a continuous purge of dinitrogen gas, and at pH 5 cyanide exists in the volatile HCN form.

At least in the case of the ascorbate procedure, Eqs. (1) and (2) represent the simplest plausible mechanism for the removal of the type 2 copper.

$$\mathrm{N(red)} + 2\mathrm{CN}^- \rightarrow \mathrm{T2D} + \mathrm{Cu(CN)_2}^- \quad (1)$$

$$\mathrm{Cu(CN)_2}^- + 2\mathrm{NN} \rightarrow \mathrm{Cu(NN)_2}^+ + 2\mathrm{CN}^- \quad (2)$$

In this scheme N(red) and NN denote reduced laccase and a copper(I) chelator, respectively. Of course the process involves many other steps, such as the protonation and deprotonation of cyanide, the reduction of native laccase, etc., which for simplicity we have omitted. It is important to note that under the reaction conditions there is no loss of copper unless the copper(I) chelator is present. Although copper(I) has a high affinity for cyanide, the steady-state concentration of the cyanide ligand in these experiments is not great enough to remove the copper on its own. Of course, as we have already shown, at higher cyanide concentrations and at a higher pH the non-selective loss of copper occurs with the production of apolaccase. The proposal that laccase releases the metal from the type 2 site as copper(I) correlates well with the results we have obtained with fungal laccase.[40]

Although the evidence for dissociation as copper(I) is not unequivocal, it seems reasonable to assume that Eq. (1) also occurs in the EDTA method. Instead of Eq. (2), we propose that Eqs. (3) and (4) are the basis of the scavenging of the free copper ion.

$$Cu(CN)_2^- + ox \rightarrow Cu^{2+} + 2CN^- \quad (3)$$

$$Cu^{2+} + EDTA^{4-} \rightarrow CuEDTA^{2-} \quad (4)$$

Here, ox denotes an oxidizing species, e.g., part of the redox buffer. *A priori*, it is evident that Eq. (3) could have an important impact on the reaction because T2D laccase reacts, at best, slowly with Cu(II).[88] This makes it easy for EDTA to compete for copper(II) in solution.

Of course, we cannot completely exclude the alternative possibility that laccase releases the type 2 copper as copper(II) from a partially reduced, i.e., mixed-valence form of the enzyme. This hypothesis would account for the need for a redox buffer, but there is no obvious explanation for the fact that the reincorporation of copper(II) is slow even with the type 3 pair in the reduced state. It is also possible that ferrocyanide attacks the type 2 copper directly, especially in light of the known adducts between ferrocyanide and copper(II).[92] However, the fact that differently charged cyanometalate ions such as $Co(CN)_6{}^{3-}$ are equally effective casts some doubt on the idea.

6. Ambiguities Regarding the Type 2 Copper

The operational definition of the type 2 copper is that it is the EPR-active copper of the type 2/type 3 cluster. Thus, when Messerschmidt and coworkers examined the electron density map of ascorbate oxidase, they assigned atom CU4 in Fig. 1 as the type 2 copper because of the oxygen (or hydroxide) bridge between CU2 and CU3.[9] Since metal–metal bonds tend to be weak between copper(II) centers, bridging ligands normally mediate antiferromagnetic coupling interactions. (The two copper centers in the Cu_A site of cytochrome *c* oxidase are somewhat exceptional in that they seem to form a direct metal–metal bond, at least in the mixed valence state.[93]) In line with this partitioning of the type 2/type 3 cluster, it is tempting to relate CU2 and CU3 to the dinuclear copper site in hemocyanin where the protein also binds each copper ion with three histidine ligands.[94] However, the analogy is a poor one because the evidence indicates that the two proteins have evolved quite differently.[11,95] Moreover, they form very different types of peroxide adducts.[20,96]

Another point to keep in mind is that the identification of the type 2 copper offered by Messerschmidt and co-workers is only relevant to the structurally characterized state of ascorbate oxidase. Many other forms of the enzyme undoubtedly exist. Indeed, in the crystallographically characterized hydroperoxide and azide adducts, the anions bind to CU2, and it no longer shares the bridging ligand with CU3.[20] In our studies of the [^{15}N]azide adduct, the protein also binds to the EPR-active copper via three nitrogen-donating ligands, vide supra.[51] Even in the absence of an inhibitor anion, the low temperature S-band EPR spectrum of T1Hg laccase suggests that the type 2 copper may have as many as three nitrogen-donating ligands.[50] To add to the possibilities, note that Cole *et al.* have postulated that there is a form of the cluster in which there is a ligand that simultaneously bridges all three copper centers in the dioxygen reduction site.[16] If antiferromagnetic coupling is possible between more than one pair of coppers in the type 2/type 3 cluster, it is not at all clear which copper dissociates from the cluster during the preparation of T2D laccase. Indeed, when Messerschmidt and co-workers applied the method to crystals of ascorbate oxidase, they found that some depletion occurred at all three copper sites within the cluster, although the greatest loss of electron density occurred at the CU4 position.[75] Since this copper center forms the fewest coordinate-covalent bonds to protein residues, it could be quite labile. But, that is not to say that it is EPR active in the resting protein. The triply bridged structure proposed by Solomon and co-workers could perhaps explain the unusual magnetic properties of the dioxygen reduction site in ceruloplasmin,[97] and it might be at the heart of the well-known temperature dependence of the EPR signal in tree laccase that we review next.

Morpurgo *et al.* first noted that the EPR signal of the type 2 copper in tree laccase varied with temperature.[59] Severns and McMillin found similar results, but both studies involved the use of a sodium phosphate buffer.[61] Thus, the drop in pH that occurs with this buffer at low temperature could have been at fault. However, the results compiled in Table 2 show that the pH has a minimal influence on the low temperature spectrum. Two distinct explanations are possible for the temperature dependence of the EPR spectrum, although both require some type of structural change. One possibility is that the coordination geometry at the EPR-active copper changes at low temperature.[59] The other possibility is that a different copper center within the trinuclear cluster becomes EPR active in frozen solution. Experiments described below are consistent with the latter view.

Results obtained with a doubly labeled form of laccase are the most relevant. To make the derivative, Li *et al.* started with ^{63}Cu-enriched tree laccase, made [^{63}Cu]T2D laccase, and then reconstituted with excess ^{65}Cu(I).[21] The remarkable result was that after reconstitution, the EPR-active site in the frozen-solution spectrum contained a ^{63}Cu nucleus. This is easy to tell from the copper hyperfine pattern in the parallel region of the spectrum because the nuclear magnetic moment of ^{65}Cu is about 7 % larger than that of ^{63}Cu. Li *et al.* did the experiments in a sodium phosphate buffer, but Miller *et al.* later obtained the same result in a potassium phosphate buffer (Fig. 10).[98] Thus, the effect is not an artifact of a pH change in frozen solution. The most informative part of the spectrum is the $M_{Cu} = -3/2$ transition of the copper hyperfine pattern in the parallel region of the spectrum, i.e., the peak that occurs at lowest flux density. The resolution of the peak is good, and the position is quite sensitive to the value of the coupling constant. The vertical line in Fig. 10 is a visual aid that facilitates comparison of the peak positions in the spectra of the various mixed-isotope derivatives as well as the ^{63}Cu-enriched and ^{65}Cu-enriched laccase controls.

Miller *et al.* also showed that copper exchange did not spoil the structure of the mixed-isotope derivative.[99] This is a potential problem because we typically treated T2D laccase with 6 eq. Cu(I) during the reconstitution step. If exchange were rapid, the product could contain as much as 66 % ^{65}Cu. In fact, however, the reconstituted protein contained only about 25 % ^{65}Cu (Fig. 10, Inset); hence, exchange was not efficient under our conditions.

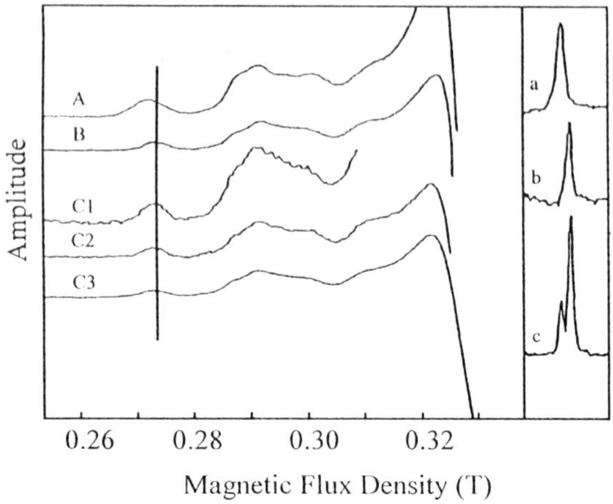

Fig. 10. X-band EPR spectra of various derivatives of tree laccase in pH 6 potassium phosphate at −150°C. A: [^{65}Cu]laccase. B: [^{63}Cu]laccase. C1–C3: Mixed-isotope derivatives made by reconstitution of [^{63}Cu]T2D laccase with ^{65}Cu(I). Inset: $M_{\mathrm{Cu}} = -3/2$ transitions of $Cu(acac)_2$ samples derived from: ^{65}Cu-enriched native laccase (a), ^{63}Cu-enriched native laccase (b), and a mixed-isotope derivative made from ^{63}Cu-enriched T2D laccase (c).

The room-temperature data that Li *et al.* obtained for the mixed-isotope derivative were equally interesting.[21] Since the X-band EPR signal is quite broad at room temperature, Li *et al.* carried out the experiments at S-band with a spectrometer at the ESR Center in Milwaukee. In contrast to the frozen-solution result, the ^{65}Cu nucleus was EPR active in fluid solution.[21] These data show that different copper centers of the cluster are EPR active under different conditions. Hence, this is a plausible explanation for the temperature dependence of the EPR spectrum.

Acknowledgment

Over the years the National Institutes of Health, the Petroleum Research Fund and Purdue University have helped fund the work carried out in our laboratory. The names of co-workers who have contributed to this effort appear in the bibliography.

References

1. B. Reinhammar, in *Copper Proteins and Copper Enzymes*, Vol. 3, ed. R. Lontie (CRC Press, Boca Raton, Florida, 1984), pp. 1–35.

2. E. T. Adman, in *Metalloproteins*, Part 1, ed. P. M. Harrison (Verlag Chemie, Deerfield Beach, Florida, 1985), pp. 1–42.
3. E. I. Solomon, M. J. Baldwin and M. D. Lowery, *Chem. Rev.* **92**, 521 (1992).
4. D. R. McMillin, C. S. Peyratout and C. Miller, in *Encyclopedia of Inorganic Chemistry*, Vol. 2, ed. R. B. King (Wiley, Surrey, UK, 1994), pp. 869–883.
5. A. M. Mayer, *Phytochemistry* **26**, 11 (1986).
6. W. Bao, D. M. O'Malley, R. Whetten and R. R. Sederoff, *Science* **260**, 672 (1993).
7. J. F. D. Dean and K. -E. L. Eriksson, *Holzforschung* **48**, 21 (1994).
8. R. Malkin and B. G. Malmström, *Adv. Enzymol.* **33**, 177 (1970).
9. A. Messerschmidt, R. Ladenstein, R. Huber, M. Bolognesi, L. Avigliano, R. Petruzzelli, A. Rossi and A. Finazzi-Agrò, *J. Mol. Biol.* **224**, 179 (1992).
10. A. Messerschmidt, *Adv. Inorg. Chem.* **40**, 121 (1993).
11. A. Messerschmidt and R. Huber, *Eur. J. Biochem.* **187**, 341 (1990).
12. P. R. LaFayette, K. -E. L. Eriksson and J. F. D. Dean, *Plant Physiol. 107*, 667 (1995).
13. S. E. Shadle, J. E. Penner-Hahn, H. J. Schugar, B. Hedman, K. O. Hodgson and E. I. Solomon, *J. Am. Chem. Soc.* **115**, 767 (1993).
14. D. R. McMillin and M. C. Morris, *Proc. Natl. Acad. Sci. U.S.A.* **78**, 6567 (1981).
15. Y. Lu, L. B. LaCroix, M. D. Lowrey, E. I. Solomon, C. J. Bender, J. Peisach, J. A. Roe, G. B. Gralla and J. S. Valentine, *J. Am. Chem. Soc.* **115**, 5907 (1993).
16. J. L. Cole, P. A. Clark and E. I. Solomon, *J. Am. Chem. Soc.* **112**, 9534 (1990).
17. B. Reinhammar, *Biophys. Biochim. Acta* **275**, 245 (1972).
18. L. Morpurgo, A. Desideri and G. Rotilio, *Biochem. J.* **207**, 625 (1982).
19. D. J. Spira-Solomon and E. I. Solomon, *J. Am. Chem. Soc.* **109**, 6421 (1987).
20. A. Messerschmidt, H. Luecke and R. Huber, *J. Mol. Biol.* **230**, 997 (1993).
21. J. -B. Li, D. R. McMillin and W. E. Antholine, *J. Am. Chem. Soc.* **114**, 725 (1992).
22. L. -E. Andréasson and B. Reinhammar, *Biochim. Biophys. Acta* **568**, 145 (1979).
23. G. B. Koudelka and M. J. Ettinger, *J. Biol. Chem.* **263**, 3698 (1988).
24. R. Tamilarasan and D. R. McMillin, *Biochem. J.* **263**, 425 (1989).
25. L. -E. Andréasson, R. Brändén and B. Reinhammar, *Biochim. Biophys. Acta* **438**, 370 (1976).
26. R. Aasa, R. Brändén, J. Deinum, B. G. Malmström, B. Reinhammar and T. Vänngård, *Biochem. Biophys. Res. Commun.* **70**, 1204 (1976).
27. J. L. Cole, D. P. Ballou and E. I. Solomon, *J. Am. Chem. Soc.* **113**, 8544 (1991).
28. J. L. Cole, G. O. Tan, E. K. Yang, K. O. Hodgson and E. I. Solomon, *J. Am. Chem. Soc.* **112**, 2243 (1990).

29. W. E. B. Shepard, R. L. Kingston, B. F. Anderson and E. N. Baker, *Acta Cryst. Sec.* **D49**, 331 (1993).
30. W. B. Church, J. M. Guss, J. J. Potter and H. C. Freeman, *J. Biol. Chem.* **261**, 234 (1986).
31. T. P. J. Garrett, D. J. Clingeleffer, J. M. Guss, S. J. Rogers, and H. C. Freeman, *J. Biol. Chem.* **259**, 2822 (1984).
32. H. Nar, A. Messerschmidt, R. Huber, M. van deKamp and G. W. Canters, *J. Mol. Biol.* **221**, 765 (1991).
33. H. Nar, R. Huber, A. Messerschmidt, A. C. Filippou, M. Barth, M. Jaquinod, M. van deKamp and G. W. Canters, *Eur. J. Biochem.* **205**, 1123 (1992).
34. J. M. Moratal, A. Romero, J. Salgado, A. Perales-Alarcón and H. R. Jiménez, *Eur. J. Biochem.* **228**, 653 (1995).
35. K. A. Blackwell, B. F. Anderson and E. N. Baker, *Acta Cryst. Sec.* **D50**, 263 (1994).
36. J. K. Burdett, *Molecular Shapes* (Wiley-Interscience, New York, 1980), pp. 170–195.
37. K. Ando, *J. Biochem.* (Tokyo) **68**, 501 (1970).
38. B. G. Malmström, L. E. Andréasson and B. Reinhammar, *The Enzymes*, 3rd ed., Vol. 12 (Academic Press, New York, 1975), pp. 507-579.
39. B. L. Hauenstein, Jr. and D. R. McMillin, *Biochem. Biophys. Res. Commun.* **85**, 505 (1978).
40. P. M. Hanna, D. R. McMillin, M. Pasenkiewicz, W. E. Antholine and B. Reinhammar, *Biochem. J.* **253**, 561 (1988).
41. P. M. Colman, H. C. Freeman, J. M. Guss, M. Murata, V. A. Norris, J. A. M. Ramshaw and M. P. Venkatappa, *Nature* **272**, 319 (1978).
42. M. C. Morris, M. M. Morie-Bebel and D. R. McMillin, *J. Am. Chem. Soc.* **106**, 3677 (1984).
43. A. S. Klemens, D. R. McMillin, H. R. Tsang and J. E. Penner-Hahn, *J. Am. Chem. Soc.* **111**, 6398 (1989).
44. P. Pyykkö, *Chem. Rev.* **88**, 563 (1988).
45. H. R. Engeseth and D. R. McMillin, *Biochemistry* **25**, 2448 (1986).
46. J. A. Larrabee and T. G. Spiro, *Biochem. Biophys. Res. Comm.* **88**, 753 (1979).
47. T. Sakurai, S. Suzuki and M. Fujita, *Inorg. Chim. Acta* **152**, 139 (1988).
48. J. -B. Li and D. R. McMillin, *Inorg. Chim. Acta* **167**, 119 (1990).
49. H. Thomann, M. Bernardo, M. J. Baldwin, M. D. Lowery and E. I. Solomon, *J. Am. Chem. Soc.* **113**, 5911 (1991).
50. M. M. Morie-Bebel, D. R. McMillin and W. E. Antholine, *Biochem. J.* **235**, 415 (1986).
51. C. S. Peyratout, J. C. Severns, S. R. Holm and D. R. McMillin, *Arch. Biochem. Biophys.* **314**, 405 (1994).
52. B. G. Malmström, B. Reinhammar and T. Vänngård, *Biochim. Biophys. Acta* **205**, 48 (1970).

53. S. S. Larsen, *Arch. Pharmacol. Chem. Soc.* **1**, 41 (1973).
54. C. Miller and D. R. McMillin, *Biophys. Chem.* **53**, 189 (1995).
55. J. Lu, C. J. Bender, J. McCracken, J. Peisach, J. C. Severns and D. R. McMillin, *Biochemistry* **31**, 6265 (1992).
56. J. Peisach and W. E. Blumberg, *Arch. Biochem. Biophys.* **165**, 691 (1974).
57. R. Barbucci and M. J. M. Campbell, *Inorg. Chim. Acta* **16**, 113 (1976).
58. J. S. Hyde and W. Froncisz, *Ann. Rev. Biophys. Bioeng.* **11**, 391 (1982).
59. L. Morpurgo, E. Agostinelli, M. Senepa and A. Desideri, *J. Inorg. Biochem.* **24**, 1 (1985).
60. R. Brändén, B. G. Malmström and T. Vänngård, *Eur. J. Biochem.* **36**, 195 (1973).
61. J. C. Severns and D. R. McMillin, *Biochemistry* **29**, 8592 (1990).
62. A. S. Yang and A. S. Brill, *Biophys. J.* **59**, 1050 (1991).
63. A. Luyet, *Ann. NY Acad. Sci.* **85**, 549 (1960).
64. J. P. Hill and P. D. Buckley, *Anal. Biochem.* **192**, 358 (1991).
65. L. Morpurgo, G. Rotilio, A. Finazzi-Agrò and B. Mondovì, *Biochim. Biophys. Acta* **336**, 324 (1974).
66. R. Malkin, B. G. Malmström and T. Vänngård, *FEBS Lett.* **1**, 50 (1968).
67. M. Geoffroy, M. Wermeille, C. O. Buchecker, J. P. Sauvage and G. Bernardinelli, *Inorg. Chim. Acta* **167**, 157 (1990).
68. M. Iwaizumi, T. Kudo and S. Kita, *Inorg. Chem.* **25**, 1546 (1986).
69. P. M. Hanna, R. Tamilarasan and D. R. McMillin, *Biochem. J.* **256**, 1001 (1988).
70. L. S. Kau, D. J. Spira-Solomon, J. E. Penner-Hahn, K. O. Hodgson and E. I. Solomon, *J. Am. Chem. Soc.* **109**, 6433 (1987).
71. P. Frank, O. Farver and I. Pecht, *J. Biol. Chem.* **258**, 11112 (1983).
72. D. T. Vu and D. M. Stanbury, *Inorg. Chem.* **26**, 1732 (1987).
73. R. Malkin, B. G. Malmström and T. Vänngård, *Eur. J. Biochem.* **10**, 253 (1969).
74. M. R. Graziani, L. Morpurgo, G. Rotilio and B. Mondovì, *FEBS. Lett.* **70**, 87 (1976).
75. A. Messerschmidt, W. Steigemann, R. Huber, G. Lang and P. M. H. Kroneck, *Eur. J. Biochem.* **209**, 597 (1992).
76. L. Morpurgo, M. T. Graziani, A. Desideri, G. Rotilio and B. Mondovì, *Biochem. J.* **187**, 367 (1980).
77. L. Morpurgo, M. T. Graziani, A. Finazzi-Agrò, G. Rotilio and B. Mondovì, *Biochem. J.* **187**, 361 (1980).
78. B. Reinhammar and Y. Oda, *J. Inorg. Biochem.* **11**, 115 (1979).
79. P. Frank and I. Pecht, *J. Phys. Chem.* **90**, 3809 (1986).
80. C. D. LuBien, M. E. Winkler, T. J. Thamann, R. A. Scott, M. So, K. O. Hodgson and E. I. Solomon, *J. Am. Chem. Soc.* **103**, 7014 (1981).
81. A. S. Klemens and D. R. McMillin, *J. Inorg. Biochem.* **38**, 107 (1990).
82. L. Morpurgo, A. Desideri, G. Rotilio and B. Mondovì, *FEBS Lett.* **113**, 153 (1980).

83. B. Reinhammar, *J. Inorg. Biochem.* **18**, 113 (1983).
84. B. Reinhammar, R. Malkin, P. Jensen, B. Karlsson, L.-E. Andréasson, R. Aasa, T. Vänngård and B. G. Malmström, *J. Biol. Chem.* **255**, 5000 (1980).
85. D. J. Spira-Solomon and E. I. Solomon, *Biochem. Biophys. Res. Commun.* **112**, 729 (1983).
86. M. M. Werst, C. E. Davoust and B. M. Hoffman, *J. Am. Chem. Soc.* **113**, 1533 (1991).
87. W. E. Antholine, P. M. Hanna and D. R. McMillin, *Biophys. J.* **64**, 267 (1993).
88. L. Morpurgo, H. J. Hartmann, A. Desideri, U. Weser and G. Rotilio, *Biochem. J.* **211**, 515 (1983).
89. J.-B. Li and D. R. McMillin, *Biochim. Biophys. Acta* **1160**, 239 (1992).
90. L. Morpurgo, M. T. Graziani, G. Marcozzi and L. Avigliano, *J. Inorg. Biochem.* **51**, 641 (1993).
91. M. K. Eggleston, C. Pecoraro and D. R. McMillin, *Arch. Biochem. Biophys.* **320**, 276 (1995).
92. L. Morpurgo, I. Mavelli, L. Calabrese, A. Finazzi-Agrò and G. Rotilio, *Biochem. Biophys. Res. Commun.* **70**, 607 (1976).
93. N. J. Blackburn, M. E. Barr, W. H. Woodruff, J. van der Ooost and S. De Vries, *Biochemistry* **33**, 10401 (1994).
94. A. Volbeda and W. G. Hol, *J. Mol. Biol.* **209**, 249 (1989).
95. K. Lerch, in *Encyclopedia of Inorganic Chemistry*, Vol. 2, ed. R. B. King (Wiley, Surrey, U.K.), pp. 850–869.
96. K. A. Magnus, H. Ton-That and J. E. Carpenter, *Chem. Rev.* **94**, 727 (1994).
97. L. Calabrese, M. Carbonaro and G. Musci, *J. Biol. Chem.* **264**, 6183 (1989).
98. C. Miller, Ph. D. Thesis, Purdue University (August 1994).
99. C. Miller, J. Fajardo, W. E. Antholine and D. R. McMillin, *Anal. Biochem.* **227**, 69 (1995).

CHAPTER 6

KINETIC STUDIES ON POLYPORUS AND TREE LACCASES

Bengt Reinhammar
Department of Biochemistry and Biophysics
University of Göteborg
Lundberg Laboratory, Medicinaregatan 9 C
S-413 90 Göteborg, Sweden
E-mail: Bengt.Reinhammar@bcbp.gu.se

ABSTRACT

This chapter summarizes results from comprehensive studies on the catalytic mechanism of *Polyporus* (fungal) and *Rhus* (tree) laccases. The properties of the three types of copper sites and their roles in the electron shuttle between substrates and molecular oxygen is reviewed. The extensive catalytic studies with only one substrate (reductants or dioxygen) at a time and the effect of pH or inhibitors on the catalytic rates have been used to formulate a catalytic mechanism of these laccases. It is suggested that this mechanism also holds for the other blue oxidases, ascorbate oxidase and ceruloplasmin.

Contents

1. Introduction

Laccase (benzenediol: oxygen oxidoreductase, EC 1.10.3.2) is a blue copper-containing oxidase which is widely distributed in higher plants and fungi. It belongs, together with ascorbate oxidase and ceruloplasmin, to the group of blue copper oxidases. They utilize the full oxidizing capacity of dioxygen, which is reduced to two water molecules, with concomitant one-electron oxidation of many different substrates. All three enzymes have many properties in common. For example, they contain three different copper ion sites with unique properties and they share the reactivity to various substrates and dioxygen. Credit goes to the recently solved crystal structure of ascorbate oxidase and to the earlier found similarity between the amino acid sequences of these blue oxidases,[1,2] that it is now possible to better understand the probable structures of, in particular, laccase. Ascorbate oxidase appears to be a dimer of two laccase-like subunits. Ceruloplasmin is more complicated but appears to have one catalytic unit in common with laccase and ascorbate oxidase besides two copper sites with unknown function.

This chapter will mainly concentrate on the catalytic function of laccase which has been extensively studied by several laboratories using different techniques. It is therefore the best understood blue oxidase as concerns the catalytic mechanism. However, as the other blue oxidases, in particular ascorbate oxidase, and also ceruloplasmin, have so many structural, spectroscopic and catalytic properties in common with laccase it is likely that the presented mechanism for laccase also holds for the other blue oxidases. There are a number of extensive reviews of the blue oxidases and the interested reader is referred to these to get a more complete picture of the current state of knowledge.[3,4]

2. Biological Distribution and Function

Laccase is widely distributed in plants and fungi. Thus, laccase-like enzymes have been found in *Pinus*[5] species, *Acer pseudoplantanus*[6] and also from plant tissues, e.g. tea-leaves and peaches.[7] Laccase from several species of lacquer trees (*Rhus*) grown in Japan, China and Vietnam have been extensively investigated. The first laccase known was discovered in 1883 by Yoshida.[8] He was able to prepare a thermolabile substance, a catalyst which could oxidize urushiol by dioxygen, from the latex of the Japanese lacquer tree. A few years later, Bertrand[9] further purified and characterized this catalyst and gave it the name "laccase". He also claimed that the laccase contained manganese and thus introduced for the first time the concept of metalloenzymes. The crude latex contains large amounts of manganese[10] which explains Bertrand's finding. Later studies by Keilin and Mann[11] and Tissières[12] showed that the enzyme contains copper, not manganese.

The function of plant laccases is not known well. However, the laccase from lacquer trees are involved in healing injuries of the bark. A white sap (lacquer) seeps out upon injury and in the presence of dioxygen the laccase oxidizes phenols present in the sap to radicals, which spontaneously polymerize to form a protective structure, somewhat similar to the blood clotting system in animals. The lacquer has been used for about 6000 years as the first sort of plastic to paint wooden articles and beautiful artistic objects have been produced over the ages. The protective plastic layer is extremely durable against various solvents, for example water, organic compounds, NaOH and HCl.[13] Laccase may also be involved in lignification in *Pinus*.[5]

Laccase is found in many fungi, for example in the basidomycetes family *Polyporus, Pleurotis, Pholiota, Agaricus* and *Phlebia* and in the ascomycetes family *Neurospora, Podospora and Aspergillus.* The enzyme is either intracellular, as in *Podospora*, or as inducible exoenzymes as in *Polyporus, Pleurotis, Pholiota* and *Neurospora*.[3] It is particularly abundant in white-rot fungi which are able to degrade lignin. It is therefore assumed that laccase plays an important role in degradation of lignin. This idea is strengthened by the recent findings that substrate analogues of lignin are oxidized by *Polyporus* laccase.[14,15] Another white-rot fungus, *Plebia radiata*, produces a laccase which is able to remove lignin-related monomeric compounds from residual lignin from the pulping process.[16] Other proposed functions are a role in sporulation[17] and pigment production.[18,19]

Laccases exhibit a rather low specificity and they can therefore oxidize a number of quite different substrates with considerable rates.[3,4] Good substrates are various phenols, for example mono-, di- and polyphenols but also aminophenols and diamines. Of the diphenols, the para-forms are the most readily oxidized. Some other substrates, for example hexacyanoferrate and ascorbate have been used in several mechanism studies of the most investigated laccases: the laccases from *Rhus* species and from *Polyporus*. The studies of these enzymes have made it possible to formulate a catalytic mechanism which is supposed to be similar for the other laccases, for ascorbate oxidase and for the catalytic unit resembling those of laccase and ascorbate oxidase which is present in ceruloplasmin.

3. Molecular Properties

3.1. *The Proteins*

Laccases are in most cases monomeric glycoproteins with varying amounts of carbohydrates. The *Polyporus* laccase has about 10 % and the *Rhus* laccase as much as 45 % carbohydrates. The enzymes consist of about 500 amino acid residues in a single polypeptide chain. The molecular weights of fungal laccases are in the range of 65,000–85,000 except one of the enzymes from *Podospora anserina* which is a tetramer with a molecular weight of 390,000. The higher molecular weights of the *Rhus* laccases, 110,000–140,000 is due to their high contents of carbohydrates.[3]

There are only very limited structural information of these laccases. No crystals suitable for X-ray analysis have yet been produced despite considerable efforts in several laboratories. It appears that the difficulty to get well ordered crystals is due to the high content of carbohydrates in these enzymes. The genes of the following fungal laccases have been cloned and sequenced: *Neurospora crassa*,[20] *Aspergillus nidulans*,[21] *Coriolus hirsutus*,[22] *Plebia radiata*[16] and *Cryphonectria parasitica*.[23] A comparison of the amino acid sequences of these laccases with ascorbate oxidase is given by Messerschmidt.[24] Recently two laccase genes from *Polyporus (Trametes, Coriolus) versicolor* have been cloned and sequenced.[25,26] The sequences are not identical. It has earlier been demonstrated that the *Polyporus* fungus produces several isoelectric forms of laccase.[27] A comparison between the sequences of several laccase genes and ascorbate oxidase[24] shows that there is homology with regard to the amino acid residues which are known to be metal ligands in ascorbate oxidase with the exception of the methionine ligand in the type 1 copper site. This amino acid residue is in some laccases replaced by leucine or phenylalanin. Studies of methionine-121 mutants of a blue copper protein, azurin from *Pseudomonas aeruginosa*, have shown that protein mutants with hydrophobic amino acid residues exhibit spectral properties of a type 1 copper site and with reduction potentials slightly higher than that of the wild type protein.[28] The very high potential of type 1 copper in *Polyporus* laccase A,[29] which has been used in the mechanism studies reported here, might then depend on the presence of a bulky hydrophobic amino acid residue, for example a phenylalanin as in *Choriolus hirsutus, Plebia radiata*, and the two *Polyporus versicolor* laccase gene sequences, instead of a methionine.

3.2. *Metal Content. The Three Types of Copper Ions*

Most laccases contain 4 copper ions per molecule but one of the three laccases from *Podospora anserina* is a tetramer and each subunit has four copper ions.[3] The metals in blue oxidases are present in at least three distinct coordination environments. They have for a long time been classified according to their spectroscopic and magnetic properties and it was proposed earlier that laccases contain three different types of copper ions.[30] However, the laccase from *Plebia radiata* is reported to contain only two copper ions, which are EPR detectable, and an organic co-factor, PQQ,[31] despite the fact that its protein sequence contains all the amino acid residues, which are homologous to the known three metal binding sites in ascorbate oxidase and several laccase gene sequences.[24–26]

(1) Type 1 Cu^{2+} shows an intense absorption in the visible region with a dominating band at about 600 nm with an extinction coefficient in the range of 4000–6000 $M^{-1}cm^{-1}$. Its EPR spectrum exhibits an unusual narrow hyperfine structure with $A_{II} < 95 \times 10^{-4}$ cm^{-1}.

(2) Type 2 Cu^{2+} shows EPR parameters similar to Cu^{2+} bound to denatured blue proteins or in low molecular weight complexes. It has only weak absorption in the visible and near ultraviolet region. This metal binds certain inhibitory anions, such as F^{-1}, $N_3{}^{-1}$ and CN^{-1} very strongly.

(3) Type 3 Cu^{2+} ions are usually not detected by EPR since they consist of a pair of copper ions that are strongly antiferromagnetically coupled.

This pair has a strong absorption band at about 330 nm with an extinction coefficient of about 3000 $M^{-1}cm^{-1}$.

3.3. *Spectroscopic Properties*

A short presentation of some relevant spectroscopic properties which are used in the studies of the mechanism of the tree and fungal laccases will be summarized below.

3.3.1. *Optical Absorption Properties*

The optical absorption spectra in the visible and near ultraviolet region of resting, oxidized blue oxidases are rather similar with a dominating band at about 600 nm, which is responsible for the strong beautiful color of all blue proteins, but also with transitions around 800, 450 and 330 nm as illustrated in Fig. 1 of *Rhus* laccase. The three first optical bands have been found to be associated with type 1 Cu^{2+}. The band at 330 nm is associated with the

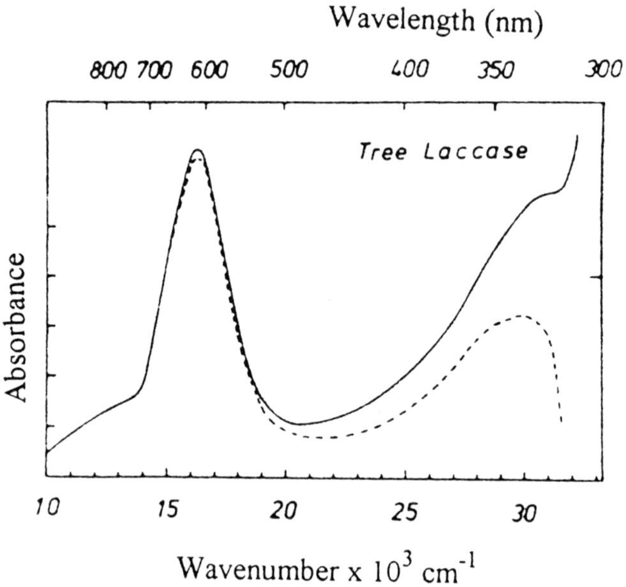

Fig. 1. Optical absorption of oxidized (full line) and oxidized-reduced difference spectrum (dashed line) of *Rhus* laccase.

oxidized copper pair. The redox reactions of these metal sites can therefore be followed by EPR and optical absorption spectroscopy.

3.3.2. *EPR Spectra*

All blue oxidases exhibit EPR spectra with superimposed signals from two different Cu^{2+} sites. Figure 2 shows the EPR spectrum of *Podospora anserina*. Spectra from *Polyporus, Neurospora* and *Podospora* laccases show very similar parameters.[3,4] The two remaining copper ions are EPR-silent under most experimental conditions (but see Sec. 8).

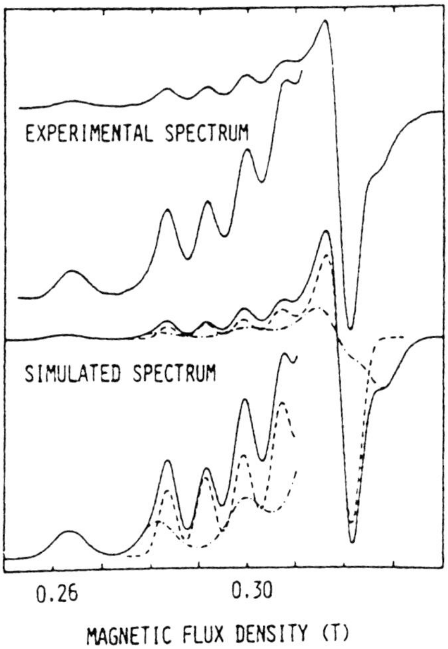

Fig. 2. Experimental and simulated EPR spectra of *Podospora anserina* laccase showing the signals from type 1 and 2 Cu^{2+} ions.

4. Oxidation-Reduction Properties

The crystal structure studies of ascorbate oxidase show that the type 2 and 3 copper ions form a trinuclear copper site.[1] It is therefore suggested that they are so intimately coupled that they should have the same reduction potential. Redox titrations using either the type 2 Cu^{2+} EPR signal or the 330 nm absorption band would give the same potential.[24] This hypothesis would then mean that these electron acceptors would titrate as a cooperative three-electron acceptor. There are several experimental studies which show that this cannot be the case. The redox titrations of tree and fungal laccases show that the 330 nm band titrates as a cooperative two-electron acceptor in both enzymes with a potential which is about 70 mV higher than the value for type 2 copper in tree laccase.[29,32] This is shown in Figs. 3 and 4. Another study of the tree laccase with various reducing substrates show, although the 330 nm band might titrate as a one- or two-electron acceptor depending on the kind of substrate, that the type 2 copper exhibits a potential about 55–80 mV lower than the copper pair.[33] Similarly, redox titrations with the fluoride inhibited fungal laccase

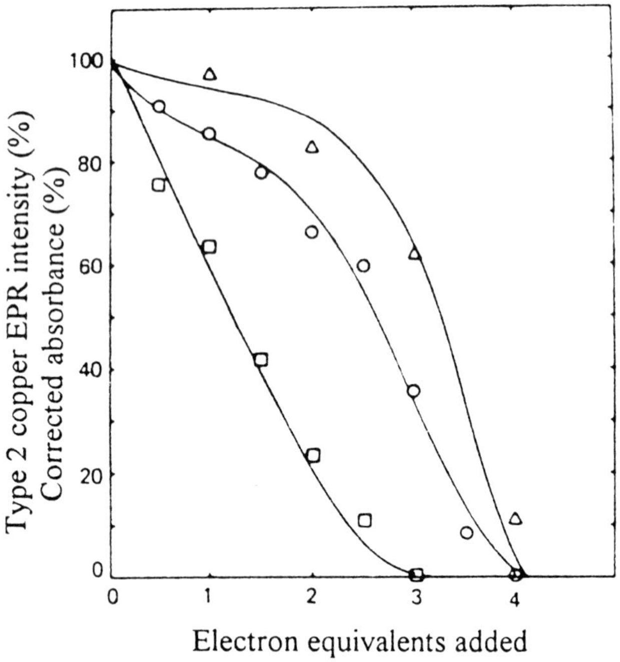

Fig. 3. Anaerobic reduction titration of *Rhus* laccase followed by optical and EPR measurements. The lines show computed reduction patterns using the reported potential values.[32]

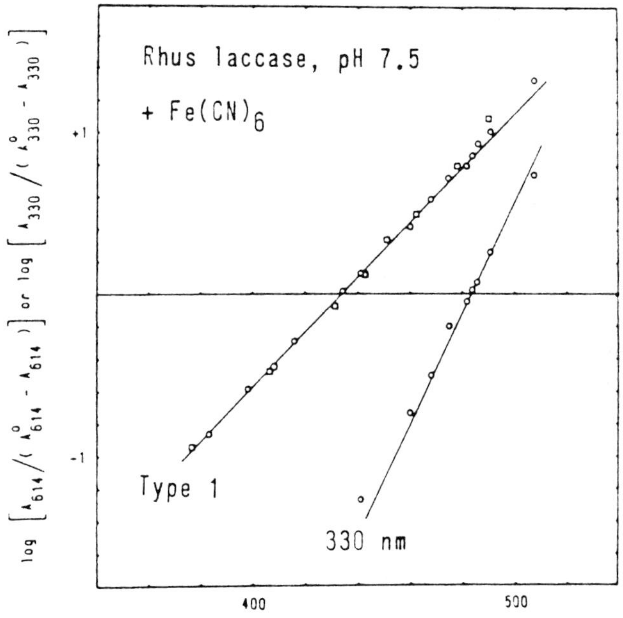

Fig. 4. Nernst plot of the reduction potentials of type 1 and 3 copper sites. The lines are computed with slopes of $n = 1$ for type 1 and $n = 2$ for the 330 nm absorption band.[29]

revealed that type 2 Cu^{2+} was only reduced after the 330 nm chromophore had completely vanished.[34] Also, the kinetic studies summarized below, show that on reduction and reoxidation of these copper sites they behave as separated electron acceptors/donators. Furthermore, the type 2 copper-depleted laccases have kept the 330 nm band in the oxidized proteins which again demonstrates that type 2 copper does not contribute significantly to this chromophore.[35,36] It is, however, not understood how these metal sites can behave as distinctly separated units despite the closeness between the three copper ions.

5. The Catalytic Reactions

There have been several studies done in different laboratories aiming at an understanding of the catalytic mechanism of laccases.[37–41] Several models have been proposed which are in conflict. This is in part due to the difference in experimental conditions which makes the interpretation of the data difficult. This will be discussed below. The most extensive studies have been performed with tree and fungal laccases. These enzymes are efficiently reduced by a number of quite different substrates. Thus, the fungal laccase type 1 Cu^{2+} is reduced with rate constants of about 10^6–10^8 M^{-1} sec^{-1} by $Fe(CN)_6{}^{4-}$, hydroquinone[42,43] and hydrated electrons.[44] A detailed mechanism of the reduction and reoxidation of the tree laccase has been formulated.[38,39] This mechanism also seems to be applicable for the fungal laccase and probably for ascorbate oxidase and ceruloplasmin. Since the reactions with reductants and dioxygen are very fast, stopped-flow and rapid-freeze EPR methods have been used to follow the changes of optical absorption bands, which are associated with types 1 and 3 copper sites together with the EPR signals of types 1 and 2 copper as well as an oxygen intermediate formed on reoxidation. The spectral changes due to the reduction and oxidation of the three types of copper sites are very complex and therefore difficult to interpret (see Fig. 9 in Sec. 6.2) in studies with both reductants and dioxygen present. It has therefore been found that it is necessary to study the reactions with only one of the reactants present at a time.

5.1. *The Kinetics of Anaerobic Reduction*

The early anaerobic studies of the fungal laccase were concentrated on the reaction of the blue chromophore. At this time it was not known that the other metal sites were associated with other absorption bands.[42] The stopped-flow results showed that type 1 Cu^{2+} was reduced in a second-order reaction of the order of 10^6 to 10^8 M^{-1} sec^{-1} with $Fe(CN)_6{}^{4-}$. This is rapid enough to involve this copper in a valence shuttle mechanism as it behaves in steady-state studies. Only one $Fe(CN)_6{}^{3-}$ was produced as rapidly as the reduction of type 1 Cu^{2+}. Two more molecules of $Fe(CN)_6{}^{3-}$ were produced in a much slower reaction indicating that there are at least two more reducible sites in addition to type 1 copper. Reduction of this laccase with pulse radiolyses generated hydrated electrons gave second order rates of about 10^8 M^{-1} sec^{-1}.[44] No reduction of the 330 nm band was observed in these studies.

Following this, it was shown in redox titration studies that an absorption band at about 330 nm is associated with electron acceptors other than types 1 and 2 copper sites, presumably with the type 3 copper ions.[34,45] This chromophore was later shown to be associated with a co-operative two-electron acceptor in both fungal and tree laccase.[29,32] Later anaerobic reduction studies of the fungal enzyme therefore also included the reactions of this band.[43,46] It showed that the 330 nm chromophore was reduced in a rate-limiting intramolecular step which is too slow (1 sec^{-1}) to be part of the catalytic mechanism. An anaerobic rapid-freeze reduction study showed that the type 2 Cu^{2+} was also reduced much slower than type 1 Cu^{2+}. Although these findings showed that these electron acceptors reduced too slowly to take part in the rapid electron shuttle it was suggested that they could take part in the electron transfer during turnover since oxygen binding at an early state of reduction might change the rates considerably.[46]

Although these earlier studies indicated that type 2 copper might not be involved in rapid electron transfer in the enzyme catalyzed reduction it was known that this metal is essential for the enzyme activity, which was demonstrated by its reversible removal[47,48] and also with its interaction with the inhibitor F^-. It was therefore suggested that this copper was not changing valence during catalysis but presumably involved in the stabilization of a putative peroxide intermediate until it was further reduced to water.[46]

The redox titrations had shown that the reduction of the two-electron acceptor requires the simultaneous electron transfer from two one-electron donors since there are no EPR or optical absorption indications for a reduction of only one of the copper ions in the type 3 copper pair.[29,32] In addition to the type 1 copper there has to be another electron acceptor-donor, which can donate an electron to the two-electron acceptor.

As mentioned above, the earlier experiments showed that type 2 copper was reduced too slowly to take part in the catalytic events. In a later rapid-freeze EPR study of fungal laccase it was demonstrated for the first time that this copper ion in fact rapidly changes valence. It is reduced subsequently to the very rapid type 1 Cu^{2+} reduction and then reoxidized simultaneously with the reduction of the type 3 copper pair.[49] Subsequent extensive studies with the tree laccase confirmed that both types 1 and 2 copper ions take part in simultaneous electron transfer to the copper pair. All three types of copper sites therefore take part in the reduction mechanism.[38,39]

The results from studies of both fungal and tree laccases, although seemingly rather different, can in fact be interpreted as rather similar if they can exist in at least two catalytically different forms, one active and the other inactive. This is most evident from studies of the tree laccase. The active enzyme is converted into an inactive form at pH higher than 6.5. The inhibition depends on the formation of a type 2 Cu^{2+}-OH complex at higher pH values. The structure of the laccases possibly resemble the known three-dimensional structure recently reported for ascorbate oxidase.[1,2] It shows that the types 2 and 3 copper ions form a trinuclear site. Two of the copper ions are bridged by an oxygen ligand, presumably OH^- or O^{2-}, and probably represents the two-electron acceptor/donor. The third copper ion would then be the EPR-detectable type 2 Cu^{2+} which is bound to two histidine residues and with a ligated water molecule. This water molecule loses a proton at higher pH and thus inhibits the reduction of type 2 Cu^{2+} (Refs. 38 and 39).

A reduction mechanism which is based on many studies of the tree laccase is shown in Scheme 1. For a full presentation of the experimental results see Refs. 38 and 39. The scheme contains two forms of the enzyme in the fully oxidized, resting state. One of the forms has an OH^- coordinated to type 2 Cu^{2+} and corresponds to the inactive state. The other form of the enzyme has probably a water molecule bound to this metal and this form represents active molecules. The results represent studies at pH values of 6.0–7.4. At pH 8.5 there are two forms of inactive molecules. The protein is represented as a box with type 1 copper in the upper, the type 3 copper pair in the next and type 2 in the lowest compartment. The distances between these copper sites are presumably the same as in ascorbate oxidase.[1,2] The copper pair is probably bridged with an O^{2-} or OH^- ion making it EPR-undetectable due to spin pairing of the unpaired electrons on both copper ions. The reason for placing the copper ions in different compartments is that they apparently behave as independent electron acceptors/donors.

Figure 5 summarizes the anaerobic reduction at pH 6.5 of the three types of copper sites in the tree laccase, as detected by stopped-flow and rapid-freeze-EPR studies. At this pH both type 1 and type 2 Cu^{2+} are reduced simultaneously in a similar fashion. There is a fast initial reduction followed by a slowing down of the rate and then a final slower reduction phase when type 3 copper is reduced. This simultaneous reduction would indicate that both metals are reduced by substrate at the same rate and

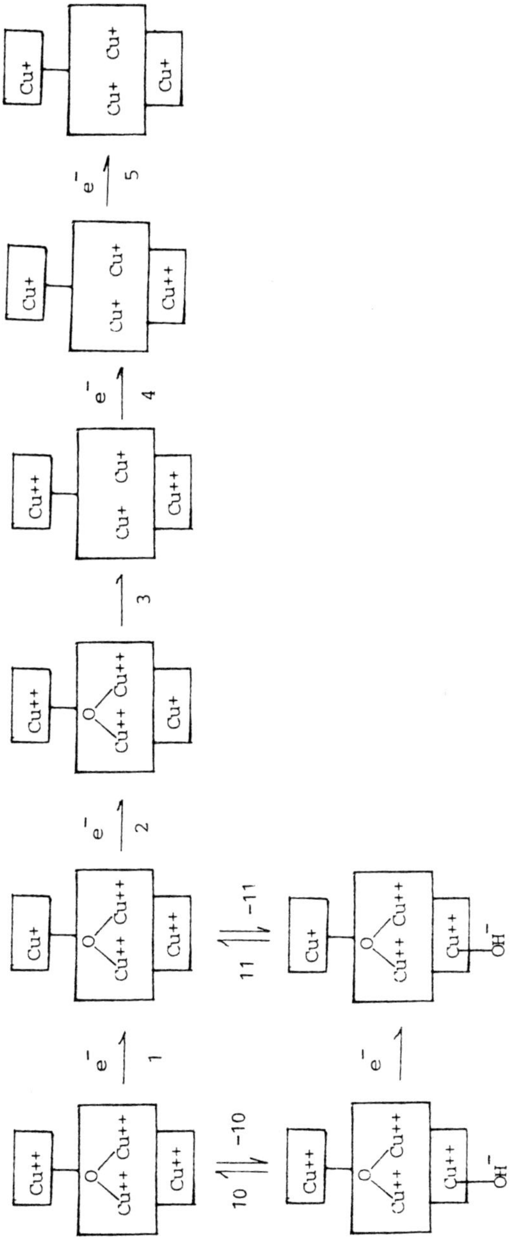

Scheme 1

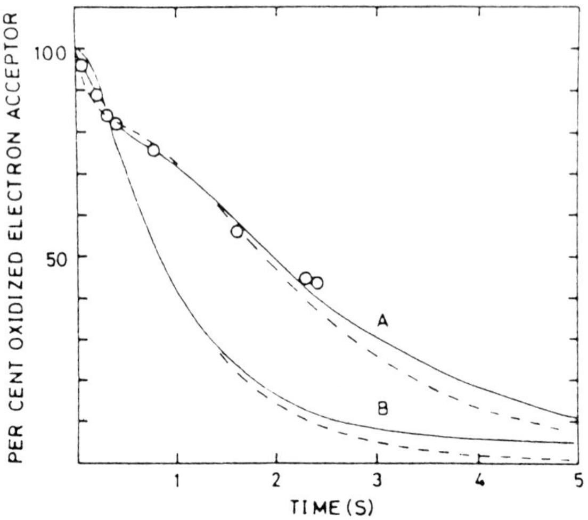

Fig. 5. Anaerobic reduction of *Rhus* laccase. Type 1 (line A), type 2 (circles), and type 3 copper (line B) at pH 6.5. Dashed lines are computed reduction courses using values for the reactions in Table 1.[38,39]

independent of each other, or that type 2 copper is reduced by type 1 Cu^{+} (or *vice versa*) in an intramolecular electron transfer. On the other hand, the latter hypothesis finds no support from the experimental results which would show that the Cu^{2+} of any of these metals exhibit a lag phase in the beginning of their reduction if they delivered an electron to each other. Another interpretation seems more likely. The simultaneous reduction could depend on a conformational interaction between these metals. Thus, the first electron from a substrate molecule reduces type 1 Cu^{2+}. When this copper is reduced it affects the reduction of type 2 Cu^{2+}, which is reduced by a second substrate molecule and not by type 1 Cu^{+}. How this proposed interaction is excerted is not known. However, the crystal structure of ascorbate oxidase shows that, although the distance between these metal sites is large, they are coordinated to one histidine each which are rather close in the amino acid sequence.[1,2]

When both type 1 and 2 copper have been reduced they simultaneously transfer one electron each to the oxidized copper pair. These reactions are consistent with the plateau phase in the reduction rates of types 1 and 2 copper ions and also of the presence of a lag phase in the reduction of the copper pair. Thereafter, the reoxidized type 1 and 2 copper ions are reduced once more and the enzyme becomes completely reduced. This reaction sequence is shown in the upper part of the scheme.

The reduction pattern of the three electron acceptors changes drastically when the pH is raised to 7.4. This is illustrated in Fig. 6. The type 1 Cu^{2+} is now reduced about five times faster than at pH 6.5. The type 2 Cu^{2+} is reduced much slower and shows an initial lag period. The type 3 copper pair is reduced in two phases. About 50 % of the absorption band at 330 nm is reduced in a second order reaction. The rest of the 330 nm absorption is reduced at a substrate concentration independent rate with hydroquinone concentrations higher than 1 mM. These reactions are consistent with the following view. At this pH about 50 % of the molecules are in the active state and reduced in the same manner as at pH 6.5. The rest of the molecules are inhibited as a result of a type 2 Cu^{2+}-OH complex that is formed at this pH. Type 1 Cu^{2+} is rapidly reduced in both active and inhibited molecules and the reduction of type 2 and type 3 copper ions are governed by the rate of reduction of type 1 copper. Although the type 1 copper is also rapidly

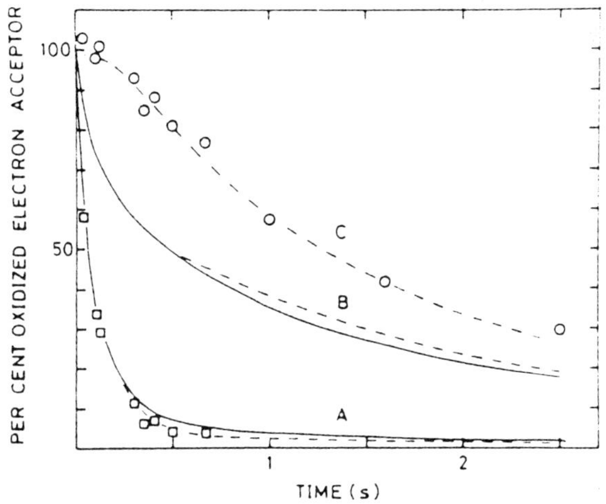

Fig. 6. Anaerobic reduction of *Rhus* laccase type 1 (line A and squares), type 3 (line B), and type 2 (circles) at pH 7.5. Dashed lines show computed reduction courses using values for the reaction rates in Table 1.[38,39]

reduced in the inactive molecules the reduction of type 2 and 3 copper are inhibited until the OH^- on type 2 Cu^{2+} is eliminated, which occurs in a slow (about 0.5 sec^{-1}) substrate concentration independent step. The reduction patterns can be simulated according to scheme and the published rate constants.[38,39]

At pH 8.5 the initial bimolecular reduction of type 1 Cu^{2+} with hydroquinone is about ten times faster than at pH 7.4. The two-electron acceptor reduction can be analyzed in terms of three reaction phases. Initially there is a reduction of its absorbance which is first order in both enzyme and substrate concentrations amounting to about 25 %. The rest of the absorbance decrease is independent of the substrate concentration and is characterized by two first order rate constants.

The earlier published kinetic results with the *Polyporus* laccase[42,43] can be explained with a similar model.[38] At pH 5.5 all the molecules are in the inactive form as they have an OH^- bound to type 2 Cu^{2+}. Type 1 Cu^{2+} is very rapidly reduced in a second order reaction. The reduction of type 2 and type 3 copper ions depend on the slow conversion to an active form (1 sec^{-1}) as the OH^- bound to type 2 Cu^{2+} is eliminated.

5.2. *The Effect of Inhibitors*

In addition to the inhibitory action caused by OH^- bound to type 2 Cu^{2+} there are also other causes for the observed inhibition. Thus the enzymes are strongly inhibited by a number of halides and pseudohalides, e.g. F^-, N_3^- and CN^-. Laccase also becomes inhibited at higher pH which is due to deprotonation of groups close to the metal sites. Studies of the tree laccase have shown that the enzyme is also inhibited at lower pH values.[38,39] Several ionizable groups therefore appear to be involved in the catalytic action of laccases. The inhibitor action is rather complex and therefore only partly understood. This is due to the fact that there are many states of the molecules, particularly during turnover, and therefore it is very difficult to determine whether the inhibitor is affecting the oxidized or reduced states of the three metal sites.

The inhibition by halides and pseudohalides has been extensively studied and is probably best understood. These anions are strongly bound to type 2 Cu^{2+} which is shown by the effect of F^-, N_3^-, and CN^- on the EPR spectrum.[30,45,47] F^- binds very strongly to this metal in the fungal laccase. It is therefore possible to obtain a one to one complex by adding stoichiometric amounts of F^-.[39] This derivative is completely inhibited. Although type 1 Cu^{2+} is rapidly reduced the type 2 Cu^{2+}-F^- complex is slowly reduced. The catalytic action increases at the same rate as the F^- is released during turnover.[50] It is therefore probable that F^- inhibits by binding to the oxidized and not to the reduced state of this copper ion.

On binding to type 2 Cu^{2+} in fungal laccase the pH increases simultaneously with the formation of the F-complex which is consistent with a displacement of an OH^- bound to this metal.[39] Azide is a stronger inhibitor than F^- in laccase. The results give strong support for the idea that these inhibitors are bound to Type 2 Cu^{2+} and that they inhibit the reduction of this ion and hence the reduction of the two-electron acceptor.

An example of the inhibitor effect in anaerobic reduction studies on the metal sites in the native and F^- inhibited tree laccase is shown in Fig. 7. Similar results are obtained with the fungal laccase.[42,43] The results show that the type 1 Cu^{2+} is reduced in a single exponential phase with a rate constant (similar as for the initial reduction of the native enzyme) which is not affected by the presence of F^-. This demonstrates that the inhibitor does not interact with this metal ion. Type 1 Cu^+ is not rapidly reoxidized and shows a plateau phase as in the native enzyme since the electron on this metal ion is only slowly transferred to the type 3 copper pair.

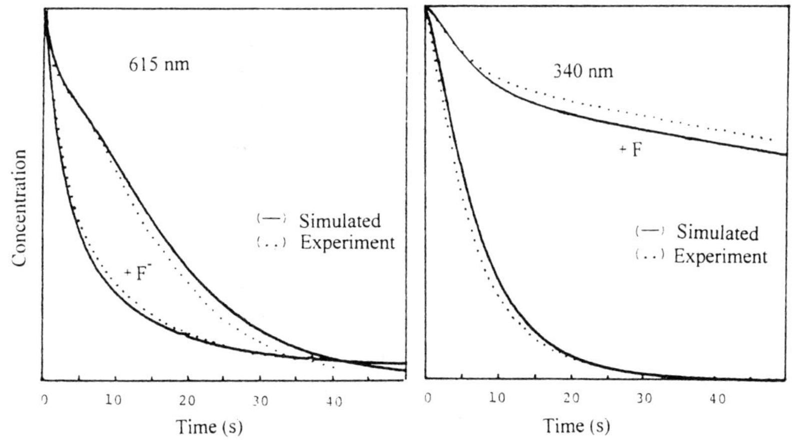

Fig. 7. Anaerobic reduction of native and fluoride-inhibited *Rhus* laccase by 0.1 mM hydroquinone at pH 7.4.

This is shown by the dramatic decrease in the reduction of the copper pair. Under these conditions about 30 % of the enzyme is not inhibited and the initial part of the reduction of type 3 copper is reduced as in the native enzyme. These results allow an estimation of the dissociation constant for F^- to be 40 mM.

As shown above, the laccases are inhibited at higher pH values which is due to the formation of a type 2 Cu^{2+}-OH^- complex. The pKa of the water molecule coordinated to this copper is about 3.5 and 7.4 in the fungal and tree laccases, respectively.[39] This Cu^{2+} cannot be reduced until the OH^- is eliminated which probably involves its protonation. The OH^- can be displaced by F^- and the inhibition at higher pH values resembles the inhibition by fluoride. These anions thus inhibit the reduction of type 2 Cu^{2+} and hence the reduction of the type 3 copper pair.

As also mentioned above the anaerobic reduction of the two-electron acceptor in tree laccase at pH 8.5 exhibits three phases. Two of these are reduced in reactions which are zero order in substrate concentration and probably represent molecules inhibited by the deprotonation of two different ionizable groups.

6. Aerobe Kinetics

6.1. *Steady-State Kinetics*

Only few studies of the steady-state reaction of laccase have been published. One study of the tree laccase reports that the kinetics is consistent with a ping-pong di Theorell-Chance mechanism.[51] With the slow substrate hydroquinone the enzyme showed no sign of saturation. The rate constant for the oxidation of this substrate (1.2×10^3 M^{-1} sec^{-1}) agrees with the value obtained in transient reduction studies[56] (see Sec. 5.1). The rate constant for the reaction between dioxygen and reduced laccase (6.1×10^6 M^{-1} sec^{-1}) is also in agreement with values obtained in transient reoxidation studies. It thus appears that some reactions of the enzyme can be studied with only one substrate at a time and that the mechanism for reduction and reoxidation[38,39] probably reflect reactions taking part during turnover. The rate constants for the oxidation of the faster substrates dimethyl-*p*-phenylenediamine and *p*-phenylendiamine showed saturation, however. The maximal turnover with these substrates are 560 electrons per second.[51] It is not possible using this technique to locate the rate-limiting steps since there are so many partial reactions revealed in the reduction and reoxidation sequences of this enzyme (see Secs. 5 and 7). However, this rate is consistent with the simulated anaerobic reduction where the rate limiting intramolecular step is proposed to be the simultaneous transfer of electrons from types 1 and 2 Cu^+ to the copper pair which is ≥ 300 per second[38,39] (see Table 1).

Table 1. Rate constants used in computer simulations of the anaerobic reduction of *Rhus* laccase by 10 mM hydroquinone at pH 6.5 and 7.4. Rate constants $k_1, -k_5$ and k_9 in M^{-1} s^{-1}. Other rate constants in s^{-1}. The initial concentrations of the two forms of the enzyme can be calculated from the rate constants k_{10} and k_{-10}.

pH	k_1	k_2	k_3	k_4	k_5	k_9	k_{10}, k_{11}	k_{-10}, k_{-11}
6.5	110	$\geq 10^4$	5	80	$\geq 10^4$	300	3.6	0.4
7.4	900	$\geq 10^4$	≥ 300	900	200	2400	0.5	0.75

From studies of the pH dependence of the oxidation of hydroquinone between pH 6.5 and 8.0, it was concluded that H^+ can be considered as an uncompetitive inhibitor with respect to dioxygen and a competitive inhibitor with respect to hydroquinone. An ionizable protein group, possibly an imidazole, with a pKa value of 7.3 was suggested to be responsible for this pH dependence.[51] As discussed in Sec. 5 the laccases become inhibited at higher pH values due to the formation of a type 2 Cu^{2+}-OH^- complex. The pKa of this complex is about 7.4 in the tree laccase. The ionizable group responsible for the observed pH dependence in steady state and transient-state kinetic studies could be the same, namely a type 2 Cu^{2+}-OH^- complex.

In a later steady state study using dimethylphenylenediamine as a reductant the influence of pH between 6.0 and 9.0 on the catalysis of tree laccase was determined. The pH dependence of both reductant-dependent and reductant-independent steps showed bell shaped curves implicating at least two protic groups in each case. The apparent pKa values for the reductant-independent steps are 8.98 and 5.91, and for the reductant-dependent steps 7.55 and 8.40. A significant solvent isotope effect was only observed on a reductant-independent step. It is suggested that the solvent isotope effect is associated with the protein protic group (pKa 5.91) required in its undissociated form.[52] The protein groups involved in these steps are not identified. However the group with a pKa of 7.55 could be the same as the one detected in the other steady state study mentioned above[51] and in the transient state reduction studies,[38,39] namely type 2 Cu^{2+}-OH^-.

In another steady-state study of the fungal laccase it was shown that maximum rate of enzyme catalyzed reaction is controlled by two ionogenic groups with pKa values of 5.0 and 6.3. Deprotonation of these groups resulted in loss of the catalytic activity.[53]

6.2. *Post Steady-State Kinetics*

Complex transient reactions are observed on aerobic reduction of laccase. For example, the reactions of the type 1 and type 3 copper sites in the *Rhus* laccase exhibits several phases before there is full reduction when dioxygen is exhausted.[38] This is shown in Fig. 8. Although these reaction courses appear to be very complicated they can in fact be explained by the reaction schemes for anaerobic reduction (see Sec. 5) and reoxidation (see Sec. 7). At pH 7.4 about 50 % of the molecules are inactive due to a type 2 Cu^{2+}-OH^- complex. During turnover this OH^- is gradually eliminated and an increasing amount of the enzyme molecules become active. The

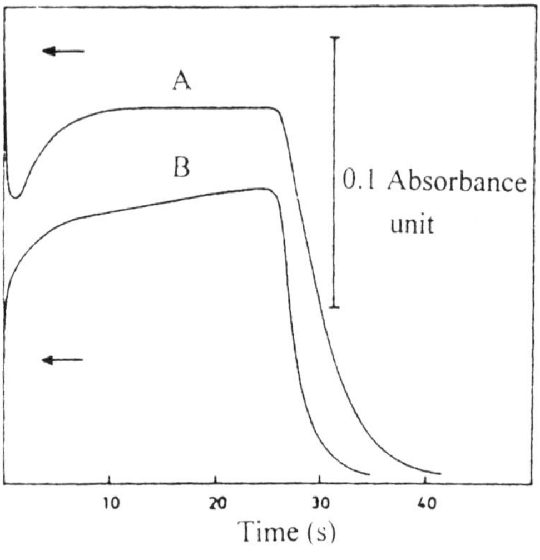

Fig. 8. Aerobic reduction courses of *Rhus* laccase at 615 nm (A; type 1 copper) and at 330 nm (B; type 3 copper).

initial absorbance decrease at 615 nm reflects the reduction of type 1 Cu^{2+} in the inactive molecules as the steady state level of this Cu^{2+} is nearly 100 % in the active molecules. This phase is dependent on the substrate concentration and the rate constant is the same as the one observed in the absence of dioxygen (see Sec. 5). The following exponential increase to a steady state occurs in a reaction that is independent of the concentration and the type of reductant used (about 0.3 sec^{-1}) This phase is associated with a conversion of inactive to active molecules as the OH^- bound to type 2 Cu^{2+} becomes eliminated during turnover. When dioxygen is consumed, type 1 copper is finally reduced in a second order reaction with the same rate as observed for the final phase (when type 3 copper is reduced) under anaerobic reduction.

The absorption at 340 nm shows an initial rapid biphasic increase. The first phase is dependent on substrate concentration and is faster than the initial reduction of type 1 Cu^{2+} and therefore possibly reflects substrate binding. The second phase of increase (within 1 sec) is due to the formation of an oxygen intermediate in the active molecules. The third phase (between 1 and 10 sec) is independent of substrate concentration and type of reductant (rate constant about 0.3 sec^{-1}) and represents the same conversion of inactive to active molecules as found for type 1 copper. The slight increase of the steady state absorbance level (between 10 and 25 sec) depend on the absorbance of the product (quinone). If ascorbate is used as reductant, the steady state level is constant. When dioxygen is consumed, the rate of decay is the same as that found for the initial, substrate concentration dependent reduction in anaerobic experiments.

Extensive series of investigations have been performed of the reduction of the metal sites in the tree laccase under similar post-steady state conditions as presented above.[37] Since type 3 copper is apparently reduced before type 1 Cu^{2+}, it was suggested that type 1 copper is not involved in the reduction of the two-electron acceptor. It was suggested that the reducing substrates would bind to type 2 Cu^{2+}. Type 3 copper would be reduced by electron transfer from reduced type 2 copper and a second substrate molecule bound to this copper ion. Type 1 Cu^{2+} was not considered to be involved in the reduction of the two-electron acceptor but itself reduced in a parallel reaction by substrate bound to type 2 copper. However, such a reduction mechanism is inconsistent with the results from anaerobic reduction studies of the native and type 2 copper depleted enzyme, presented in

Secs. 5 and 9. These studies demonstrate that the type 1 Cu^{2+} is reduced with similar rates by hydroquinone in the native and type 2 copper-depleted laccase which excludes the proposed role that type 2 copper reduces type 1 Cu^{2+} (Ref. 35).

7. The Reactions with Dioxygen

In reoxidation studies with ceruloplasmin it was observed that a new transient species, which absorbs at 420 nm, appeared.[54] It was formed at the same time as the rapid reoxidation of type 1 copper and apparently also the type 3 copper site. This species decayed within 2 seconds to the resting state absorbance. Similar optical intermediates were later observed on reoxidation of fungal laccase and also during turn over of tree laccase[56,58] showing that it represents a common intermediate in the reoxidation of blue oxidases. The optical absorption and EPR properties of the intermediate state in the *Polyporus* laccase show similarities with the complex between type 2 Cu^{2+} and added H_2O_2.[57] It was therefore suggested that the optical intermediate represents peroxide which is formed after dioxygen has received two electrons from the reduced two-electron acceptor. This peroxide would be bound to type 2 Cu^{2+} until it is reduced to water in another two-electron transfer step. However, the results from the reoxidation of ceruloplasmin and *Polyporus* laccase show that the type 1 Cu^{+} and the two-electron acceptor are reoxidized at the same time. This observation is obviously inconsistent with the idea that only two electrons are rapidly transferred to dioxygen to form the intermediate.

In a later study of the reduced tree laccase it was observed that the reaction with dioxygen occurs in a more complex mechanism.[58] On reoxidation, both the 615 and 330 nm chromophores are developed simultaneously with the formation of an optical intermediate, which has an absorption maximum at 360 nm. These rapid reactions are dependent on the oxygen concentration and occur with a rate constant of 5×10^6 M^{-1} sec^{-1}. The type 2 Cu^{+} is reoxidized in a slow (half-time about 20 sec) intramolecular reaction. Since the optical intermediate decays in a reaction, which is first order with respect to the enzyme concentration, with the same half-time, it appears that these two reactions are coupled. The results are consistent with a reaction in which dioxygen rapidly receives three electrons from the reduced type 1 copper and the two-electron acceptor and that an oxygen intermediate is bound to the enzyme until it is reduced by a fourth electron from type 2 Cu^{+}. This last step completes the reoxidation reactions.

These reactions lead to the development of some paramagnetic intermediates. No new EPR signals were observed at 77 K but studies at much lower temperatures (about 10 K) led to the discovery of a new paramagnetic intermediate in *Polyporus* and *Rhus* laccase[59,60] (see Fig. 9). Since the EPR signal of this intermediate shows unusual relaxation properties similar to those of the free O^- radical it was suggested that the intermediate could be an O^- radical which is probably bound to the type 3 copper pair. That the reaction intermediate is derived from oxygen was also demonstrated in reoxidation studies with $^{17}O_2$ (Ref. 60). This oxygen intermediate is also formed during turnover condition with ascorbate and hydroquinone as reductants, which shows that it is not an artificial species formed only under the conditions used in the reoxidation experiments.[39]

As the electron transfer between type 2 Cu^+ and the oxygen intermediate is rather slow in the different proteins (half-times 1 to 20 sec) it cannot be of catalytic importance.[3,4] The decay of the oxygen intermediate is limited by the rate of reduction of type 1 Cu^{2+} and it is therefore suggested

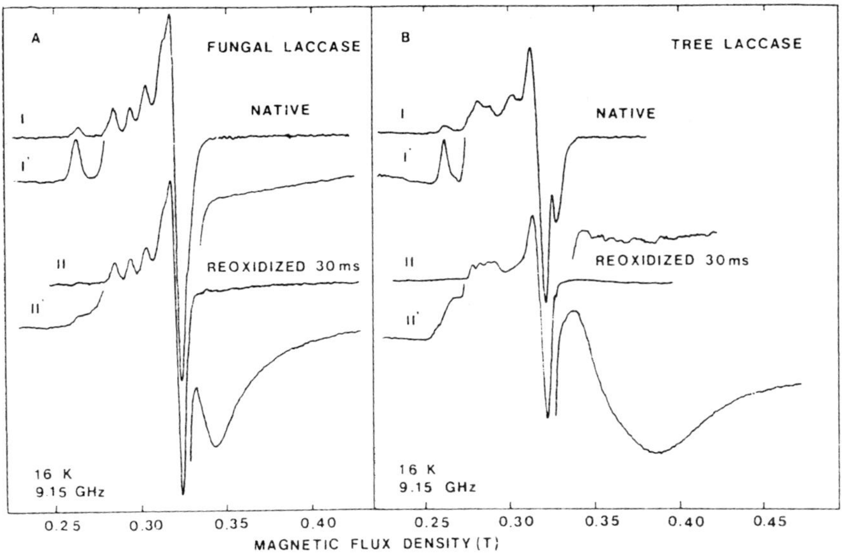

Fig. 9. Rapid-freeze EPR spectra of *Polyporus* (A) and *Rhus* (B) laccase. Spectra I shows the oxidized enzymes. Spectra II shows the 4 electron equivalents of ascorbate reduced enzymes after reoxidation with dioxygen for 30 ms.

that the electron transfer during catalysis occurs via the re-reduced type 1 copper.[38,39]

The EPR detectable oxygen intermediate is also formed in the type 2 copper-depleted tree enzyme[35] (Fig. 10). This copper ion can therefore not be necessary in the reoxidation reactions. If the intermediate is bound to metal ions, it is probably then bound to the type 3 copper ions.

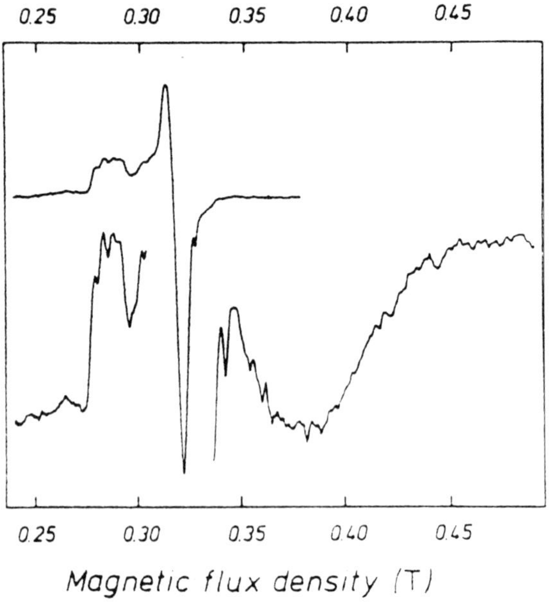

Fig. 10. EPR spectrum at 9 GHz of type 2 copper-depleted *Rhus* laccase after reoxidation showing type 1 Cu^{2+} signal (lower field) and the oxygen intermediate signal (higher field).

Our knowledge about the detailed reoxidation mechanism is limited and not so well understood as the reduction mechanism. However, a few intermediate steps appear likely to occur. This is summarized in the upper part of Scheme 2. Although both type 1 copper and the two-electron acceptor are reoxidized simultaneously, it is possible that this reaction occurs in at least two steps. The first step is probably the binding of dioxygen as a peroxide to the reoxidized type 3 copper pair. Then, an electron from type 1 Cu^{+} is rapidly transferred to this metal-oxygen complex and the

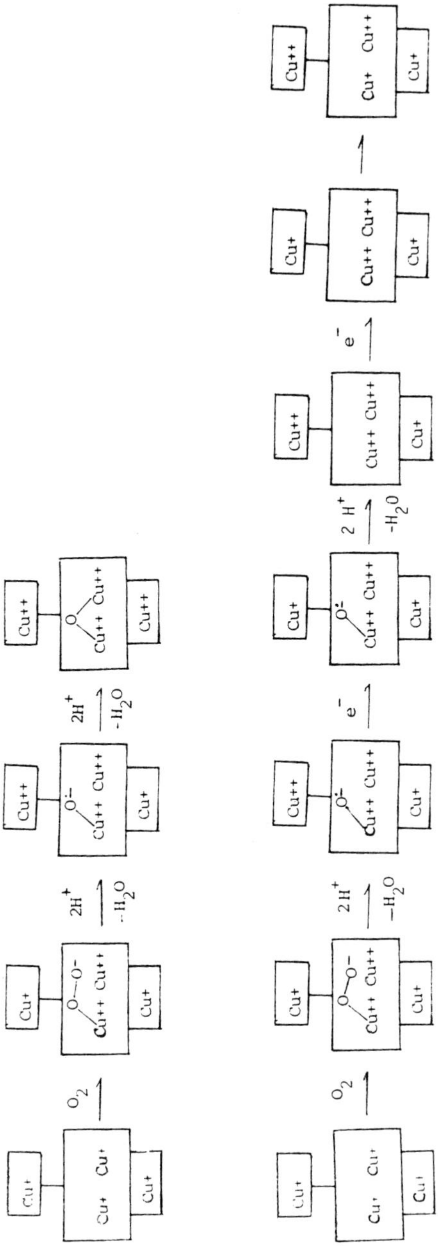

Scheme 2

paramagnetic oxygen intermediate is formed. The O–O bond is probably broken at this stage according to a study of proton consumption upon reoxidation of tree laccase.[61] It showed that two protons are consumed in the very rapid reoxidation phase in which the paramagnetic oxygen intermediate is formed. This result indicates that one water molecule is probably formed and the oxygen intermediate would therefore probably be an O^- radical. Two more protons are then consumed as the oxygen intermediate decays through electron transfer from the type 2 Cu^+, which reaction completes full reoxidation of the enzyme.

It is not known at what reduction states the enzyme reacts with dioxygen during catalysis. Results from pulse-radiolysis reduction studies of tree laccase indicate that the singly reduced enzyme is rapidly reoxidized by dioxygen. It was suggested that dioxygen reacts either with type 2 Cu^+ or the half-reduced copper pair after it had received an electron from the type 1 Cu^+ (Ref. 41). Neither of these assumptions seem likely. First, Type 2 Cu^+ cannot be the oxygen binding site as this copper is not necessary even for the oxidation reactions. Also, the type 1 Cu^+ is very slowly reoxidized by dioxygen in the 2 copper depleted tree laccase which shows that type 1 copper cannot rapidly donate an electron to dioxygen or the copper pair. These results show that the dioxygen-reducing site should be the reduced type 3 copper pair. It seems probable that dioxygen will react as soon as the copper pair is reduced and that neither type 1 nor type 2 copper has to be reduced for oxygen binding to occur to the enzyme. This hypothesis has recently gained support from studies between the reduced mercury derivate of tree laccase and dioxygen (see Sec. 9).

8. The Half-Reduced Copper Pair

The studies of redox titration and of the reduction and reoxidation discussed above show that the type 3 copper pair appears to act as a strictly cooperative two-electron acceptor/donor. No new EPR signals, which could originate from a half-reduced type 3 copper pair, have been observed in these types of experiments. A new type of EPR signal can, however, be produced under certain conditions in both native[62] and type 2 copper depleted tree and *Polyporus* laccase.[35,36,62–64] Since neither type 1 nor type 2 Cu^{2+} ions can be responsible for this signal it must represent one of the type 3 copper ions. The new signal can be formed both on a re-reduction after reoxidation with dioxygen or on reoxidation by peroxide of the completely reduced tree enzyme.[62]

Some special conditions appear to be involved in the reduction of only one of the type 3 copper ions. It is only possible immediately after reoxidation by dioxygen of the native or type 2 copper depleted laccase which was first reduced with a small excess of reductant.[35,62] The signal appears after the re-reduction of the reoxidized type 1 copper and after the reduction of the oxygen intermediate.

The new signal has also been prepared with native tree laccase in experiments where the reduced enzyme was treated with small excesses of H_2O_2 at low temperatures for a few seconds before freezing and EPR analysis.[62] Under these conditions type 2 Cu^+ is not reoxidized but the type 1 Cu^{2+} EPR signal is fully developed. The new EPR signal intensity corresponds to 0.6 spin per molecule. Figure 11 shows the Q-band EPR spectra of the enzyme containing only the type 1 Cu^{2+} signal (A) and the type 3 Cu^{2+} signal (B). The símulated EPR spectrum of this new signal is also shown (C).

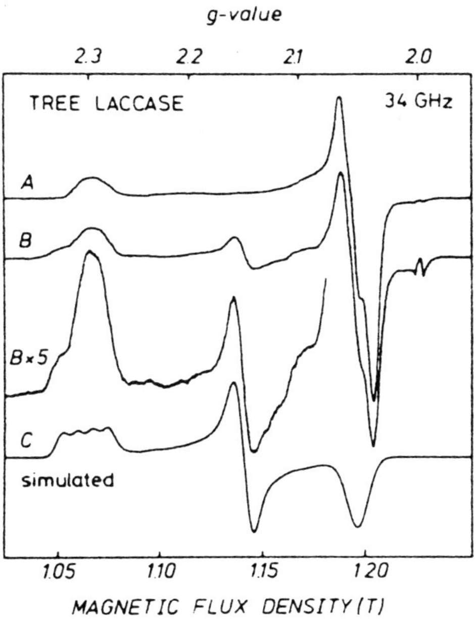

Fig. 11. EPR spectra at 34 GHz of *Rhus* laccase type 1 Cu^{2+} (A) and the new Cu^{2+} signal (B). The simulated spectrum new Cu^{2+} signal is also shown (C).

It seems that the reactions leading to the half-reduced copper pair is only possible in enzyme molecules where the type 2 copper is either reduced or absent. It has never been observed in anaerobic reduction studies or on reoxidation by dioxygen of reduced enzyme. It has also not appeared on turnover reactions. It is therefore possible that it is not formed during the normal catalytic reactions. Under turnover conditions, it appears that type 2 copper is mainly oxidized.[39] This copper ion is also the last to be reduced in transient-state reduction studies or during redox titrations mentioned above.[29,32,38,39] The formation of the half-reduced copper pair can occur in two ways depending on the experimental conditions. On reoxidation of the native enzyme with H_2O_2, the peroxide probably binds to the reduced type 3 copper site and reoxidizes both metal ions. The type 1 Cu^+ is then quickly reoxidized by electron transfer to one of the Cu^{2+} in the copper pair and the other Cu^{2+} therefore becomes EPR-detectable. With another peroxide molecule added the only EPR signals observed are the type 1 Cu^{2+} and the oxygen intermediate while type 2 copper is still reduced. The second peroxide molecule thus reoxidizes the half-reduced type 3 copper pair and then decays to a water molecule and a bound oxygen intermediate.[62]

In the lower part of Scheme 2 the reoxidation by dioxygen of native and type 2 copper depleted tree laccase, which is reduced with a small excess of reductant, is summarized. Type 2 copper does not seem to take part in the reaction with dioxygen and the formation of the half-reduced copper pair, as it can be absent in these reactions. It can therefore be excluded. On reoxidation, the oxygen intermediate state is rapidly formed. In the presence of excess reductant, the type 1 Cu^{2+} is re-reduced and then rapidly reoxidized by electron transfer to the oxygen intermediate which is reduced to water. Another substrate molecule re-reduces type 1 Cu^{2+} which donates its electron to the type 3 Cu^{2+} pair and the half-reduced copper pair is formed. This reaction sequence is consistent with the observation that the oxygen intermediate is reduced before the type 3 Cu^{2+} EPR signal appears.[35,63]

9. The Copper-Depleted Laccase

9.1. *The Type 2 Copper-Depleted Laccases*

This copper ion can be reversibly removed in laccases from *Polyporus*[48,64] and *Rhus* laccases.[35,65] Most studies report that this enzyme form is obtained with a reduced copper pair. This is shown by the loss of the 330 nm absorption band and by EXAFS studies.[66] It is, however, possible to isolate this enzyme form with an oxidized copper pair[35] (Fig. 12). In fact, the first report on the removal of this copper ion showed an absorption spectrum consistent with the presence of the 330 nm band.[65] Also a recent report on the type 2 copper-depleted laccase from *Coliolus consors* shows a 330 nm band similar to the native enzyme.[67] The different results are probably due to some differences in the preparation of the type 2-depleted enzymes. On removal of type 2 copper the enzymes become inactive which shows that this copper ion has a functional role in the catalytic reaction.

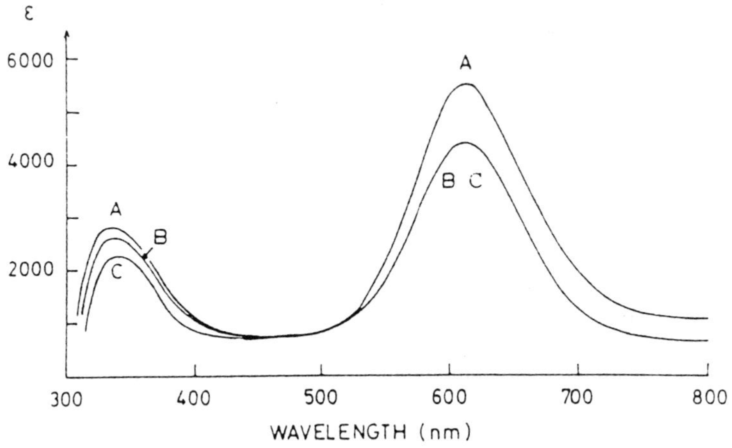

Fig. 12. Oxidized-reduced difference spectra of native (A) and type 2 copper-depleted samples (B,C) of *Rhus* laccase.

There have been many suggestions about the role of this metal ion in the reduction and reoxidation reactions. Since type 2 Cu^{2+} is reduced slowly in the *Polyporus* laccase it was suggested that it is not involved in the reduction of the enzyme but probably in the stabilization of a putative peroxide intermediate.[55] This idea gained further support from studies of the reaction between *Rhus* laccase and dioxygen.[33,66] In these studies it is thought that type 2 copper is involved in the breaking of the O–O of a peroxide intermediate. The peroxide is bound between type 2 and type 3 copper sites which requires a close proximity between these sites. It has therefore been stated in several reports (see Ref. 66 and references therein) that type 2 copper is necessary for the reaction between reduced enzyme and dioxygen.

It was also shown that one of the water molecules formed during reoxidation by dioxygen is bound to type 2 Cu^{2+}. These results suggested that this metal ion plays an important role in the reduction of dioxygen to water by laccase.[68]

With the type 2 copper depleted *Rhus* laccase it has been possible to understand the role of this metal in the catalytic mechanism.[35] It was shown that the rate of reduction of type 1 Cu^{2+} is not changed on removal of type 2 copper, but the reduction of the two-electron acceptor is very slow. This result demonstrates that type 1 Cu^{2+} is the primary electron acceptor that is reduced by substrates independent of the states of the other electron acceptors and that type 2 copper is involved in the reduction of the two-electron acceptor. The results also contradict the proposal that type 2 copper is the primary site for electron transfer from substrate to the two-electron acceptor and type 1 copper.[37]

During reoxidation with dioxygen both types 1 and 3 copper sites are reoxidized with the same rate as in the native enzyme. A paramagnetic oxygen intermediate with the same properties is formed in both native and type 2 copper depleted enzyme (see Fig. 10). It is therefore obvious that type 2 copper is not involved in the reoxidation mechanism and that it is not necessary in the formation and the stabilization of any oxygen intermediate.[35]

Reduction studies of type 2 depleted enzyme has also helped to clarify another point of debate. It was proposed that dioxygen can react with the enzyme at any stage of reduction. For example, dioxygen would react with either type 2 Cu^+ or the half-reduced type 3 copper pair after it had received an electron from the reduced type 1 copper.[69] However, the studies of the type 2 copper-depleted enzyme shows that type 1 Cu^+ is reoxidized very slowly in the one-electron reduced enzyme.[35] Furthermore, an O_2^- radical has never been detected in reoxidation or turnover studies of native and type 2 copper depleted enzyme.

9.2. *The Mercury Derivative of Rhus Laccase*

Type 1 copper can be replaced with the redox inactive metal mercury without altering the other metal sites.[70] This protein derivative is inactive and has the advantage that it is now possible to study the reaction of the reduced protein with dioxygen without the interference of this metal site. In the earlier studies with native and type 2 copper depleted enzymes it was not possible to isolate a two-electron reduced oxygen intermediate since the reoxidations of both type 1 and 3 copper sites occur in similar very rapid reaction phases. With the mercury derivative the only redox sites left are the type 2 and 3 copper sites. On reoxidation with dioxygen an optical intermediate, which is different from the one found in the native enzyme, is rapidly formed. It shows optical absorption bands at 340 and 470 nm but no signs of an intermediate EPR-detectable radical as observed in the native enzyme. The peroxide intermediate is suggested to be bound between one of the oxidized type 3 copper ions and type 2 Cu^+. The intermediate decays slowly ($t/2 = 4.3$ min.), the rate of which is faster than the reoxidation of type 2 Cu^+ ($t/2 = 6.4$ min.). This electron transfer would result in the formation of an oxygen radical but there are no reports of such a species.[71,72] These studies suggest that the dioxygen is probably reduced in two two-electron transfer steps in the native enzyme. However, as discussed above the reactions with the type 2 copper-depleted enzyme shows that this metal ion is not necessary for the reaction with dioxygen.

References

1. A. Messerschmidt, R. Ladenstein, R. Huber, M. Bolognesi, L. Avigliano, R. Petruzzelli, A. Rossi and A. Finazzi-Agrò, *J. Mol. Biol.* **224**, 170 (1992).
2. A. Messerschmidt and R. Huber, *Eur. J. Biochem.* **187**, 341 (1990).
3. B. Reinhammar, in *Copper Proteins and Copper Enzymes*, ed. R. Lontie (CRC Press, Boca Raton, Fl., 1984), Vol. 3, p. 1.
4. B. Reinhammar and B. G. Malmström, in *Copper Proteins, Metals in Biology*, ed. T. Spiro (Wiley & Sons, New York, 1981), Vol. 3, p. 109.
5. W. Bao, D. M. O'Malley, R. Whetten and R. R Sederoff, *Science* **260**, 672 (1993).
6. R. Bligny, J. Gaillard and R. Douce, *Biochem. J.* **237**, 583 (1986).
7. A. M. Mayer and E. Harel, *Phytochemistry* **7**, 1253 (1968).
8. H. Yoshida, *J. Chem. Soc.* (Tokyo) **43**, 472 (1883).
9. G. Bertrand, C. H. *Hebd. Acad. Sci.* (Paris) **118**, 1215 (1894).
10. J. -M. Lehn, B. G. Malmström, E. Selin and M. Öblad, *Trends Biochem. Sci.*, **11**, 228 (1986).
11. D. Keilin and T. Mann, *Nature* (London) **143**, 23 (1939).
12. A. Tissières, *Nature* (London) **162**, 340 (1948).
13. H. Binghuan, C. Wending and L. Jinhuo, *Int. Symp. On Oriental Lacquers*, May 2–5 (Fuzhou, China, 1993), p. 94.

14. R. Bourbonnais and M. G. Paice, *FEBS Lett.* **267**, 99 (1990).
15. P. J. Kersten, B. Kalyanaraman, K. E. Hammel, B. Reinhammar and T. K. Kent, *Biochem J.* **268**, 475 (1990).
16. M. Saloheimo, M. -L. Niku-Paavlova and J. K. C. Knowles, *J. Gen. Microbiol.* **137**, 1537 (1991).
17. G. F. Leatham and M. A, J. Stahtmann, *J. Gen. Microbiol.* **125**, 147 (1981).
18. T. J. Leonhard, *J. Bacteriol.* **106**, 162 (1971).
19. A. J. Clutterbuck, *J. Gen. Microbiol.* **70**, 423 (1972).
20. U. A. German, G. Muller, P. E. Hunziker and K. Lerch, *J. Biol. Chem.* **263**, 885 (1988).
21. R. Aramayo and W. E. Timberlake, *Nucleic Acids Res.* **18**, 3415 (1990).
22. Y. Kosima, Y. Tsukuda, Y. Kawai, A. Tsukamota, J. Segiura, M. Sakaino and Y. Kita, *J. Biol. Chem.* **265**, 15244 (1990).
23. G. H. Choi, T. G. Larson and D. I. Nuss, *Mol. Plant-Microbe Interact.* **5**, 119 (1992).
24. A. Messerschmidt, *Advances in Inorganic Chemistry* **40**, 121 (1993).
25. Y. Iimura, K. Takenouchi, M. Nakamura, S. Kawai, Y. Katayama and N. Morohoshi, *5th Int. Conf. on Biotechnology in the Pulp and Paper Industry,* May 27-30 (Kyoto Japan 1992) p. 427.
26. L. Jönsson, K. Sjöström, I. Häggström and P. O. Nyman, *Biochim. Biophys. Acta* **1251**, 210 (1995).
27. M. Jonsson, E. Pettersson and B. Reinhammar, *Acta. Chem. Scand.* **22**, 2135 (1968).
28. T. Pascher, B. G. Karlsson, M. Nordling, B. G. Malmström and T. Vänngård, *Eur. J. Biochem.* **212**, 289 (1993).
29. B. Reinhammar, *Biochim. Biophys. Acta* **275**, 245 (1972).
30. B. G. Malmström, B. Reinhammar and T. Vänngård, *Biochim. Biophys. Acta* **156**, 67 (1968).
31. E. Karhunen, M. -L. Niku-Paavlova, L. Viikari, T. Haltia, R. A. van der Meer and J. A. Duine, *FEBS Lett.* **267**, 6 (1990).
32. B. Reinhammar and T. Vänngård, *Eur. J. Biochem.* **18**, 463 (1971).
33. O. Farver, M. Goldberg, S. Wherland and I. Pecht, *Proc. Natl. Acad. Sci. USA* **75**, 5245 (1978).
34. R. Malkin, B. G. Malmström and T. Vänngård, *Eur. J. Biochem.* **10**, 324 (1969).
35. B. Reinhammar and Y. Oda, *J. Inorg. Biochem.* **11**, 115 (1979).
36. B. Reinhammar, *J. Inorg. Biochem.* **18**, 113 (1983).
37. R. A. Holverda and H. B. Gray, *J. Am. Chem. Soc* **96**, 6008 (1974).
38. L. -E. Andréasson and B. Reinhammar, *Biochim. Biophys. Acta* **445**, 579 (1976).
39. L. -E. Andréasson and B. Reinhammar, *Biochim. Biophys. Acta* **568**, 568 (1979).
40. F. B. Hansen, R. W. Noble and M. J. Ettinger, *Biochemistry* **23**, 2049 (1984).
41. O. Farver, M. Goldberg and I. Pecht, *Eur. J. Biochem.* **104**, 71 (1980).

42. B. G. Malmström, A. Finazzi-Agrò and E. Antonini, *Eur. J. Biochem.* **9**, 383 (1969).
43. L. -E. Andréasson, B. G. Malmström and T. Vänngård, *Eur. J. Biochem.* **34**, 434 (1973).
44. I. Pecht and M. Faraggi,*Nature New Biology* **233**, 116 (1971).
45. J. A. Fee, R. Malkin, B. G. Malmström and T. Vänngård, *J. Biol. Chem.* **244**, 4200 (1969).
46. L. -E. Andréasson, R. Brändén, B. G. Malmström, C. Strömberg and T. Vänngård, in *Oxidases and Related Redox Systems*, eds. T. E. King, H. S. Mason and M. Morrison (Univ. Park Press, Baltimore, 1973), p. 87.
47. R. Malkin, B. G. Malmström and T. Vänngård, *FEBS Lett.* **1**, 50 (1968).
48. R. Malkin, B. G. Malmström and T. Vänngård, *Eur. J. Biochem* **7**, 253 (1969).
49. R. Brändén and B. Reinhammar, *Biochim. Biophys. Acta* **405**, 236 (1975).
50. R. Brändén, B. G. Malmström and T. Vänngård, *Eur. J. Biochem.* **36**, 195 (1973).
51. L. C. Petersen and H. Degn, *Biochem. Biophys. Acta* **526**, 85 (1978).
52. G. B. Koudelka, F. B. Hansen and M. J. Ettinger, *J. Biol. Chem.* **260**, 15561 (1985).
53. A. Naqui and S. D. Varfolomeev, *FEBS Lett.* **113**, 157 (1980).
54. N. Manabe, K. Manabe and H. Hatano, *FEBS Lett.* **23**, 268 (1972).
55. L. -E. Andréasson, R. Bränden, B. G. Malmström and T. Vänngård, *FEBS Lett.* **32**, 187 (1973).
56. L. -E. Andréasson, R. Brändén and B. Reinhammar, *Biochim. Biophys. Acta* **445**, 579 (1976).
57. R. Brändén, B. G. Malmström and T. Vänngård, *Eur. J. Biochem.* **18**, 238 (1971).
58. L. -E. Andréasson, R. Brändén and B. Reinhammar, *Biochim. Biophys. Acta* **438**, 370 (1976).
59. R. Aasa, R. Brändén, J. Deinum, B. G. Malmström, B. Reinhammar and T. Vänngård, *FEBS Lett.* **61**, 115(1976).
60. R. Aasa, R. Brändén, J. Deinum, B. G. Malmström, B. Reinhammar and T. Vänngård, *Biochem. Biophys. Res. Commun.* **70**, 1204 (1976).
61. B. Reinhammar, *Chemica Scripta* **25**, 172 (1985).
62. B. Reinhammar, *J. Inorg. Biochem,* **15**, 27 (1981).
63. B. Reinhammar, R. Malkin, P. Jensen, B. Karlsson, L. -E. Andréasson, R. Aasa, T. Vänngård and B. G. Malmström, *J. Biol, Chem.* **255**, 5000 (1980).
64. P. M. Hanna, D. R. McMillin, M. Pasenkiewicz-Gierula, W. E. Antholine and B. Reinhammar, *Biochem. J.* **253**, 561 (1988).
65. M. T. Graziani, L. Morpurgo, G. Rotilio and B. Mondovi, *FEBS Lett.* **70**, 87 (1976).
66. E. I. Solomon and M. D. Lowery, *Science* **259**, 1575 (1993).
67. T. Sakurai, *J. Inorg. Biochem.* **41**, 277 (1991).
68. R. Brändén and J. Deinum, *FEBS Lett.* **73**, 144 (1977).

69. I. Pecht, O. Farver and M. Goldberg, *Bioinorg. Chem. II: Adv. Chem. Sem., Am. Chem. Soc.* **100**, 179 (1977).
70. M. M. Morie-Bebel, M. C. Morris, J. L. Menzie and D. R. McMillin, *J. Am. Chem. Soc.* **106**, 3677 (1984).
71. J. L Cole, G. O. Tan, E. K. Yang, K. O. Hodgson and E. I. Solomon, *J. Am. Chem. Soc.* **112**, 2243 (1990).
72. J. L. Cole, D. P. Ballou and E. I. Solomon, *J. Am. Chem. Soc.* **113**, 8544 (1991).

CHAPTER 7

FUNGAL LACCASES: ROLE IN DELIGNIFICATION AND POSSIBLE INDUSTRIAL APPLICATIONS

Melvyn Smith, Christopher F. Thurston
Microbiology Group, Division of Life Sciences
King's College London
Campden Hill Road, London W8 7AH, U.K.
E-mail: udba009@hazel.cc.kcl.ac.uk

and

David A. Wood
Microbiology and Plant Pathology Department
Horticulture Research International
Wellesbourne, Warwick CV35 9EF, U.K.
E-mail: david.wood@bbsrc.ac.uk

ABSTRACT

Laccase enzymes are produced by many fungi, particularly those that degrade lignin in woody plant material — the white rot fungi. Fungal laccases are typically extracellular glycoproteins. Their function in lignin degradation remains controversial, but is made more likely by the recent demonstration of the mediating activity of some artificial reducing substrates. In some other fungi laccase is a key agent of pathogenesis against the plant host. Cloning of laccase genes has shown that laccase and ascorbate oxidase sequences are similar, particularly in relation to copper liganding amino acid residues and domain structure, although laccase is usually monomeric in contrast to the ascorbate oxidase homo-dimer. Industrial applications of laccase are foreseen in biobleaching of paper pulp, adhesive formation from lignin wastes, in the construction of biosensors for industrial pollutants, in removal of phenolics from foodstuffs and in bioremediation.

Contents

1. Introduction

Laccase (Benzenediol: oxygen oxidoreductase, EC 1.10.3.2) is a member of the blue multi-copper family of enzymes, containing four copper atoms comprising one Type I, one Type II and an antiferromagnetically-coupled Type III pair of Cu^{2+}. The enzyme was first discovered in the sap of the Japanese lacquer tree *Rhus vernicifera*.[1] Subsequently it was also demonstrated to be a fungal enzyme.[2,3] To date only one species of bacterium, *Azospirillum lipoferum* has been shown to exhibit laccase-like phenol oxidase activity.[4,5] Laccase is regarded as the simplest of the large blue copper oxidases, because in some examples, it is the smallest of these proteins and contains the least copper. This apparent simplicity is both superficial and misleading. It is superficial in the sense that laccase and ascorbate oxidase are equivalent (as shown by structural comparisons, see Chapter 2), except that the well-studied examples of ascorbate oxidase are all dimers of identical subunits, whilst many of the laccases apparently function as monomers (see Table 1).

Table 1. Properties of some purified fungal laccases.

Organism	No. of enzymes	Mol. mass (kDa)	Carbohydrate content (%)	No. of Cu atoms/mol.	References
Agaricus bisporus‡	1(2)	100(65)	15	2	23, 32
Armillaria mellea	2	59	n.d.	n.d.	98
Aspergillus nidulans (conidial)	1	110	12	n.d.	21
Basidiomycete PM1	1	64	6.5	4	99
Botrytis cinerea	1	74	49	n.d.	102
Ceriporiopsis subvermispora	2	71 68	15 10	4	103
Cryphonectria parasitica	2?	77	n.d.	n.d.	82
Lentinus edodes	1	66	n.d.	n.d.	104
*Neurospora crassa**	1	65	11(12)	3(4)	28
Phlebia radiata	1	64	n.d.	2	105
Pleurotus ostreatus	1	59	n.d.	n.d.	106
Podospora anserina	4	70 80(2) 390	25 23 22	3.3	39
Trametes (Polyporus) pinsitus	2	63	14	2–4	34
Rhizoctonia solani+	1(4)	66–100	—	2	107
Schizophyllum commune	1	62–64	n.d.	2	22

‡*A. Bisporus* is shown as having 1(2) laccases because the two laccase genes *lcc* 1 and *lcc* 2 are both expressed but encode only slightly different proteins. As mentioned in the text the copper content of 2 is likely to be an underestimate.
*The figures in brackets are from later work, see Ref. 29.
+*R. solani* is shown as having 1(4) laccases because only one of the four laccases (from the four distinct genes found) was isolated from the culture medium under the conditions employed.

Taking the view that laccases are simple is also quite misleading. The several reasons for this are considered in this chapter with emphasis on laccases of fungal origin, together with the controversy as to their possible function in lignin degradation and the prospect of their application in novel biotechnologies.

2. Reactions Catalysed by Laccase

As with ascorbate oxidase and ceruloplasmin, laccases carry out a one-electron oxidation of a reducing substrate coupled with a four-electron reduction of dioxygen to water (for details of electron transfer reactions see elsewhere in this book). However, there are a number of difficulties in attempting to define laccases by their reducing substrate requirements. First, laccases are capable of oxidising a wide range of compounds, but the range is not always the same comparing one laccase to another. Reducing substrates include mono-phenols, *ortho* and *para* di-phenols, methoxyphenols, ascorbic acid and diamines, such as *p*-phenylene diamine.[6–11] A second complication arises due to an overlap in the substrate range with tyrosinase, another copper-containing phenol oxidase, (EC 1.14.18.1), often referred to as catechol oxidase.[12] A typical reaction, carried out by all laccases described to date, is the one-electron oxidation of diphenols forming an oxygen-centred free radical. This can then be converted to the quinone, either by a second catalytic step, or by spontaneous disproportionation. The quinone and free radical products can then undergo polymerisation.[13,14]

Catechol oxidase or tyrosinase catalyses two distinct reactions. The first, whereby an oxygen is inserted *ortho* to an existing hydroxyl group, is typically followed by the oxidation of the diphenol to the corresponding quinone, often referred to as cresolase activity. The second reaction is the oxidation of *o*-dihydric phenols, such as catechol, with hydrogen abstraction to the corresponding *o*-quinone, referred to as catecholase activity.[13] It is this reaction which is superficially similar to that carried out by laccase. However, laccases cannot oxidise tyrosine and are not inhibited by carbon monoxide, but their ability to oxidise *p*-diphenols is considered diagnostic.[7] In contrast, tyrosinase activity is inhibited by carbon monoxide and its inability to oxidise *p*-diphenols under any circumstances is a typical property of the enzyme. Syringaldazine, [N,N′-bis(3,5-dimethoxy-4-hydroxybenzylidene) hydrazine] is considered a specific substrate for laccase.[15] It is commonly used to detect the presence of laccase (see for example Ref. 16). In the reaction between laccase and ethanolic syringaldazine the initial product, formed by a one-electron oxidation, is a free radical.[14] The second one-electron oxidation step (either enzymatic and/or disproportionation), forms the quinone with a characteristic deep purple colour. It is important to exclude hydrogen peroxide from this reaction because syringaldazine is also oxidised by manganese-dependent peroxidases, produced by many lignin degrading basidiomycetes. The presence of endogenous H_2O_2 produced by laccase itself does not interfere, because if it is produced at all, it is reduced to H_2O on the enzyme and not released from it.[13,17] ABTS [2,2′-azinobis(3-ethylbenzthiazoline-6-sulphonate)] has increasingly been used as an experimental substrate for laccase assays.[9,18] The remarkable ability of ABTS to act as a mediating substrate is considered in a later section of this chapter. Finally, some laccases will accept ascorbate as a reducing substrate[19,13] and consequently the distinction between laccase and ascorbate oxidase is less than clear.

3. Studies on Purified Fungal Laccases

3.1. *Size and Copper Content*

Table 1 shows a list of some of the fungal laccases that have been purified over the last 20 years. All these enzymes are extracellular glycoproteins. However, as can be seen from this brief survey, there are differences in the structural components of laccases from different sources. The copper content appears to vary between two and four copper atoms per enzyme molecule or subunit. It is now generally accepted that all laccases contain four copper atoms.[13,14,20] There are a number of reasons why the lower figures have been rejected. Firstly, many purified enzyme preparations contain nicked molecules that are shown as a number of cleavage products upon electrophoretic analysis.[21–23] This is an inevitable consequence of the extracellular location of the great majority of laccases, because fungal media often also contain proteolytic and glycolytic activities.[24] Consequently, it is unlikely that such shortened and/or fragmented molecules retain the full copper complement of the intact enzyme. Another feature of laccases and one that has been exploited in attempts to unravel the reaction mechanism, is that copper can be selectively depleted from the enzyme.[13,17,25,26] It is possible that copper is lost accidentally from the enzyme during purification in a similar manner. This is a problem particular to fungal laccases, which typically lack a liganding methionine to one of the copper sites (type I) compared with ascorbate oxidases and other blue oxidases. Work carried out on a *Pseudomonas* azurin, in which a methionine ligand had been mutated to a leucine residue, showed this specific site to be relatively unstable and that some of the copper was lost during purification.[27] It is relevant to note that the lower figure of two copper atoms for *Neurospora crassa* laccase[28] shown in Table 1 was later revised to four copper atoms per molecule.[29] The main difference in their approach was to simplify the purification procedure from five steps to two. This resulted in an increase in enzyme yield from 19 % to 50 % and presumably an increase in the number of intact enzyme molecules recovered. Further indirect evidence in support of the four copper atoms/molecule model for laccase has come from gene sequence and deduced amino acid comparisons with ascorbate oxidase from zucchini squash.[30,31] The crystal structure and complete amino acid sequence are known for ascorbate oxidase. In solution ascorbate oxidase exists as a dimer, each subunit consisting of 552 amino acid residues (mol. mass 70 kDa) and four copper atoms. By comparing amino acid sequences from *Neurospora crassa* laccase and human ceruloplasmin with ascorbate oxidase it was found that all ligands, (except methionine for the type I copper in laccase) were strictly conserved.[31] Laccase (deduced) amino acid sequence homology is discussed further in the next section.

The variations in the molecular masses reported for laccases are probably in part related to the variations in the methods of their determination. The 100 kDa figure for *Agaricus bisporus* laccase was determined by sedimentation analysis,[32] compared with the 65 kDa value from electrophoresis under denaturing conditions.[23] Also the higher figure is not in accord with values subsequently deduced from gene sequence analysis.[33] However, it is possible that *Agaricus* laccase may exist as a dimer and that the 100 kDa figure is an underestimate related to the attached carbohydrate chains. At least two of the laccases from *Trametes pinsitus* are dimers.[34]

3.2. *Carbohydrate Content*

Molecular mass variations in glycoproteins related to their carbohydrate moiety are referred to as microheterogeneities.[35] It has also been found that the biosynthesis of the carbohydrate units is not completed on all protein molecules and that quantitative alterations in the carbohydrate sequence may occur.[35] Detailed analyses of the glycan moiety have been carried out on very few laccases, but for those that have, it seems that they share two common features, namely a high mannose content and an *N*-linked structure, although the laccase of *A. bisporus* is thought to contain equal amounts of *N*-and *O*-linked carbohydrate.[23] The two laccases of *Botrytis cinerea* were found to contain mannose, galactose, glucose, arabinose and xylose;[36] *Podospora anserina* laccase II was found to contain mannose, galactose and glucose attached to a number of potential *N*-linked sites.[35] The laccase of *Neurospora crassa* was also found to contain high levels of *N*-linked mannose.[28] Work on laccase III of *Coriolus versicolor*, which contains at least four *N*-linked carbohydrate chains high in mannose, has provided some insights into the function of these structures.[37] By using endoglycosidases and the *N*-asparagine specific inhibitor tunicamycin, it was found that the carbohydrate moiety was not necessary for excretion or enzyme activity. However, the carbohydrate-depleted laccase was found to be more sensitive to both proteolytic attack, (from proteases in the culture medium) and to elevated temperatures. It seems therefore that the glycan moiety of the laccase is involved in stabilising and protecting the enzyme from raised temperatures and proteolysis, conditions expected to be present during active growth of a white rot fungus. Similar experiments, using tunicamycin, were carried out on the intra-and extracellular laccases of *Botrytis cinerea*.[36] Both activities were reduced, but the extracellular activity was suppressed to a greater extent, indicating a possible role for the carbohydrate chain in secretion of the enzyme.

The formation of oligomeric units accounts for much of the variation in molecular masses in the laccases reported for *Podospora anserina*.[38,39] It is thought that laccase I (mol. mass 390 kDa) is composed of four units of the smaller laccase molecules. Later work by Durrens resulted in the discovery of a further laccase activity, laccase IV, with the same molecular mass as laccase III (mol. mass 80 kDa).[39] The separation of the two activities was based on their differences during isoelectric focusing experiments. Durrens proposed that the native protein, laccase I, was formed from four laccase IV subunits, each with two associated small molecules, which when separated under increasing pH gave rise to laccase III. Therefore, it seems likely that laccases are a heterogeneous group of glycoproteins with a carbohydrate content of 10–30 % and a basic structural protein unit of between 50–70 kDa that in some instances function as a dimer or higher aggregate.

4. Studies on Laccase cDNA and Gene Sequences

4.1. *Organisation and Similarity of Laccase and Ascorbate Oxidase Sequences*

During recent years laccase gene and/or DNA sequences have been described from a number of fungi (see Table 2). Generally, these laccase

Table 2. Cloned fungal laccase genes.

Species	No. of a/a residues	No. of introns	No. of a/a residues leader sequence	Reference
Agaricus bisporus	520	15	19	33
Aspergillus nidulans	609	5	n.d.	108
Coriolus hirsutus	499	10	21	109
Cryptococcus neoformans	624	14	20	110
Neurospora crassa	619	1	49	8
Basidiomycete PM1	517	10	21	99
Phlebia radiata	527	9	21	11
Pleurotus ostreatus	506	19	23	111
Trametes (Polyporus) pinsitus	520(*lcc* 1)	8	21	34
Rhizoctonia solani	572 (*lcc* 1)	n.d.	19	107
	599 (*lcc* 2)	n.d.	19	
	572 (*lcc* 3)	n.d.	19	
	529 (*lcc* 4)	n.d.	19	

genes show a common pattern in that they all encode polypeptides of approximately 520–600 amino acid residues, including an N-terminal secretion peptide. In sequence comparisons the overall amino acid similarity is generally not very high, but importantly they all show strict conservation in and around the single cysteine and ten histidine residues proposed as copper ligands. In all cases these residues are distributed in the N-terminal domain (two pairs of histidines) and the C-terminal region for the cysteine and remaining histidines. Early work in the 1980's had begun to identify these residues as ligands. In studies on laccase B from *Polyporus versicolor*, the peptide sequence from a number of cysteine-containing, C-terminal fragments of the enzyme was obtained.[20] In comparisons with plastocyanin, human ceruloplasmin and azurin (all blue type-I proteins) it was determined that the cysteine residue provided at least one of the ligands for the type I copper. Using magnetic circular dichroism and EPR techniques on Japanese lacquer tree laccase, it was demonstrated that the anionic inhibitor azide is able to bridge the type II site and one of the type III coppers, greatly decreasing the rate of reduction of the type III copper.[40] These results suggested that these three coppers were closely associated in a tri-nuclear active site. Sequence alignments using the deduced amino acid sequence of *Neurospora crassa* laccase and human ceruloplasmin also identified putative copper ligands in the carboxy-terminal part of the fungal sequence.[41] These earlier findings are in accord with more recent comparisons with sequence and crystal structure data from ascorbate oxidase.[30,31,42] There is one notable difference between the laccases and ascorbate oxidase. Both proteins bind a type I copper with one cysteine and two histidines, but in ascorbate oxidase there is a fourth ligand, methionine, that is absent in all but one fungal laccase sequenced so far (the ascomycete *A. nidulans*). There is a significant parallel with copper liganding in the small blue copper proteins that contain only a single type I copper. Of these, stellacyanin is the only one that lacks a methionine ligand, whilst retaining the two histidine and single cysteine ligands. It has also been found that when the methionine ligand is lost from *Pseudomonas* azurin by mutation, electron transferase activity is retained, but the protein exhibits a higher redox potential.[27]

Modelling studies of the amino acid sequence of *Neurospora crassa* laccase[31] and the *Phlebia radiata* laccase[11] both indicate that the laccase sequences can fold into the three-domain β-barrel structure described for the ascorbate oxidase subunit. There is however quite limited similarity among the ascorbate oxidase and laccase primary sequences.[33] The aligned sequences of five laccase and two ascorbate oxidases show only 7 % identical residues. The *Aspergillus nidulans* and *Neurospora crassa* sequences are somewhat surprisingly no more similar to the basidiomycete laccases than are the cucumber and squash ascorbate oxidases (around 50 % similar amino acid residues in paired alignments). As more fungal laccase sequences have been described, it is becoming clear that there is significantly higher homology amongst those isolated from white-rot basidiomycetes (see Fig. 1). What remains uncertain is whether this has much functional significance. It may be only that these fungi are more closely related (i.e., their laccase sequences have had less opportunity to diverge). At present, there is insufficient information on other gene sequences from these organisms to be able to determine whether e.g., the similarity between *Agaricus bisporus* laccase sequence and those from *Phlebia radiata*, *Coriolus hirsutus* and *Polyporus pinsitus* reflects only evolutionary distance or is (in part) because the enzymes have different functions in some way. It is nevertheless intriguing that the *Aspergillus nidulans* enzyme is not very similar to the other

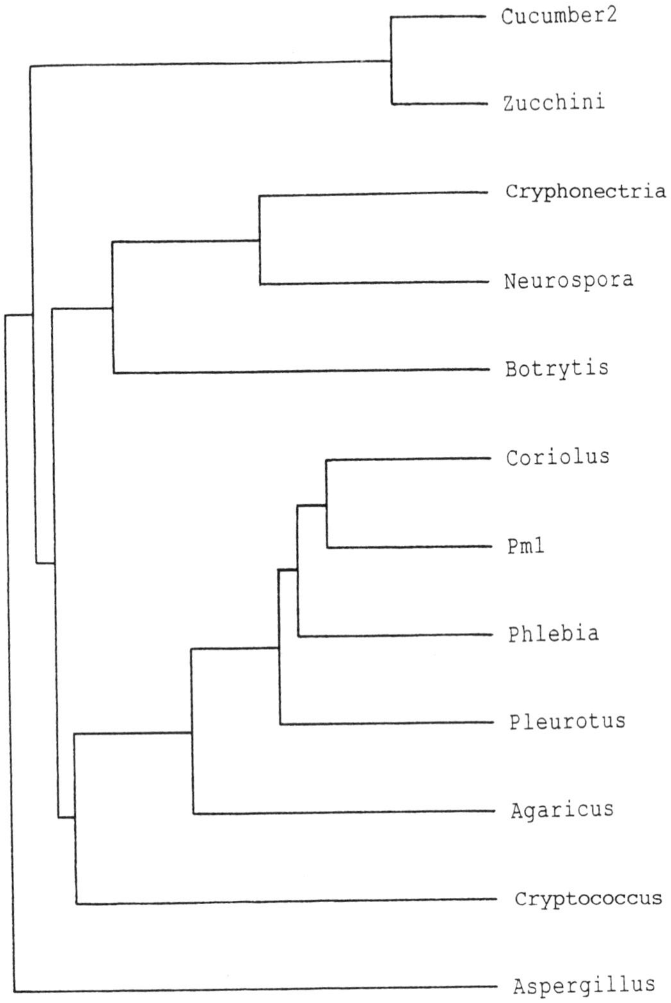

Fig. 1. Dendrogram compiled using the GCG program PILEUP showing the similarities between deduced amino acid sequences of different fungal laccases and plant ascorbate oxidases. The origin of the fungalsequence data, with the exception of *Botrytis*, may be found in the references cited in Table 2. The position of laccase from *Botrytis* is provisional as the sequence employed is incomplete, see GCG Accession No. U20192. For the sequences of ascorbate oxidase from cucumber and zucchini see GCG Accession Nos. S11027 and A30066 respectively. Where two alleles of the laccase gene have been sequenced and in the case of *A. bisporus* where two very similar non-allelic genes have been sequenced, only one of each was used to produce the dendrogram. Vertical lines represent degree of similarity, with similarity increasing from left to right.

ascomycete laccases (*Botrytis, Neurospora* and *Cryphonectria*) and is the one enzyme known to be involved in pigment formation (see Sec. 6.1).

4.2. *Do all Laccases Contain Four Cu Atoms?*

The presence of a new combination of prosthetic groups in the laccase from *Phlebia radiata* has been claimed.[10,11] This laccase is reported to contain only two copper atoms/molecule.[10] It is proposed that the co-factor pyrroloquinoline quinone (PQQ) acts in an analogous manner to the type III coppers of other laccases. However, as discussed previously there are problems associated with copper quantitation in these enzymes; also the *P. radiata* sequence shows the presence of ligands for all four coppers. A more recent re-evaluation of the work on diamine oxidases has also cast doubt on whether PQQ is the co-factor in these enzymes.[43] It appears that not only do enzyme preparation/purification methods lead to the formation of PQQ-like compounds, but that also, fungal culture media often contain compounds with similar spectroscopic properties to PQQ. It may be of relevance that galactose oxidase (a one-copper oxidase) has an active-site tyrosine that is modified and involved in radical formation.[44] The second copper binding motif in the N-terminal domain of laccase and ascorbate oxidase (His-Ser-His) is immediately preceded by a tyrosine residue. This could conceivably be modified giving some spectral properties similar to PQQ, but again as this residue is conserved in all the AO and laccase sequences available to date, it is unclear why the structure/mechanism of the *Phlebia* laccase might be different.

5. Laccase and the Degradation of Lignin

5.1. *Lignin Degradation—The Background*

It has been considered for many years that the laccases from a variety of fungi play a direct role in the oxidation of lignin.[13,45] Significantly, it has recently been reported that laccase also participates in the lignification of plant tissues.[46,47] Laccase has been purified from the cell walls of differentiating xylem tissues of loblolly pine (*Pinus taeda*), its appearance coinciding with the formation of lignin.[46] Further, *in vitro* studies showed that this enzyme was able to oxidise sinapyl and coniferyl alcohols (both monolignin precursors) to form polymeric dehydrogenitive products by free radical coupling.

Lignin is found in all higher plants and with the exception of polysaccharides is the most abundant cell wall component, particularly in supporting tissues such as xylem.[48,49] Lignin forms a network that permeates the matrix of the cell wall, enhancing structural support and also creating a barrier to attack from pathogenic organisms on the cellulose and hemicellulose components within the cell wall.

Biosynthetically, lignin is typically made up of three precursor alcohols: *p*-hydroxycinnamyl (coumaryl) alcohol, which gives rise to *p*-hydroxyphenyl units in the polymer; 4-hydroxy-3-methoxy cinnamyl (coniferyl) alcohol, the guaiacyl units and 3,5-dimethoxy-4-hydroxycinnamyl (sinapyl) alcohol, the syringyl units. By a process of free-radical co-polymerisation, these units form the heterogeneous, optically inactive, cross-linked, highly polydisperse polymer.[48] The mechanism of lignin polymerisation is still not fully understood however, because the identity of the full complement of enzymes responsible for polymerisation remains unclear.[47]

The biodegradation of lignin is of interest for three reasons. This process is a major component in terrestrial carbon cycling,[50] it is an unusual biochemical process because of the chemical structure of lignin as described above, and the removal of lignin from wood pulp during paper manufacture is a difficult and polluting chemical process that might be improved or replaced by a biotechnological alternative.[48]

Although lignin biodegradation has been the subject of much study over many years, it is far from completely understood. It is probable that some bacteria (other than Actinomycetes, see below) can degrade lignin to a greater or lesser extent, but this has not been convincingly demonstrated with laboratory cultures of pure strains. It is known that some Actinomycetes are capable of extensive modification of lignin, but this does not apparently involve the complete mineralisation (conversion to CO_2) of a large proportion of lignin.[51–53]

In contrast, some fungi are able to mineralise lignin completely, or to a much larger extent than has presently been demonstrated with other organisms. This process is a characteristic property of the basidiomycete fungi that are known as "white rot fungi"[48,54,55] and may occur in a similar fashion in (some ?) leaf litter degrading species.[56] Studies of the enzymology of lignin degradation have presupposed the following: the initial attack on lignin must be carried out by extracellular enzymes because lignin is a large molecule, intimately integrated with other insoluble polymers in plant cell wall material. The enzymes are unlikely to be hydrolases because the carbon–carbon bonds and ether linkages of lignin are not amenable to hydrolytic cleavage and changes in lignin structure observed during white-rot degradation result from oxidation reactions (e.g., aromatic ring cleavage; see Refs. in 48). Given that the process of wood and leaf litter rotting with which these fungi are associated is aerobic (water-logged, and hence anaerobic systems do not show rapid lignin breakdown), the enzymatic process is likely to be directly, or indirectly, oxidative.[48]

5.2. *Lignin Degradation by White Rot Fungi*

Until the early 1980's laccase was the only enzyme known to be secreted by these fungi that had at least some of the requisite properties. The problem was that, although laccase can cleave some of the bonds found in lignin, there are others that it cannot directly attack[9,17,57] and purified laccase will polymerise lignin model compounds, but was not apparently able to depolymerise lignin or artificially made lignin-like structures.[58,59] Experiments designed to test whether laccase was directly involved in ligninolytic growth were equivocal — some supporting a role for laccase[45,60–62] and others suggesting that laccase was not required.[63,64] This paradox, together with the strongly oxidative nature of lignin degradation gave impetus for searches that led to the discovery of the haem-containing peroxidases — lignin peroxidase (LiP)[65,66] and manganese peroxidase (MnP).[66,67] These enzymes were discovered in strains of *Phanerochaete chrysosporium* that are powerfully lignolytic, but which do not make measurable amounts of laccase under the conditions used to study LiP and MnP, although conditions under which *P. chrysosporium* BKM-F 1767 **does** secrete laccase have very recently been established[112] confirming the earlier work of Ander and Eriksson.[60] For some time, therefore, it looked likely that the peroxidases were responsible for lignin breakdown in white rot fungi and that laccase was not directly involved.

This view is now less easy to sustain. As more information has accumulated on the properties of LiP and MnP, it has emerged that these enzymes like laccase, tend to polymerise rather than depolymerise lignin *in vitro.* Recently conditions under which LiP will affect some net depolymerisation of DHP have been established,[68] but equivalent studies of MnP and laccase have not been reported. Secondly, in surveys of white rot fungi it appears that whilst some organisms make LiP, MnP and laccase, more often two out of the three can be detected.[69] Thirdly, MnP acts by producing chelated Mn^{3+} which is a mediator of oxidative degradation/damage of lignin[70] and LiP may well also act via a reactive mediating product (veratryl alcohol radical cation).[66,71] This is most significant because of the recently discovered ability of laccases to act via "mediating" substrates/products.[9] When ABTS is present as a substrate, other compounds are oxidised. Using purified laccase from *C. versicolor* and ABTS as co-substrate, it was demonstrated *in vitro* that laccase could oxidatively cleave $C\alpha$–$C\beta$ linkages of non-phenolic components of kraft pulp lignin and oxidise veratryl alcohol to veratraldehyde, neither of which can be done by laccase on its own.[9,18] The presence of ABTS also prevented the repolymerisation of the phenoxy radicals produced by oxidation of the lignin phenolic groups. Whether ABTS is a true mediator (being oxidised by the enzyme and reduced back to its original form by interaction with a second substrate) is not clear. The amount of ABTS present in these experiments was greater than the the amount of second substrate oxidised,[9] consequently ABTS may act as a co-substrate which is not regenerated.

In summary, laccase is produced by many if not the majority of lignolytic fungi, together with either MnP or LiP and sometimes both. All three enzymes may therefore degrade lignin employing low molecular weight substrates which give reactive diffusible oxidation products that can in turn oxidatively cleave the lignin polymer. This proposal is most speculative in the case of laccase because no naturally occurring component of a fungal culture medium during lignolytic growth has yet been shown to have the mediating activity demonstrated for ABTS *in vitro*.

Laccase may also interact with other enzymes during lignin degradation to allow quinone cycling. It has been shown that the lignolytic activity of *Trametes* (*Coriolus*) *versicolor* laccase is greatly enhanced by the presence of glucose oxidase.[72] The action of laccase alone on high molecular weight lignosulphonates produced quinonoid intermediates resulting in low rates of depolymerisation. However the addition of glucose oxidase, which is thought to reduce the reactive quinone radicals, greatly improved the efficiency of lignin depolymerisation. Similar studies have shown that the white-rot Basidiomycete, *Rigidoporus lignosus* excretes both a laccase and a manganese-dependent peroxidase, neither of which was able to degrade radioactively-labelled lignins on its own.[57] It was found that together the enzymes acted synergistically, however, resulting in extensive lignin-degradation. It was also found that the presence of glucose oxidase enhanced degradation rates, again possibly by preventing re-polymerisation of radicals formed by the two oxidative enzymes. A similar effect had been observed with cellobiose oxidase during lignin degradation by *P. chrysosporium* peroxidases.[73] Cellobiose oxidase is able to reduce radicals and quinones, resulting in a decrease in polymerisation of lignin-degradation products, which may increase the rate of both lignin and cellulose biodegradation.

It has been shown that the laccase of *C. versicolor* can produce manganic (MnIII) chelates from manganous (MnII) ions in the presence of phenolic 'accessory' substrates, suggesting an overlap of specificity between laccases and manganeses peroxidases.[74] However, more recent work on the laccase, manganese peroxidase and lignin peroxidases of *Phlebia radiata* have shown that with preferred reducing substrates, such as syringaldazine and ABTS, no laccase-coupled oxidation of Mn(II) occurred.[75]

As a number of authors have pointed out, laccase could well have a function during lignolytic growth other than a direct involvement in lignin cleavage.[24,76] All the enzymes thought to participate in lignin degradation produce reactive (and hence toxic) species from which the fungal mycelium must be protected. Laccase may function under these circumstances by polymerising such compounds before they can enter the mycelium. In species that do not produce laccase, such as *P. chrysosporium*, this protective function would need to be carried out by alternative means and could conceivably be achieved by secretion of glucan slime.[77]

5.3. *Laccase as an Agent of Pathogenesis*

Whilst the role(s) of fungal laccase in lignin degradation have yet to be fully evaluated, one aspect that is becoming more clearly defined is its importance in pathogenicity. *Botrytis cinerea*, which causes soft-rot infections in many crop species including beans, carrots, cucumbers and grapes,[78] produces both a constitutive laccase and a number of inducible, extracellular, forms of the enzyme. These are produced in response to pectins, phenolics and/or their quinone oxidation products.[79,80] When *Botrytis* begins the infective process in cucumber fruit, it produces a number of "attack" enzymes, including cellulase, polygalacturonase, proteinase and laccase(s). In response, the infected plant produces a range of cucurbitacins (tetracyclic triterpenoid compounds). These have been shown to have a specific inhibitory effect only on laccase. Cucurbitacins are highly active in repressing the appearance of extracellular laccase and may act directly on gene expression. Tests on both intra-and extracellular laccases showed cucurbitacins reduced enzyme activity by up to 70 %. In the presence of cucurbitacins the electrophoretic mobility of the extracellular laccase was altered, indicating that the compounds may cause structural or conformational changes. Laccase may also play a role in the pathogenesis of the chestnut blight fungus, *Cryphonectria parasitica*.[81,82] This fungus produces two forms of laccase and is susceptible to infection by a double-stranded RNA virus which represses both enzymes, by preventing laccase mRNA accumulation. A number of other gene products are also repressed, the overall result of which leads to the formation of a hypovirulent strain of the fungus with reduced sporulation and pigmentation. The specificity of laccase in infection is not yet known, but chestnut bark does have high concentrations of tannins. This suggests that a successful pathogen must be able to tolerate high tannin levels. Laccase may well play a role in the tolerance of host phenolics in virulent strains of the fungus.

6. Industrial Applications of Laccase

6.1. *Adhesives*

In Sec. 5.2 we have considered laccase functions that depend at least in part on depolymerisation of polyphenolic compounds, but this is not the only type of function laccase can exhibit *in vivo*, or that can form the basis of an industrial application. Laccase has a well characterised function in pigment formation in *Aspergillus nidulans*[83] and is postulated to be involved in rhizomorph formation by Armillaria species.[84] The laccase-catalysed polymerisation of phenolic compounds may be the mechanism of hyphal adherence giving rhizomorphs their rigidity and dehydration resistance. Laccase may also effect the binding of hyphae by polymerising phenols in the intercellular spaces in *Lentinus edodes* fruit bodies.[7] For a full discussion of adherence in fruit body formation, see Ref. 85.

These adhesive properties of laccase-polymerised phenolic compounds have been investigated for potential industrial application. It has been demonstrated that wood adhesives can be produced using laccase and lignin wastes (from pulping) on a commercial scale, for use in industrial processes such as the manufacture of chip-board.[86] This process has not, however, come into use at the present time. High volume production of any derivative of lignin-containing wastes from pulping would be advantageous because of the problems disposal of these wastes present.

6.2. *Biosensors*

Environmental damage resulting from industrial pollution, whether accidental or as a by-product of industrial processes, requires monitoring. The substrate range of laccases include some of the components in the waste products from the coal, oil, gas and paper industries; also a range of phenolics and catechols are present in wines, tea and fruit juices. Laccase is a good candidate for use as a biosensor transducer because it can readily be coupled to an oxygen electrode, which allows amperometric measurements of relatively high sensitivity.

A bienzyme system has been developed using laccase (from *Coriolus hirsutus*) and glucose dehydrogenase co-immobilised onto plastic discs, coupled with an oxygen electrode.[87] This system has been used for the detection of *p*-aminophenols (PAP), whereby laccase catalyses its oxidation to *p*-iminoquinone. The latter compound is the electron acceptor in the glucose oxidation reaction catalysed by glucose dehydrogenase. In this way an excess of glucose enables low levels of PAP to be detected by recycling it, which results in an increase in oxygen depletion. This system has been shown to detect levels of neuromediators (dopamine, adrenalin and noradrenalin) up to ten times lower than those detected by tyrosine/ascorbic acid enzymatic chemical recycling methods.[87] Another use for the laccase from *C. hirsutus*, that of an enzyme marker in immunoassays, has been developed.[88] The assay is based on laccase-antibody conjugates and is simpler than using peroxidase-antibody conjugates because the laccase system uses air-oxygen as the second substrate of the enzymatic reaction.

The laccase bi-enzyme sensor has also been developed to detect levels of ferrocene down to concentrations as low as 0.2nM.[89] An amperometric biosensor employing laccase and tyrosinase co-immobilised by adsorption onto a graphite electrode has been used to detect phenols for environmental analysis.[90] The combined use of laccase and tyrosinase (see Sec. 2) enables a wider range of phenols to be detected. The catalytic activity in non-aqueous environments of the laccase from *Coriolus hirsutus* has been exploited in the development of an organic-phase amperometric biosensor.[91] The enzyme is immobilised onto carbon electrodes by entrapment with a polymeric film to increase stability of the enzyme in organic media. The resulting biosensor responds rapidly and is able to detect low concentrations of catechols and hydroquinones in various alcohols. The detection limit for hydroquinone is reported to be approximately 6×10^{-7}M. Another application of the enzymatic electrode is in the detection of phenolic compounds in tea. Oxidation of catechols by oxygen in the presence of immobilised laccase results in variations in the oxygen concentration of the reaction mixture which can be monitored by an oxygen electrode. Two versions of the biosensor were tested: batch and flow injection which can be used to assay leaf catechols at various stages in tea-leaf production and quality control.[92]

6.3. *Phenolics in Foodstuffs*

Immobilised laccase from *Coriolus* (*Trametes*) *versicolor* has been employed in the enzymatic removal of phenolic compounds in must and wines.[93] Musts and wines are complex mixtures of ethanol, organic acids, salts and phenolic compounds. The colour and flavour of wine depends significantly on the phenolic components in the mixture and these are subject to various changes during storage. Problems can arise from the modification of these phenolics and subsequent chemical reactions, adversely affecting the flavour and colour of the wine. Various methods have been used to prevent the phenol-dependent discoloration and/or off-flavour formation, including removal of phenolics with cross-linked polyvinylpyrrolidone and hyperoxidation. In the laccase system the enzyme was immobilised in agarose-based matrix (AffiGel-10 and 15) column reactors. Variations in the flow rate and bed volume showed that a low flow rate and large bed volume in the reactor were most efficient at total polyphenol removal. It was found that the immobilised enzyme could be reused up to eight times in successive experiments over a six-month period with negligible loss in activity.[93]

6.4. *Bioremediation*

One potential area of investigation for the use of laccase is in the biological bleaching of kraft pulps (industrially produced, chemically delignified/derivatised wood fibres for the paper industry). The abilities of a range of white rot fungi, especially *Coriolus* (*Trametes*) *versicolor*, to delignify kraft pulps have been reviewed.[94] Whilst the enzymatic treatments were slow, the enzymes thought to be involved, laccase and/or MnP, were able to demethylate and partially solubilise the lignin in pulps. Studies on biopulping wheat-straw by members of the genus *Pleurotus*, particularly *P. eryngii*, has shown them capable of extensive delignification, without significant attack on the cellulose fibres.[95] The possibilities of using laccases from *Coriolus versicolor* in the dechlorination of chlorophenolic compounds in wood-pulp mill bleachery effluent have been investigated.[96] Chlorine has been a widely used bleaching agent for chemically produced wood pulps. Chlorination and subsequent alkaline extraction of the pulp produces large volumes of effluent containing substantial levels of adsorbable organic halogens (AOX). Whilst chlorolignin production is diminishing, paper pulping will continue to produce some lignin-derived by-product which will require elimination. Lignin peroxidase from *Phanerochaete chrysosporium* has also been extensively studied in this context (see for example Ref. 97). Whilst the peroxidase was slightly better in overall terms of substrate consumption, laccase offered a number of advantages over lignin peroxidase: (i) laccases are more stable and can function effectively when immobilised, (ii) no H_2O_2 is required (iii) a number of laccases are constitutively produced in large quantities and (iv) laccases were more effective in dechlorinating the compounds tested.[96] The performance of soluble and immobilised laccase from *Rhizoctonia praticola* has been evaluated for its ability to decrease the toxicity of a range of substituted phenols.[76,100] This is based on the enzyme's ability to oxidatively couple these compounds to syringic acid, a naturally occurring phenol. Thus, laccase in both free and immobilised forms was able to oxidise a range of methoxy- and methyl-substituted phenols. Unlike the laccase of *C. versicolor* however, the *R. praticola* enzyme showed a poor ability at removing chlorinated phenols.

The use of laccase in pulp bleaching has taken on renewed interest since the discovery of its broadened substrate range in the presence of mediating substrates (see Sec. 5.2). A process employing 1-hydroxybenzotriazole and related compounds as mediators is presently under development.[101]

From the foregoing it is clear that there are a number of applications for laccase(s) in industrial processes that could prove to be useful. Specifically adapted immobilised enzyme systems tailored to a particular waste product might prove more effective. For example the use of a range of laccases from different fungal species, or laccases in combination with LiPs and/or MnPs might be more efficient in some applications. Similarly the use of immobilised mycelium would allow for the continuous production of a range of enzymes.[100] The use of heterologous production methods or engineered/induced overproducing fungal strains to give economically viable quantities of laccase will also be a useful area of investigation. Such investigations would also help to further our understanding of the biological processes in which laccase is involved.

Acknowledgments

We acknowledge financial support from the Biotechnology and Biological Sciences Reseach Council of the UK. We thank Dr. Phil Cunningham for his expert assistance with sequence alignment. We are most grateful to Dr. Debbie Yaver and Dr. Jill Wahleithner for giving us preprints of recent work.

References

1. H. Yoshida, *J. Chem. Soc.* **43**, 472 (1883).
2. G. Bertrand, *C. R. Hebd Seances Acad. Sci.* **123**, 463 (1896).
3. J. Laborde, *C. R. Hebd Seances Acad. Sci.* **123**, 1074 (1896).
4. A. Givaudan, A. Effosse, D. Faure, P. Potier, M. Bouillant and R. Bally, *FEMS Microbiol. Lett.* **108**, 205 (1993).
5. D. Faure, M. L. Bouillant and R. Bally, *Appl. Environmental Microbiol.* **60**, 3414 (1994).
6. D. A. Wood, *J. Gen. Microbiol.* **117**, 339 (1980).
7. G. F. Leatham and M. A. Stahmann, *J. Gen. Microbiol.* **125**, 147 (1981).
8. U. A. Germann, G. Muller, P. E. Hunziker and K. Lerch, *J. Biol. Chem,* **263**, 885 (1988).
9. R. Bourbonnais and M. G. Paice, *FEBS Lett.* **267**, 99 (1990).
10. E. Karhunen, M. -L. Niku-Paavola, L. Viikari, T. Haltia, R. A. Van Der Meer and J. A. Duine, *FEBS Lett.* **267**, 6 (1990).
11. M. Saloheimo, M. -L. Niku-Paavola and J. K. C. Knowles, *J. Gen. Microbiol.* **137**, 1537 (1991).
12. A. M. Mayer, *Phytochem.* **26**, 11 (1987).
13. A. M. Mayer and E. Harel, *Phytochem.* **18**, 193 (1979).
14. C. F. Thurston, *Microbiol.* **140**, 19 (1994).
15. J. M. Harkin, M. J. Larsen and J. R. Obst, *Mycologia* **66**, 469 (1974).
16. C. Grassin and D. Dubourdieu, *J. Science Food Agricul.* **48**, 369 (1989).
17. J. Li and D. R. McMillin, *Biochim. Biophys. Acta* **1160**, 239 (1992).
18. R. Bourbonnais, M. G. Paice, I. D. Reid, P. Lanthier and M. Yaguchi, *Appl. Environmental Microbiol.* **61**, 1876 (1995).

19. R. Brändén and J. Deinum, *FEBS Lett.* **73**, 144 (1977).
20. C. Briving, E. -K. Gandvik and P. O. Nyman, *Biochem. Biophys. Research Comm.* **93**, 454 (1980).
21. M. B. Kurtz and S. P. Champe, *J. Bacteriol.* **151**, 1338 (1982).
22. O. H. M. De Vries, W. H. C. F. Kooistra and J. G. H. Wessels, *J. Gen. Microbiol.* **132**, 2817 (1986).
23. C. R. Perry, S. E. Matcham, D. A. Wood and C. F. Thurston, *J. Gen. Microbiol.* **139**, 171 (1993).
24. D. A. Wood, in *Developmental Biology of Higher Fungi*, eds. D. Moore, L. A. Casselton, D. A. Wood and J. C. Frankland (Cambridge Univ. Press, Cambridge 1985), pp. 375–387.
25. B. Reinhammar and B. G. Malmström, in *Metal Ions in Biology*, 2nd ed., ed. G. Spiro (Wiley, New York, 1981), pp. 109–149.
26. B. Reinhammar, *J. Inorganic Biochem.* **18**, 113 (1983).
27. B. G. Karlsson, R. Aasa, B. G. Malmström and L. G. Lundberg, *FEBS Lett.* **253**, 99 (1989).
28. S. C. Froehner and K. -E. Eriksson, *J. Bacteriol.* **120**, 458 (1974).
29. K. Lerch, J. Deinum and B. Reihnhammar, *Biochim. Biophys. Acta* **534**, 7 (1978).
30. A. Messerschmidt, A. Rossi, R. Ladenstein, R. Huber, M. Bolognesi, G. Gatti, A. Marchesini, R. Petruzzelli and A. Finazzi-Agro, *J. Mol. Biol.* **206**, 513 (1989).
31. A. Messerschmidt and R. Huber, *Eur. J. Biochem.* **187**, 341 (1990).
32. D. A. Wood, *J. Gen. Microbiol.* **117**, 327 (1980).
33. C. R. Perry, M. Smith, C. Britnell, D. A. Wood and C. F. Thurston, *J. Gen. Microbiol.* **139**, 1209 (1993).
34. D. S. Yaver, X. Feng, E. J. Golightly, K. M. Brown, S. H. Brown, M. W. Rey, P. Schneider, T. Halkier, K. Mondorf and H. Dalbøge, The purification, characterisation, molecular cloning and expression of two laccase genes from the white-rot basidiomycete *Trametes pinsitus* (1995), in press.
35. K. Esser and W. Munith, *Eur. J. Biochem.* **23**, 484 (1971).
36. I. Marbach, E. Harel and A. M. Mayer, *Phytochem.* **23**, 2713 (1984).
37. A. Yoshitake, Y. Katayama, M. Nakamura, Y. Iimura, S. Kawai and N. Morohoshi, *J. Gen. Microbiol.* **139**, 179 (1993).
38. H. P. Molitoris and K. Esser, *Arch. Microbiol.* **72**, 267 (1970).
39. P. Durrens, *Arch. Microbiol.* **130**, 121 (1981).
40. M. D. Allendorf, D. J. Spira and E. I. Solomon, *Proc. Nat. Acad. Sci. USA* **82**, 3063 (1985).
41. U. A. Germann and K. Lerch, *Proc. Natl. Acad. Sci. USA* **83**, 8854 (1986).
42. A. Messerschmidt, R. Ladenstein, R. Huber, M. Bolognesi, L. Avigliano, R. Petruzzelli, A. Rossi and A. Finazzi-Agro, *J. Mol. Biol.* **224**, 179 (1992).
43. J. P. Klinman, D.M. Dooley, J. A. Duine, P. F. Knowles, B. Mondovi and J. J. Villafranca, *FEBS Lett.* **282**, 1 (1991).

44. A. J. Baron, S. C. Stevens, C. Wilmot, K. D. Seneviratne, V. Blakeley, D. M. Dooley, P. F. Knowles and M. J. McPherson, *J. Biol. Chem.* **269**, 25095 (1994).
45. E. M. Turner, *Transactions of the British Mycological Society* **63**, 541 (1974).
46. W. Bao, D.M. O'Malley, R. Whetten and R. R. Sederoff, *Science* **260**, 672 (1993).
47. D. M. O'Malley, R. Whetten, W. Bao, C. -L. Chen and R. Sederoff, *The Plant Journal* **4**, 751 (1993).
48. T. K. Kirk and R. L. Farrell, *Ann. Rev. Microbiol.* **41**, 465 (1987).
49. T. W. Goodwin and E. I. Mercer, in *Introduction to Plant Biochemistry*, 2nd ed. (Pergamon Press, Oxford, 1985), pp. 55–91.
50. E. Srebotnik, K. A. Jensen Jr and K. Hammel, *Proc. Nat. Acad. Sci. USA* **91**, 12794 (1994).
51. A. J. McCarthy and P. Broda, *J. Gen. Microbiol.* **130**, 2905 (1984).
52. A. J. McCarthy, *FEMS Microbiol. Rev.* **46**, 145 (1987).
53. C. Trigo and A. S. Ball, *Appl. Microbiol. Tech.* **41**, 366 (1994).
54. R. L. Crawford, in *Lignin Biodegradation and Transformation* (Wiley, New York, 1981), pp. 38–60.
55. D. H. Jennings, in *The Physiology of Fungal Nutrition* (Cambridge Univ. Press, Cambridge, 1995).
56. A. J. Durrant, D. A. Wood and R. B. Cain, *J. Gen. Microbiol.* **137**, 751 (1991).
57. H. Galliano, G. Gas, J. L. Seris and A. M. Boudet, *Enzyme and Microbial Technology* **13**, 478 (1991).
58. T. Ishihara, in *Lignin Biodegradation: Microbiology, Chemistry and Potential Applications*, Vol. 2, eds. T. K. Kirk and H. M. Chang (CRC Press, Inc., Boca Raton, Florida, 1980), pp. 17–31.
59. T. Higuchi, *ACS Symposium Series* **399**, 482 (1989).
60. P. Ander and K. -E. Eriksson, *Arch. Microbiol.* **109**, 1 (1976).
61. M. W. Platt, Y. Hadar and I. Chet, *Appl. Microbial Biotech.* **20**, 150 (1984).
62. B. Breuer, M. Bergbauer, H. Teepe and G. Kraepelin, in *Abstracts of the Fourth Int. Mycological Congress*, eds. A. Reisinger and A. Breinsky (Univ. of Regensburg, Regensburg, 1990), p. 224.
63. C. S. Evans, *FEMS Microbiol. Lett.* **27**, 339 (1985).
64. A. Haars and A. Huttermann, *Arch. Microbiol.* **125**, 233 (1980).
65. M. Tien and T. K. Kirk, *Science* **221** 661 (1983).
66. M. H. Gold and M. Alic, *Microbiol. Rev.* **57**, 605 (1993).
67. J. K. Glenn and M. H. Gold, *Arch. Biochem. Biophys.* **243**, 329 (1985).
68. K. E. Hammel, K. A. Jensen, M. D. Mozuch, L. L. Landucci, M. Tien and E. A. Pease, *J. Biol. Chem.* **268**, 12274 (1993).
69. E. De Jong, F. P. De Vries, J. A. Field, R. P. Van De Zwan and J. A. M. De Bont, *Mycological Research* **96**, 1098 (1992).
70. M. H. Gold, H. Wariishi and K. Valli, *ACS Symposium Series* **389**, 127 (1989).

71. L. Marquez, H. Wariishi, H. B. Dunford and M. H. Gold, *J. Biol. Chem.* **263**, 10549 (1988).
72. G. Szklarz and A. Leonowicz, *Phytochem.* **25**, 2537 (1986).
73. P. Ander, *FEMS Microbiol. Rev.* **13**, 297 (1994).
74. F. Archibald and B. Roy, *Appl. Environmental Microbiol.* **58**, 1496 (1992).
75. T. Lundell and A. Hatakka, *FEBS Lett.* **348**, 291 (1994).
76. J. -M. Bollag, k. L. Shuttleworth and D. H. Anderson, *Appl. Environmental Microbiol.* **54**, 3086 (1988).
77. B. Bes, B. Pettersson, H. Lennholm, T. Iversen and K. -E. Eriksson, *Biotech. Appl. Biochem.* **9**, 310 (1987).
78. A. Viterbo, B. Yagen and A. M. Mayer, *Phytochem.* **34**, 47 (1993).
79. I. Marbach, E. Harel and A. M. Mayer, *Phytochem.* **24**, 2559 (1985).
80. A. Viterbo, B. Yagen and A. M. Mayer, *Phytochem.* **32**, 61 (1993).
81. D. Rigling and N. K. Van Alfen, *J. Bacteriol.* **173**, 8000 (1991).
82. D. Rigling and N. K. Van Alfen, *Appl. Environmental Microbiol.* **59**, 3634 (1993).
83. A. J. Clutterbuck, *J. Gen. Microbiol.* **70**, 423 (1972).
84. J. J. Worrall, I. Chet and H. Huttermann, *J. Gen. Microbiol.* **132**, 2527 (1986).
85. J. G. H. Wessels, in *Development of Fruit Bodies in Homobasidiomycetes*, eds. J. G. H. Wessels and F. Minehardt (Springer Verlag, Berlin, 1994), pp. 351–361.
86. A. Haars and A. Huttermann, Process for producing a binder for wood materials. United States Patent No. 4432921 (1984).
87. A. L. Ghindilis, A. Makover, C. G Bauer, F. F. Bier and F. W. Scheller, *Anal. Chim. Acta* **304**, 25 (1995).
88. O. V. Skorobogatko, A. L. Ghindilis, E. N. Troitskaya, A. M. Shuster and A. I. Yarpolov, *Anal. Lett.* **27**, 2997 (1994).
89. A. L. Ghindilis, A. Makower and F. W. Scheller, *Anal. Lett.* **28**, 1 (1995).
90. A. I. Yaropolov, A. N. Kharybin, J. Emneus and L. Gorton, *Anal. Chim. Acta* **308**, 137 (1995).
91. J. Wang, Y. H. Lin, A. V. Eremenko, A. L. Ghindilis and I. N. Kurochkin, *Anal. Lett.* **26**, 197 (1993).
92. A. I. Yaropolov, O. V. Skorobogogatko, S. S. Vartanov and S. D. Varfolomayev, *Appl. Biochem. Biotech.* **49**, 257 (1994).
93. O. Brenna and E. Bianchi, *Biotech. Lett.* **16**, 35 (1994).
94. I. D. Reid and M. G. Paice, *FEMS Microbiol. Rev.* **13**, 369 (1994).
95. A. T. Martinez, S. Camarero, F. Guillen, A. Gutierrez, C. Munoz, E. Varela, M. J. Martinez, J. M. Barrasa, K. Ruel and J. M. Pelayo, *FEMS Microbiol. Rev.* **13**, 265 (1994).
96. L. Roy-Arcand and F. S. Archibald, *Enzyme and Microbial Technology* **13**, 194 (1991),
97. J. A. Bumpus and S. T. Aust, *Appl. Environmental Microbiol.* **53**, 2001 (1987).

98. A. U. Rehman and C. F. Thurston, *J. Gen. Microbiol.* **138**, 1251 (1992).
99. P. M. Coll, C. Tabernero, R. Santamaria and P. Perez, *Appl. Environmental Microbiol.* **59**, 4129 (1993).
100. K. L. Shuttleworth and J. -M. Bollag, *Enzyme Microbial Tech.* **8**, 171 (1986).
101. H. P. Call, Process for modifying, breaking down or bleaching lignin, materials containing lignin or like substances. World Patent Application WO 94/29510 (1994).
102. D. Slomczynski, J. P. Nakas and S. W. Tanebaum, *Appl. Environmental Microbiol.* **61**, 907 (1995).
103. Y. Fukushima and T. K. Kirk, *Appl. Environmental Microbiol.* **61**, 872 (1995).
104. H. Kofujita, T. Ohta, Y. Asada and M. Kuwahara, *Mokuzai Gakkaishi* **37**, 562 (1991).
105. M. -L. Niku-Paavola, E. Karhunen, P. Salola and V. Raunio, *Biochem. J.* **254**, 877 (1988).
106. G. Sannia, M. Giardina, M. Luna, M. Rossi and V. Buonocore, *Biotech. Lett.* **8**, 797 (1986).
107. J. A. Wahleithner, F. Xu, K. M. Brown, S. H. Brown, E. J. Golightly, T. Halkier, S. Kauppinen, A. Pederson and P. Schneider, The identification and characterisation of four laccases from the plant pathogenic fungus *Rhizoctonia solani* (1995), in press.
108. O. R. Aramayo and W. E. Timberlake, *Nucleic Acids Research* **18**, 3415 (1990).
109. Y. M. Kojima, Y. Tsukuda, Y. Kawai, A. Tsukamoto, J. Sugiura, M. Sakaino and Y. Kita, *J. Biol. Chem.* **265**, 15224 (1990).
110. P. R. Williamson, *J. Bacteriol.* **176**, 656 (1994).
111. P. Giardina, R. Cannio, L. Martirani, L. Marzullo, G. Palmieri and G. Sannia, *Appl. Environmental Microbiol.* **61**, 2408 (1995).
112. C. Srinivasan, T. M. D'souza, K. Boominathan and C. A. Reddy, Laccase activity in the white-rot fungus *Phanerochaete chrysosporium*. Abst. 7th Int. Symp. on Microbial Ecology (Santos-Sao Paulo, Brazil, 1995), p. 2–20.34.

CHAPTER 8

SPECTROSCOPY OF CUCUMBER ASCORBATE OXIDASE AND FUNGAL LACCASE

Takeshi Sakurai
National Institute for Molecular Science
Myodaiji, Okazaki, Aichi 444, Japan
E-mail: tsakurai@ims.ac.jp
and
Shinnichiro Suzuki
Faculty of Science, Osaka University,
Machikaneyama 1-16, Toyonaka, Osaka 560, Japan
E-mail: bic@ch.wani.osaka-u.ac.jp

ABSTRACT

Cucumber ascorbate oxidase and fungal laccase give the five absorption bands due to the charge transfers, σS^-(Cys) $\rightarrow Cu^{2+}$, πS^-(Cys) $\rightarrow Cu^{2+}$, N(His) $\rightarrow Cu^{2+}$, S(Met) $\rightarrow Cu^{2+}$, and d–d transitions in the visible region coming from the type 1 copper, and a charge transfer band at 330 nm coming from the EPR undetectable met form of the type 3 coppers (Cu^{2+}–OH^-–Cu^{2+}). The latter band, which is completely lost or partially reduced of its intensity during the procedure to deplete the type 2 copper, is characteristic of the multicopper oxidase family. The type 1 and type 2 coppers are EPR detectable, while type 3 coppers are EPR undetectable because of the antiferromagnetic interaction between them. However, inhibitors such as N_3^-, SCN^-, OCN^-, and F^- bind to the trinuclear site, presumably making the type 3 copper to be EPR detectable and the type 2 copper to be autoreduced. These type 3 coppers bound by exogenous inhibitors give the narrower hyperfine splitting compared to those of the type 2 coppers in the resting multi-copper oxidases. Other spectroscopies such as XANES and resonance Raman are also reviewed.

Contents

1. Introduction

Ascorbate oxidase (L-ascorbate: oxygen oxidoreductase, EC 1.10.3.3) was first extracted from cabbage leaves by Szent-Györgi in 1931. Although ceruloplasmin also shows ascorbate oxidase activity, ascorbate oxidase is found in plants and some bacteria. Higher plants such as cucumber (*Cucumis sativus*), green zucchini squash (*Cucurbita pepo medullosa*), and crook-neck squash (*C. pepo condensa*) are the good sources for enzyme purification.[1] The texts for dietetics write that vitamin C might be decomposed by dietary cucumber.

The biological role of ascorbate oxidase has not been known. However, the content of ascorbate oxidase is high in growing parts and is usually isolated from the early harvest of cucumber and zucchini peeling in the early summer. The seedling is also suitable for the isolation of the enzyme. Ascorbate oxidase is utilized to remove interfering ascorbate from analytical system. Although the cloning of cucumber ascorbate oxidase has already been performed,[2] the enzyme with activity has not been obtained yet.

Laccase (Benzenediol: oxygen oxidoreductase, EC 1.10.3.2) is a type of polyphenol oxidase which has been isolated from different sources such as higher plants and fungi. The enzyme was first discovered from latex of lacquer tree by Yoshida in 1883 and was subsequently demonstrated as a fungal enzyme in 1896.[3,4] A laccase-type phenol oxidase from a bacterium,

Azospirillum lipoferum has been recently reported.[5] The main sources for fungal laccase are the Ascomycetes and the Basidomycetes (see Tables 7.1 and 7.2). These enzymes are extracellular acidic glycoproteins. The typical laccase is a 60–80 kDa molecule of which 15 – 25 % is carbohydrate, while tree laccases have the larger content (45 %) of carbohydrate. Moreover, many fungal laccases show heterogeneity after purification. For example, on sodium dodecyl sulfate-PAGE *Monocillium indicum* laccase (100 kDa) is resolved into three proteins of 72, 56, and 24 kDa.[6] The ratio of type 1 Cu:type 2 Cu:type 3 Cu in laccases from *Coriolus consors, Neurospora crassa, Podospora anserina*, and *Polyporus versicolor* is 1:1:2 per protein molecule. The extensive spectroscopic studies on fungal laccases have not been studied in detail except for the four enzymes above.

2. Cucumber Ascorbate Oxidase

2.1. *Absorption and Circular Dichroism (CD) Spectra of Native and Type 2 Copper-Depleted (T2D) Cucumber Ascorbate Oxidases*

Absorption and CD spectra of native cucumber ascorbate oxidase are shown in Fig. 1. The resting enzyme exhibits two absorption maxima at 607 nm and 760 nm with a shoulder at 330 nm.[7] The extinction coefficient of the most peculiar absorption band at 607 nm is 10400 M^{-1} cm^{-1} per dimer. The extinction coefficients of the 330- and 760-nm bands are 8600 and 4800 M^{-1} cm^{-1}, respectively. The CD spectrum exhibits the absorption maxima at 310 ($\Delta\varepsilon = -1.1$ M^{-1} cm^{-1}), 413 (+1.1), 470 (−3.8), 593 (+8.3), and 745 nm (−15), of which the 593-nm band has a positive shoulder at ca. 540 nm.[8] These absorption and CD spectra are nearly identical with those of the zucchini enzyme[9] but are slightly different from *Acremonium* ascorbate oxidase, of which the 330-nm band is more intense and the corresponding negative CD band is not observed.[10] Nevertheless, whole spectral feature of ascorbate oxidase is apparently similar regardless of its source. In the single crystal polarized absorption spectrum for green zucchini ascorbate oxidase, the feature at around 330 nm changed depending on the orientation, although the feature at around 600 nm did not change prominently.[11]

The molar absorptivity at 607 nm is constant at pH 4–8. However, it gradually decreases with increasing pH, being accompanied by the decrease in the total amount of the EPR detectable Cu^{2+}.[12]

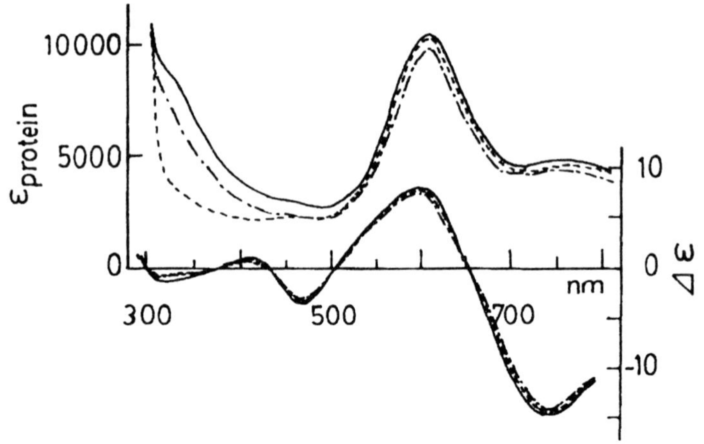

Fig. 1. Absorption and CD spectra of native cucumber ascorbate oxidase (——), the type 2 copper-depleted derivative (– – ––), and hydrogen peroxide-treated T2D (— - —).

The CD spectral pattern in the visible region is similar to those of many blue copper proteins. Since the ligand groups around the type 1 Cu in ascorbate oxidase are the same with those of the typical blue copper proteins, ascorbate oxidase shows the absorption and CD spectral features similar to those of simple blue copper proteins. Nevertheless the curve analysis has not been successfully performed yet, because the contribution from the type 2 and type 3 Cu's has not been identified yet.

In analogy with the CD spectra of small blue copper proteins, the band at around 750 nm originates in the *d–d* transitions, whose absorption intensities are unusually strong. Two positive bands at 596 nm and ca. 550 nm come from the charge transfers, $\pi S^-(Cys) \rightarrow Cu^{2+}$ and $\sigma S^-(Cys) \rightarrow Cu^{2+}$, respectively.[13] The former with π character exhibits the higher ellipticity compared with the latter with σ character. Blue copper proteins exhibit an absorption band and two CD bands at around 450 nm. They are assumed to come from the charge transfers, $N(His) \rightarrow Cu^{2+}$ and $S(Met) \rightarrow Cu^{2+}$. However, stellacyanin exhibits only one CD band because Met is substituted by Gln which is more electron-donating.[14] The fact that the weak absorption at ca. 450 nm is composed of two oppositely signed CD bands, is in accordance with 1Cys-1Met-2His coordination of the type 1 Cu site as represented by the X-ray crystal structure analysis.[15,16]

The shoulder at ca. 330 nm is characteristic of multi-copper oxidases. The presence of this band differentiates multi-copper oxidase from blue copper protein. This band is supposed to come from the coupled type 3 Cu's in the EPR undetectable met form. In the resting form of the enzymes the type 3 Cu's are antiferromagnetically coupled through a bridging group, OH^- (or O^{2-}).[15,16]

By the selective removal of the type 2 Cu using hexacyanoferrate(II), EDTA, and dimethylglyoxime, the band at 330 nm disappears (Fig. 1).[8] When this T2D ascorbate oxidase is treated with hydrogen peroxide, the 330-nm band is restored considerably. The presence or absence of the 330-nm band in T2D laccase has been a matter of discussion.[17,18] The confusion would have been brought because the oxidation state of the type 3 Cu may change in preparations by each researcher. According to our experience, the 330-nm band of ascorbate oxidase and *Rhus* laccase disappears during the procedure to deplete the type 2 Cu, making the absence of this band a marker of the successful type 2 Cu depletion.[8] However, it is also true that we occasionally obtained T2D ascorbate oxidase having the rather strong 330-nm band in spite of the absence of the type 2 Cu EPR signal. Therefore, the type 3 Cu is considered to exist as a mixture of the reduced form (Cu^+ Cu^+) and the EPR undetectable met form (Cu^{2+}–OH^-–Cu^{2+}) after the selective removal of the type 2 Cu.

2.2. *EPR Spectra of Native and T2D Ascorbate Oxidases*

The EPR spectrum of native cucumber ascorbate oxidase at 77 K is shown in Fig. 2. Of the three types of coppers the type 1 and type 2 Cu's are observable.[12,19,20] The spectral feature of plant ascorbate oxidase is similar regardless of its source. The spin Hamiltonian parameters of the type 1 Cu are $g_{\|} = 2.230$, $g_{\perp} = 2.050$, $A_{\|} = 6.2$ mT (6.6×10^{-3} cm^{-1}), and $A_{\perp} = 1.8$ mT and those for the type 2 Cu are $g_{\|} = 2.240$, $g_{\perp} = 2.050$, $A_{\|} = 19.8$ mT (20.5×10^{-3} cm^{-1}), and $A_{\perp} = 2.0$ mT at 77 K.

The mystery about the EPR spectrum of ascorbate oxidase is the amounts of the type 1 Cu and the type 2 Cu detectable by EPR. As shown in Fig. 2, the satisfactory simulation cannot be performed for 2 type 1 Cu and 2 type 2 Cu but can be performed for 3 type 1 Cu and 1 type 2 Cu.[12,21] Since the total amount of the EPR detectable Cu is about 4, ascorbate oxidase was once believed to have 3 type 1 Cu, 1 type 2 Cu, and 4 type 3 Cu per dimer. However, the recent X-ray crystallography[15,16] showed that

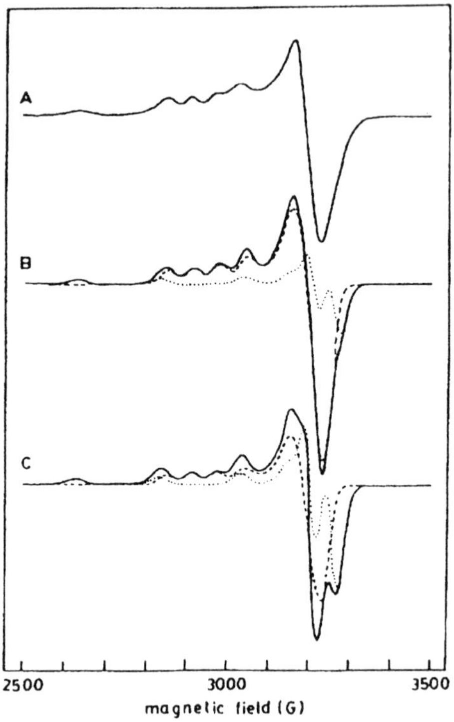

Fig. 2. Experimental (A) and simulated (B for three type 1 Cu's and one type 2 Cu, (C) for two type 1 Cu's and two type 2 Cu's) spectra of cucumber ascorbate oxidase at 9.21 GHz and 77 K. (Taken from Ref. 12. With permission.)

zucchini ascorbate oxidase is composed of the identical 2 subunits, each of which contains 1 type 1 Cu, 1 type 2 Cu, and 2 type 3 Cu. As for the amount of the EPR detectable Cu, it has been repeatedly reported to be 4, while it is also true that it has dispersed from 3 to 4 per dimer according to our experiments.[22] In the estimations using very fresh enzyme, the total amount of the EPR detectable Cu was apparently less than 4, although zucchini ascorbate oxidase obtained by dissolving a crystal gave 4.[23] In the end the discrepancy might show a limit for the double integration method. Although we usually use an external standard to compensate the difference in tuning condition, error of ca. 5 % should be anticipated in determining the amount of the EPR detectable Cu^{2+}.

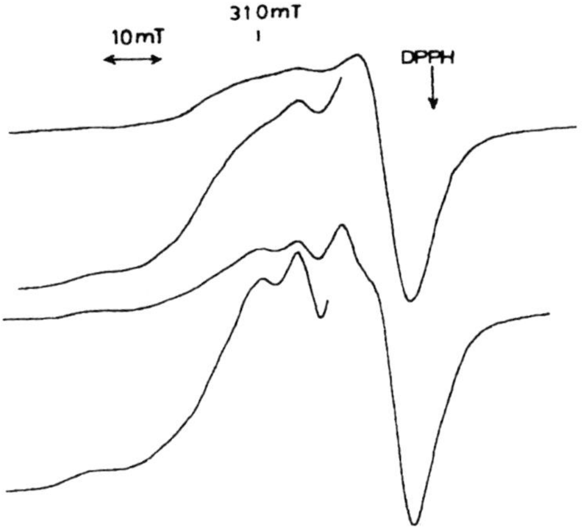

Fig. 3. Room temperature EPR spectra of two forms of cucumber ascorbate oxidase. In the upper spectrum 70 % of the type 2 Cu are EPR undetectable (the total amount of the EPR detectable Cu is 2.6), while in the lower spectrum the type 2 Cu is almost fully EPR detectable (the total amount of the EPR detectable Cu is 3.8).

Contrary to the confusion above, both the type 1 Cu and type 2 Cu in *Acremonium* ascorbate oxidase are fully EPR detectable.[10] Further, we recently found that only ca. 30 % of the type 2 Cu both in cucumber ascorbate oxidase and in *Rhus* laccase are ESR detectable at room temperature, although the type 2 Cu is almost fully EPR detectable at 77 K.[22,24] The room temperature EPR spectrum of ascorbate oxidase ($g_{\|} = 2.21$, $A_{\|} = 5.7$ mT for the type 1 Cu and $g_{\|} = 2.26$, $A_{\|} = 14.8$ mT for the type 2 Cu) apparently differs from that at 77 K (Fig. 3). This fact represènts that the type 2 Cu is in equilibrium between cupric and cuprous forms at room temperature but is mainly in the cupric form in a frozen solution. Alternatively, the trinuclear center might have two forms in which the type 2 Cu becomes both EPR detectable and undetectable (very broadened) because of the delicate difference in the antiferromagnetic interaction between the type 3 Cu's depending on temperature. Since the spin Hamiltonian parameters of the type 1 and type 2 Cu's differ at room temperature and at 77 K, it appears that a change is induced on the protein structure by freezing. This is not apparently the effect of the change in pH on freezing. In line with this, fresh ceruloplasmin isolated promptly by affinity chromatography lacks the type 2 Cu signal.[25]

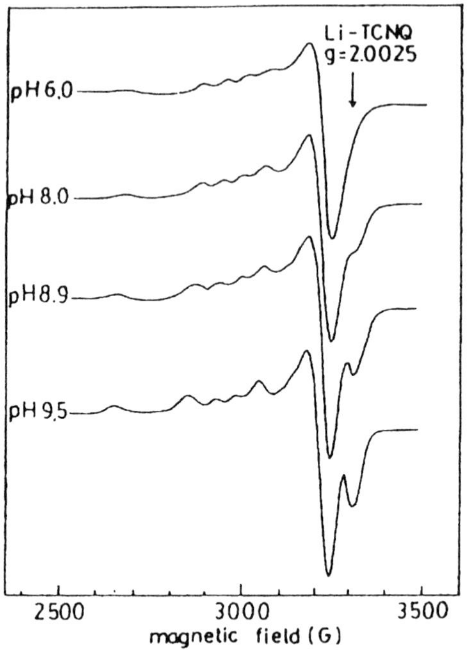

Fig. 4. EPR spectra of cucumber ascorbate oxidase at 77 K in different pH buffers (Taken from Ref. 12. With permission.)

As shown in Fig. 4,[12] the type 2 Cu signal becomes prominent with increasing pH because of a change on the protein structure or the deprotonation from the coordinated water. Although the total amount of the EPR detectable Cu decreases with increasing pH, it is not certain whether the type 1 Cu or type 2 Cu is autoreduced or both. The clear type 2 Cu EPR signal is shown by changing the type 1 Cu with Hg.[26] The ESR signals of the type 1 Cu and type 2 Cu are not saturated at room temperature.[22] In contrast, the half saturation power ($P_{1/2}$) of both types of coppers are about 10 mW at 10 K.[12]

2.3. *Effect of Hexacyanoferrate(II) on the Selective Removal of the Type 2 Copper from Cucumber Ascorbate Oxidase*

In order to selectively deplete the type 2 Cu, ascorbate oxidase is anaerobically reacted with hexacyanoferrate(II), followed by the treatments with EDTA and dimethylglyoxime. During this procedure, the absorption at

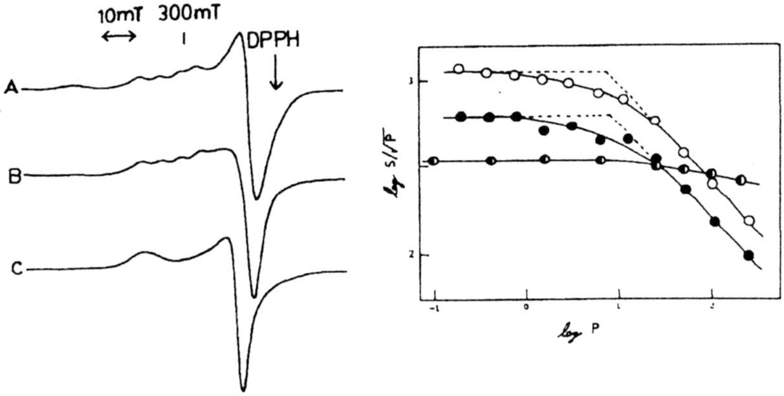

Fig. 5. The effect of hexacyanoferrate(II) on the EPR spectra of ascorbate oxidase. Left: the EPR spectra of the native enzyme (A), hexacyanoferrate(II)-treated ascorbate oxidase (B), and ascorbate-reduced as (B) (C). The amounts of the EPR active spins are 3.7, 3.5, and 1.8, respectively. Right: the power saturation curves of the type 1 (open circle), and the type 2 coppers (closed circle) of the native enzyme and the ascorbate-reduced hexacyanoferrate(II)-treated enzyme (half closed circle) at 4 K. S and P denote signal intensity and microwave power (mW), respectively. (Taken from Ref. 29. With permission.)

607 nm characteristic to the type 1 Cu is decreased. In the case of *Rhus* laccase, the blue color is completely bleached probably because the redox potential of the type 1 Cu in laccase[27] is more positive than that in ascorbate oxidase.[7] In the corresponding EPR spectrum, the type 2 Cu signal disappears and a new signal superimposes on the type 1 Cu signal. The anaerobic reduction of this hexacyanoferrate(II)-treated ascorbate oxidase with a reducing agent such as ascorbate gives a signal with no hyperfine structure in the parallel region ($g_{\parallel} = 2.28$, $g_{\perp} = 2.08$) (Fig. 5).[28,29] This signal is considerably sharp as derived from a transition metal ion, and is not saturated even by 200 mW microwave at 4 K in contrast to the type 1 and 2 Cu signals. The corresponding absorption spectrum shows a broad band at 400–500 nm. This novel species is assumed to come from Cu-CN-Fe with the exchange narrowing. The radical center might be present on the Fe ion rather than the Cu ion. *Rhus* laccase also shows quite similar behavior and a small molecule study supports the formation of the Cu-CN-Fe species.[30] Furthermore, the similar behavior is also observed for the T2D ascorbate oxidase, indicating that the novel signal is not derived from the depleted Cu but from a sort of adduct involving the type 3 Cu and hexacyanoferrate(II).

2.4. *Effect of Exogenous Small Inhibitors on the Absorption, CD and EPR Spectra of Cucumber Ascorbate Oxidase*

Small anions such as azide, thiocyanate, cyanate, and fluoride ions have been used to reveal the structure of the copper centers in multi-copper oxidases and proteins containing only the type 2 or type 3 Cu by modulating the Cu-EPR signal or by making the EPR undetectable Cu to be EPR detectable. These anions are effective inhibitors of multi-copper oxidases that interfere with the access of dioxygen to the trinuclear site.

Upon treatment of ascorbate oxidase with N_3^-, the band at 330 nm disappears and broad bands appear at around ca. 480 nm and 420 nm (Fig. 6).[8,24,31,32] The increase in the absorption with increasing the amount of N_3^- is biphasic,[24] indicating that 2 N_3^- bind to the trinuclear site, although three binding modes of N_3^- to the trinuclear site are proposed for zucchini ascorbate oxidase.[33] In the case of *Rhus* laccase, the effect of N_3^- is also biphasic. The absorptions at ca. 480 nm and 420 nm increase

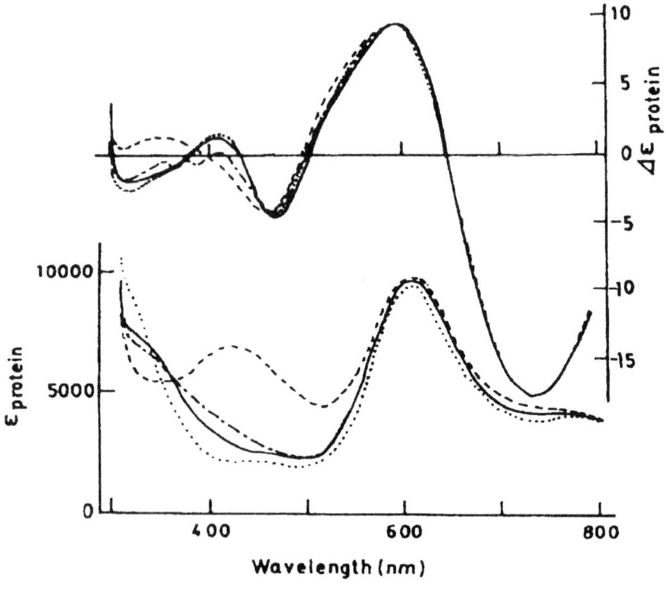

Fig. 6. Absorption and CD spectra of native ascorbate oxidase (——), N_3^- (x370) -ascorbate oxidase (– – – – –), SCN^- (×370) -ascorbate oxidase (— - —), and F^- (×1000)- ascorbate oxidase (· · ·). (Taken from Ref. 32. With permission.)

by the high affinity N_3^- and the low affinity N_3^-, respectively.[34] In the corresponding CD spectrum of ascorbate oxidase, the feature in the range 300–450 nm prominently changes by the action of N_3^-, although that at the longer wavelength region is scarcely modified. Contrary to this, the CD spectrum of *Rhus* laccase treated with large excess of N_3^- is significantly modified over the wide range between 300 and 800 nm, indicating that the binding of 2 N_3^- to the trinuclear site modifies the structure of the remote type 1 Cu site [24,32] as in the case of ceruloplasmin.[35] The X-ray crystal structure of zucchini ascorbate oxidase showed $2N_3^-$ bind to one of type 3 Cu's.[36,37] The elimination of the bridging group (OH^- or O^{2-}) between the type 3 Cu's makes the N_3^- -bound type 3 Cu to be EPR detectable.

When T2D ascorbate oxidase is treated with N_3^-, a charge transfer band appears at about 450 nm irrespective of treating the derivative with hydrogen peroxide.[8,38] This fact might also support that the type 3 Cu site in the T2D enzyme exists as a mixture of the oxidized and the reduced forms.

SCN^- and F^- perturb the absorption spectrum in the near UV region. The effect of these inhibitors on cucumber ascorbate oxidase is basically the same as that on *Rhus* laccase.[31,32] The absorption at ca. 400 nm slightly increases and that at 330 nm decreases by the action of SCN^-. On the other hand, the band at 330 nm slightly blue-shifts with the concomitant increase in its intensity by the action of F^-.

EPR spectra of ascorbate oxidase treated with small exogenous anions are informative of the structure of the trinuclear site. When N_3^- is acted on native ascorbate oxidase, the type 2 Cu signal superficially changes to give a narrower $A_{\|}$ value of 14.0 mT (Fig. 7).[31,32] In the comparable study of *Rhus* laccase, N_3^- has been believed to bind to the type 2 Cu and also to bridge between the type 2 Cu and type 3 Cu.[39] In the case of zucchini ascorbate oxidase, three binding forms are proposed from the low temperature MCD study: one is that an azide ion bridges between the type 2 Cu and type 3 Cu and the others are those that an azide anion binds only to the type 2 Cu or type 3 Cu.[33] However, the only one N_3^- -Cu species with $g_{\|} = 2.247$ and $A_{\|} = 14.0$ mT has been observed, although the absorption change shows the biphasic process for the binding of N_3^-. Even taking into account the difference in the binding constant of N_3^- at room temperature and 77 K, only one N_3^- species has been identified. Contrary to this, two Cu species are detected from the N_3^- study of *Rhus* laccase both at room temperature

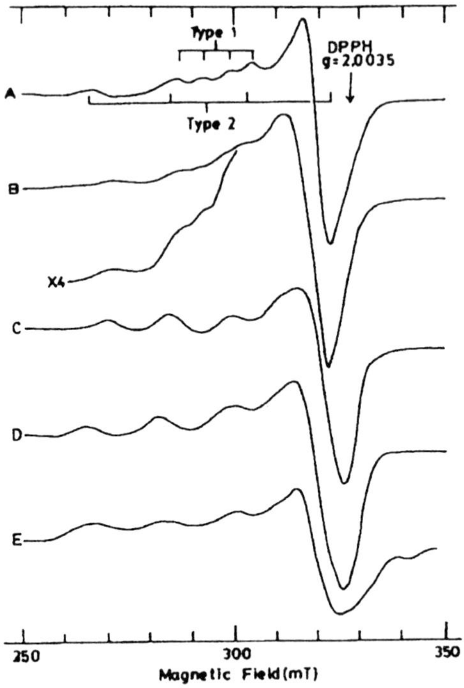

Fig. 7. EPR spectra of native ascorbate oxidase (A), N_3^- (×370) -ascorbate oxidase (B) (The EPR signals of the type 1 Cu and N_3^- -type 3 Cu are detected), B anaerobically reacted with L-ascorbate (C), SCN^- (×370) -ascorbate oxidase anaerobically reacted with a slight excess of L-ascorbate (D), F^- (×1000) -ascorbate oxidase anaerobically reacted with a slight excess of L-ascorbate (E). (Taken from Ref. 32. With permission.)

($g_\parallel = 2.28$, $A_\parallel = 8.1$ mT and $g_\parallel = 2.26$, $A_\parallel = 13.8$ mT) and at 77 K ($g_\parallel = 2.272$, $A_\parallel = 9.2$ mT and $g_\parallel = 2.248$, $A_\parallel = 13.9$ mT).[24,31,32] (The parameters for the former species seem to be near to those for the highly hindered type 3 Cu species detected as the intermediates during the reoxidation process of laccase and the Cu_B in cytochrome *c* oxidase.[40]) Considering from the X-ray crystallography for N_3^--zucchini ascorbate oxidase,[36,37] these signals are not supposed to come from the N_3^--type 2 Cu but from the N_3^--type 3 Cu. SCN^- also gives the EPR signal with narrow hyperfine as shown in Fig. 7. When excess F^- is acted on ascorbate oxidase, each of the four components parallel to magnetic field is split in the ratio of 1:2:1, indicating that two F^- ions bind to one of type 3 Cu's (After an anaerobic reduction Fig. 7E is obtained).

The parameters of these inhibitor-bound type 3 Cu signals are tabulated in Table 1. The Blumberg-Peisach plots ($g_{||}$ vs. $A_{||}$ correlation) indicate that the EPR signals which emerged by acting inhibitors on multi-copper oxidases are at the intermediate regions between those for the type 1 Cu and type 2 Cu in resting multi-copper oxidases. Some of the type 3 Cu signals are in the signal region of the type 2 Cu (non-blue copper) contained in proteins such as superoxide-dismutase (SOD) and nitrite reductase. Nevertheless, they have the narrower hyperfine splittings than those of the native type 2 Cu's in multi-copper oxidases, indicating that the structure around the type 3 Cu site is highly tetrahedrally hindered.

Table 1. Spin Hamiltonian parameters of Cu-species in cucumber ascorbate oxidase.

Cu-Species	gz	Az (mT)
Type 1 Cu in native ascorbate oxidase	2.230	6.2
Type 2 Cu in native ascorbate oxidase	2.240	19.8
Type 3 Cu in N_3^- -treated ascorbate oxidase	2.247	14.0
Type 3 Cu in SCN^- -treated ascorbate oxidase	2.259	16.7
Type 3 Cu in F^- -treated ascorbate oxidase	2.248	17.0

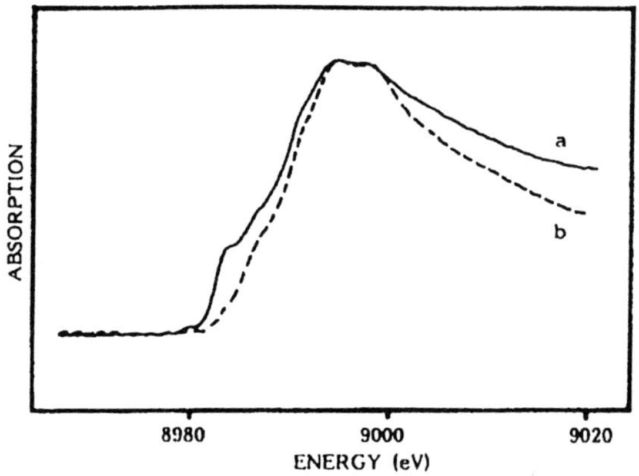

Fig. 8. Normalized Cu edge spectra for T2D ascorbate oxidase (a) and its hydrogen peroxide (x28)-treated derivative (b). (Taken from Ref. 41. With permission.)

2.5. *X-ray Absorption of Cucumber Ascorbate Oxidase*

The K-edge spectra of T2D ascorbate oxidase and its H_2O_2-treated derivative have been measured in order to reveal the oxidation state of the type 3 Cu.[41] A very weak pre-edge due to the $1s \rightarrow 3d$ transition, which is characteristic of Cu(II) species, is observed at 8980 eV (Fig. 8). The distinct transition at approximately 8984 eV for T2D ascorbate oxidase shows the presence of Cu(I) species $1s \rightarrow 4s$ transition). The peculiar transition for the cuprous ion disappears by treating T2D ascorbate oxidase with hydrogen peroxide, indicating that the reduced type 3 Cu's are oxidized. However, the detailed XANES and EXAFS studies have not been performed for ascorbate oxidase.

2.6. *Vibration and Resonance Raman Spectra of Cucumber Ascorbate Oxidase*

We have measured the resonance Raman spectra of cucumber ascorbate oxidase, and found it was identical with that of zucchini.[24] Resonance Raman spectra of multi-copper oxidases are more complex than those of simple blue copper proteins.[42] However, since the resonance Raman lines of blue copper proteins are now assigned in detail, the precise inspection of the spectra might give the missing structural information of the type 1 Cu site in multi-copper oxidases and also of other types of coppers. The successful detection of the resonance Raman spectra irradiating the near UV region has not been reported yet. No detailed information for the trinuclear site is available from resonance Raman and the related vibrational spectroscopies. The ν_{as} band of N_3^- bound to one of type 3 coppers was observed at 2054 cm^{-1} by FT-IR spectroscopy, suggesting the end-on coordination.[22]

3. Fungal Laccase

3.1. *Absorption, CD and Magnetic Circular Dichroism (MCD) Spectra of Fungal Native and T2D Laccases*

The absorption and CD spectra of fungal laccase (*Polyporus versicolor*) are compared with those of tree laccase (*Rhus vernicifera*) in Fig. 9.[43] Those of *C. consors* native and T2D laccases are also shown in Fig. 10.[44] Absorption and CD spectral data of several fungal laccases are tabulated in Table 2. With respect to all fungal laccases the blue bands are observed at 595–610 nm and shoulders are around 330 and 750 nm. The blue

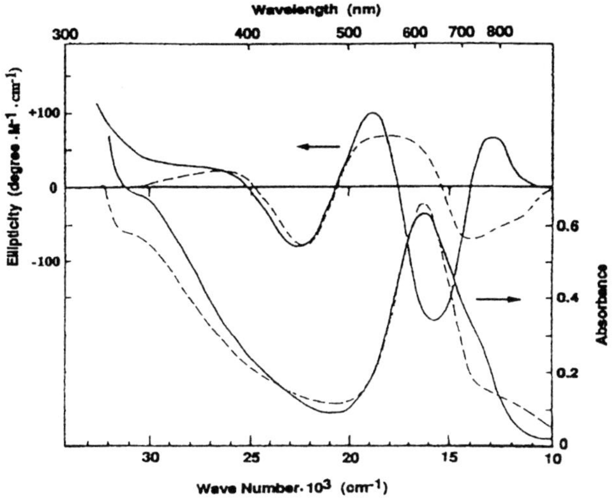

Fig. 9. Absorption (lower curves) and CD (upper curves) spectra of fungal (*P. versicolor*, pH 6.0) (——) and tree (*R. vernicifera*, pH 7.5) (– – ––) laccases. (Taken from Ref. 43. With permission.)

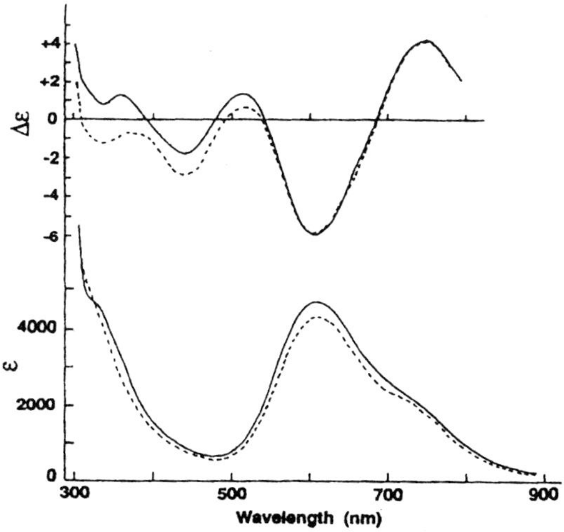

Fig. 10. Absorption (lower curves) and CD (upper curves) spectra of native (——) and T2D (– – ––) laccases from *C. consors* at pH 6.0. (Taken from Ref. 44. With permission).

Table 2. Electronic absorption maxima and CD absorption extrema of fungal and tree laccases at room temperature.

Electronic absorption spectra fungal laccase (kDa)	Blue band, nm (ε: M^{-1} cm^{-1})	Ref.
Coriolus consors (58)	608 (4,700)	44
Neurospora crassa (65)	595 (3,300)	45
Podospora anserina II and III (70 and 80)	610 (3,700)	46
Polyporus versicolor (65)	610 (4,900)	43
Tree laccase		
Rhus vernicifera (lacquer tree)	615 (5,700)	43
sycamore (*Acer pseudoplatanus* L.) cell	615	48

CD spectra fungal laccase	λ_{ext}, nm (sign)	
Coriolus consors	360(+) 440(−) 510(+) 605(−) 750(+)	44
Neurospora crassa	306(+) 370(+) 426(−) (+)*¹ (+)*¹ 690(−)	45
Podospora anserina II–III	307(−) 357(+) 441(−) 556(+) (+)*¹ 725(−)	46
Polyporus versicolor	ca. 360(+) 442(−) 526(+) 625(−) 769(+) 909(−) 1380(+)	47
Tree laccase		
Rhus vernicifera	ca. 380(+) 435(−) 526(+)*² 620(+) 769(−) 855(−) 1053(−) 1670(+)	47

*¹ The positive bands are obtained around 600 nm by fitting the experimental spectra to a number of Gaussian components (see text).
*² Ref. 43.

bands shift about 5 nm to the shorter wavelength than those of tree laccase. The shoulder bands around 750 nm are more peculiar compared with that of *R.vernicifera* laccase. The intense visible absorption bands due to the type 1 Cu in *P. versicolor* laccase disappear upon specifically reducing the Cu site at pH > 6.5.[49] In general, the relatively intense shoulders around 330 nm assigned to the type 3 Cu site do not have any CD activity. Considering the remarkable similarity of the visible CD spectra of proteins containing the type 1 Cu site, it is important to note that the spectra of

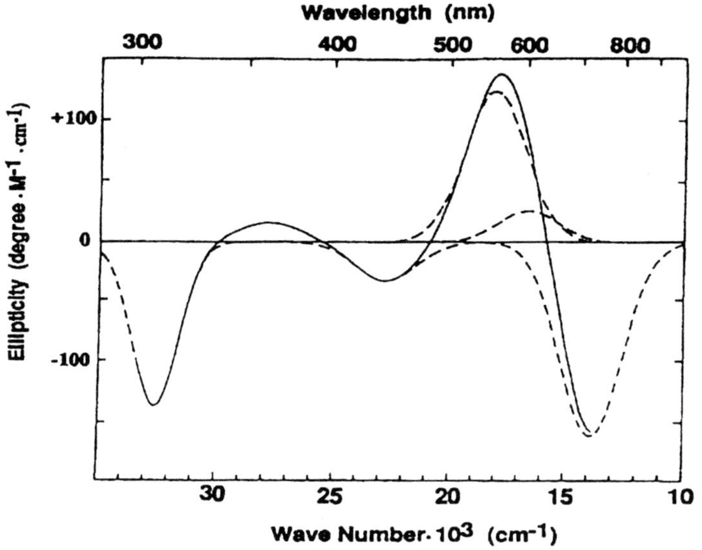

Fig. 11. CD spectrum (——) of *P. anserina* laccase II or III at pH 7.0 and simulated Gaussian bands (– – –) obtained from the experimental curve. (Ref. 46. With permission.)

two fungal enzymes, *P. versicolor* and *C. consors* are obviously different from those of the other type 1 Cu-containing proteins. There are usually five CD bands in the region 350–800 nm and most of proteins containing the type 1 Cu show the positive and negative CD bands near 350–420 nm and 430–470 nm, respectively.[50] Tree laccase gives at least two positive and one negative CD bands (+, +, −) between 500 nm and 800 nm as observed for many copper proteins containing the type 1 Cu, while two fungal laccases (*C. consors* and *P. versicolor*) display three CD bands with signs (+, −, +).[51] On the other hand, the CD spectra of *N. crassa* and *P. anserina* laccases in this region show three bands with signs (+, +, −) as in many proteins containing the type 1 Cu, although the positive bands occur at 532 and 575 nm (*N. crassa*) and 606 nm (*P. anserina*) (Fig. 11) by the simulation analysis of the experimental spectra.

According to the Kuhn anisotropy factor evaluated from the CD and absorption bands of *R. versicolor* laccase,[47] reasonable assignments for the bands at 442 and 625 nm are πN(His) $\rightarrow$ Cu (II) and σS^-(Cys) $\rightarrow$ Cu (II) charge-transfer transitions, respectively. The near-infrared bands at

909 and 1380 nm are attributed to *d–d* transitions of the type 1 Cu (II). The 526- and 769-nm bands are assigned to the charge-transfer transitions σS(Met) $\rightarrow$ Cu (II) and πS^- (Cys) $\rightarrow$ Cu (II), respectively, but the two corresponding transitions of a blue copper protein, plastocyanin have been recently assigned to pseudo-S^- (Cys) $\rightarrow$ Cu (II) and *d–d*, respectively.[13]

R. versicolor laccase exhibits a room-temperature MCD pattern unique among the copper proteins containing the type 1 Cu. In the range 500–800 nm, a negative peak is observed at 660 nm ($\Delta\varepsilon = -0.40$) and a positive peak at 560 nm ($\Delta\varepsilon = +0.10$), while the MCD spectra of tree laccase, ceruloplasmin, and blue small copper proteins display the only negative maxima near 700 nm.[47]

The observation of the absorption and CD spectra of the types 2 and 3 Cu in oxidized laccase is generally difficult except the 330-nm region because of the type 1 Cu. However, under the reduction of the type 1 Cu of *P. versicolor* laccase at high pH, the types 2 and 3 Cu display two shoulders near 330 nm and 600 nm. In the CD spectrum three peaks and a shoulder are resolved at 444(−), 513(+), and 667(−) nm and 351(+) nm, respectively.[47]

The type 2 Cu can be selectively removed from laccase by the cyanide treatment.[52] The total Cu content of *C. consors* T2D laccase in Fig. 10 was reported to be 2.8 per protein molecule and about one Type 2 Cu was removed from the native enzyme (total Cu 4.0); the EPR spectrum supports the depletion of the type 2 Cu (*vide infra*).[44] The electronic absorption and CD spectra of T2D laccase from *P. versicolor* are also similar to those of the native enzyme except the ultraviolet region of 300–400 nm[53] like *C. consors* native and T2D laccases: the features in the visible region are hardly changed by removing the type 2 Cu (Fig. 10). In addition, the intensity of the 330-nm shoulder band is not drastically decreased in contrast to that of *R. vernicifera* tree laccase which disappears completely by the reduction of the type 3 Cu during the depletion of the type 2 Cu.[13,53] Therefore, the type 3 Cu sites of these fungal laccases seem to maintain the oxidized form or to be reduced partially. The prominent alteration on the CD feature in the region 300–400 nm might occur from that the type 2 Cu, together with the type 3 Cu in the oxidized form, has originally contributed to this region and/or the steric structure around the type 3 Cu site is significantly modified by the removal of the type 2 Cu adjacent to the type 3 Cu's.

Finally the difference of optical behaviors (> 550 nm) between *P. versicolor* and *R. vernicifera* laccases and the reduction of the type 1 Cu in the former enzyme at high pH are probably associated with the more positive redox potential of the fungal enzyme (+785 mV at pH 5.5)[54] than that of the tree enzyme (+394 mV at pH 7.5). According to the structurally based amino-acid sequence alignment of fungal laccases, ascorbate oxidases, and small blue copper proteins, there is very often a Met as the fourth ligand of the type 1 Cu in *Aspergillus nidulans* laccase, ascorbate oxidases, and small blue copper protein, as demonstrated in the X-ray structures of type 1 Cu sites in zucchini ascorbate oxidase[37] and several blue copper proteins.[55] However, a Leu or Phe which cannot serve as a ligand, is substituted for the Met residue in four laccases from *Phlebia radiata*, *Coriolus hirsutus*, *Neurospora crassa*, and *Cryphonectria parasitica*.[37] This fourth non-Met ligand probably relates to the spectral and electrochemical properties of these fungal laccases, e.g., the CD feature and the redox potential of the type 1 Cu.

3.2. *EPR of Fungal Native and T2D Laccases*

EPR spectra of all laccases reported so far are composed of superimposed signals from two Cu(II) ions in different coordination environments (types 1 and 2 Cu), which originate from equal amounts of two coppers. Only ca. 50 % of the total Cu ions are detected by EPR and the two remaining Cu ions (type 3 Cu) are EPR silent. Figure 12 shows typical spectra of laccases from *P. anserina* and *C. consors* at 77 K. According to a simulation of the EPR spectrum of the former laccase (Fig. 12A), the signal is apparently separated into two signals of Cu(II) species, the types 1 and 2 Cu which have hyperfine constants (Az or $A_{||}$), 4–9 mT and 12–20 mT, respectively. The EPR parameters of both types of coppers in four fungal laccases and two tree laccases are shown in Table 3. The $A_{||}$ values of the type 1 Cu's in four fungal laccases are similar to that of *R. succedanea* laccase but are much greater than that of *R. vernicifera* laccase. Anaerobic rapid-freeze reduction of the type 1 Cu in *P. versicolor* laccase B by ascorbic acid at pH 5–6 is so fast that the contribution from this Cu is not observed in the EPR spectrum even at 10 ms.[60] The low-field part of the spectra demonstrates that the type 2 Cu signal rapidly disappears and an intermediate signal corresponding to 0.3 Cu^{2+} per enzyme molecule is formed at 30 ms. After about 150 ms most of the type 2 Cu signal with superhyperfine lines in the

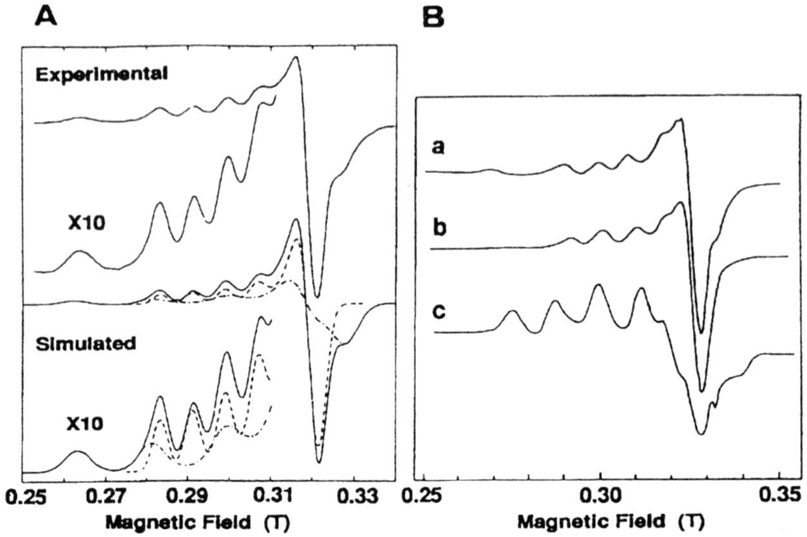

Fig. 12. (A) Experimental and simulated X-band EPR spectra of *P. anserina* laccase at pH 6.0 and 77 K (– – – – –, type 1 Cu; — - — , type 2 Cu); (B) X-band EPR spectra of native (a) and T2D (b) laccases from *C. consors* at pH 6.0 and 77 K, and the T2D enzyme (c) at pH 9.5 and 77 K. (A: Taken from Ref. 46; B: Taken from Ref. 44. With permission.)

region of $g_\perp$ is recovered and the intermediate signal decreases to less than half of its intensity at 30 ms.

Selective removal of the type 2 Cu from *C. consors* shows the EPR signal of the typical type 1 Cu (Fig. 12B(b)). In this T2D enzyme, the amount of the EPR detectable Cu content decreased from 2.0 to 1.0. The EPR parameters of the signal are quite similar to those of the type 1 Cu in the native enzyme as well as in the case of *P. versicolor* laccase, as presented in Table 3. In the EPR spectrum of ^{63}Cu-enriched T2D laccase from *P. versicolor*, the type 1 Cu signal presents the well-resolved ligand hyperfine splitting structure in the $g_\perp$ region, which is simulated on the basis of contributions from two N(His) and two H nuclei: the H atoms are assumed to be bound to the β-carbon of the covalently bonded Cys ligand.[52,59] It also requires a substantial delocalization of an unpaired electron density onto the S(Cys) atom. On the basis of the second integration of the signal, it is reported that 40 % of the type 3 Cu pairs are in a partially reduced state which is remarkably stable, and neither is readily oxidized by H_2O_2

Table 3. EPR parameters of Cu(II) signals in fungal and tree laccases at 77 K.

Source		$g_{\|\|}$	$g_{\perp}$	$A_{\|\|}$ (mT)	Ref.
Native fungal laccase					
Coliolus consors	T1	2.19	2.05	8.8	44
	T2	2.25	2.05	18.2	
Neurospora crassa	T1	2.19	2.04	9	45
	T2	2.23	2.04	18.5	
Podospora anserina	T1	2.209	2.042	7.8	46
	T2	2.246	2.046	16.8	
Polyporus versicolor	T1	2.190	2.042	8.8	56
	T2	2.262	2.036	18.5	
Native tree laccase					
Rhus vernicifera	T1	2.298	2.047	4.0	57
	T2	2.237	2.053	19.2	
Rhus succedanea	T1	2.204	2.045	7.4	58
	T2	2.217		17.8	
T2D fungal lacasse					
Coliolus consors	T1	2.185	2.050	9.7	22
Polyporus versicolor	T1	2.193	2.047	8.9	59

	g_x	g_y	g_z	A_z (mT)	Ref.
New Cu (II) signal of fungal laccase					
Coliolus consors	2.012	2.067	2.272	12.5	44
Polyporus versicolor	2.025	2.148	2.268	12.5	40
New Cu (II) signal of tree laccase					
Rhus vernicifera	2.03	2.15	2.27	12.5	3

and $IrCl_6{}^{2-}$ nor is reduced by $Fe(CN)_6^{4-}$.[52] On the other hand, the type 3 Cu site, which is reduced during the procedure to deplete the type 2 Cu in tree laccase, is easily oxidized by H_2O_2 and the 330-nm shoulder band is recovered as that of the native enzyme.[13,53]

The new Cu (II) EPR signal with respect to the type 3 Cu is exhibited in Fig. 12B (c). The T2D laccase gives the signal with the highly rhombic character after seven days incubation at pH 9.5 (Table 3). This finding implies that a portion of type 3 coppers becomes EPR detectable in accord with the autoreduction of the type 1 Cu at high alkaline pH's. The spectrum is very similar to those of type 3 Cu sites in tree T2D laccase and half met hemocyanin ($g_z = 2.30$, $A_z = 12.3$ mT (13.2×10^{-3} cm^{-1})).[3]

3.3. *Resonance Raman (RR) Spectra of Fungal Native and T2D Laccases*

Figure 13 compares the room-temperature RR spectra of the type 1 Cu site of *P. versicolor* laccase B with that of tree laccase (*R. vernicifera*) and also shows the effect of removing the type 2 Cu from these laccases.[61] The RR spectrum of the native laccase B nearly agrees with that of laccase A from the same fungus[62] and appears to correlate with the tree laccase spectrum; all the frequencies are slightly higher (2–8 cm^{-1}). For tree laccase the 380-, 405-, and 419-cm^{-1} bands are of comparable intensity, with the 405-cm^{-1} band being strongest, whereas for fungal laccase most of the intensity resides in the 427-cm^{-1} band. Since it is plausible that one of the prominent bands in the 400-cm^{-1} region of blue copper RR

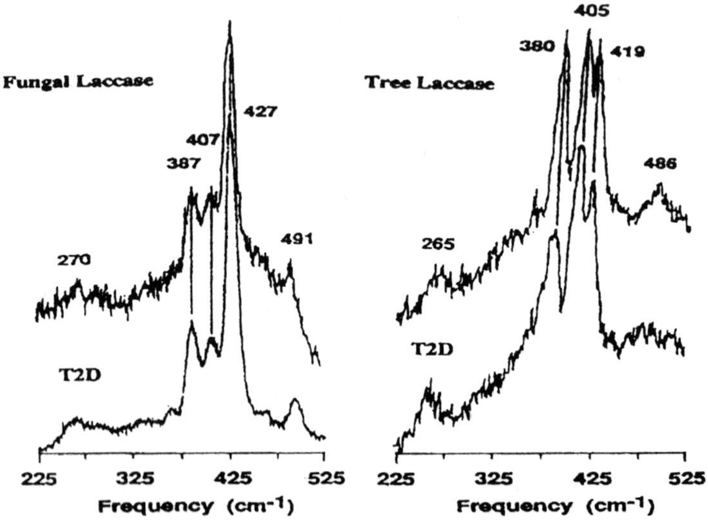

Fig. 13. 647.1-nm RR spectra of tree and fungal laccases, showing the effect to type 2 Cu removal (T2D) at pH 6.0. (Taken from Ref. 61. With permission.)

spectra is accountable to Cu-S^- (Cys) stretching, the higher frequency (427 cm^{-1}) suggests to elucidate the more positive redox potential of the type 1 Cu in fungal laccase (+785 mV) than that in tree laccase (+394 mV). The binding of F^- ion to the type 2 site in *P. versicolor* laccase A[63] does not induce a detectable change on the RR spectrum of the type 1 Cu site.[62] This observation implies that the structural perturbations at the type 2-type 3 Cu triangle site induced by F^- binding are not propagated to the type 1 Cu site.

The depletion of the type 2 Cu from laccase does not substantially change the type 1 Cu EPR spectrum and slightly diminishes in the intensity of the 600-nm absorption band. Removal of the type 2 Cu lowers the intensity of one of the tree laccase RR-bands (380 cm^{-1}), inducing a further lowering of both the 387- and 407- cm^{-1} band intensities of fungal laccase. These RR-spectral changes clearly show that the removal of the type 2 Cu is accompanied by a structural change at the type 1 Cu site.[62]

$^{63/65}$Cu and H_2O/D_2O isotope shifts in the RR spectrum are important for determining the contribution of Cu-ligand stretching (especially Cu-S^-(Cys) and Cu-N(His)). The N3 proton of imidazole exchanges on contact with D_2O and the C2 proton exchanges more slowly. Low-temperature Raman difference technique was used to obtain the isotope shifts for the three prominent bands of *P. versicolor* laccase A, which are compared with those of the corresponding prominent bands of tree laccase, stellacyanin, and azurin.[64] Although the RR spectra of fungal and tree laccases look very different as shown in Fig. 13, the frequencies and the isotope shifts are nearly the same, differing from stellacyanin and azurin. The strong 428-cm^{-1} (427-cm^{-1} in Fig. 13) band and the moderate intensity 408-cm^{-1} (407-cm^{-1} in Fig. 13) and 387-cm^{-1} bands of the fungal enzymes show small $^{63/65}$Cu isotope shifts (0.5–0.6 cm^{-1}),[64] which are substantially less than that expected for an isolated Cu-S(Cys) stretch. The finding suggests a high degree of mixing of this coordinate with two or more internal modes of the ligands. In the case of stellacyanin, the combined shifts of the 385- and 347-cm^{-1} bands are equal to the calculated $^{63/65}$Cu isotope shift (2.8–3.3 cm^{-1}), so that it is reasonable to suppose that this pair of bands arises from strong coupling between the Cu-S^-(Cys) mode and another mode of similar energy like the SCC bent of the Cys side chain. H_2O/D_2O shifts of 1.1 and ca. 0.3 cm^{-1} are observed for the 387- and 428-cm^{-1} bands of the fungal enzymes, respectively. The largest shifts among the major ca. 400-cm^{-1} bands are associated with the feature at 380–390 cm^{-1} for *P. versicolor* laccase (at 387 cm^{-1}), *R. vernicifera* laccase (0.8 cm^{-1} at 381 cm^{-1}), and stellacyanin (0.6 cm^{-1} at 385 cm^{-1}), but a shift of this magnitude for azurin (1.4 cm^{-1}) is observed at the 428 cm^{-1} band. However, none of these shifts is large enough to be associated with a Cu-N(His) stretching vibration directly. The 2-cm^{-1} H_2O/D_2O shift identifies the RR band of stellacyanin at 273 cm^{-1} as a Cu-N(His) mode, but fungal and tree laccases show no moderate intensity band near 270 cm^{-1}. The 741 cm^{-1} band of fungal laccase at 77 K is assigned to the C-S stretching mode and the lower C-S frequency compared with the corresponding bands of tree laccase, stellacyanin, and azurin (746–753 cm^{-1}) may be attributable to a Cu-S-C angle somewhat closer to 90 degree.

References

1. B. Mondovi and L. Avigliano, in *Copper Proteins and Copper Enzymes*, Vol. 3, ed. R. Lontie (CRC Press, Boca Raton, 1984), p. 101.
2. J. Ohkawa, N. Okada, A. Shinmyo and M. Takano, *Proc. Natl. Acad. Sci. USA* **86**, 1239 (1989).
3. B. Reinhammar, in *Copper Proteins and Copper Enzymes*, Vol. 3, ed. R. Lontie (CRC Press, Boca Raton, 1984), p. 1.
4. C. F. Thurston, *Microbiology* **140**, 19 (1994).
5. A. Givaudan, A. Effosse, D. Faure, P. Potier, M. L. Bouillant and R. Bally, *FEMS Micro. Lett.* **108**, 205 (1993).
6. D. D. Thakker, C. S. Evans and K. K. Rao, *Appl. Microbiol. Biotechnol.* **37**, 321 (1992).
7. K. Kawahara, T. Sakurai, S. Suzuki and A. Nakahara, *Inorg. Chim. Acta* **92**, L33 (1984).
8. T. Sakurai, S. Sawada, S. Suzuki and A. Nakahara, *Biochim. Biophys. Acta* **915**, 238 (1987).
9. D. Dooley, J. H. Dawson, P. Philip and H. B. Gray, *Biochemistry* **20**, 2024 (1981).
10. J. Hirose, T. Sakurai, K. Imamura, H. Watanabe, H. Iwamoto, K. Hiromi, H. Itoh, T. Shin and S. Murao, *J. Biochem.* **115**, 811 (1994).
11. A. Merli, G. L. Rossi, M. Bolognesi, G. Gatti, L. Morpurgo and A. Finazzi-Agro, *FEBS Lett.*, **231**, 89 (1988).
12. T. Sakurai, S. Suzuki and M. Chikira, *J. Biochem.* **107**, 37 (1990).
13. E. I. Solomon, M. J. Baldwin and M. D. Lowery, *Chem. Rev.* **92**, 521 (1992).
14. B. A. Fields, J. M. Guss and H. C. Freeman, *J. Mol. Biol.* **222**, 1053 (1991).
15. A. Messerschmidt, A. Rossi, R. Ladenstein, R. Huber, M. Bolognesi, G. Gatti, A. Marchesini, R. Petruzzelli and A. Finazzi-Agro, *J. Mol. Biol.* **206**, 513 (1989).

16. A. Messerschmidt, R. Ladenstein, R. Huber, M. Bolognesi, L. Avigliano, R. Petruzzelli, A. Rossi and A. Finazzi-Agro, *J. Mol. Biol.* **224**, 179 (1992).
17. B. Reinhammar and Y. Oda, *J. Inorg. Biochem.* **11**, 115 (1979).
18. J. E. Penner-Hahn, B. Hedman, K. O. Hodgson, D. J. Spira and E. I. Solomon, *Biochem. Biophys. Res. Comm.* **119**, 567 (1984).
19. T. Nakamura, N. Makino and Y. Ogura, *J. Biochem.* **64**, 189 (1968).
20. V. Ts. Aikazyan and R. M. Nalbandyan, *FEBS Lett.* **104**, 127 (1979).
21. A. Marchesini and P. M. Kroneck, *Eur. J. Biochem.* **101**, 65 (1979).
22. T. Sakurai and J. Takahashi, unpublished data.
23. L. Morpurgo, I. Savini, G. Gatti, M. Bolognesi and L. Avigliano, *Biochem. Biophys. Res. Commun.* **152**, 623 (1988).
24. T. Sakurai and J.Takahashi, *Biochem. Biophys. Res. Comm.*, **215**, 235 (1995).
25. L. Calabrese, M. Carbonaro and G. Musci, *J. Biol. Chem.* **263**, 6480 (1988).
26. T. Sakurai, *Inorg. Chim. Acta* **157**, 117 (1989).
27. T. Nakamura, *Biochim. Biophys. Acta* **30**, 44 (1958).
28. K. Kawahara, S. Suzuki, T. Sakurai and A. Nakahara, *Arch. Biochem. Biophys.* **241**, 179 (1985).
29. T. Sakurai, *Inorg. Chim. Acta* **195**, 255 (1992).
30. T. Sakurai, *Biochem. J.* **284**, 681 (1992).
31. T. Sakurai, *Biochem. (Life Sci. Adv.)* **14**, 17 (1995).
32. T. Sakurai and J. Takahashi, *Biochim. Biophys. Acta* **1248**, 143 (1995).
33. J. L. Cole, L. Avigliano, L. Morpurgo and E. I. Solomon, *J. Am. Chem. Soc.* **113**, 9080 (1991).
34. D. J. Spira-Solomon, M. D. Allendorf and E. I. Solomon, *J. Am. Chem. Soc.* **108**, 5318 (1986).
35. T. Sakurai and A. Nakahara, *J. Inorg. Biochem.* **27**, 85 (1986).
36. A. Messerschmidt, H. Luecke and R. Huber, *J. Mol. Biol.* **230**, 997 (1993).
37. A. Messerschmidt, in *Advances in Inorganic Chemistry*, Vol. 40, ed. A. G. Sykes (Academic Press INC., San Diego, 1993), p. 121.
38. L. Casella, M. Gullotti, G. Pallanza, A. Pinter and A. Marchesini, *Bio Metals* **4**, 81 (1991).
39. D. J. Spira-Solomon and E. I. Solomon, *J. Am. Chem. Soc.* **109**, 6421 (1987).
40. B. Reinhammar, R. Malkin, P. Jensen, B. Karlsson, L. -E. Andreasson, R. Aasa, T. Vänngard and B. G. Malmström, *J. Biol. Chem.* **225**, 5000 (1980).
41. T. Sakurai, S. Suzuki and M. Sano, *Inorg. Chim. Acta* **152**, 3 (1988).
42. O. Siiman, N. M. Young and P. R. Carey, *J. Am. Chem. Soc.* **98**, 744 (1976).
43. K. -F. Falk and B. Reinhammar, *Biochem. Biophys. Acta* **285**, 84 (1972).
44. T. Sakurai, J. *Inorg. Biochem.* **41**, 277 (1991).
45. K. Lerch, J. Deinum and B. Reinhammar, *Biochim. Biophys. Acta* **534**, 7 (1978).
46. H. P. Molitori and B. Reinhammar, *Biochim. Biophys. Acta* **386**, 493 (1975).
47. D. M. Dooley, J. Rawlings, J. H. Dawson, P.J. Stephens, L. -E. Andreasson, B. G. Malmström and H. B. Gray, *J. Am. Chem. Soc.* **101**, 5038 (1979).

48. R. Bligny, J. Gaillard and R. Douce, *Biochem. J.* **237**, 583 (1986).
49. J. A. Fee, B. G. Malmström and T. Vänngard, *Biochim. Biophys. Acta* **197**, 136 (1970).
50. H. B. Gray and E. I. Solomon, in *Copper Proteins*, ed. T. G. Spiro (John Wiley & Sons Inc., New York, 1981), p. 1.
51. S.-P. W. Tang, J. E. Coleman and Y. P. Myer, *J. Biol. Chem.* **243**, 4286 (1968).
52. P. M. Hanna, D. R. McMillin, M. Pasenkiewicz-Gierula, W. E. Antholine and B. Reinhammar, *Biochem. J.* **253**, 561 (1988).
53. C. D. LuBien, M. E. Winkler, T. J. Thamann, R. A. Scott, M. S. Co, K. O. Hodgson and E. I. Solomon, *J. Am. Chem. Soc.* **103**, 7014 (1981).
54. B. R. M. Reinhammar, *Biochim. Biophys. Acta* **275**, 245 (1972).
55. S. K. Chapman, in *Perspectives on Bioinorganic Chemistry*, Vol. 1, eds. R. W. Hay, Dilworth and K. B. Nolan (JAI Press LTD., London, 1991), p. 95.
56. B. G. Malmström, B. Reinhammar and T. Vänngard, *Biochim. Biophys. Acta* **156**, 67 (1968).
57. B. G. Malmström, B. Reinhammar and T. Vänngard, *Biochim. Biophys. Acta* **205**, 48 (1970).
58. R. Bränden and J. Deinum, *Biochim. Biophys. Acta* **524**, 297 (1978).
59. S. K. Wrigley and J. F. Gibson, *Biochim. Biophys. Acta* **916**, 259 (1987).
60. R. Bränden and B. Reinhammar, *Biochim. Biophys. Acta* **405**, 236 (1975).
61. L. Nestor, J. A. Larrabee, G. Woolery, B. Reinhammar and T. G. Spiro, *Biochemistry* **23**, 1084 (1984).
62. D. F. Blair, G. W. Campbell, V. Lum, C. T. Martin, H. B. Gray, B. G. Malmström and S. I. Chan, *J. Inorg. Biochem.* **19**, 65 (1983).
63. R. Bränden, B. G. Malmström and T. Vänngard, *Eur. J. Biochem.* **36**, 195 (1973).
64. L. Nestor, B. Reinhammar and T. G. Spiro, *Biochim. Biophys. Acta* **869**, 286 (1986).

CHAPTER 9

BIOLOGICAL FUNCTION AND ENZYME KINETICS OF ASCORBATE OXIDASE

Luciana Avigliano and Alessandro Finazzi-Agrò
Dipartimento di Medicina Sperimentale e Scienze Biochimiche, Università di Roma "Tor Vergata", via di Tor Vergata, 135. 00133 Roma, Italy
E-mail: avigliano@utovrm.it

ABSTRACT

Ascorbate oxidase is a copper enzyme found exclusively in plants. It oxidizes ascorbate to semidehydroascorbate with the concomitant reduction of oxygen to water. Despite large amounts of structural and kinetic information on the purified enzyme the proposed physiological roles of ascorbate oxidase are largerly circumstantial. In particular this enzyme has been involved either in growth-related processes or in antioxidative defense, due to the well-known chemistry of ascorbate.

The kinetics of ascorbate oxidase, substantiated by X-ray derived 3-D structure, show that the blue, type 1 copper is the entrance gate for electrons which afterwords are transferred very efficiently to a trinuclear copper cluster where oxygen is bound and reduced. The efficiency of substrate oxidation and the relative specificity for ascorbate make ascorbate oxidase an appropriate tool in analytical and biomedical applications.

Contents

1. Introduction

Ascorbate oxidase (L-ascorbate:oxygen oxidoreductase, EC 1.10.3.3) is a multicopper glycoenzyme belonging to (along with ceruloplasmin and laccase) the group of "blue oxidases".[1] As its name implies, ascorbate oxidase catalyzes the one-electron oxidation of four ascorbate molecules to the semidehydroascorbate radical with the concomitant four-electron reduction of molecular oxygen to water. Ascorbate oxidase is present only in higher plants, the most abundant sources being the cucurbits such as cucumber, pumpkin and green marrows. To date no definite biological function has been established for this enzyme. However ascorbate oxidase appears to be of special interest because of the high concentration of ascorbate in plant cells and the reactivity of the ascorbate/dehydroascorbate redox pair.

Comprehensive reviews about structural, spectroscopic and kinetic properties of ascorbate oxidase[2–4] or its physiological role[5] appeared during the 1980s. Recently, knowledge about this topic has significantly increased with the determination of the three-dimensional structure of ascorbate oxidase from green zucchini marrows (*Cucurbita pepo medullosa*).[6,7] Furthermore, the cDNAs of cucumber (*Cucumis sativus*)[8], pumpkin (*Cucurbita sp.*)[9] and green zucchini (A. Rossi, personal communication) have been sequenced. Altogether this information stimulates new interest in ascorbate oxidase, as it provides the required structural basis to understand the electron transfer process and the mechanism by which the reactive oxygen intermediates formed during the reduction are not released free in solution. It also provides a basis for studying enzyme expression, which is preliminary to understanding its biological role in plants.

The enzyme purified from green zucchini is a homodimer composed of 70 kDa subunits, each containing four copper ions. These ions belong to three different types, classified as type 1, 2 and 3 according to their spectroscopic features.[1] In the oxidized form, type 1 Cu is intensely blue (ε at 610 nm = 10000 M^{-1} cm^{-1}), paramagnetic and shows a low hyperfine splitting ($A_{\|} = 5.8$ mT) in the $g_{\|}$ region of the EPR spectrum. Type 2 Cu is colorless and paramagnetic with a broader $A_{\|}$ (19 mT) i.e. in the usual range of low molecular weight Cu(II) complexes. Type 3 is formed by a pair of copper atoms which, due to antiferromagnetic coupling, are diamagnetic and is associated with an optical absorption band at 330 nm ($\Delta\,\varepsilon$ = 3300 M^{-1} cm^{-1}). Upon reduction all the spectroscopic absorptions related to copper are bleached; this allows to follow the redox process involving the different copper types.

Each ascorbate oxidase subunit is formed by three domains all showing a β-barrel folding. Domain 2 alone is not involved in copper binding. Type 1 Cu is bound inside domain 3 by 2 His, Met and Cys. Type 2 Cu and type 3 Cu are situated close to each other forming a trinuclear cluster, with an average copper distance of 3.7 Å, and are located at an interface between domains 1 and 3, which equally supply the ligands for them. The two paired ions are each coordinated by three His and bridged by an OH^- or O^{2-}, the type 2 Cu is coordinated by two His and a H_2O or OH^-. The amino acid sequence His-Cys-His links type 1 and type 3 coppers. In this sequence the cystein binds type 1 Cu while the two histidines bind the type 3 coppers, making approximately 12 Å the shortest distance between type 1 Cu and the trinuclear coppers.[7] As shown by the X-ray structure, the geometry of the reduced type 1 Cu remains the same as the oxidized form,[10] thus requires little reorganization in the redox process, as does the single copper in the small blue copper proteins plastocyanin[11] and azurin.[12] Instead considerable structural changes occur at the trinuclear copper site where, upon reduction, the hydroxyl bridge is released and the copper-copper distance increases.[10] Type 2 Cu is the most accessible to exogenous ligands and inhibitors such as F^- and N_3^-.[13] It can be removed from ascorbate oxidase in solution[14] and from ascorbate oxidase crystals[15] by chelating agents under mild reducing conditions producing an inactive derivative, with bleached 330 nm absorption. This absorption is restored by reoxidation with excess hydrogen peroxide.[16] In both cases, type 3 loss is observed after prolonged depletion treatment.[15,16]

The carbohydrate content of ascorbate oxidase from green zucchini amounts to 3 % (w/w) of the protein. The primary structure Man_3-Xyl-$GlcNAc_2$ of the N-linked moiety (two identical chains per subunit) seems to be a common element of many plant N-glycoproteins (Fig. 1).[17]

Manα1
|
6
Manα1-3Manβ1-4GlcNAcβ1-4GlcNAc
2
|
Xylβ1

Fig. 1. Primary structure of N-linked carbohydrate chain of ascorbate oxidase from green zucchini (from Ref. 17).

2. Biological Aspects of Ascorbate Oxidase

2.1. *Distribution in Plant Kingdom*

Ascorbate oxidase has been found only in higher plants, although enzymes showing significant ascorbate oxidase activity have been isolated from microorganisms. Ascorbate oxidase from *Myrothecium verrucaria* has been termed "atypical" due to the absence of the characteristic absorbance in the visible region and EPR signal and is unaffected by inhibitors of heavy metal catalysis.[18] Recently, an ascorbate oxidase was isolated from the fungus *Acremonium* sp. HI-25.[19] It is a monomeric multi-copper protein of molecular weight 80 kDa, containing four different copper atoms of types 1, 2 and 3. The optical, CD and EPR spectroscopic features closely resemble those of ascorbate oxidase from cucurbits (ε at 600 nm = 4270 M^{-1} cm^{-1}). The enzyme had no laccase activity. It was maximally active towards ascorbate at pH 4.5 where, as indicated by Km value (0.29 mM) and by kcat (3100 s^{-1}), the fungal enzyme had the same catalytic efficiency of the monomeric unit of the plant ascorbate oxidase at neutral pH.[20]

Ascorbate oxidase is widely diffused in different tissues and plants. The highest level of ascorbate oxidase is known to be in cucurbit fruits. Thus ascorbate oxidase was isolated and characterized from cucumber,[21] squash[22,23] and green zucchini.[24,25] Recently ascorbate oxidase from muskmelon (*Cucumis melo*)[26] and from satsuma mandarin (*Citrus inshiu*),[27] popular fruit and the richest source of ascorbate in Japan, were purified and characterized. These enzymes exist in multiple forms unlike the enzyme extracted from cucurbits where the presence of two forms was occasionally reported only for the green zucchini enzyme.[25] The enzyme purified from melon is a glycoprotein of 137 kDa, composed of two subunits of 68 kDa and formed by six isoenzymes with isoeletric points in the range of pH 7.7 to 8.3. It contains eight copper atoms whose spectroscopic and catalytic features were reported to be similar to those of analogous enzymes. In satsuma mandarin three ascorbate oxidase forms with different behavior towards cation and anion exchangers were observed. The basic form (pI = 7.0) is a dimer of 141 kDa composed by two subunits with an apparent mass of 74 kDa and 62 kDa. Metal ligands such as diethyldithiocarbamate, cyanide or fluoride inhibited enzyme activity.

Ascorbate oxidase enzymatic activity was also found in shoot apices of rice,[28] pea leaves,[29] potatoes,[30] apple buds[31] and fresh herbs[32] including thyme, coriander, fennel, parsley, oregano, spearmint and basil.

The presence of ascorbate oxidase in higher plants can be detected through several methods, but mainly through enzymatic and immunological tests. A rapid method for detection of ascorbate oxidase activity in electrophoretic gels is based upon the photo-induced reduction of nitroblue tetrazolium to the chromophore formazan. Ascorbate acted as an electron donor; the presence of ascorbate oxidase prevented this reduction through catalytic removal of ascorbate and therefore the enzyme can be detected as an achromatic band on a blue background.[33,34]

Ascorbate oxidase is a glycoprotein whose glycans seem to be strongly immunogenic. Indeed polyclonal antibodies directed against the native form of ascorbate oxidase recognize poorly the deglycosylated protein.[35,36] On the other hand a glycan chain of glycoproteins in the plants possibly shares antigenic determinants. Polyclonal antibodies were produced against the enzyme purified from zucchini and used as probes.[37] Such heterologous antibodies lack specificity since ascorbate oxidase proteins from different sources may differ in structure. An immunological cross-reactivity with the green zucchini enzyme and ascorbate oxidase activity was found in carrot roots, parsley leaves, egg plant fruit, capsicum fruit and fennel leaf stalks, though proteins were not identical to the green zucchini enzyme as they showed a different electrophoretic mobility and a typical non-identity pattern in immunodiffusion experiments. Celery and spinach showed significant ascorbate oxidase activity but no immunological cross-reactivity with green zucchini ascorbate oxidase.[37]

2.2. *Tissue and Cell Localization*

The immunological methods were however fundamental to the study of ascorbate oxidase distribution in various plant tissues and inside the cells. The immunohistochemical localization demonstrated the presence of ascorbate oxidase in both the vegetative and reproductive organs of green zucchini (Figs. 2A and B). The enzyme was found in stem, branches, leaves,

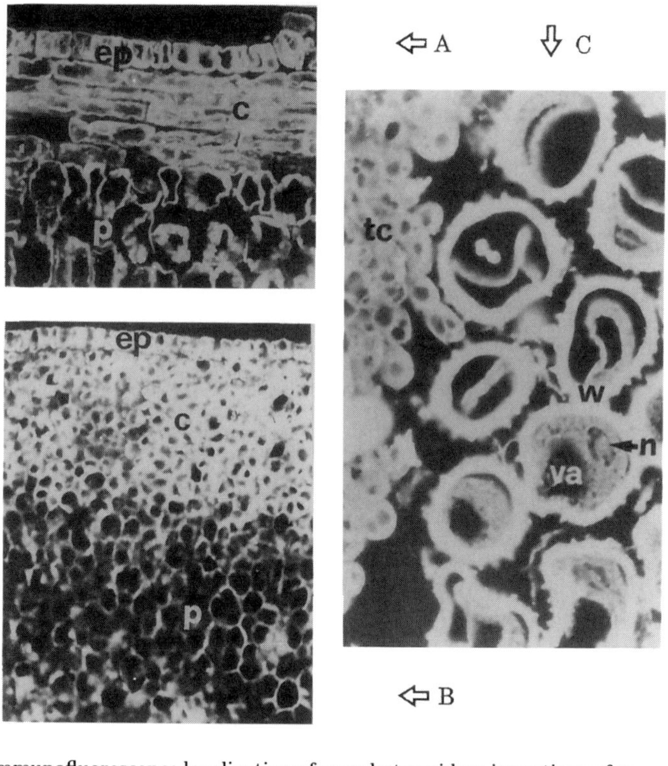

Fig. 2. Immunofluorescence localization of ascorbate oxidase in sections of green zucchini (from Ref. 38). A: leaf midrib in longitudinal section (225 x); B: mature fruit in cross-section (245 x); C: anther in longitudinal section (300 x). c–collenchyma; ep–epidermis; n–nucleolus; p–parenchyma; tc–tapetal cells; va–vacuole; and w, wall. White is positive.

flowers, fruits and seeds, both at early and advanced developmental stages, and both in undifferentiated cells, such as those of cambium and organ primordia, or in differentiated ones, including hair cells. It was also detected in dead cells, which undergo wall lignification, such as vessels and seed integuments.[38]

At the cellular level, the enzyme was localized in the cytoplasm, mostly at the cell wall, but it was not present in the vacuoles and the nucleus (Fig. 2C).[38]

In the apoplast of epicotyls from *Vigna angularis* almost all ascorbate oxidase is present in wall-bound form, about two-thirds of which was solubilized by suspension of cell walls in 0.5 M KCl, suggesting that some of ascorbate oxidase is non-covalently, and perhaps electrostatically bound to walls.[39] This pattern of cellular distribution supports early findings on the biochemical localization of ascorbate oxidase both in the soluble fraction and cell wall of cabbage leaves and barley roots.[40]

2.3. *Gene and Expression*

2.3.1. *Gene*

The cDNA sequences coding for ascorbate oxidase from cucumber,[8] pumpkin[9] and zucchini (A. Rossi, personal communication) has been established. The cDNAs encode for proteins of 554 (cucumber), 549 (pumpkin) and 552 (zucchini) aminoacid residues respectively. The three sequences are very similar, and the overall sequence identity is more than 80 %. NH_2-terminal signal peptides of 33 and 30 residues are present in the cDNAs coding for cucumber and pumpkin enzymes respectively, while that of zucchini has not yet been determined. The two signal sequences, rich in hydrophobic amino acid residues, and the bonds presumably cleaved, as shown by cDNA sequences, are consistent with the specificity of the signal peptidase which may direct the nascent polypeptide chains across the rough endoplasmic reticulum to the periplasmic space. A genomic clone containing the complete gene coding for cucumber ascorbate oxidase is also known.[41] It shows the presence of four exons and three introns and, at the 5'-flanking region, the putatives TATA and CAAT boxes; the site of transcription was determined by primer extension. The 5'-flanking region, placed upstream the reporter gene coding for β-glucuronidase (GUS), was able to direct the synthesis of the GUS protein in cucumber and tobacco plants.

Northern blot analysis was developed to check the presence of mRNA coding for ascorbate oxidase in plant homogenates. The probe was a ^{32}P-labelled RNA synthesized by T7 RNA polymerase runoff transcription of a linearized plasmid containing an EcoRI cDNA fragment coding for the last 180 amino acids of zucchini ascorbate oxidase.[34]

2.3.2. *Expression in Plants*

The expression of ascorbate oxidase has been extensively studied in cucurbits. Positive correlation between growth rate and enzyme level was found in both fruit and leaves of zucchini squash (*Cucurbita pepo*, black zucchini).[42] In the developing fruit, the specific activity, protein and mRNA were highest between 4 and 6 days after anthesis, i.e. during the period of highest growth rate of the fruit. Within the fruit there was a gradient in the enzyme level moving outward with the highest activity in epidermis and the lowest in the central placental region. In leaf tissue, ascorbate oxidase activity decreased with age. Within the leaf the enzyme decreased from the lower third of the blade to the apex, i.e. from fast-growing to slow-growing regions. Ascorbate oxidase activity and protein were also traced during the development of melon fruit.[26] The highest activity of ascorbate oxidase was found during anthesis and during specific periods of melon fruit development corresponding to the first days of fruit growth and to early events of fruit ripening in 30 to 35 day old fruits. Ethylene production paralleled the ascorbate oxidase activity while the ascorbate content showed an inverse correlation. Ascorbate oxidase expression in pumpkin fruits was found to be quite similar to that in squash and melon.[43] Low ascorbate oxidase level was present in pumpkin seeds but gradually increased during early seedling growth.[44] Correlation between growth rate and enzyme level was also reported in a study on the activity of ascorbate oxidase in apple buds (*Malus domestica*) during dormancy and budbreak induced by thidiazuron (N-phenyl-N'-1,2,3,-thidiazol-5-yl-urea), a plant bioregulator.[45] In dormant buds the ascorbate oxidase activity was low compared to buds treated with thidiazuron and that had resumed growth. An increase in ascorbate oxidase activity began at the time of metabolic transition. The highest level was reached ten days after thidiazuron induction during the expansion growth phase. The ascorbate content was positively correlated to ascorbate oxidase activity.[31]

Ascorbate oxidase expression appears to be regulated by phytohormones. The activity of cell wall-bound ascorbate oxidase was significantly increased by treatment with the auxin IAA (indole-3-acetic acid) in epicotyls from *Vigna angularis.* The increase in ascorbate oxidase activity was correlated to increased levels of ascorbate and dehydroascorbate, while the activity of cell wall peroxidase was not affected.[39] Ethephon (2-chlorethylphosphonic acid) slowly releases ethylene by a chemical reaction, allowing the hormone to exert its effect. The spraying of apple fruits advanced fruit ripening and increased respiration rate, ethylene production and ascorbate oxidase activity.[46]

Ascorbate oxidase synthesis also seems to be controlled by phytochromes. In mustard (*Sinapis alba*) cotyledons ascorbate oxidase activity and protein level were markedly increased by irradiation with continuous far-red light. Pulse-labeling experiments indicated an increased rate of enzyme synthesis. The involvement of phytochrome in this light-mediated response was demonstrated by red/far-red reversibility experiments.[36,47,48] This finding differs from a report of another research group claiming a post-translational control for ascorbate oxidase by phytochrome in mustard cotyledons.[49]

2.3.3. *Expression in Cultured Cells*

Cell cultures of tissue from pumpkin,[50,51] cucumber[52] and green zucchini[53] were developed to study the expression of ascorbate oxidase. The sarcocarp tissue of fruit was induced to produce callus on Murashige-Skoog basal medium with plant hormones (2,4-dichlorophenoxyacetic acid and kinetin; Fig. 3).

Ascorbate oxidase activity increased during callus formation, reaching the maximum on day 5, and during growth of callus (maximum on day 10–15), and then declined. Maximum enzyme activity was found on day 10–16 in callus obtained from fruit peel tissue of cucumber.[52] Similar results were found using green zucchini cells where a maximum in enzymatic activity, protein level and specific mRNA were determined at day 10 of subculturing.[53] The ascorbate oxidase level also rapidly increased after transferring the callus to fresh culture media.

Ascorbate oxidase was secreted into the culture medium from cell suspensions, supporting the above reported observation that its mRNA also codes for a signal sequence to a secretory pathway. The activity in the

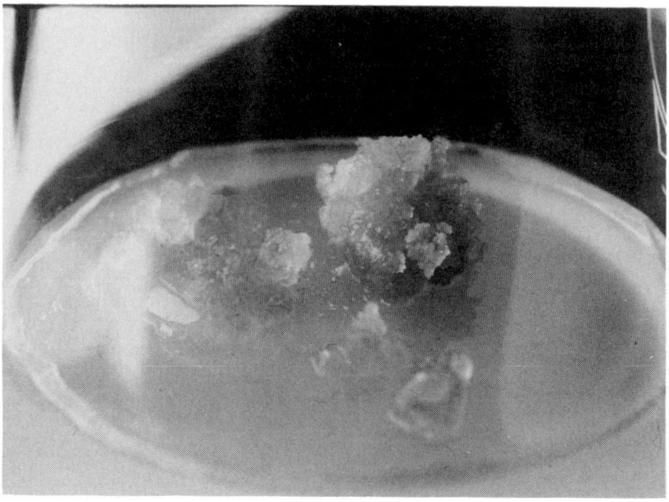

Fig. 3. Callus induced from sarcocarp of green zucchini fruit.

medium as well as that in the cell increased soon after transfer of cells to fresh medium and reached a maximum at about day 5 when 50–70 % of ascorbate oxidase activity was present in the medium.[52–54]

The biological significance of the oligosaccharide chains present in ascorbate oxidase was investigated in green zucchini cultured cells treated with tunicamycin, a specific inhibitor of N-glycosylation which prevents the binding of the first N-acetylglucosamine residue to dolichol phosphate.[55] A sugar-free active enzyme was produced by treatment of zucchini cell suspension with 2–4 mg/l tunicamycin at day 10 culturing.[56] Although the cells treated with tunicamycin had a low rate of protein synthesis, they showed an ascorbate oxidase specific activity 3–4 times higher than that of the controls. This results seems to be in line with the data obtained by *in vitro* enzymatic deglycosylation, i.e., that carbohydrates are not essential for catalytic function. In fact, fully deglycosylated ascorbate oxidase, obtained by peptide N-glycosidase F treatment in the presence of a detergent solution made of 0.1 % (w/v) sodium dodecyl sulphate plus 0.5 % (v/v) Nonidet P-40, showed the same kinetic properties as the native enzyme.[35] However, ascorbate oxidase, deglycosylated up to 50 % by sequential exoglycosidase treatment under non-denaturing conditions, showed a slight increase both in substrate affinity and catalytic activity with respect to the native enzyme.[57]

The expression of ascorbate oxidase in cultured cells was shown to be modulated by a number of factors such as copper, calcium, chemicals and plant hormones.

The addition of copper 100 times greater than the basal level of the culture medium (0.1 μM $CuSO_4$) resulted in a significant increase in enzyme activity.[42,51–53] In pumpkin callus the ascorbate oxidase activity and the amount of protein were increased approximately fivefold by copper supplementation, although the amount of mRNA coding for ascorbate oxidase was not affected.[43] Therefore, it was suggested that copper might regulate ascorbate expression at the translational or post-translational level. These results somewhat differ from those found for green zucchini cells. The intracellular activity of ascorbate oxidase increased threefold in response to copper with no change in the specific mRNA and protein levels; therefore, it was concluded that the metal cofactor may reconstitute the active enzyme from the inactive apoenzyme present in the cell.[53] The effect of copper deficiency on Cu-containing enzymes has been previously investigated in plants grown in low-copper media. Under copper-deficient conditions, the activity of ascorbate oxidase decreased by about 50 % in the first generation and by 80 % in the second generation of pea plants.[29] In clover (*Trifolium subterraneum*) copper deficiency had little effect on protein levels but markedly decreased the activity of the copper-containing enzymes ascorbate oxidase and diamine oxidase. The activities were not restored upon copper addition.[58] Altogether, these reports indicate that the expression of ascorbate oxidase is controlled by copper, although the mechanism can vary from plant to plant.

Since it is well-known that calcium plays a pivotal role in cell functions, including exocytosis and protein secretion, the effect of calcium on ascorbate secretion was also investigated. Addition of $CaCl_2$ and $MgCl_2$, but not of KCl, markedly stimulated the ascorbate oxidase secretion from cultured pumpkin cells.[59] Ascorbate oxidase activity in the culture medium containing 40 mM $CaCl_2$ was about 50 times the level attained in the absence of $CaCl_2$. Since a high concentration of calcium was required, the effect on membrane permeability seems to be non-specific. It seemed, however, to be specific for ascorbate oxidase since the secretion of other secretory enzymes such as peroxidase was only slightly increased. The secretion of ascorbate oxidase from cucumber was stimulated by the addition of calcium but not of magnesium ions.[59]

The secretion of ascorbate oxidase from cultured pumpkin cells was also markedly stimulated by salycilate and by eosine yellowish which are known to be inducers of phatogenesis-related proteins, i.e. soluble proteins induced by pathogens and secreted into intercellular spaces.[60] Such behavior does not necessarily imply that ascorbate oxidase itself is a pathogenesis-related protein, though it is suggestive to recall that the cognate protein ceruloplasmin is an acute-phase reactant protein in mammals.

Plant hormones were found to regulate ascorbate oxidase expression. The auxin 2,4-dichlorophenoxyacetic acid, which acts mainly as a growth promotor, markedly increased ascorbate oxidase activity and mRNA in pumpkin cultured cells.[43] Abscisic acid, which instead is an inhibitor of growth and metabolic processes, caused a reduction of ascorbate oxidase activity, protein and mRNA levels in green zucchini cells.[53]

2.4. *Physiological Role*

Since ascorbate oxidase is highly active in the degradation of ascorbate *in vitro*, it is reasonable to look at its physiological role from this point of view.

Ascorbate is involved in a variety of cell processes in plants, mainly related to cell growth and detoxification. A complete picture of the complex ascorbate system in plants has been recently reviewed.[61,62] The level of ascorbate, its redox state and its specific function depend on several enzyme activities related to biosynthesis and metabolism. Ascorbate is synthesized *ex novo* from glucose and used in cell metabolism as an electron donor in two successive one-electron steps. The first oxidation product, regardless of the oxidative reaction that takes place, is the semiquinone-like free radical semidehydroascorbate. The ascorbate radical can act either as an electron acceptor or donor, being reconverted to ascorbate by NAD(P)H-dependent semidehydroascorbate reductase, or spontaneously undergoing disproportionation thus generating ascorbate and dehydroascorbate.[63] Dehydroascorbate is either reduced back to ascorbate by glutathione-dependent dehydroascorbate reductase or further oxidized to produce oxalate. Many metabolic systems are known to involve ascorbate. Ascorbate peroxidases utilize ascorbate and H_2O_2 to catalyze the polymerization of phenolic compounds that commonly make up lignins[64] and the cross-linking of cell wall constituents[65] and therefore it is thought to participate in the inhibition of elongation growth. Ascorbate is used by peptidyl-proline hydroxylase for biosynthesis of the hydroxyproline-rich proteins required for cell division.[66] Ascorbate and its free radical are involved in a plasma membrane redox reaction that induces a high vacuolization which is responsible for elongation.[67] *In vitro* ascorbate appears to act as a co-substrate of the enzyme 1-aminocyclopropane-1-carboxylate oxidase which catalyzes the final step in the biosynthesis of the plant hormone ethylene.[68] Ascorbate also exerts an antioxidant action by eliminating harmful oxygen reactive species such as $^1O_2$, O_2^-, $OH^\cdot$, H_2O_2.[69] In plants, activated states of oxygen are produced both in physiological conditions such as in photosynthetic activity or by cytotoxic compounds such as air pollutants or herbicides. Ascorbate peroxidase represents the main system for the removal of H_2O_2; nevertheless, the tissue level of ascorbate is high enough to cause chemical scavenging of oxygen radicals. Ascorbate can regenerate some membrane-bound radical chain-breakers such as α-tocopherol, cooperating in order to protect lipids and lipid-containing structures against peroxidation.[70,71]

As yet the proper role of ascorbate oxidase in the complex net of reactions involving ascorbate has not been defined. The finding that ascorbate oxidase is a cell wall-localized protein, that is: (a) highly expressed at a stage when rapid growth occurred, (b) mainly concentrated in fruit epidermis where cells are under greatest tension during rapid growth and (c) inducible by auxins, altogether support the idea that this enzyme might be involved in reorganization of the cell wall. One hypothesis is that ascorbate oxidase and ascorbate peroxidases, both present in the plant cell wall, participate together in the control of the synthesis of cell-wall components.[39,64] It is important to note that complexes between ascorbate oxidase and peroxidase may occur in solution causing problems in the purification procedure.[25] Ascorbate and ascorbate oxidase might modulate cell elongation by a redox chain occurring at the plasma membrane.[62,72] The extracellular ascorbate free radical, supplied by ascorbate oxidase, can function as an electron acceptor in a transmembrane electron transport chain, from cytosolic NADH through an intramembrane b-type cytochrome.[67] Electron flux along this chain induces hyperpolarization of plasma membrane leading to vacuolization and to other growth-associated events such as increased uptake of nutrients into the cell.[73] A further mechanism has been proposed which utilizes dehydroascorbate, the final product of ascorbate oxidase activity, for cell-wall loosening during cell expansion.[42] According to this suggestion, at least a couple of mechanisms, based on the known chemistry of dehydroascorbate, might be operative. One involves the modification of lysine and arginine residues of wall proteins. This modification can prevent both covalent cross-linking of wall proteins and non-covalent ionic interactions of the positively charged residues of proteins with the negatively charged pectin (polygalacturonate). Another mechanism is related to the enhancement of oxalate formation with consequent removal of calcium from the calcium-pectin complex in the wall, resulting in a less rigid structure.

The involvement of ascorbate oxidase in cell metabolism reactions other than growth has been suggested. Fruit ripening requires a series of metabolic changes eventually leading to fruit senescence that include the softening of cell walls.[74] Ascorbate oxidase may play a role in such process possibly through its role in ethylene biosynthesis.[26] Ascorbate oxidase may also participate in a redox chain, involving ascorbate, glutathione and NADPH, responsible for the redox balance of pyridine nucleotides.[69]

Finally ascorbate oxidase may have both a direct antioxidant role in the plant by scavenging organic radicals, since it is able to accept their extra electrons and to transfer them to oxygen with formation of water (see Sec. 3.3) or an indirect role, by producing the relatively harmless ascorbate radical which in turn may quench free radical chains. Ascorbate free radical is effective in preventing chain propagation as it is neither a strong oxidant or reductant (E°, semidehydroascorbate/ascorbate = +282 mV; dehydroascorbate/semidehydroascorbate = −174 mV, at pH 7.0). Furthermore it reacts poorly with oxygen, producing very little, if any, superoxide, which is a species responsible for the initiation of free-radical chain reactions.[75]

The apoplast compartment has been reported to be a site of production of oxygen free radicals if the air pollutant ozone interacts with the cellular surface. Ascorbate oxidase was shown to reduce the *in vitro* production of singlet oxygen under treatment with O_3.[76] It has been reported that herbicides paraquat, amitrole or acilfluorfen led to oxidative stress resulting in a drastical increase of monodehydroascorbate, probably due to the protective activity of ascorbate oxidase.[77]

Some of the proposed roles for ascorbate oxidase are resumed in Fig. 4.

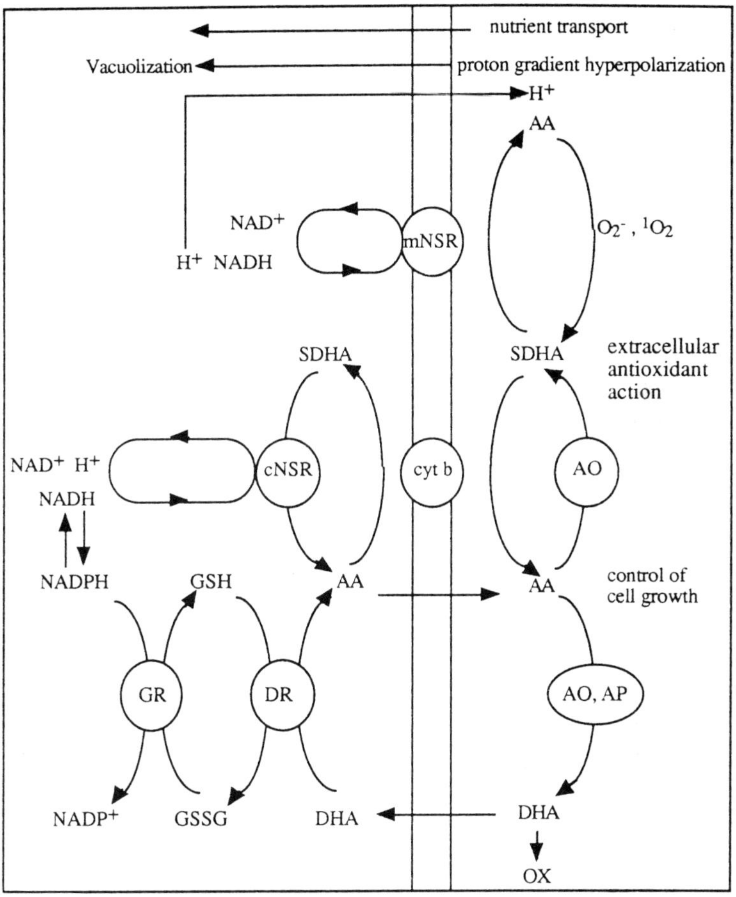

Fig. 4. Ascorbate and ascorbate oxidase system in plant cell (according to Ref. 62). AA–ascorbate; AO–ascorbate oxidase; AP–ascorbate peroxidase; cNSR–cytosolic NADH semidehydroascorbate reductase; DHA–dehydroascorbate; DR–dehydroascorbate reductase; GR–glutahione reductase; mNSR–membrane NADH semidehydroascorbate reductase; OX–oxalate; SDHA–semidehydroascorbate. On the left: cytosol; on the right: apoplast.

3. Functional Properties of Ascorbate Oxidase

Ascorbate oxidase oxidizes L-ascorbate quite efficiently. The most active samples of ascorbate oxidase extracted from green zucchini show a K_m for ascorbate of 0.2 mM, a K_m for oxygen of 0.2 mM, a k_{cat} of 7500 s^{-1}, and hence a k_{cat}/K_m value of 3.8×10^7 M^{-1} s^{-1} at 25°C. The activity as a function of pH follows a bell-shaped curve, nearly flat between pH 5 and pH 7 but with maximum at pH 6. The maximal activity was found at 35°C. The Ea value was 10 kJ mol^{-1}, in accordance with the high efficiency of ascorbate oxidation.[78]

3.1. *Substrate Specificity*

The specificity of the reduction site of ascorbate oxidase and the steric and electronic requirement for a substrate was studied in the presence of L-ascorbate analogs and derivatives.[79] Structural modifications at C-6 position were uninfluential on the activity. The enzyme turnover remained almost unchanged with the compounds tested (6-amino-; 6-deoxy-; 6-deoxy-6-bromo-; 6-O-phenyl-; 6-S-phenyl-L-ascorbic acid). The enzyme affinity remained unchanged for the two deoxyderivatives while it was decreased four times for the aminoderivative, possibly due to the presence of a positively charged group. The affinity was increased six times in the two phenyl derivatives, possibly due to the presence of a hydrophobic region in the active site of the enzyme. Removal or blocking of C-5 OH (5,6-ene-L-ascorbic acid; 5,6-isopropylidene-L-ascorbic acid; D-isoascorbic acid) significantly decreased the affinity but not the turnover number. Instead, a structural modification at C-2 position had a drastic effect since 2-amino-6-S-phenyl-L-ascorbic acid and imino-D-glucoascorbic acid were neither substrates nor inhibitors of ascorbate oxidase. The most significant difference between ascorbic acid and 2-amino ascorbic acid was the ionization of C-3 OH in the natural compound ($pK_a = 4.2$) which occurs at higher pH values in the amino derivative ($pK_a = 6.8$), indicating that the enzyme preferentially interacts with the anionic form of the substrate. The lack of reactivity of imino-D-glucoascorbic acid further supports the idea that the presence of a lactone ring carrying a vicinal enediol adjacent to carbonyl group is required for reaction with ascorbate oxidase.[80] Interestingly, the specificity of the ascorbate oxidase is remarkably similar to the specificity of the mammalian copper enzyme ascorbate-dependent dopamine β-monooxygenase,[81] suggesting that the ascorbate binding sites of these two enzymes might be similar, although there is no known sequence identity.

Besides ascorbate, several one-electron reductants may interact with ascorbate oxidase. Substituted hydroquinones and leuco-2,6-dichloroindophenol were oxidized to the respective p-quinones and to the blue quinoid dye.[80] Compounds containing the functional o-diphenolic group, such as the naturally occurring (+)-catechin, were also oxidized *in vitro*. *In vivo* o-diphenolic compounds normally do not come in contact with ascorbate oxidase unless the normal cellular architecture is upset by wounding or senescence. This reaction may be responsible for the darkening process in fruits and vegetables and is prevented by competitive reaction with ascorbate.[82]

Ascorbate oxidase can also oxidize redox compounds like organic radical and ferrocyanide (see below).

3.2. *Ascorbate Oxidase Entrapped in Liposomes*

Since many biological effects of ascorbate involve its interaction with cell or vesicle membranes both in plants (i.e. apoplasts, vacuoles) and animals (i.e. chromaffin granules, synaptosomes),[72] ascorbate oxidase-loaded vesicles might be a good model for monitoring such interactions. The enzyme was entrapped in dipalmitoyl phosphatidylcholine/cholesterol unilamellar liposomes by using two different methods. Liposomes loaded with ascorbate oxidase obtained by controlled detergent dialysis were found to be more homogenous than those obtained by reverse phase evaporation.[83,84] Ascorbate oxidase was mostly embedded in the lipid bilayer, as indicated by transmission electron microscopy and freeze-fracture and by labeling with specific polyclonal antibodies. The size of liposomes formed in the presence of ascorbate oxidase was one-half of that of empty vesicles, possibly due to the localization of ascorbate oxidase across the membrane. The affinity of ascorbate for the entrapped enzyme was similar to that for the enzyme free in solution. However, an increased oxidase activity after detergent treatment indicated that about 50 % of the activity was masked in intact liposomes. The relative lipophily of ascorbate oxidase may explain its ability either to cross the cell membrane or to stick to it.

3.3. *Electron Transfer Inside Ascorbate Oxidase*

The known three-dimensional structure of ascorbate oxidase makes this protein highly suitable for the elucidation of the individual roles of the different copper centers in blue oxidases. It also provides a model for studying the different factors which control long-range (>10 Å) electron transfer. Recently ascorbate oxidase has been the subject of several transient kinetic studies. Although there have been conflicting reports from different research groups, these studies have allowed researchers to elucidate many steps of the electron-transfer process involved in the catalytic cycle.

3.3.1. *Reaction with Ascorbate and Reductate*

Static anaerobic reduction of ascorbate oxidase showed that the blue type 1 Cu was first reduced, then in a few minutes an intramolecular electron transfer occurred with some reduction of the 330 nm chromphore and a partial reappearance of the blue color. Upon reoxidation, the process took place in the opposite direction.[85] These early observations suggest that the blue type 1 Cu is the entrance site for electrons into the enzyme and that the type 3 Cu, which is associated with the 330 nm band, is the oxygen reduction site. In addition, these observations support the hypothesis of a linear array of redox centers which undergo sequential reduction.

Pre-steady-state kinetics of the reaction between ascorbate oxidase and ascorbate was investigated by stopped-flow method. The type 1 Cu reduction was very rapid being completed in the dead-time of the instrument used (3.5 ms). The reoxidation process appeared to be biphasic, with an initial rapid step followed by a slower one.[3]

Reductate was found to be a more convenient substrate than ascorbate for rapid kinetic studies as the reaction could be followed by stopped-flow technique using a wide range of substrate concentrations. However, complex multiphasic kinetic behavior was observed by monitoring either the reduction of type 1 Cu at 610 nm or the reduction of type 3 copper at 330 nm. The rate and amplitude of each phase depended on the reductate concentration and on the presence or the absence of oxygen.[86]

Other one-electron reductants such as ferrocyanide, transient radicals produced by pulse radiolysis or flash photolysis were used in rapid kinetic reactions with ascorbate oxidase, on the assumption that they might provide more information on the reaction mechanism.

3.3.2. *Reaction with Ferrocyanide*

Ferrocyanide behaves as a substrate for ascorbate oxidase and shows K_m value of 3 mM and K_{cat} value of 250 s^{-1} at pH 6.0 and 25°C. Unlike ascorbate, the oxidation of ferrocyanide is strongly affected by pH, showing a two-order higher turnover number at pH 5.5 than at pH 7.5. This difference may be due to a different site of interaction with the enzyme. In fact, in anaerobiosis ferrocyanide selectively reduced type 2 Cu, as detected by EPR spectroscopy.[87] Type 1 Cu was only partially reduced, with an asymptotic value of 30 %, even in the presence of a large excess of the reductant, as detected by EPR,[87] static titration and rapid kinetic studies.[3] This behavior may be explained by the fact that ferrocyanide is a mild reductant with a redox potential (E° = +420 mV) higher than that of type 1 Cu (E° = +344 mV at pH 7.0). The redox potentials of type 1 Cu and type 3 Cu are almost identical at 25°C.[2] The redox potential of type 2 Cu is not known but the coordination number of the histidyl residues suggest a value not very different from that of the similarly coordinated type 3 Cu. However, under the above experimental conditions, the redox potential of type 2 Cu must be higher than that of ferrocyanide, and lower than that of type 3 Cu since electron transfer most likely took place from type 2 to type 3 copper in this catalytic cycle.

As a matter of fact, redox potential of the three types of copper may be affected by several factors. At a lower temperature, i.e. 10°C, the redox potential of type 1 Cu did not change but that of type 3 copper became more positive by 40 mV.[2] Interaction of ferrocyanide made the redox potential of the type 3 Cu in the parent protein tree laccase more positive.[88] The binding of ligands to type 2 may instead decrease its reduction potential by increasing the coordination number.[16]

Ascorbate oxidase is stable towards chelators either in the fully-oxidized or the fully-reduced forms. Selective removal of type 2 Cu was obtained only in anaerobiosis, in the presence of ferrocyanide and of chelators with high affinity for oxidized copper (EDTA, nitrilotriacetate, dimethylglioxime).[16] The experimental conditions suggest that the reactive species was the type 3 Cu (I)/type 2 Cu (II) derivative. It appears that different redox states may occur at the type 2 Cu/type 3 Cu site since redox potentials may be modulated as described above.

3.3.3. *Reaction with Radicals*

The intermolecular electron transfer from the reducing substrate to the type 1 Cu was studied by reacting ascorbate oxidase with radicals generated by pulse radiolysis. The reaction was monitored by following absorption changes at 400–410 nm for the radicals and at 610 nm for type 1 Cu.

Radiolysis of deoxygenated buffer produced e^-_{aq} or CO_2^-. The interaction of e^-_{aq} with ascorbate oxidase resulted in a non-specific reduction of the protein moiety, which, in turn, transferred electrons to type 1 Cu through an intramolecular pathway (first-order rate constant $\approx 10^4$ s^{-1}).[89] This sequence of events is likely to occur because of the low redox potentials of the reducing species (E° $H_2O/e^-_{aq} = -2870$ mV at pH 7.0[75]), which may induce reduction of aminoacyl residues. Ascorbate oxidase has three disulphide bridges.[6] Disulphide radical anions, which strongly absorb at $\approx$ 400 nm ($\varepsilon = 10000$ M^{-1} cm^{-1}), do not seem to be produced, unlike tree laccase[90] and ceruloplasmin[91] in which such radicals have been produced in analogous experiments. The blue copper reduction never exceeded 50 % even in the presence of a large excess of the reductant indicating saturation of the primary acceptor.

The interaction of CO_2^- with ascorbate oxidase also caused the reduction of type 1 Cu. The time-course of the bleaching of the blue band was analyzed either in terms of two first-order reactions for which a non-specific reduction of the protein by formate radical was suggested,[89] or in terms of a second-order reaction for which a direct bimolecular interaction with rate constant 1.2×10^9 M^{-1} s^{-1} at pH 5.5–7.5 was indicated.[92,93]

Superoxide, formed by the radiolysis of an oxygenated buffer, was shown to be able to reduce ascorbate oxidase by direct interaction with type 1 Cu at a rate of $< 5 \times 10^6$ M^{-1} s^{-1}. Type 1 Cu was reduced by up to 30 % in the holoenzyme and to a greater extent in the type 2 Cu-depleted derivative.[94]

Table 1. Second-order rate constants (M^{-1} s^{-1}) of the reduction of the type 1 Cu of ascorbate oxidase by different radical reductants (E° = mV at pH 7.0) at 25°C (* 19°C). Values for ascorbate are reported for comparison (# k_{cat}/K_m).

Radical	k	E°	Ref.
O_2^-	<5x10^6	−330	94
CO_2^-	1.2x10^9	−2000	92,93
Metronidazole	3x10^7	−486	89
Lumiflavin*	3.8x10^7	−240	93
Lumiflavin	2.7x10^7		95
Deazaflavin*	1.5x10^8	−650	93
Methyl viologen^{+*}	2.6x10^7	−450	93
Ascorbate	3.8x10^7#	+110	3

One-electron nitroaromatic radical anions were generated by pulse radiolysis of nitroaromatic compounds, such as metronidazole and misonidazole, which are generally used as sensitizers, since their redox potentials are considerably more positive than those of e^-_{aq} or CO_2^- and may be more selective in their interaction with ascorbate oxidase (Table 1).[89,94]

Upon reaction, a fast electron transfer to type 1 Cu occurred in a bimolecular second-order reaction ($k = 3 \times 10^7$ M^{-1} s^{-1} at pH 6.0–7.5) comparable to the k_{cat}/K_m ratio obtained using ascorbate as the reducing substrate. Ascorbate oxidase reduction was almost stoichiometric as six to seven equivalents of the nitroaromatic radicals were required for the complete bleaching of type 1 Cu in the native enzyme. In type 2 Cu depleted derivative, the reduction of type 1 Cu occurred at the same rate as it did in the holoenzyme. However, only three reducing equivalents were needed for stoichiometric reduction of type 1 Cu.

Lumiflavin semiquinone produced by laser flash photolysis in anaerobiosis was used as a reducing substrate.[95] In this case too, the pseudo-first-order rate constant was found to be dependent on the protein concentration and the calculated second-order constant was 2.7×10^7 M^{-1} s^{-1} at pH 7.0 for both native and type 2 Cu depleted ascorbate oxidase. A fast bimolecular reduction of the type 1 Cu by lumiflavin, deazaflavin and methyl viologen (1,1'-dimethyl-4-4'-bipyridinium) radicals generated by pulse radiolysis (Table 1) was also observed.[93]

In view of these results, it can be concluded that ascorbate oxidase may interact with a variety of radicals which reduce type 1 Cu by different mechanisms depending on their redox potential. A specific reaction seems operative only with some less powerful redox species. In this case, the rate constants are independent of the one-electron redox potentials of the radical. Type 2 Cu is not required for efficient type 1 reduction but appears important for redistribution of electrons within the different copper sites.

3.3.4. *Intramolecular Electron Transfer*

The intramolecular electron transfer from type 1 Cu to the trinuclear copper center may be observed in the one-electron reduced enzyme by radicals produced by laser flash photolysis[95] and pulse radiolysis[92,93] under anaerobic conditions.

Unlike studies on the reduction mechanism, in which a large excess of reducing equivalents with respect to the protein concentration were used, the measurements regarding intramolecular electron transfer were carried out by using a large molar excess of protein with respect to the amount of the radical. This procedure should ensure that no enzyme molecule was reduced by more than one electron.

Following fast reduction by lumiflavin semiquinone produced by flash photolysis, the type 1 Cu was slowly reoxidized in a reaction independent of enzyme concentration ($k = 160\ s^{-1}$).[95] Type 1 Cu was reoxidized by only about 55 % and then approached equilibrium, presumably reflecting the electron distribution between different coppers depending on their redox potentials. This observation is consistent with electron transfer to an approximately equipotential one-electron acceptor and thus the observed rate constant may be the sum of the forward and reverse constants for this equilibration.

The rate constant of 160 s^{-1} for the intramolecular electron transfer was only observed in ascorbate oxidase samples showing maximum turnover in steady-state kinetics. On the contrary, many samples which appeared homogenous from the standpoint of protein biochemistry were found to be heterogenous with respect to their catalytic activity. Enzyme preparations which had about 80 % less activity had a rate constant of 2.6 s^{-1}. Other samples showed biphasic rate constants for intramolecular electron transfer. Some of the less active enzyme samples could be reactivated using ascorbate to induce oxidative turnover or through direct oxidation by ferricyanide or hydrogen peroxide, resulting in an increase in the intramolecular electron transfer rate constant to a value comparable to that of the native enzyme.[96]

Reduction at the type 2 Cu/type 3 Cu trinuclear site should occur together with type 1 Cu reoxidation. However, the absorbance of flavin did not allow direct monitoring of spectral changes at 330 nm where type 3 Cu absorbs. In addition, no new EPR signal was detected as a result of uncoupling of the type 3 Cu pair. Still, the reoxidation process was interpreted as an intramolecular electron transfer from type 1 Cu to type 3 Cu on the basis that the process occurred at the same rate in the absence of type 2 Cu, as in the presence of inhibitors which bind type 2 Cu.[95] In fact, the removal of type 2 Cu from ascorbate oxidase did not significanly affect the reoxidation process rate, although the extent of reoxidation in type 2 Cu depleted enzyme was significatively larger than that in the native enzyme (73 % vs 55 %), possibly due to a greater difference in the redox potentials of type 1 Cu and type 3 Cu in type 2 Cu-depleted derivative. Fluoride and azide did not cause any significant changes in the kinetics of reduction and intramolecular electron transfer, although, both (a) a bond with type 2 Cu has been proposed for F^- (Ref. 13) and (b) either a bond with one of two type 3 Cu (Ref. 10) or a bridge between type 2 Cu and one of the type 3 Cu (Refs. 13 and 97) have been proposed for N_3^-. In contrast, cyanide caused a concentration-dependent decrease in the rate of the type 1 Cu reduction and in the extent of electron transfer with no change in rate constant, both in holo and type 2 Cu-depleted enzymes. A change in redox potentials upon binding of CN^- was suggested.[95]

This early report apparently disagreed with subsequent findings in which either two or three unimolecular steps of electron-tranfer were observed upon reoxidation of the reduced type 1 Cu.

In one such report,[92] following the fast-reduction phase related to reaction of type 1 Cu with formate radical produced by pulse radiolysis, either two or three reoxidation steps were monitored depending on the extent of type 1 Cu reduction. A concomitant reduction of the 330 nm band was observed, again proceeding through two or three phases. The respective rate constants were independent of protein and CO_2^- concentration. These observations confirm that an electron transfer process occurs from reduced type 1 Cu to oxidized type 3 Cu. At pH 5.5 and at 25°C, the rate constants observed for the three unimolecular electron transfer steps were 201, 20 and 2.3 s^{-1}, respectively. At pH 7.0, only a 20 % decrease in rate constant value was observed, which is in line with the kinetic studies showing almost constant enzymatic activity between pH 5.5 and 7.0. The activation parameters for the fastest ($\Delta H = 9.1$ kJ mol^{-1}, $\Delta S = -170$ J K^{-1} mol^{-1}) and the slowest phases ($\Delta H = 6.8$ kJ mol^{-1}, $\Delta S = -215$ J K^{-1} mol^{-1}) were determined.

In another report,[93] a biphasic reoxidation process following the fast reduction of type 1 Cu by CO_2^-, lumiflavin, deazaflavin and methyl viologen radicals produced by pulse radiolysis was observed. The rate constant had an average value of 122 s^{-1} for fast step and of 2.3 s^{-1} for slow step and was independent of both protein and radical concentrations (Table 2). Activation parameters of the two steps were ΔH=29.8 kJ mol^{-1} and $\Delta S = -108.3$ J K^{-1} mol^{-1} and $\Delta H = 53.3$ kJ mol^{-1} and $\Delta S = -58.8$ J K^{-1} mol^{-1} respectively.

On the basis of the three-dimensional structure of ascorbate oxidase, it has been proposed that shortest pathway for electron transfer from type

Table 2. First-order rate (s^{-1}) for the reoxidation of type 1 Cu in ascorbate oxidase (*pH 5.5, 25°C; ** pH 7.0, 19°C, # pH 7.0, 25°C).

Radical	k_1	k_2	k_3	Ref.
CO_2^{-*}	201	20	2.3	92
CO2−**	120	2.0	–	93
Lumiflavin#	160	–	–	95
Lumiflavin**	97	2.4	–	93
Deazaflavin**	121	2.5	–	93
Methyl viologen+**	127	2.3	–	93

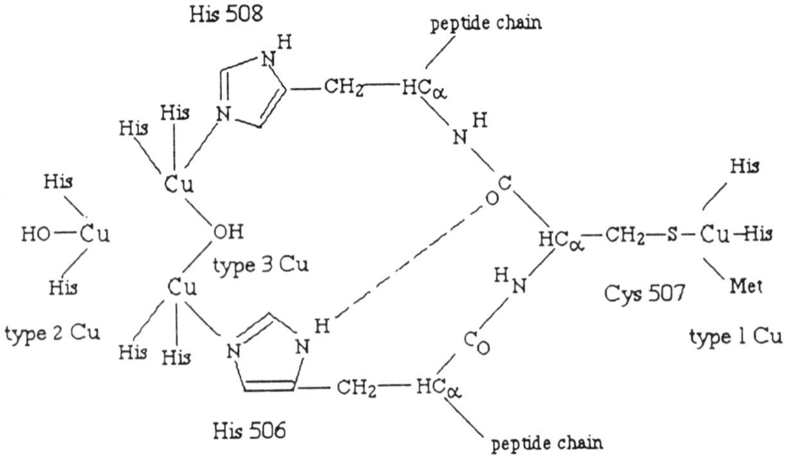

Fig. 5. Coordination of type 1 Cu, type 2 Cu and type 3 Cu in fully oxidized ascorbate oxidase and possible pathways for electron transfer (according to Refs. 7, 93).

1 Cu to type 3 Cu is via Cys 507 and either His 506 or His 508 (Fig. 5). Both routes consist of nine covalent bonds. An alternative route may be provided by a hydrogen bond between the carbonyl oxygen of the Cys 507 and Nδ of His 506. The three pathways occur with equal probability[93,98] as calculated by using an algorithm developed to identify the most favourable long-range electron-transfer route in metalloprotein.[99]

The fast process might correspond to the intramolecular electron transfer of the first electron from the reduced type 1 Cu to an equipotential electron acceptor, which was probably one of the type 3 Cu. Since multiphasic behavior related to different electron pathways seems to be excluded, the presence of the two steps was explained by different reactivity of ascorbate oxidase reduced to different extents, possibly due to structural changes occurring at the trinuclear site after the acceptance of the first electron which slowed down the transfer of the second one.[93] Alternatively, the presence of more than one step was attributed to the presence of distinct reactive forms of ascorbate oxidase. Large structural changes at the trinuclear copper center of ascorbate oxidase crystals occurred upon reduction and upon binding of ligands such as hydrogen peroxide and azide.[10] Limited variations at this site are possibly present in the native fully-oxidized enzyme due to, for example, different protonation of the oxygen ligands of type 2 Cu and type 3 Cu.[92,98] A third hypothesis regarding the existence of a "resting" ascorbate oxidase and of a "pulsed" one activated by catalytic turnover as described for cytochrome oxidase[100] was examined and ruled out.[101]

In any case, the rate of the fast intramolecular process is considerably lower than the turnover number (7500 s^{-1} at pH 6.0). This indicates that the electron transfer should occur faster in multi-reduced enzyme and in the presence of oxygen than it should in the single-reduced enzyme.

The hypothesis that the number of electrons taken up ascorbate oxidase may modulate intramolecular electron transfer has been supported by complete time-resolved titrations by formate radical. When the 610 nm and the 330 nm bands were almost fully bleached, the rate constant of the fast intramolecular electron transfer phase was 1000 s^{-1}, which was five fold larger than the corresponding fast phase for the one-electron reduced enzyme.[102] Intramolecular electron transfer, subsequent to the one-electron reduction of type 1 Cu by formate radical, was also followed in the presence of a slight excess of oxygen over ascorbate oxidase. Under this condition the rate constant for intramolecular electron transfer was 1100 s^{-1}, which was about five fold times that found for electron transfer under anaerobic conditions. Comparison of the spectral changes monitored at 330 nm for the reaction in the presence and in the absence of oxygen suggested that the one-electron reduced trinuclear center may already interact with oxygen.[101]

These findings are in line with the results of a study on intramolecular electron redistribution in three-electron reduced ascorbate oxidase.[103] The enzyme, fully reduced by ascorbate, was reoxidized by laser-generated triplet state of deazariboflavin. This reaction should be very rapid, as suggested by bimolecular rate constant in the diffusion-controlled range (10^9 M^{-1} s^{-1}) for the redox proteins cytochrome c[104] and plastocyanin.[105] The concentration of the oxidant was considerably smaller than the concentration of the reduced enzyme to ensure that only a single electron was removed during the reaction. Following reaction with the radical, oxidation of the reduced type 1 Cu occurred in two phases of approximately equal amplitute and with rate constant of 9500 s^{-1} and 1400 s^{-1} as monitored at 610 nm. Monitoring at 330 nm was not possible due to the flavin absorbance. It was hypothesized that the initial oxidation reaction removed an electron from the reduced type 2 Cu/type 3 Cu trinuclear center, probably via parallel reactions of two of the three copper atoms. Then intramolecular electron transfer from reduced type 1 Cu to two different oxidized coppers, possibly type 2 Cu and one of the type 3 Cu, followed at a rate of one to two order faster than that observed for electron transfer in the one-electron reduced enzyme. Interestingly, the observed rate constant for fast reaction is comparable to the turnover number obtained in steady-state kinetics.

Altogether, the above results provide convincing evidence that the intramolecular electron transfer in ascorbate oxidase is dependent upon the overall redox state, being more effective in greatly reduced enzyme, and upon the binding of oxygen. In fact, the presence of copper-oxygen complexes may affect the redox properties of the trinuclear copper site with the consequent increase in the driving force.

3.3.5. *Reaction with Dioxygen*

The catalytic cycle of ascorbate oxidase may be envisaged as a sequential mechanism. Single electrons are provided to the type 1 Cu by the substrate. The trinuclear type 2 Cu/type 3 Cu center may be the site of dioxygen binding and of its reduction to water. The sequential intramolecular electron transfer from type 1 Cu to the trinuclear copper center, via type 3 Cu, provides the necessary four electrons. At the moment, it is not clear how many electrons must be transferred to the trinuclear site before the site can bind the oxygen, nor is the nature of the intermediates of the action well known.

A mode of dioxygen binding has been proposed based on the kinetic and spectroscopic properties of ascorbate oxidase, on the known three-dimensional structures of fully-oxidized, fully-reduced and peroxide-bound forms of the enzyme and on a comparison with the catalytic and spectroscopic properties of the related "blue" oxidase laccase. The full amount of evidence and hypotheses concerning this issue are reported in Ref. 7, creating a unified picture. According to this proposal, dioxygen binds to the fully-reduced enzyme and may form bridges between any copper of the trinuclear center. A classical ping-pong mechanism for ascorbate oxidase has been described on the basis of steady-state experiments (L. Avigliano, unpublished results). This allows rapid electron transfer from each of the three reduced copper ions, leading to the formation of three electron reduced species $O_2^{\cdot 3-}$. This species must accept protons from the bulk solvent in order to release the first molecule of water. At this point it is possible that, in analogy with laccase as suggested by EPR experiments,[106] an intermediate containing an oxygen radical $O^{\cdot -}$ may be formed. It is then reduced by the remaining electron. The proposed model implies that the oxygen ligand of type 2 Cu is removed upon reduction. This release may be a slow process and may correspond to the slow phase observed in the reoxidation of type 1 Cu. This hypothesis is in line with EPR analysis of rapid-freeze-quenched reaction showing a slow reduction of type 2 Cu in ascorbate oxidase.[2]

Another possibility is that molecular oxygen binds to the completely reduced form of the enzyme to give a peroxide derivative,[10] analogous to that described by the X-ray analysis, where peroxide is bound end-on as hydroperoxide to one of the type 3 copper pair, i.e. the one coordinated by His 506. According to this hypothesis, the transfer of the two electrons from type 3 Cu to dioxygen may lead to the formation of a hydroperoxide intermediate. At this point, two further electrons may be provided by reduced type 2 Cu and type 1 Cu with the consequent rupture of the O-O bond. The first water molecule is liberated and an intermediate containing an oxygen radical is formed. The next step leads to the formation of the second water molecule.

Not all the reported kinetic studies prove the existence of a "resting" and of a "pulsed" form of ascorbate oxidase. It is possible, however, that during the turnover the trinuclear site maintains a structure close to that of the reduced enzyme. This requires only minor rearrangements of the site, facilitating the subsequent electron transfer. This may be a further explanation for the large difference between the calculated rate of electron transfer in the rapid kinetics and the turnover number of steady-state kinetics.[7,10]

3.4. *Reaction with Cytochrome c*

Ascorbate oxidase is involved in reactions with small reductants and has not known proteins as natural redox partners. However, ascorbate oxidase has been found to interact and to accept an electron from reduced horse heart cytochrome *c* ($E° = +250$ mV) with second-order rate constant $k = 820\ M^{-1}\ s^{-1}$ in 0.2 M phosphate buffer at pH 6 and 25°C. The activation parameters were estimated to be $\Delta H = 5$ kJ M^{-1} and $\Delta S = -188$ J $K^{-1}\ M^{-1}$. The rate of reaction was increased by both decreasing pH and ionic strength, suggesting that the electron transfer between cytochrome *c* and ascorbate oxidase was carried out through local electrostatic interactions.[107] The reported redox system, due to the known structure of both cytochrome *c* and ascorbate oxidase, may be a useful model for studying the mechanisms of macromolecule recognition and of intermolecular long-range electron transfer.

4. Biotechnological Applications of Ascorbate Oxidase

Many methods have been developed for the determination of ascorbate (vitamin C) in clinical and food analysis. Most of them are based on redox properties of ascorbate and therefore lack specificity. A more specific although time consuming method is high-performance liquid chromatography.[108]

Ascorbate oxidase can be used as rapid and specific method for the determination of ascorbate in biological samples due to the relatively high affinity and the high catalytic power towards ascorbate. The enzyme is also widely used for the removal of interferences from ascorbate in common clinical chemistry assays such as the estimation of urinary oxalate[109] or serum creatine,[110] urate and cholesterol.[111]

The lability of ascorbate oxidase may limit its use. The enzyme may be inactivated by physical and chemical factors, such as temperature, pH and UV irradiation. The $t_{1/2}$ of inactivation at 55°C is about 20 min. Inactivation also occurs at a pH lower than 4.5 and higher than 9.0. Ascorbate oxidase is sensitive to UV irradiation which causes a 90 % drop in activity in 30 min, possibly due to photooxidation of aromatic aminoacyl residues.[78] The enzyme undergoes inactivation when frozen for long periods.[96] Freeze-drying also reduces enzyme activity. In this case, an effective stabilization was obtained through treatment with lyoprotectants.[112] Sucrose, mannose, glucose and polyvinylpyrrolidone increased the recovery of activity in rehydrated samples. The protective effect was observed at high ascorbate oxidase concentration only.[113]

Immobilization of ascorbate oxidase on various supports markedly improves the stability of the enzyme. In general the kinetic properties of the immobilized enzyme were found to be similar of those of the free form. However, it was more stable to heat and pH than the free enzyme. It also remained fully active for at least one to three months at 4°C. Therefore, the immobilized ascorbate oxidase was used to set up methods for the determination of ascorbate. The rationale is to utilize both the high specificity of the enzyme and the high sensitivity of the detector, i.e. voltammetric or amperometric detectors. Some recent methods are reported below.

Ascorbate oxidase was physically entrapped within calcium-alginate gel beads. The stability of the entrapped enzyme was enhanced by treatment with glutaraldehyde. The system was static and the ascorbic acid concentration was monitored by measuring the change in enzyme activity spectrophotometrically at 265 nm.[114]

Alternatively ascorbate oxidase was immobilized on CH-Sepharose 4B via carbodiimide with the formation of amide bonds between the -COOH groups of CH-Sepharose and the $-NH_2$ group of the enzyme. The immobilized enzyme was used in a flow-through system equipped with a polarographic detector which monitors oxygen depletion due to its reaction with ascorbate. This method allowed for a rapid determination of ascorbate as low as 1–10 μM in biological samples, such as human plasma and fruit juices.[115]

In a different set-up, ascorbate oxidase was immobilized on cyanogen bromide-activated Sepharose 4B and incorporated in a flow-injection system with amperometric detection at a glassy carbon electrode at + 600 mV. Upon passage through the immobilized enzyme, ascorbate was converted into dehydroascorbate which is not electroactive and the decrease in signal compared to the blank was measured. The detection limit was lower than 1 μM in fruit and vegetable juices. Fresh fruits contain a number of organic acids and sugars, such as tartrate, citrate, malate, fructose and glucose, which did not seriously interfere with ascorbate oxidase activity even in much larger quantities than ascorbate. Therefore, manipulation of the biological samples before the measurements was not required.[116]

These examples are only a few of the many possible applications of ascorbate oxidase as a detecting device for measurement of ascorbate, as well as its removal from biological fluids.

References

1. J. A. Fee, *Struct. Bond.* **23**, 1 (1975).
2. P. M. H. Kroneck, F. A. Armstrong, H. Merkle and A. Marchesini, in *Advances in Chemistry. Ascorbic Acid: Chemistry, Metabolism and Uses*, eds. P. A. Seib and B. M. Tolbert (Am. Chem. Soc., Washington, 1982), p. 223.
3. B. Mondovi and L. Avigliano, in *Copper Proteins and Copper Enzymes*, Vol. 3, ed. R. Lontie (CRC Press Inc., Boca Raton, Florida, 1984), p. 101.
4. A. Finazzi-Agrò, *Life Chem. Rep.* **5**, 199 (1987).
5. V. S. Butt, in *The Biochemistry of Plants. A Comprehensive Treatise. Metabolism and Respiration*, Vol. 2, ed. P. J. Davies (Acad. Press, 1980), p. 85.
6. A. Messerschmidt, A. Rossi, R. Ladenstein, R. Huber, M. Bolognesi, G. Gatti, A. Marchesini, R. Petruzzelli and A. Finazzi-Agrò, *J. Mol. Biol.* **206**, 513 (1989).
7. A. Messerschmidt, R. Ladenstein, R. Huber, M. Bolognesi, L. Avigliano, R. Petruzzelli, A. Rossi and A. Finazzi-Agrò, *J. Mol. Biol.* **224**, 179 (1992).
8. J. Ohkawa, N. Okada, A. Shinmyo and M. Takano, *Proc. Natl. Acad. Sci. USA* **86**, 1239 (1989).
9. M. Esaka, T. Hattori, K. Fujisawa, S. Sakajo and T. Asahi, *Eur. J. Biochem.* **191**, 537 (1990).
10. A. Messerschmidt, H. Luecke and R. Huber, *J. Mol. Biol.* **230**, 997 (1993).
11. J. M. Guss, P. R. Harrowell, M. Murata, V. A. Norris and H. C. Freeman, *J. Mol. Biol.* **192**, 361 (1986).
12. W. E. B. Shepard, B. F. Anderson, D. H. Lewandoski, G. E. Norris and E. N. Baker,*J. Amer. Chem. Soc.* **112**, 7817 (1990).
13. J. Cole, L. Avigliano, L. Morpurgo and E. I. Solomon, *J. Am. Chem. Soc.* **11**, 9080 (1991).
14. L. Avigliano, A. Desideri, S. Urbanelli, B. Mondovi and A. Marchesini, *FEBS Lett.* **100**, 318 (1979).
15. A. Messerschmidt, W. Steigemann, R. Huber, G. Lang and P. M. H. Kroneck, *Eur. J. Biochem.* **209**, 597 (1992).
16. L. Morpurgo, M. T. Graziani, G. Marcozzi and L. Avigliano, *J. Inorg. Biochem.* **51**, 641 (1993).
17. G. D'Andrea, J. B. Bouwstra, J. P. Kamerling and J. F. G. Vliegenthart, *Glycoconjugate J.* **5**, 151 (1988).
18. G. A. White and R. M. Krupka, *Arch. Biochem. Biophys.* **110**, 448 (1965).
19. S. Murao, H. Itoh, T. Yajima, Y. Ozaki, S. Fukuyasu and T. Shin, *Biosci. Biotech. Biochem.* **56**, 847 (1992).

20. J. Hirose, T. Sakurai, K. Imamura, H. Watanabe, H. Iwamoto, K. Hiromi, H. Itoh, T. Shin and S. Murao, *J. Biochem.* **115**, 811 (1994).
21. T. Nakamura, N. Makino and Y. Ogura, *J. Biochem.* **64**, 189 (1968).
22. M. H. Lee and C. R. Dawson, *Meth. Enzymol.* **62**, 30 (1979).
23. L. B. Carvalho Jr, C. J. Lima and P. H. Medeiros, *Phytochemistry* **20**, 2423 (1981).
24. L. Avigliano, P. Gerosa, G. Rotilio, A. Finazzi-Agrò, L. Calabrese and B. Mondovi, *Ital. J. Biochem.* **21**, 248 (1972).
25. A. Marchesini and P. M. H. Kroneck, *Eur. J. Biochem.* **101**, 65 (1979).
26. O. Moser and A. K. Kanellis, *J. Exp. Bot.* **45**, 717 (1994).
27. N. B. Saari, S. Fujita, K. Haraguchi and T. Tono, *J. Sci Food Agric.* **65**, 153 (1994).
28. K. P. Reddy, P. A. Khan, S. Patnaik, G. B. Mohanty and K. B. Kumar, *Plant Cell Physiol.* **27**, 725 (1986).
29. M. Baron Ayala and G. Sandmann, *Plant Physiol.* **72**, 801 (1988).
30. Y. Kida, N. Honda, M. Uchida, Y. Kunisada and M. Fukuda, *J. Japan. Soc. Food Sci. Tecnol.* **38**, 160 (1991).
31. S. Y. Wang and M. Faust, *HortScience* **27**, 1102 (1992).
32. K. Yamawaki, N. Morita, K. Murakami and T. Murata, *J. Japan. Soc. Food Sci. Technol.* **40**, 636 (1993).
33. C. Beauchamp and I. Fridovich, *Anal. Biochem.* **44**, 276 (1971).
34. M. Maccarrone, A. Rossi, G. D'Andrea, G. Amicosante and L. Avigliano, *Anal. Biochem.* **188**, 101 (1990).
35. G. D'Andrea, M. L. Salucci, G. Pitari and L. Avigliano, *Glycobiology* **3**, 563 (1993).
36. R. Hayashi and Y. Morohashi, *Plant Physiol.* **102**, 1237 (1993).
37. G. D'Andrea, A. Oratore, L. Avigliano, A. Rossi and A. Finazzi-Agrò, *Plant Sci.* **56**, 107 (1988).
38. G. Chicchiriccò, M. P. Cerù, A. D'Alessandro, A. Oratore and L. Avigliano, *Plant Sci.* **64**, 61 (1989).
39. U. Takahama and T. Oniki, *Plant Cell Physiol.* **35**, 257 (1994).
40. M. Hallaway, P. D. Phethean and J. Taggart, *Phytochemistry* **9**, 935 (1970).
41. J. Ohkawa, T. Ohya, T. Ito, H. Nozawa, Y. Nishi, N. Okada, K. Yoshida, M. Takano and A. Shinmyo, *Plant Cell Rep.* **13**, 481 (1994).
42. L. -S. Lin and J. E. Varner, *Plant Physiol.* **96**, 159 (1991).
43. M. Esaka, K. Fujisawa, M. Goto and Y. Kisu, *Plant Physiol.* **100**, 231 (1992).
44. M. Esaka, M. Uchida, K. Suzuki and K. Kubota, *Agric. Biol. Chem.* **53**, 1181 (1989).
45. S. Y. Wang, G. L. Steffens and M. Faust, *Phytochemistry* **25**, 311 (1986).
46. R. M. Ray, J. D. Tewari and N. Pant, *Prog. Hortic.* **15**, 276 (1985).
47. G. J Acton, H. D Drumm and H. Mohr, *Planta* **121**, 39 (1974).
48. T. H. Attridge, *Biochim. Biophys. Acta* **362**, 258 (1974).
49. L. Leaper and H. J. Newbury, *Plant Sci.* **64**, 79 (1989).

50. M. Esaka, J. Imagi, K. Suzuki and K. Kubota, *Plant Cell Physiol.* **29**, 231 (1988).
51. M. Esaka, M. Uchida, H. Fukui, K. Kubota and K. Suzuki, *Plant Physiol.* **88**, 656 (1988).
52. H. J. Cho, T. Aimi, S. -Y. Paik and Y. Murooka, *J. Fermentation Bioengineering* **68**, 193 (1989).
53. G. Pitari, G. Chicchiriccò, G. Marcozzi, A. Rossi, M. Maccarrone and L. Avigliano, *Plant Physiol. Biochem.* **31**, 593 (1993).
54. M. Esaka, H. Fukui, K. Suzuki and K. Kubota, *Phytochemistry* **28**, 117 (1989).
55. H. Lis and N. Sharon, *Eur. J. Biochem.* **218**, 1 (1993).
56. G. Pitari, G. D'Andrea, M. L. Salucci and L. Avigliano, *It. Biochem. Soc. Trans.* **5**, 163 (1994).
57. G. D'Andrea, M. Maccarrone, A. Oratore, L. Avigliano and A. Messerschmidt, *Biochem. J.* **264**, 601 (1989).
58. E. Delhaize, J. F. Loneragan and J. Webb, *Plant Physiol.* **78**, 4 (1985).
59. M. Esaka, I. Nishitani, H. Fukui, K. Suzuki and K. Kubota, *Phytochemistry* **28**, 2655 (1989).
60. M. Esaka, K. Suzuki and K. Kubota, *Phytochemistry* **29**, 1547 (1990).
61. O. Arrigoni, *J. Bioenerg. Biomembrane* **26**, 407 (1994).
62. F. Cordoba and J. A. Gonzales-Reyes, *J. Bioenerg. Biomembrane* **26**, 399 (1994).
63. I. Yamazaki and L. H. Piette, *Biochim. Biophys. Acta* **50**, 62 (1961).
64. U. Takahama, *Plant Cell Physiol.* **34**, 809 (1993).
65. J. B. Cooper and J. E. Varner, *Plant Physiol.* **76**, 414 (1984).
66. O. Arrigoni, R. Arrigoni-Liso and G. Calabrese, *FEBS Lett.* **82**, 135 (1977).
67. N. Horemans, H. Asard and R. J. Caubergs, *Plant Physiol.* **104**, 1455 (1994).
68. J. G. Dong, J. C. Fernandez-Maculet and S. -F. Yang, *Proc. Natl. Acad. Sci. USA* **89**, 9789 (1992).
69. C. H. Foyer, P. Descourvieres and K. J. Kunert, *Plant Cell Environment* **17**, 507 (1994).
70. H. Wefer and H. Sies, *Eur. J. Biochem.* **174**, 353 (1988).
71. R. H. Bisby and A. W. Parker, *Arch. Biochem. Biophys.* **317**, 170 (1995).
72. B. Rubinstein, *J. Bioenerg. Biomembranes* **26**, 385 (1994).
73. J. A. Gonzales-Reyes, A. Hidalgo, J. A. Caler, R. Palos and P. Navas, *Plant Physiol.* **104**, 271 (1994).
74. R. E. Christoffersen, M. L. Tucker and G. G. Laties, *Plant. Mol. Biol.* **3**, 385 (1984).
75. G. R. Buettner, *Arch. Biochem. Biophys.* **300**, 535 (1993).
76. J. R. Kanofsky and P. D. Sima, *J. Biol. Chem.* **270**, 7850 (1995).
77. H. B. Stegmann, P. Schuler, S. Westphal and E. Wagner, *Z. Naturforsch. (Biosci.)* **48**, 766 (1993).
78. M. Maccarrone, G. D'Andrea, M. L. Salucci, L. Avigliano and A. Finazzi-Agrò, *Phytochemistry* **32**, 795 (1993).

79. K. Wimalasena and S. Dharmasena, *Biochem. Biophys. Res. Commun.* **203**, 1417 (1994).
80. Dayan and C. R. Dawson, *Biochem. Biophys. Res. Commun.* **73**, 451 (1976).
81. K. Wimalasena, S. Dharmasena and D. S. Wimalasena, *Biochem. Biophys. Res. Comm.* **200**, 113 (1994).
82. A. Marchesini, P. Capelletti, L. Canonica, B. Danieli and S. Tollari, *Biochim. Biophys. Acta* **484**, 290 (1977).
83. G. Mossa, A. DiGiulio, L. Dini and A. Finazzi-Agrò, *Biochim. Biophys. Acta* **986**, 310 (1989).
84. L. Dini, A. Di Giulio, A. Pavan, G. Ravagnan and G. Mossa, *Biochim. Biophys. Acta* **1062**, 108 (1991).
85. L. Avigliano, G. Rotilio, S. Urbanelli, B. Mondovi and A. Finazzi-Agrò, *Arch. Biochem. Biophys.* **185**, 419 (1978).
86. A. Baici, P.L. Luisi, S. Palmieri and A. Marchesini, *J. Mol. Cat.* **6**, 87 (1979).
87. K. Kawahara, S. Suzuki, T. Sakurai and A. Nakahara, *Arch. Biochem. Biophys.* **241**, 179 (1985).
88. B. Reinhammar, *Biochim. Biophys. Acta* **275**, 245 (1972).
89. P. O'Neill, E. M. Fielden, A. Finazzi-Agrò and L. Avigliano, *Biochem. J.* **209**, 167 (1983).
90. I. Pecht and M. Faraggi, *Nature* **233**, 116 (1971).
91. M. Faraggi and I. Pecht, *J. Biol. Chem.* **248**, 3146 (1973).
92. O. Farver and I. Pecht, *Proc. Natl. Acad. Sci. USA* **89**, 8283 (1992).
93. P. Kyritsis, A. Messerschmidt, R. Huber, G. A. Salmon and A. G. Sykes, *J. Chem. Soc. Dalton Trans.*, 731 (1993).
94. P. O'Neill, E. M. Fielden, L. Avigliano, G. Marcozzi, A. Ballini and A. Finazzi-Agrò, *Biochem. J.* **222**, 65 (1984).
95. T. E. Meyer, A. Marchesini, M. A. Cusanovich and G. Tollin, *Biochemistry* **30**, 4619 (1991).
96. G. Tollin, T. E. Meyer, M. A. Cusanovich, P. Curir and A. Marchesini, *Biochim. Biophys. Acta* **1183**, 309 (1993).
97. L. Casella, M. Gullotti, G. Pallanza, G. Pintar and A. Marchesini, *Biochem. J.* **251**, 441 (1988).
98. O. Farver and I. Pecht, *Biophys. Chem.* **50**, 203 (1994).
99. (a) J. N. Onuchic and D. N. Beratan, *J. Chem. Phys.* **92**, 722 (1990); (b) D. N. Beratan, J. N. Betts and J. N. Onuchic, *Science* **252**, 1285 (1991).
100. F. Malatesta, P. Sarti, G. Antonini, B. Vallone and M. Brunori, *Proc. Natl. Acad. Sci. USA* **87**, 7410 (1990).
101. O. Farver, S. Wherland and I. Pecht, *J. Biol. Chem.* **269**, 22933 (1994).
102. O. Farver, S. Wherland and I. Pecht, *J. Inorg. Biochem.* **51**, 14 (1993).
103. J. T. Hazzard, A. Marchesini, P. Curir and G. Tollin, *Biochim. Biophys. Acta* **1208**, 166 (1994).
104. M. Roncel, M. Hervas, J. A. Navarro, M. A. DeLaRosa and G. Tollin, *Eur. J. Biochem.* **191**, 531 (1990).

105. J. A. Navarro, M. A. DeLaRosa and G. Tollin, *Eur. J. Biochem.* **199**, 239 (1991).
106. (a) R. Aasa, R. Bränden, J. Deinum, B. G. Malmström, B. Reinhammar and T. Vänngard, *FEBS Lett.* **61**, 115 (1976); (b) L -E Andreasson, R. Bränden and B. Reinhammar, *Biochim. Biophys. Acta* **438**, 370 (1976).
107. T. Sakurai, *J. Inorg. Biochem.* **55**, 193 (1993).
108. P. W. Washko, R. W. Welch, K. R. Dharival, Y. Wang and M. Levine, *Anal. Biochem.* **204**, 1 (1992).
109. G. A. Crawford, J. F. Mahoney and A. Z. Györy, *Clin. Chim. Acta* **147**, 51 (1985).
110. M. Suzuky and M. Yoshida, *Clin. Chim. Acta* **140**, 289 (1984).
111. J. Freemantle, M. J. Freemantle and T. Badrik, *Clin. Chem.* **40**, 950 (1994).
112. J. C. Lee and S. N. Timasheff, *J. Biol. Chem.* **256**, 7193 (1981).
113. G. D'Andrea, L. Salucci and L. Avigliano, *Process Biochem.* **31**, 173 (1996).
114. M. Esaka, K. Suzuki and K. Kubota, *Agric. Biol. Chem.* **49**, 2955 (1985).
115. R. Stevanato, L. Avigliano, A. Finazzi-Agrò and A Rigo, *Anal. Biochem.* **149**, 537 (1985).
116. G. M. Greenway and P. Ongomo, *Analyst* **115**, 1297 (1990).

CHAPTER 10

THE BIOLOGY OF HUMAN CERULOPLASMIN

Z. Leah Harris
Edward Mallinckrodt Department of Pediatrics
Washington University School of Medicine
St. Louis, Missouri, USA

Hiroshi Morita
Department of Medicine
Shinshu University School of Medicine
Matsumoto, Japan

Jonathan D. Gitlin
Edward Mallinckrodt Department of Pediatrics
Washington University School of Medicine
St. Louis, Missouri, USA
E-mail: gitlin@kidsa1.wustl.edu

ABSTRACT

Ceruloplasmin, the "sky-blue protein" is a multi-copper ferroxidase which plays an essential role in iron metabolism in humans. This protein is encoded by a single 45 kB gene located on chromosome 3q23-25 which is expressed in the fetal liver by the 5th week of embryonic development. Six atoms of copper are incorporated into ceruloplasmin during biosynthesis resulting in the secretion of a holoprotein which contains more than 95 % of the copper found in the plasma. In hepatocytes, copper incorporation into ceruloplasmin is dependent upon the activity of the Wilson disease P-type ATPase which transports copper into the secretory pathway where it is available for holoceruloplasmin synthesis and biliary copper excretion. Aceruloplasminemia is an autosomal recessive disorder of iron metabolism characterized by a complete deficiency of ceruloplasmin ferroxidase activity due to mutations in the ceruloplasmin gene. Affected individuals develop a triad of diabetes, retinal degeneration and neurologic disease in association with excess tissue iron accumulation. These recent studies on the biology of human ceruloplasmin provide new insights into the structure and function of this multi-copper oxidase and reveal a remarkable evolutionary conservation of the role of such proteins in copper and iron metabolism.

Contents

1. Introduction

The development of a method to separate proteins by electrophoresis earned Tiselius the Nobel prize and ushered in an era of investigation which eventually resulted in the identification and characterization of the proteins of human plasma. In the course of such studies, Holmberg and Laurell observed that pig serum precipitated with 40 % ammonium sulfate had a bluish green color resulting from the abundant copper content of this fraction. From this fraction, these investigators isolated and purified a protein containing "most if not all serum copper" which, because of the beautiful blue color of the purified protein, was given the name "caeruloplasmin".[1] In the same year Wilson disease was shown to be due to copper accumulation and subsequently Scheinberg and Gitlin demonstrated a deficiency of ceruloplasmin in the sera of patients with Wilson disease.[2] Soon thereafter the ferroxidase activity of ceruloplasmin was discovered and this finding along with studies in copper-deficient pigs established a role for ceruloplasmin in iron metabolism.[3,4]

Since this initial work, many studies have appeared in the literature detailing the structure and function of ceruloplasmin. In this chapter we review recent studies on ceruloplasmin gene expression and discuss the current understanding of the mechanisms of holoceruloplasmin biosynthesis and the potential role of the Wilson disease P-type ATPase in this process.

In addition, we discuss aceruloplasminemia, a newly identified autosomal recessive disorder of iron metabolism resulting from a deficiency of ceruloplasmin ferroxidase activity due to mutations in the ceruloplasmin gene. Earlier studies on ceruloplasmin are discussed in detail elsewhere and the reader is referred to these reviews for detailed information.[5–8]

2. Gene Structure and Expression

The complete amino acid sequence of human ceruloplasmin was obtained by Putnam and colleagues in 1984.[9] These data definitively demonstrated the single chain structure of this molecule revealing internal triplication of the sequence. Delineation of the multi-copper oxidase signature sequences was accomplished following solution of the X-ray structure of ascorbate oxidase and amino acid sequence alignment with ceruloplasmin.[10] Subsequently, human ceruloplasmin cDNA clones were isolated and characterized and the derived amino acid sequence was found to be consistent with the earlier protein work.[11–13] RNA blot analysis using these cDNA clones identified two ceruloplasmin specific transcripts of 4.2 and 3.7 kB in RNA from liver and HepG2 cells. These initial data revealed abundant, tissue-specific expression of the ceruloplasmin gene in the liver.

Studies using classical genetic techniques identified linkage of the human transferrin and ceruloplasmin genes.[14] Subsequent molecular analysis using ceruloplasmin cDNA probes confirmed this finding and localized the human ceruloplasmin gene to chromosome 3q23-25.[15] Further genetic analysis has localized the genes for transferrin, the transferrin receptor, lactotransferrin and melanotransferrin to this same region of chromosome 3 suggesting the possibility of an iron regulatory locus in this region.[16] In support of this concept, synteny has been demonstrated for each of these genes in a number of different mammalian species. A processed pseudogene for human ceruloplasmin has also been isolated and localized to chromosome 8 by *in situ* hybridization.[17,18] Although not expressed, the presence of this sequence in the human genome has important methodologic implications in the identification of ceruloplasmin gene mutations in patients with aceruloplasminemia (see below).

The human ceruloplasmin gene has been isolated and characterized and consists of 19 exons spanning approximately 45 kB[19] (Fig. 1). The two transcripts of 3.7 and 4.2 kB arise from alternative usage of two polyadenylation sites in the 3′ untranslated region. Translation results in a 132 kDa

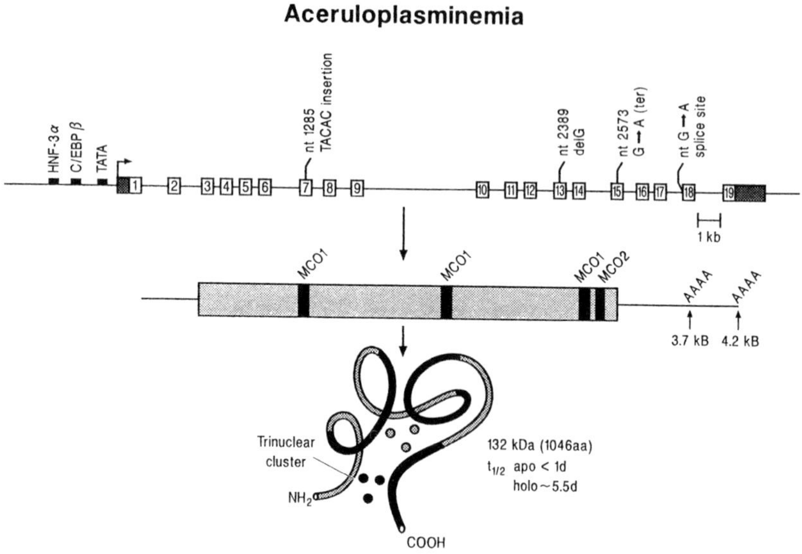

Fig. 1. Structure of the ceruloplasmin gene. The exon structure of the ceruloplasmin gene, putative *cis*-acting regulatory elements and currently identified mutations in patients with aceruloplasminemia are shown. The mRNA transcript sizes are depicted along with the alternative polyadenylation sites. A model of the translated, six-domain structure of ceruloplasmin is shown including the trinuclear copper cluster encompassed in domains 1 and 6 of the molecule (closed circles). The three mononuclear coppers (open circles) are contained in the second domains of the two-domain building block of the triplication.

protein with 6 copper atoms including a trinuclear copper cluster with ligands contributed to by amino acids encoded in domains 1 and 6 (Fig. 1). Cloning of the ceruloplasmin cDNA from the rat revealed 93 % amino acid homology with the corresponding human sequence and permitted a careful analysis of the mechanisms of ceruloplasmin gene expression in a number of species.[20,21]

Ceruloplasmin is an acute phase reactant and the serum concentration increases during infection, inflammation and trauma. The mechanism of acute phase regulation of the ceruloplasmin gene has been examined in human hepatoma cell lines and primary hepatocytes.[22,23] These studies reveal that interleukin-6 is the predominant cytokine resulting in the increase in ceruloplasmin expression during the acute phase response. An example of this regulation is shown in Fig. 2 which demonstrates the IL-6 mediated

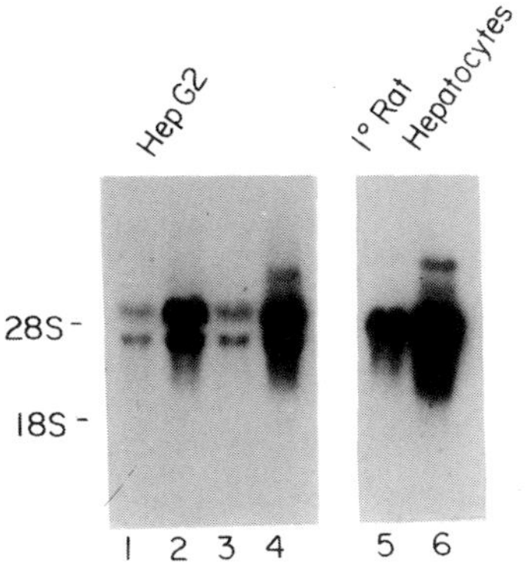

(A)

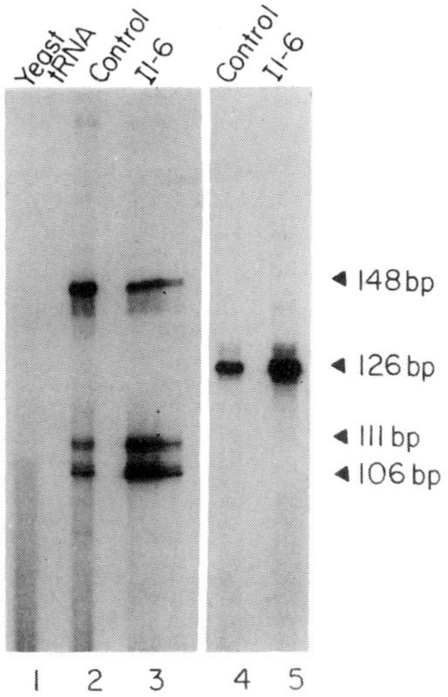

(B)

Fig. 2. Acute phase regulation of ceruloplasmin gene expression. (A) RNA blot analysis of ceruloplasmin gene expression in HepG2 cells or primary rat hepatocytes following treatment with saline (lanes 1,3,5), 10 ng (lanes 2) or 100 ng (lanes 4,6) of recombinant IL-6. RNA blots were probed with antisense cRNA probes for human (lanes 1–4) or rat (lanes 5 and 6) ceruloplasmin. (B) Primer extension analysis of HepG2 (lanes 1–3) or primary rat hepatocyte (lanes 4 and 5) RNA, following treatment of cells with saline (lanes 2, 4) or 100 ng of recombinant IL-6 (lanes 3,5). The sizes of extended products is shown.

increase in the abundance of hepatocyte ceruloplasmin-specific mRNA. Although interleukin-1 and TNF-α also result in an increase in hepatic ceruloplasmin gene expression, these cytokines mediate this response through a secondary increase in interleukin-6.

The molecular mechanisms regulating these changes in ceruloplasmin gene expression were further characterized using a rodent model. Turpentine-induced inflammation in rats was shown to increase the content of hepatic ceruloplasmin mRNA.[20] When this response was examined in detail, hepatic ceruloplasmin mRNA content was shown to increase approximately 6–10 fold over control values within 24 hours following the induction of inflammation.[24] Nuclear run-on assays demonstrated an increase in the relative rate of transcription of the ceruloplasmin gene within 2–3 hours following induction with maximum values reached by 18 hours. These changes were paralleled by an increase in hepatic ceruloplasmin mRNA content which increased 2-fold within 12 hours of induction, and reached a maximum by 24 hours, returning to control values within 72 hours. In contrast, the serum ceruloplasmin concentration in turpentine treated animals did not increase until 36 hours and reached maximum levels 4 days following the induction of an inflammatory response. These data demonstrate that inflammation leads to a rapid increase in hepatic ceruloplasmin mRNA content predominantly as a result of increased ceruloplasmin gene transcription. A careful comparison of the relative rate of transcription and mRNA accumulation also suggests that changes in ceruloplasmin mRNA turnover and/or post-translational mechanisms may account for some of the observed changes in serum ceruloplasmin concentrations seen during inflammation.[24]

Ceruloplasmin accounts for more than 95 % of the copper found in the plasma of vertebrate species. For this reason, several investigators have examined the role of copper in the regulation of ceruloplasmin gene expression. In primary mouse hepatocytes an elevation in intracellular copper content had no effect on ceruloplasmin mRNA abundance despite a marked elevation of metallothionein mRNA in these experiments.[25] Similar observations were made in primary rat hepatocytes from normal and copper deficient rats. Although copper deficiency resulted in a complete loss of serum ceruloplasmin oxidase activity, the abundance of hepatic ceruloplasmin mRNA was equivalent in these animals.[26] Interleukin-1 and lipopolysaccharide (LPS) increased hepatic ceruloplasmin mRNA content to a similar degree in both normal and copper deficient animals without a concomitant increase in ceruloplasmin oxidase activity in the copper deficient group.[26] These studies indicate that hepatocyte copper content has no effect on hepatic ceruloplasmin gene expression and that the acute phase mediated increase in hepatic ceruloplasmin mRNA is independent of the copper status of the animal.

Although the liver is the major site of ceruloplasmin production, early studies revealed the unexpected finding of ceruloplasmin biosynthesis in synovial tissue from patients with rheumatoid arthritis.[27] Subsequently, ceruloplasmin biosynthesis has been detected in sertoli cells of the testis and murine embryonic fibroblasts.[28,29] A detailed analysis in the rat revealed ceruloplasmin gene expression in multiple extrahepatic tissues including chorid plexus, yolk sac, placenta and testis.[20] A developmental study of ceruloplasmin gene expression in the rat demonstrated ceruloplasmin mRNA in fetal rat liver and lung by 15 days of gestation.[21] During fetal development the abundance of the ceruloplasmin mRNA increases selectively such that at birth the mRNA content in these tissues is approximately 60 % of that in the adult liver. Primary extension analysis revealed a single, identical start site of ceruloplasmin gene transcription in both of these tissues. Ceruloplasmin mRNA is also detected in human fetal lung explants and human lung adenocarcinoma cell lines suggesting a similar pattern of expression can be studied using *in situ* hybridization (Fig. 3).

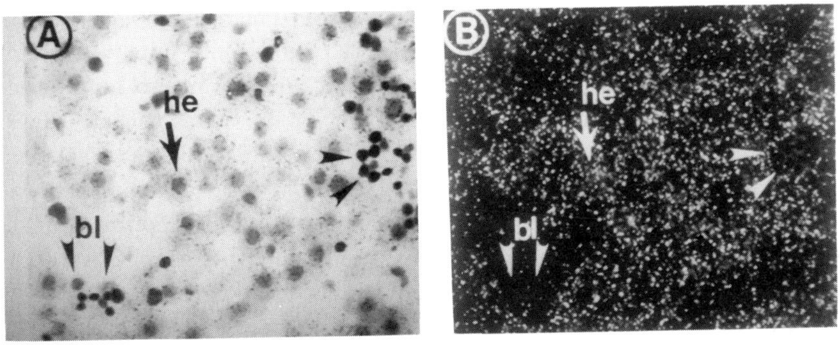

Fig. 3. Cell specific expression of the ceruloplasmin gene. Light (A) and dark field (B) photomicrograph of newborn rat liver following *in situ* hybridization with antisense rat ceruloplasmin cRNA probe. Silver grains are seen over the hepatocytes (he) but are absent over hematopoetic cells (arrowheads) within blood islands (bl).

In mammary gland this technique localized ceruloplasmin gene expression to the epithelium lining the alveolar ducts with no expression in the surrounding mesenchyme.[30] Consistent with this observation, ceruloplasmin gene expression is also detected in breast adenocarcinoma cell lines and in biopsy tissue from women with invasive ductal carcinoma.[30]

To examine the effect of inflammation on extrahepatic ceruloplasmin gene expression, the ceruloplasmin mRNA content of adult rat tissues was examined following endotoxin injection.[31] Within 8 hours of a single inflammatory dose, ceruloplasmin specific transcripts increased as expected in the liver but were also detectable in the lung. These effects were tissue-specific because ceruloplasmin mRNA was not detected in other extrahepatic tissues under these conditions. The kinetics of ceruloplasmin mRNA accumulation was shown to be equivalent in lung and liver tissue with a maximum 7–9 fold increase in ceruloplasmin mRNA content in each tissue. The relative rate of ceruloplasmin gene transcription was increased in lung tissue within three hours of endotoxin administration suggesting similar mechanisms of regulation of ceruloplasmin gene expression in hepatic and extrahepatic tissues during inflammation. To determine the cellular sites of ceruloplasmin production in the inflamed lung, alveolar macrophages were isolated and examined. Ceruloplasmin mRNA was increased in alveolar macrophages isolated from endotoxin treated animals and a similar increase was seen in macrophages from normal animals treated *in vitro* with tumor necrosis factor or LPS.[31]

Ceruloplasmin mRNA content in lung was also increased following hyperoxic exposure. Exposure of adult rats to 95 % oxygen results in a 6-fold induction of ceruloplasmin mRNA content in lung within 46 hours and this response is time-dependent, reaching maximum values within 5 days of exposure. Hyperoxic induction of ceruloplasmin mRNA is tissue-specific since hepatic ceruloplasmin mRNA abundance does not change under these conditions. The lung is therefore a prominent site of ceruloplasmin gene expression during development, inflammation and hyperoxia. Consistent with these findings, a number of clinical studies have detected an increase in ceruloplasmin in bronchioalveolar lavage fluid from patients with acute respiratory disease.[32]

In an effort to understand the mechanisms determining tissue-specific ceruloplasmin gene expression during development and inflammation, the rat ceruloplasmin gene was isolated and characterized.[33] Nucleotide sequence analysis of the promoter revealed a typical eukaryotic TATA-box and several putative *cis*-acting elements (Fig. 1). Although these elements do not fit a tight consensus sequence, studies utilizing transient transfection assays have revealed a role for both HNF-3α and CEBP-β in the regulation of ceruloplasmin gene expression in HepG2 cells.[34] In addition, a nucleotide region was detected approximately 400bp upstream from the transcription start site which appears to direct tissue-specific ceruloplasmin expression in lung and liver during development.[33] An analysis of human ceruloplasmin gene expression in cell lines revealed the presence of an alternatively spliced transcript by RT-PCR, but the relevance of this finding to ceruloplasmin expression *in vivo* is unclear since translation products were not identified.[35]

3. Metabolism

Ceruloplasmin is first detected during human fetal development early in the 5th week of gestation in both the fetal liver and yolk sac.[36] The concentration of ceruloplasmin increases steadily throughout gestation and turnover rates observed at birth indicate that the majority of ceruloplasmin present in neonatal plasma is derived from endogenous synthesis.[36] Consistent with this observation, little or no ceruloplasmin is transported across the human placenta. The levels of ceruloplasmin in the developing infant remain low for the first several weeks and then slowly reach adult levels in the first years of life. Several electrophoretic variants of human ceruloplasmin, have been identified which result in a 2–3 fold variation in serum ceruloplasmin concentrations.[37,38] Although these polymorphic variants presumably arise from amino acid sequence differences, the mechanisms resulting in changes in serum ceruloplasmin concentration are unknown. In addition to these polymorphisms, heterogeneity of the carbohydrate chains of human ceruloplasmin has been demonstrated and variations in carbohydrate structure have been observed during malignancy and the acute phase response.[39–41] Serum ceruloplasmin concentrations in normal individuals are dependent upon the changes associated with puberty and pregnancy and in this regard, exogenous administration of estrogen results in an increase in the ceruloplasmin concentration in the plasma.[42,43]

Turnover studies utilizing radioactive copper or ^{131}I demonstrate a $t_{\frac{1}{2}}$ of 5.5 days for human ceruloplasmin.[44,45] Kinetic studies on the turnover of the copper and protein moieties of ceruloplasmin demonstrate similar half-lives for each component indicating little or no exchange of ceruloplasmin copper with tissues.[46] Although previous studies have demonstrated that the copper present in ceruloplasmin ultimately will be found in tissues, this occurs following ceruloplasmin degredation and does not represent a transport function. About 10 % of circulating ceruloplasmin exists as the apoprotein which is synthesized and secreted from the liver without copper and turns over rapidly in circulation.[47,48] Holoceruloplasmin is rapidly synthesized *in vivo* following copper supplementation in copper deficient rats and this is consistent with the observation that the stimulation of ceruloplasmin synthesis results in the rapid appearance of newly incorporated copper into ceruloplasmin.[44,48] Stimulation of ceruloplasmin biosynthesis with estrogen maintains the normal plasma ratio of holo to apoprotein despite a 10-fold increase in total ceruloplasmin synthesis suggesting that copper is not rate limiting in holoceruloplasmin production.[44] This is supported by biosynthetic studies in primary hepatocytes from normal and copper deficient rats which demonstrate that the biosynthesis and secretion of apoceruloplasmin is independent of copper availability.[26] Taken together these data suggest a model for copper incorporation into ceruloplasmin where copper availability serves only to determine the ratio of holo to apoprotein secreted from the cell. At steady state, the serum ceruloplasmin concentration represents the sum of apo and holoprotein determined by intracellular copper availability and the observed differences in the half lives of these two forms. As expected from such a model, a rapid increase in the hepatic copper pool of normal individuals results in a marked increase in holoceruloplasmin in the serum.[49]

To examine the mechanisms of copper incorporation in greater detail, methods were developed to identify holoceruloplasmin by oxidase activity staining and immunoprecipitation.[50] When human liver and lung cell lines were metabolically labeled with ^{67}Cu, newly synthesized holoceruloplasmin was detected in the culture media as two proteins with apparent molecular masses of 84 kDa and 79 kDa. Pulse-chase studies demonstrated that exogenous copper is readily available for incorporation into newly synthesized holoceruloplasmin and that the kinetics of apo and holoceruloplasmin biosynthesis and secretion were identical confirming, previous studies in hepatocytes from copper deficient animals. Pretreatment with tunicamycin did not affect the rate or amount of copper incorporated into holoceruloplasmin indicating that copper incorporation is independent of N-linked glycosylation. Tunicamycin treatment resulted in secretion of a single 68 kDa protein and thus the 84 kDa and 79 kDa proteins arise from glycosylation differences detected by these techniques. The apparent size differences between apo and holoceruloplasmin suggest a conformational change in ceruloplasmin due to copper incorporation which does not affect the secretory rate of the protein. Presumably it is this conformational change determined by metal incorporation which results in stabilization of the secreted protein. In support of this, substitution of silver for copper during biosynthesis results in an inactive protein due to the lack of electron exchange in this metal which is nevertheless stable in the plasma with a $t_{1/2}$ equivalent to that of copper-ceruloplasmin.[51]

Radiocopper studies in patients with Wilson disease support the mechanism of copper incorporation noted above and suggest that the basic defect in this disease results from impaired copper transfer into a common pool for holoceruloplasmin production and a biliary excretion.[52,53] This has recently been confirmed following cloning of the Wilson disease gene.[54–56]

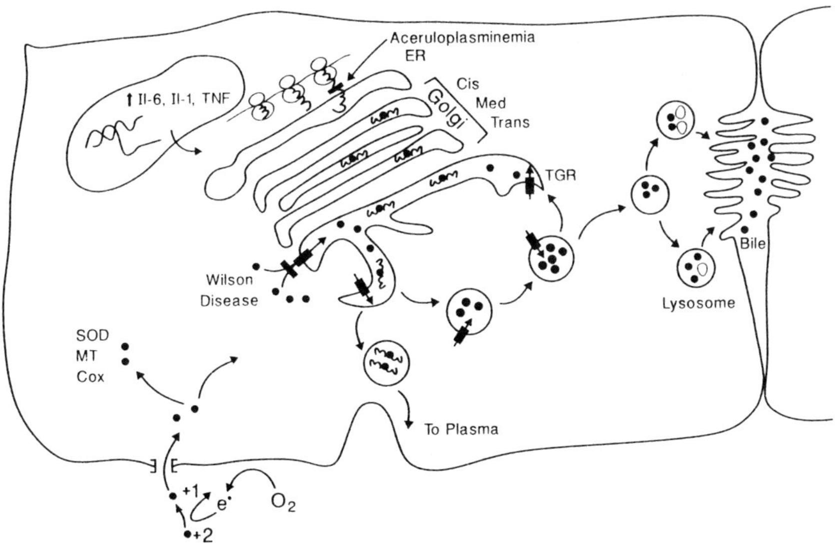

Fig. 4. Mechanisms of holoceruloplasmin biosynthesis. A model of copper traffic into the hepatocyte secretory pathway is shown along with the putative sites of copper transport by the Wilson disease P-type ATPase.

This gene encodes a copper transporting P-type ATPase which is localized in the secretory pathway of hepatocytes. Loss of function of this protein results in impaired intracellular hepatocyte copper transport with copper accumulation in the cytosol and copper depletion from compartments essential for holoceruloplasmin biosynthesis and copper excretion (Fig. 4). The details of this process have been critically examined in the Long-Evans Cinnamon (LEC) rat which is a rodent model of Wilson disease with deficient expression of the Wilson disease gene due to a deletion in the rat Wilson P-type ATPase gene.[57–59] Analysis of copper incorporation into ceruloplasmin in normal and LEC rat hepatocytes confirms the hypothesis of a primary defect in copper transport into the secretory pathway on Wilson disease.[60] Although a recent study suggests a second pathway of copper incorporation into ceruloplasmin in LEC rats these data are complicated by a failure to demonstrate deficiency of the P-type ATPase in the cell lines used and by quantitative abnormalities in the recovery of holoceruloplasmin from transfected cells.[61] Pulse chase experiments and fractionation studies suggest that the Golgi or TGN is the site of copper incorporation into ceruloplasmin[52,62,63] (Fig. 4). The incorporation of copper late in the secretory pathway suggests that ceruloplasmin either remains unfolded or is unfolded again following transport from the endoplasmic reticulum (ER). Although this would be unusual for a secreted protein, the oxidizing environment of the ER may preclude this compartment as a site of copper incorporation since reduced copper is involved in this process.[64]

4. Aceruloplasminemia

Previous studies demonstrated that 20 % of obligate heterozygotes for Wilson disease will have decreased serum ceruloplasmin concentrations in association with a 2–3 fold increase in hepatic copper content.[65–67] The mechanism for this decrease in ceruloplasmin is unclear but may reflect the effect of specific mutations in the Wilson disease gene in such families. In the course of these early studies, individuals were detected with inherited differences in serum ceruloplasmin concentrations. The lack of symptoms in such individuals as well as the inheritance pattern and the incidence suggested that such differences could not be attributed to the heterozygote state of Wilson disease.[42,65] Careful analysis of one such family lead to the designation of "hereditary hypoceruloplasminemia" and indicated that a decrease in serum ceruloplasmin concentration to 50 % of normal was not associated with any clinical signs or symptoms.[68]

The signs and symptoms of Wilson disease are the result of copper accumulation and are not related to the secondary deficiency in serum ceruloplasmin concentration. Nevertheless, recent reports in the literature have described individuals with diabetes, retinal degeneration and neurologic symptoms who have a total absence of serum ceruloplasmin.[69,70] The cloning of the Wilson disease gene prompted further investigation in these families and this has revealed the presence of inherited mutations in the ceruloplasmin gene in these patients[71–74] (Fig. 1). The presence of these mutations in conjunction with the clinical and pathologic features has revealed an essential role for ceruloplasmin in human biology and identifies aceruloplasminemia as an autosomal recessive disorder of iron metabolism.

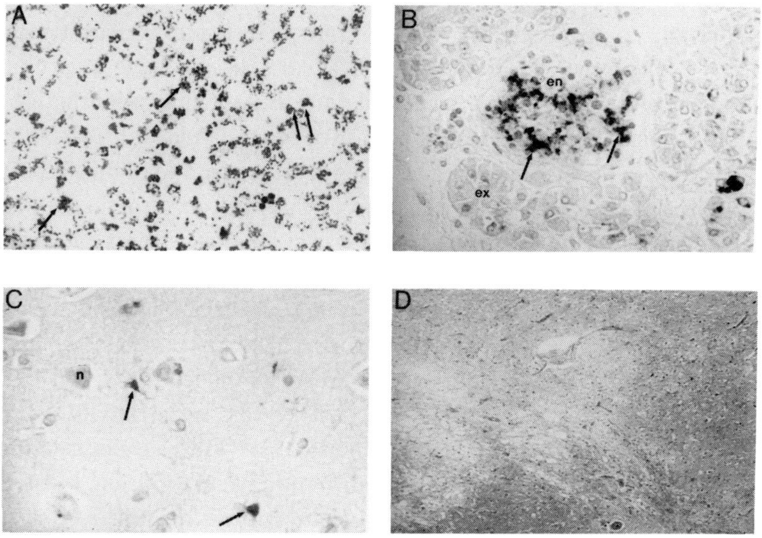

Fig. 5. Histopathology of aceruloplasminemia. (A) Liver section from biopsy of a patient with aceruloplasminemia stained with Prussian blue. Arrows indicate accumulation of iron within the hepatocytes and Kupffer cells. (B) Prussian blue staining of the pancreas revealing iron accumulation (arrows) within the endocrine islets (en) as well as paucity of iron staining in exocrine tissue (ex). (C) Section of prefrontal gyrus stained with Prussian blue revealing iron accumulation within microglia (arrows) but not neurons (n). (D) H & E section from the caudate nucleus revealing gliosis and neuronal degeneration.

Affected patients have normal hepatic parenchyma on biopsy and normal hepatic copper content but a marked elevation of hepatic and pancreatic iron.[69–75] T_2 weighted magnetic resonance imaging reveals an increased iron content in basal ganglia and examination of tissues reveals abundant parenchymal and reticuoendothelial (RE) system iron. In the brain there is microglial and neuronal iron accumulation in association with spongyform degeneration and neuronal loss in a number of basal ganglia structures[75] (Fig. 5).

The ceruloplasmin gene mutations identified thus far all result in a truncation of the open reading frame prior to the carboxyl-terminal trinuclear copper cluster (Fig. 1). Without this region, copper incorporation during biosynthesis cannot occur and any translated protein would be devoid of oxidase activity and subject to rapid degregation in the plasma. The lack of detectable ceruloplasmin in the sera of patients with aceruloplasminemia supports this concept. Under normal circumstances only 5 % of the plasma ceruloplasmin is required for iron flow into transferrin and iron removal from the RE system.[76,77] Patients with Wilson disease have a ceruloplasmin concentration above this level and thus do not have abnormalities of iron metabolism. However, a total absence of ceruloplasmin results in an insidious long-term accumulation of parenchymal iron due to uptake of non-transferrin bound serum iron. The initial flux of iron into the iron cycle is dependent upon the ability to transfer iron into transferrin and any disease with either a decreased ability to transfer iron into transferrin or over saturation of transferrin will result in an increase in non-transferrin bound serum iron which is rapidly deposited in parenchymal tissues (Fig. 5). The presumed role of ceruloplasmin in this process is dependent upon the ferroxidase activity which catalyzes iron transfer into transferrin and mobilizes iron from the RE system[76,77] (Fig. 6). Patients with aceruloplasminemia are only mildly anemic, because nonenzymatic iron oxidation will permit some iron transfer into transferrin.[76] In other forms of iron overload, RE system iron storage is minimal despite a marked increase in non-transferrin bound iron because ceruloplasmin is functioning normal. The finding of normal copper metabolism in these patients and off spring confirms that ceruloplasmin has no essential role in copper transport in man.

The uniqueness of aceruloplasminemia resides in the neurodegeneration which distinguishes this disease from other inherited and acquired iron storage diseases (Table 1). The implication of the neurologic picture is that

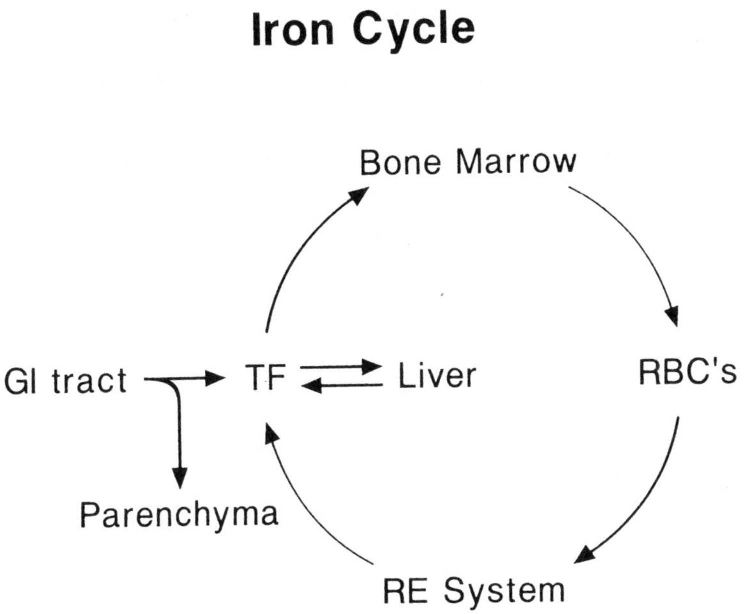

Fig. 6. Normal flow of iron into transferrin and the iron cycle.

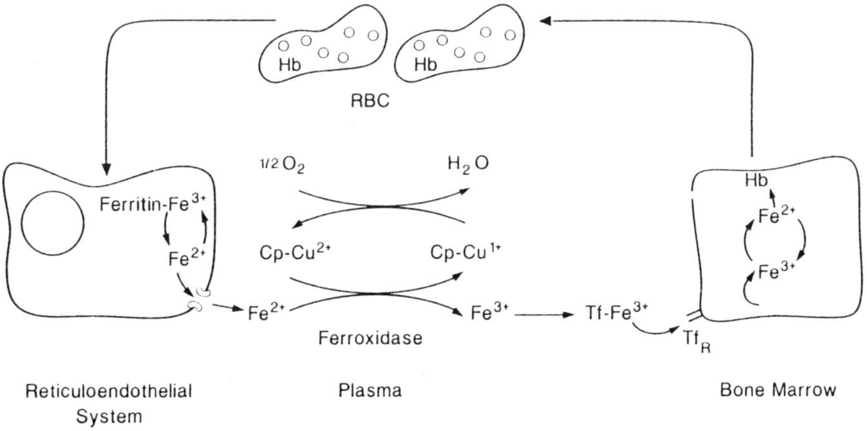

Fig. 7. Potential role for ceruloplasmin in the delivery of iron to transferrin.

Table 1. Hereditary disorders of iron metabolism.

	Aceroloplasminemia	1°Hemochromatosis	Atransferrinemia
Genetics	3q 23-25 (CP) Autosomal recessive Unknown	6p22 (unknown) Autosomal recessive 1:400	3q22-25 (Tf) Autosomal recessive 1:250,000
Clinical	Adult onset Dementia Dystonia Diabetes Retinal degeneration	Adult onset Cirrhosis Heart failure Diabetes Bronze skin	Childhood onset Severe anemia Heart failure
Laboratory	Serum Fe ↓; Ferritin ↑ Liver Fe ↑; Brain Fe↓ Serum Cu, and Cp ↓	Serum Fe↑; Ferritin↑ Liver Fe ↑; Brain Fe nl Serum Cu and Cp nl	Serum Fe↓; Ferritin↑ Liver Fe ↑; Brain Fe nl Serum Cu and Cp nl
Defect	↓ Fe export from RE cells	↑ Fe absorbtion	↓ Fe delivery to erythron
Treatment	Unknown	Desferrioxamine Phlebotomy	Transferrin

ceruloplasmin plays an essential role in brain iron metabolism. It would seem reasonable to conclude that the brain accumulation of iron is not derived from non-transferrin bound iron but is instead due to accumulation in the brain RE (microglial) system. If this is not the case, iron deposition in the brain and neurologic symptoms would be a major component of all forms of iron overload (Table 1). It seems unlikely that plasma ceruloplasmin is essential for brain RE (microglia) system iron removal since this protein does not cross the blood brain barrier. The implication is that brain iron metabolism is compartmentalized and any role of ceruloplasmin in this process is derived from ceruloplasmin production within the central nervous system. This concept is supported by previous studies on extrahepatic ceruloplasmin production as well as by existing data on the known mechanisms of iron regulation in the brain.[21]

The mechanisms of tissue injury in aceruloplasminemia are unknown but presumably relate directly to iron accumulation. Such injury is most likely the result of oxidant induced free-radical generation due to the accumulation of ferrous iron and consistent with this concept recent studies indicate a marked increase in plasma lipid peroxidation in patients with aceruloplasminemia.[78] These data are supported by the proposed role of iron in neurotoxicity and with studies indicating that the antioxidant role of ceruloplasmin is related to ferrous iron oxidation.[79,80] Such findings may have broader implications for our understanding of the role of transition metals and free-radical tissue balance in brain injury in a variety of neurodegenerative diseases. Prior studies have revealed age related changes in human ceruloplasmin resulting from oxidative modification.[81] Although these studies revealed only a modest decrease in ceruloplasmin oxidase activity, it is tempting to speculate that such changes in ceruloplasmin oxidase activity in association with other environmental or genetic factors may contribute to certain forms of neurodegenerative disease observed with ageing.

5. Epilogue

This review has focused on the molecular and genetic factors regulating expression of the ceruloplasmin gene and the metabolism of this protein in humans. Although the genetic basis for Wilson disease and aceruloplasminemia are distinct, these studies illustrate how each disorder provides fundamental insight into the cellular and molecular mechanisms involved in iron and copper metabolism. A final note should be made of the remarkable evolutionary conservation of this process. Genetic studies analyzing the mechanisms of ferrous iron transport in *S. cerevisiae* have defined a novel multi-copper oxidase (Fet3).[82] This protein has limited amino acid sequence homology to ceruloplasmin and has been shown to function as a copper-oxidase with an essential role in plasma membrane iron transport in *S. cerevisiae.* Deletion of the Wilson disease P-type ATPase homologue in *S. cerevisiae* results in the biosynthesis of apo fet3 which, is unable to function in iron metabolism.[83] The uniform preservation of this process in two species separated by millions of years of evolution suggest that multi-copper oxidases identified in diverse species be examined for a role in iron metabolism.

Acknowledgments

We gratefully acknowledge the collaboration of our colleagues, Drs. Yoshitomo Takahashi, Hiroaki Miyajima, Ross MacGillivray, Rob Steiner, Anne Hughes, John Logan, Godfrey Gillett and John Walshe for sharing ideas and information and for the referral of patients and clinical samples. Work reported in this review from the author's laboratory was supported in part by funds from National Institutes of Health grants DK44464 and HL41536.

References

1. C. G. Holmberg and C. B. Laurell, *Clin. Chem. Acta.* **2**, 550 (1948).
2. I. H Scheinberg and D. Gitlin, *Science* **116**, 484 (1952).
3. S. Osaki, D. A. Johnson and E. Frieden, *J. Biol. Chem.* **241**, 2746 (1966).
4. G. R. Lee, S. Nacht, J. N. Lukens and G. E. Cartwright, *J. Clin. Invest.* **47**, 2058 (1968).
5. J. M. Gutteridge and J. Stocks, *Crit. Rev. Clin. Lab. Sci.* **14**, 257 (1981).
6. P. L. Wolf, *Crit. Rev. Clin. Lab. Sci.* **17**, 229 (1982).
7. R. J. Cousins, *Physiol. Rev.* **65**, 238 (1985).
8. E. Frieden, *Clin. Physiol. Biochem.* **4**, 11 (1986).
9. N. Takahashi, T. L. Ortel and F. W. Putnam, *Proc. Natl. Acad. Sci.* **390**, 394 (1984).
10. A. Messerschmidt and R. Huber, *Eur. J. Biochem.* **187**, 341 (1990).
11. F. Yang, S. L. Naylor, J. B. Lum, S. Cutshaw, J. L. McCombs, K. H. Naberhaus, J. R. McGill, G. S. Adrian, C. M. Moore, D. R. Barnett and B. H. Bowman, *Proc. Natl. Acad. Sci.* **83**, 3257 (1986).
12. M. L. Koschinsky, W. D. Funk, B. A. van Oost and R. T. MacGillivray, *Proc. Natl. Acad. Sci.* **83**, 5086 (1986).
13. J. F. Mercer and A. Grimes, *FEBS Lett.* **203**, 185 (1986).
14. L. R. Weitkamp, *Ann. Hum. Genet.* **47**, 293 (1983).
15. N. J. Royle, D. M. Irwin, M. L. Koschinsky, R. T. MacGillivray and J. L. Hamerton, *Somat. Cell Mol. Genet.* **13**, 285 (1987).
16. P. Aisen in *Iron Metabolism*, ed. J. H. Brock *et al.*, (W. B. Saunders, 1994), pp. 1–30.
17. M. L. Koschinsky, B. K. Chow, J. Schwartz, J. L. Hamerton and R. T. MacGillivray, *Biochemistry* **26**, 7760 (1987).
18. H. Wang, M. Koschinsky and J. L. Hamerton, *Cytogenet. Cell Genet.* **47**, 230 (1988).
19. M. Daimon, K. I. Yamatani, M. Igarashi, N. Fukase, T. Kawanami, T. Kato, M. Tominaga and H. Sasaki, *Biochem. Biophy. Res. Comm.* **208**, 1028 (1995).
20. A. R. Aldred, A. Grimes, G. Schreiber and J. F. Mercer, *J. Biol. Chem.* **262**, 2875 (1987).
21. R. E. Fleming and J. D. Gitlin, *J. Biol. Chem.* **265**, 7701 (1990).
22. G. Ramadori, J. Van Damme, H. Rieder and K. H. Meyer zum Buschenfelde, *Eur. J. Immunol.* **18**, 1259 (1988).
23. A. Mackiewicz and I. Kushner, *Scand. J. Immunol.* **29**, 265 (1989).
24. J. D. Gitlin, *J. Biol. Chem.* **263**, 6281 (1988).

25. H. J. McArdle, J. F. Mercer, A. M. Sargeson, and D. M. Danks, *J. Nutr.* **120**, 1370 (1990).
26. J. D. Gitlin, J. J. Schroeder, L. M. Lee-Ambrose and R. J. Cousins, *Biochem. J.* **282**, 835 (1992).
27. J. D. Gitlin, J. I. Gitlin and D. Gitlin, *Arthritis Rheum.* **20**, 1491 (1977).
28. M. K. Skinner and M. D. Griswold, *Biol. Reprod.* **28**, 1225 (1983).
29. F. F. Chu and K. Olden, *Biochem. Biophys. Res. Comm.* **126**, 15 (1985).
30. J. L. Jaeger, N. Shimizu and J. D. Gitlin, *Biochem. J.* **280**, 671 (1991).
31. R. E. Fleming, I. P. Whitman and J. D. Gitlin. *Am. J. Physiol.* **260**, L68 (1991).
32. M. G. Lykens, W. B. Davis and E. R. Pacht. *Am. J. Physiol.* **262**, L169 (1992).
33. R. E. Fleming and J. D. Gitlin, *J. Biol. Chem.* **267**, 479 (1992).
34. C. D. Bingle, R. E. Fleming and J. D. Gitlin, *Biochem. J.* **294**, 473 (1993).
35. F. M. Yang, W. E. Friedrichs, R. L. Cupples, M. J. Bonifacio, J. A. Sanford, W. A. Horton and B. H. Bowman, *J. Biol. Chem.* **265**, 10780 (1990).
36. D. Gitlin and A. Biasucci, *J. Clin. Invest.* **48**, 1433 (1969).
37. D. C. Shreffler, G. J. Brewer, J. C. Gall and M. S. Honeryman, *Biochem. Genet.* **1**, 101 (1967).
38. H. W. Mohrenweiser and R. S. Decker, *Hum. Hered.* **32**, 369 (1982).
39. M. Endo, K. Suzuki, KI. Schmid, B. Fournet, Y. Karamanos, J. Montreuil, L. Dorland, H. Van Halbeek and J. F. Vliegenthart, *J. Biol. Chem.* **257**, 8755 (1982).
40. J. E. Hansen, J. Iversen, A. Lihme and T. C. Bog-Hansen, *Cancer* **60**, 1630 (1987).
41. A. Mackiewicz, M. K. Ganapathi, D. Schultz *et al.*, *J. Exp. Med.* **166**, 253 (1987).
42. D. W. Cox, *J. Lab. Clin. Med.* **68**, 893 (1966).
43. L. A. Tovey and G.H. Lathe, *Lancet* **2**, 596 (1968).
44. I. Sternlieb, A. G. Morell, W. D. Tucker, M. W. Greene and I. H. Scheinberg, *J. Clin. Invest.* **40**, 1834 (1961).
45. M. Kekki, P. Koskelo and E. A. Nikkila, *Nature* **209**, 1252 (1966).
46. D. Gitlin and C. A. Janeway, *Nature* **185**, 693 (1960).
47. N. A. Holtzman and B. M. Gaumnitz, *J. Biol. Chem.* **245**, 2350 (1970).
48. N. A. Holtzman and B. M. Gaumnitz, *J. Biol. Chem.* **245**, 2354 (1970).
49. N. A. Holtzman, D. A. Elliott and R. H. Heller, *New Engl. J. Med.* **275**, 347 (1966).
50. M. Sato and J. D. Gitlin, *J. Biol. Chem.* **266**, 5128 (1991).
51. F. Hirasawa, M. Sato, and Y. Yakizawa, *Toxicol. Lett.* **70**, 193 (1994).
52. I. Sternlieb and I. H. Scheinberg, *Gastroenterology* **77**, 138 (1979).
53. J. M. Vierling, R. Shrager, W. F. Rumble, R. Aamodt, M. D. Berman and E. A. Jones, *Gastroenterology* **74**, 652 (1978).

54. Y. Yamaguchi, M. E. Heiny and J. D. Gitlin, *Biochem. Biophys. Res. Commun.* **197**, 271 (1993).
55. P. C. Bull, G. R. Thomas, J. M. Rommens, J. R. Forbes and D. W. Cox, *Nature Genet.* **5**, 327 (1993).
56. R. E. Tanzi, K. Petrukhin, I. Chernov, J. L. Pellequer, W. Wasco, B. Ross, D. M. Romano, E. Parano, L. Pavone, L. M. Brzustowicz, M. Devoto, J. Peppercorn, A. I. Bush, I. Sternlieb, M. Pirastu, J. F. Gusella, O. Evgrafov, G. K. Penchaszadeh, B. Honig, I. S. Edelman, M. B. Soares, I. H. Scheinberg and T. C. Gilliam, *Nature Genet.* **5**, 344 (1993).
57. Y. Li, Y. Togashi, S. Sato, T. Emoto, J. H. Kang, N. Takeichi, H. Kobayashi, Y. Kojima, Y. Une and J. Uchino, *J. Clin. Invest.* **87**, 1858 (1991).
58. Y. Yamaguchi, M. E. Heiny, N. Shimizu, T. Aoki and J. D. Gitlin, *Biochem. J.* **301**, 1 (1994).
59. J. Wu, J. R. Forbes, H. S. Chen and D. W. Cox, *Nature Genet.* **7**, 541 (1994).
60. T. Yamada, T. Agui, Y. Suzuki, M. Sato and K. Matsumoto, *J. Biol. Chem.* **268**, 8965 (1993).
61. K. Nakamura, F. Endo, T. Ueno, H. Awata, A. Tanoue and I. Matsuda, *J. Biol. Chem.* **270**, 7656 (1995).
62. K. Terada, Y. Kawarada, N. Miura, O. Yasui, K. Koyama and T. Sugiyama, *Biochem. Biophys. Acta* **1270**, 58 (1995).
63. Y. Murata, E. Yamakawa, T. Iizuka, H. Kodama, T. Abe, Y. Seki and M. Kodama, *Biochem. Biophys. Res. Commun.* **209**, 349 (1995).
64. J. T. Sawyer, T. Lukaczyk and M. Yilla, *J. Biol. Chem.* **269**, 22440 (1994).
65. G. E. Cartwright, H. Markowitz, F. S. Shields and M. M. Wintrobe, *Am. J. Med.* **28**, 555 (1960).
66. I. Sternlieb and I. H. Scheinberg, *New Engl. J. Med.* **278**, 352 (1968).
67. K. Gibbs and J. M. Walshe, *Q. J. Med.* **48**, 447 (1979).
68. C. Q. Edwards, D. M. Williams and G. E. Cartwright, *Clin. Genet.* **15**, 311 (1979).
69. H. Miyajima, Y. Nishimura, K. Mizoguchi, M. Sakamoto, T. Shimizu and N. Honda, *Neurology* **37**, 761 (1987).
70. J. I. Logan, K. B. Harveyson, G. B. Wisdom, A. E. Hughes and G. P. Archbold, *Q. J. Med.* **87**, 663 (1994).
71. Z. L. Harris, Y. Takahashi, H. Miyajima, M. Serizawa, R. T. A. MacGillivray and J. D. Gitlin, *Proc. Natl. Acad. Sci.* **92**, 2539 (1995).
72. K. Yoshida, K. Furihata, S. Takeda, A. Nakamura, K. Yamamoto, H. Morita, S. Hiyamuta, S. Ikeda, N. Shimizu and N. Yanagisawa, *Nature Genet.* **9**, 267 (1995).
73. Y. Takahashi, H. Miyajima, S. Shirabe, S. Nagataki, A. Suenaga and J. D. Gitlin, *Hum. Molec. Genec.* **5**, 81 (1996).
74. L. Harris, M. Migas, A. Hughes, J. I. Logan and J. D. Gitlin, *Q. J. Med.* **86**, 355 (1996).

75. H. Morita, S. Ikeda, K. Yamamoto, S. Morita, K. Yoshida, S. Nomoto, M. Kato and N. Yanagisawa, *Ann. Neurol.* **37**, 646 (1995).
76. S. Osaki and D. A. Johnson, *J. Biol. Chem.* **244**, 5757 (1969).
77. H. P. Roeser, G. R. Lee, S. Nacht and G. E. Cartwright, *J. Clin. Invest.* **49**, 2408 (1970).
78. H. Miyajima, Y. Takahashi, M. Suizawa, E. Kaneko and J. D. Gitlin, *Free Rad. Biol. Med.* **20**, 757 (1996).
79. J. R. Connor and A. Benkovic, *Ann. Neurol.* **32**, S51 (1992).
80. V. M. Samokyszyn, D. M. Miller, D. W. Reif and S. D. Aust, *J. Biol. Chem.* **264**, 21 (1989).
81. G. Musci, M. C. Bonaccorsi di Patti, U. Fagiola and L. Calabrese, *J. Biol. Chem.* **268**, 13388 (1993).
82. C. Askwith, D. Eide, A. VanHo, P. S. Bernard, L. Li, S. Davis-Kaplan, D. M. Sipe and J. Kaplan, *Cell* **76**, 403 (1994).
83. D. S. Yuan, R. Stearman, A. Dancis, T. Dunn, T. Beeler and R. D. Klausner, *Proc. Natl. Acad. Sci.* **92**, 2632 (1995).

CHAPTER 11

MOLECULAR PROPERTIES OF CERULOPLASMIN FROM DIFFERENT SPECIES

Lilia Calabrese
Department of Biology, Third University of Rome
via Ostiense, 173, 00154 Rome, Italy
E-mail: calabrese@axcasp.caspur.it

and

Giovanni Musci
Department of Organic and Biological Chemistry, University of Messina Salita Sperone, 31, 98166 Messina, Italy
E-mail: musci@axcasp.caspur.it

ABSTRACT

Ceruloplasmin, isolated from several species encompassing various classes of vertebrates, turned out to be a highly conserved system, as far as protein structure, copper-binding sites and ligand-binding properties were concerned. A common three-domain structure was evidenced in all species by Differential Scanning Calorimetry. The three domains are asymmetrical with respect to the homology domains inferred, at least on the human protein, from the amino acid sequence, and are stabilized by the presence of copper bound at the constitutive sites. The three-domain architecture was retained even after limited proteolysis of the polypeptide chain, although the pattern of fragmentation differed among the various species of ceruloplasmin.

The stoichiometry of copper was confidently established as six metal ions per ceruloplasmin molecule in human ceruloplasmin. The spectroscopic analysis of the protein under different redox conditions allowed to conclude that the metal sites are distributed into three type 1 copper sites and a type 2-type 3 trinuclear cluster. A considerable fraction of blue copper was found to be in the reduced state in the resting human or sheep proteins, and could be oxidized only by treatment with strong oxidants like ferricyanide, or through intramolecular electron transfer to the trinuclear cluster. This latter phenomenon was observed in a number of conditions, including limited proteolysis and treatment of the protein with chloride. Chicken ceruloplasmin constituted an exception, since copper stoichiometry was evaluated as five atoms per molecule, and the protein was essentially insensitive to chloride. Nevertheless, all ceruloplasmins examined showed the same reactivity towards other ligands, including nitric oxide and the divalent cations Ca(II) and Mg(II).

Contents

1. Introduction

In the forties, Holmberg and Laurell made the discovery that a copper-containing glycoprotein, which they named ceruloplasmin after its strong blue color, is present in the human plasma.[1–3] It has been then ascertained that ceruloplasmin circulates in the blood of all vertebrates, while its presence in lower organisms remains to be established.

Ceruloplasmin binds most of the plasma copper. The copper content varies widely in the plasma of different vertebrates, and this is usually reflected in a change of the oxidase activity toward aromatic amines, a parameter which has been taken as representative of the concentration of ceruloplasmin. A comprehensive report of the levels of oxidase activity of plasma from a large number of vertebrates has been published more than 30 years ago.[4] The data clearly evidenced that aves and reptiles have lower enzymatic activities when compared to mammals, fishes and amphibians. Surprisingly, in all classes there are one or more species which apparently lack measurable levels of oxidase activity. The relationship between plasma copper and plasma oxidase activity, however, is not always straightforward. The most obvious explanation is that the amount of copper that is non-associated to ceruloplasmin varies among species. The pig stands as a striking example, with non-ceruloplasmin copper accounting for as much as 45 % of total copper.[5] A second reason could be the specific activity of ceruloplasmins in the plasma of various animals. At this regard, it is interesting to recall that the catalytic parameters of four ceruloplasmins, the mammalian sheep and dolphin, the avian chicken and the reptilian turtle, display considerable variations in both K_m (ca. 10-fold) and V_{max} (ca. 4-fold) values for the oxidase activity vs. *p*PD (*p*-phenylendiamine).[6,7] Furthermore, it is known that the pH optimum for the oxidase activity of ceruloplasmin varies among animals.[8] As a last possible reason, the presence of inhibitors of ceruloplasmin at different levels in the various plasma may also contribute to the discrepancies between plasma copper and oxidase activity. The case of frog plasma is worth to be mentioned, since the oxidase activity could be detected in that case only on dialyzed plasma.[9]

The biological function(s) of ceruloplasmin is (are) still unknown, although an enzymatic activity as an oxidase toward a broad range of substrates has been widely documented.[10] From a structural point of view, the human protein comprises 1046 amino acid residues, aligned in three homologous domains, each constituted by two subdomains.[11] According to evolutionary studies, the three domains arose from triplication of an ancestral gene coding for a primitive oxidase.[12] In turn, this gene was the product of a duplication event, stemming from the gene which has nowadays evolved to code for simple blue copper proteins like plastocyanin and phytocyanin. This pathway would account for the overall six-domains structure of ceruloplasmin. The other two multi-copper blue oxidases ascorbate oxidase and laccase, as well as the functionally unrelated coagulation factors V and VIII, have their origin from the same "simple blue copper" gene, although through different duplication events.[12] As stated above, the presence of ceruloplasmin in invertebrates has never been reported. It is worth recalling, however, the recent discovery of a yeast protein named Fet3p, having a high degree of homology with multi-copper blue oxidases, especially with laccase.[13,14] The yeast protein is required for ferrous ion uptake by the cell,[13] a function compatible to that proposed for ceruloplasmin.[15]

The presence of multiple metal sites in ceruloplasmin and in the other multi-copper oxidases is expected, since they have originated from successive duplication of a gene that coded for an ancestral simple blue protein. However, this constitutes an intriguing problem, since to the six-domain and to the three-domain structures of ceruloplasmin and of laccase or ascorbate oxidase do not correspond respectively six or three blue copper sites. Laccase and monomeric ascorbate oxidase actually have only one blue "type 1" site, while human ceruloplasmin, as it will be better outlined below, bears three blue sites and an integrated trinuclear cluster. This latter unit cannot structurally arise from the rearrangement of the ancestral copper sites,[12] suggesting that several events involving either loss or acquirement of a copper binding locus have occurred at different stages of evolution, leading to an unbalanced evolution of the polypeptide chain and of the metal sites.

On this basis, it becomes of fundamental importance a comparative investigation of ceruloplasmins extracted from the blood of different species. In this chapter we will review our results on some molecular aspects of different ceruloplasmins, carried out in the effort to answer the following questions: Is ceruloplasmin a highly conserved protein? Can its function be carried out with a broad range of structural arrangements? Is there any flexibility in the stoichiometry of the copper sites for the protein to be functionally active?

2. The Stability of Ceruloplasmin

2.1. *Limited Proteolysis Studies*

Ceruloplasmin has been long recognized to be extremely susceptible to the proteolytic cleavage, such that the single chain nature of the protein was unconfutably proven only in 1982, a quarter of a century after its first isolation.[16] Direct analysis of ceruloplasmin in the human plasma by means of immunoelectrophoretic techniques reveals that a fraction of the protein circulates proteolytically cleaved.[17] This finding suggests that limited proteolysis phenomena might be relevant for the function of ceruloplasmin *in vivo*. It is peculiar that the first group to report the primary structure of intact ceruloplasmin has speculated that "the fragments may have different specific activities because of the multifunctional nature of ceruloplasmin".[18] In this regard, a recent study has shown that the well recognized antioxidant properties of the protein[19] are exerted only by proteolyzed ceruloplasmin, while the intact protein rather has a prooxidant ability, at least on the copper ion-mediated oxidation of plasma LDL.[20]

The nature of the protease(s) responsible for cleavage of ceruloplasmin is still under debate. It is clear from a number of studies carried out in vitro that serine proteases, including trypsin, chymotrypsin and plasmin, easily hydrolyze ceruloplasmin into a 116 and a 19 kDa components, the same pattern observed in the plasma. This has led to hypothesize that either plasmin or thrombin may be the physiologically relevant proteases for ceruloplasmin. However, Ehrenwald and Fox[21] have recently provided evidence that a metalloproteinase is involved in the cleavage of ceruloplasmin, since under their conditions the purified protein is rapidly degraded unless well established metal chelators like EDTA or 1,10-phenantroline are present.

2.1.1. *Patterns of Proteolytic Fragmentation of Ceruloplasmin*

The extreme susceptibility of ceruloplasmin to proteolysis can be exploited to comparatively investigate the structure of ceruloplasmin from different sources. We have therefore studied the behavior of ceruloplasmin extracted from the blood of several species with respect to the susceptibility towards serine proteases, and the effects of the cleavage on the spectroscopic properties of the protein.[22] Figure 1A, which reports the SDS-PAGE pattern of sheep ceruloplasmin before (lane 1) and after limited proteolysis by trypsin (lanes 2 and 3) or plasmin (lanes 4 and 5), shows that, as expected, both proteases generate the two major fragments with M_r 116,000 and 19,000, but with quite different kinetics of hydrolysis. Trypsin further cleaves the 116 kDa component into other fragments with M_r 75,000, 45,000 and 25,000, while plasmin essentially cleaves at a single site, giving rise only to the 116 and the 19 kDa components. The fragmentation pattern is very similar to that obtained by others with human ceruloplasmin, where three fragments with M_r 67,000, 50,000 and 19,000, the former two stemming from a primary 116 kDa fragment, have been identified as the N-terminal, the middle and the C-terminal regions of the molecule.[11] The six N-terminal residues of the 19 kDa fragment of sheep ceruloplasmin are Val-Phe-Asn-Pro-Arg-Arg, a sequence identical to that of the corresponding

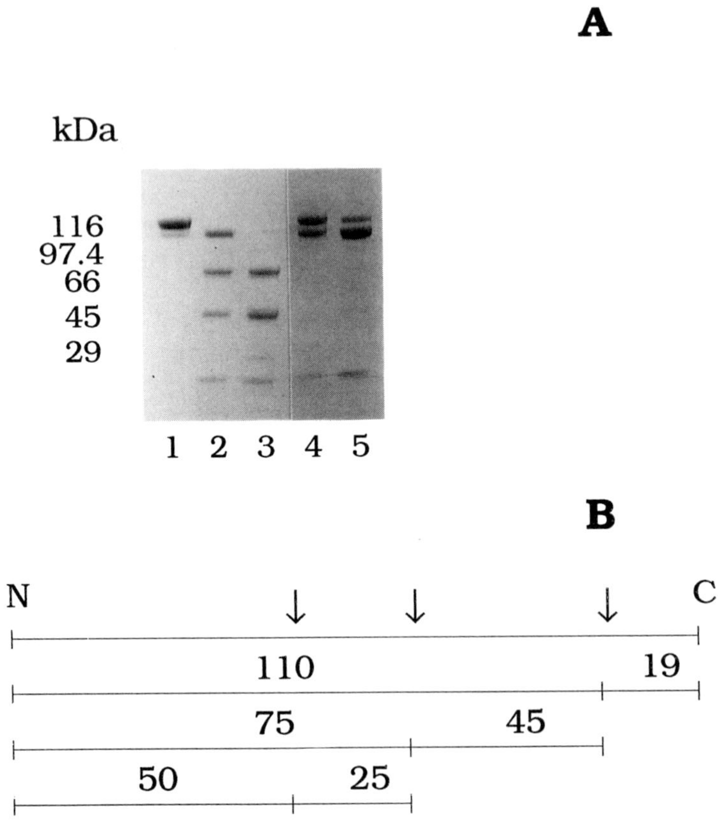

Fig. 1. (A) SDS-PAGE analysis of sheep ceruloplasmin after limited proteolysis. The native protein (lane 1) was incubated for 5 min. and 30 min. with trypsin (lanes 2, 3) or plasmin (lanes 4, 5). (B) Diagram of the major fragments obtained for the sheep protein.

human fragment. This suggests that a common cleavage site for trypsin and plasmin, which can be identified as Lys 887 (human numbering), exists in human and sheep ceruloplasmin. The 75 and the 45 kDa components arise directly from the 116 kDa fragment, and are homologous to the 67 and 50 kDa fragments obtained with human ceruloplasmin[11] and also detectable with trypsin-treated, but not with plasmin-treated rat ceruloplasmin.[7,23] Cleavage of the 75 kDa component could be the source of the 25 kDa fragment, since the appearance of this latter fragment is accompanied by the intensification of the electrophoretic band around 45–50 kDa (Fig. 1A, lanes 2, 3). Figure 1B summarizes the formation pathway of the major fragments of sheep ceruloplasmin. For this protein, the cleavage patterns obtained with other proteases such as chymotrypsin are similar to those observed with trypsin or plasmin.

The proteolysis of chicken ceruloplasmin by trypsin and plasmin follows a pathway qualitatively similar to that observed with the sheep protein, although the fragments have notably different values of M_r (Fig. 2A). The first proteolytic event leads in this case to two chains with M_r 95,000 and 45,000. Also in this case the digestion with trypsin goes on very rapidly (lanes 2, 3) while it stops with plasmin (lanes 4, 5). Owing to the lack of

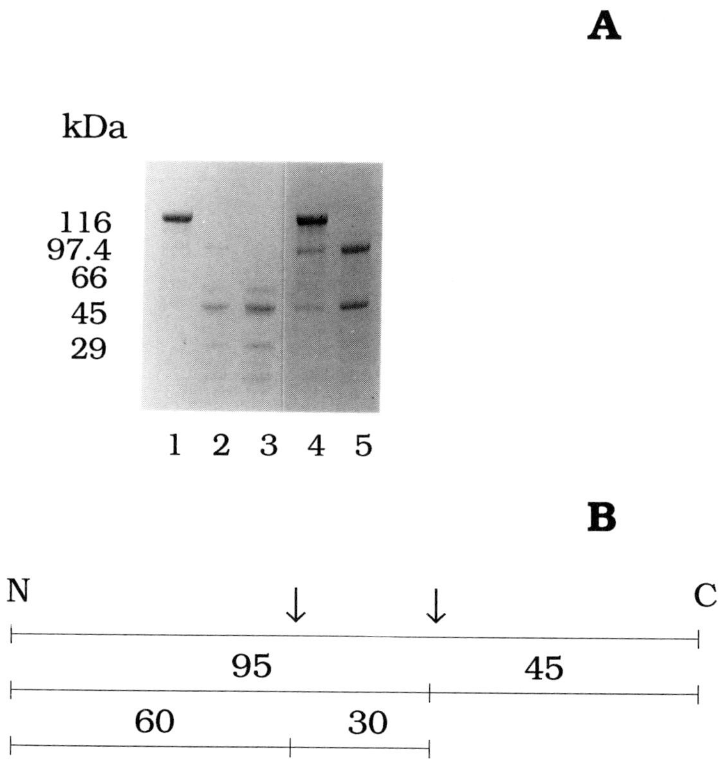

Fig. 2. (A) SDS-PAGE analysis of chicken ceruloplasmin after limited proteolysis. The native protein (lane 1) was incubated for 5 min. and 30 min. with trypsin (lanes 2, 3) or plasmin (lanes 4, 5). (B) Diagram of the major fragments obtained for the chicken protein.

information on the primary structure of bird ceruloplasmin, including the chicken protein, the diagram shown in Fig. 1B is only tentative.

It is clear from the data reported above that all mammalian ceruloplasmins contain an extremely accessible loop which is initially attacked, since proteases with different specificities lead to fragments with similar molecular weights. This also seems to be the case with chicken ceruloplasmin, however the loop would be located in this case on a different region of the molecule.

A noticeable exception is constituted by ceruloplasmin extracted from the plasma of the dolphin *Stenella stenella*, a marine mammal.[7] This protein is totally insensitive to the attack of trypsin, plasmin or chymotrypsin, as exemplified in Fig. 3B. The absence of the proper sequence forming the loop accessible to proteases can obviously explain these results. However, as

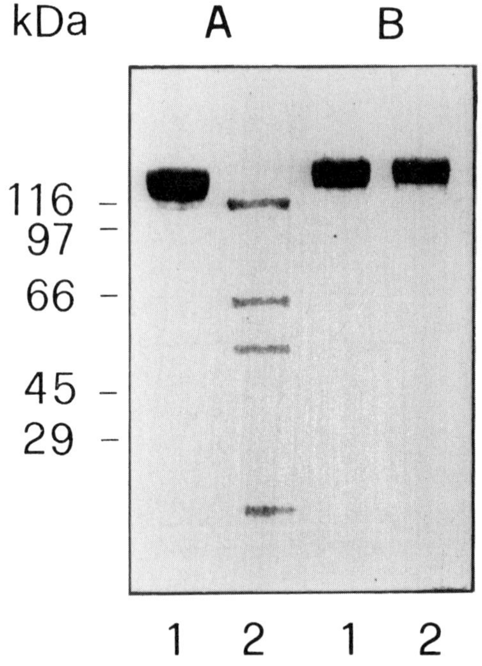

Fig. 3. SDS-PAGE analysis of rat (A) or dolphin (B) ceruloplasmin after limited proteolysis by trypsin. The native proteins (lanes 1) were incubated for 30 min. with the protease (lanes 2).

just mentioned, chicken ceruloplasmin is readily cleaved by trypsin or plasmin although with a pattern different from that of mammalian proteins. This suggests that dolphin ceruloplasmin carries multiple local sequence variations with respect to other mammalian ceruloplasmins. In an effort to correlate sensitivity to proteases to local sequence variations, another mammalian ceruloplasmin, that from the rat, has been examined. Rat ceruloplasmin was chosen because it is the only ceruloplasmin, apart the human, for which the primary structure is known, and this protein has been suggested to have a different sensitivity to proteases.[24] Sequence analyses have revealed that while the region of the first cleavage site (Lys 887) is completely conserved, there are considerable differences around the second cleavage site: Arg 481 is substituted by Lys 480 and there is a five amino acid gap in the sequence of the rat protein.[24] Despite this sequence variation, the digestion pattern by trypsin of rat ceruloplasmin (Fig. 3A) is very similar to that obtained with other mammalian ceruloplasmins. Differences in the secondary structure and in the accessibility of Lys 480 might explain the relatively low content of the resulting fragments (M_r 63,000 and 50,000) even after 30 min of digestion. It should be noted that under the same conditions the 116 kDa fragment completely disappears in all other mammalian ceruloplasmins tested.

Dolphin ceruloplasmin is so far the only case among mammalian ceruloplasmins found to be resistant to proteolysis. However, the protein from another sea organism, the turtle *Caretta caretta*, has earlier been reported to be insensitive to the enzymatic action of trypsin and plasmin.[6] This unusual resistance to proteolysis has led to postulate that the key bond(s) to be attacked by the protease is(are) shielded, for instance by the carbohydrate moiety. As a matter of fact, although the glucidic content has not been measured for turtle ceruloplasmin, it should be recalled that this protein has an apparent molecular weight of ca. 145,000, one the highest among ceruloplasmins. However, in a later report, we have found that turtle ceruloplasmin is extensively degraded by chymotrypsin.[7] Since the studies with other ceruloplasmins have shown that the site of attack of chymotrypsin must be very close to that of trypsin or plasmin, we now believe that the protein extracted from *Caretta caretta* lacks the residues constituting the site of recognition by these two latter proteases.

The limited proteolysis data suggest that ceruloplasmin is a labile protein. Indeed ceruloplasmin is a structurally compact system, since the fragments generated by the proteolytic cleavage remain tightly associated in a sample that is enzymatically active and with spectroscopic properties not far from those typical of the native one (see below). As a matter of fact, not only proteolyzed ceruloplasmin runs as a single band in non-denaturing PAGE, but the fragments do not show up in SDS-PAGE unless the sample is boiled before the electrophoretic run. SDS does not perturb the structure of ceruloplasmin at room temperature, which remains active and retains the native optical and EPR properties in the presence of the detergent (unpublished results from our laboratories). The calorimetric data that will be shown in Sec. 2.2 are also consistent with a compact folding of ceruloplasmin, since the three-domain structure is retained in the proteolyzed protein, unless an extensive fragmentation has occurred. Thus, the ceruloplasmin molecule can be viewed as a tightly packed protein bearing multiple flexible loops, which tolerate, as expected for an exposed polypeptide segment, diverse sequence variations. It is clear that copper plays a major role in the stability of ceruloplasmin. In particular, the ligands of the type 2-type 3 trinuclear cluster are in both the C- and the N-terminal portion of the protein, which implies that the metals hold together the extreme regions of the polypeptide chain, thus playing a stabilizing effect. That the conformation of ceruloplasmin depends on the presence of copper is also clear from the electrophoretic behavior of the single-chain protein in SDS-PAGE. Sato and Gitlin[17] were first to report that apo- and holo- human ceruloplasmin can be distinguished in non-denaturing SDS-PAGE, i.e. without pre-boiling the sample, with the holo form running as a doublet of bands with apparent M_r of 78,000 and 84,000 and the apo as a 130 kDa band. We have evidence that, on one hand, ceruloplasmins from different species behave differently, and that the rat and, to some extent the sheep protein, for instance, run at 130 kDa even when copper-saturated. On the other hand, in a recent study on the mechanism of copper reincorporation into apoceruloplasmin we have further circumstantiated the role of copper and found that it is not the metal status of the protein, but rather the native geometries of the copper sites that make the holoprotein running with anomalous molecular weights.[25]

2.1.2. *Spectroscopic Properties of Ceruloplasmin after Proteolysis*

A careful analysis of the effects of proteolysis on the spectroscopic properties of ceruloplasmin has been carried out on the sheep and the chicken protein

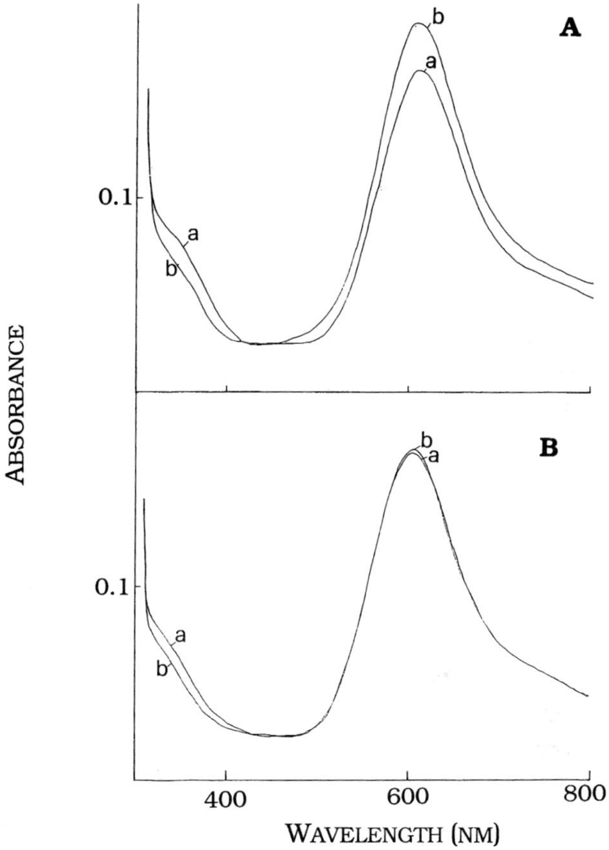

Fig. 4. Optical spectra of sheep (A) and chicken (B) ceruloplasmin after plasmin digestion. (a) Native protein (2×10^{-5} M in 0.1 M phosphate buffer, pH 7); (b) after 30 min. incubation with 1×10^{-6} M protease.

treated with either trypsin or plasmin.[22] Figure 4 shows the optical spectra of sheep and chicken ceruloplasmin after digestion with plasmin for 30 min at room temperature (protease/ceruloplasmin ratio 1:20). The chromophore at 330 nm typical of the type 3 pair, diminishes in intensity in both cases –15 % for the sheep and 11 % for the chicken protein — while the band of the type 1 copper sites at around 600 nm is stable in the chicken, and increases ca. 10 % in the sheep protein. The corresponding EPR spectra, reported in Figs. 6 and 7, show that digestion with plasmin

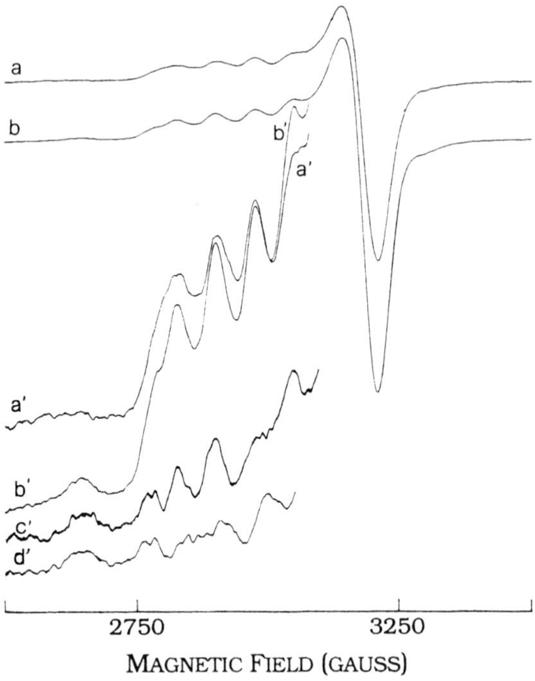

Fig. 5. EPR spectra of sheep ceruloplasmin after plasmin digestion. (a) Native protein ($8x10^{-5}$ M in 0.1 M phosphate buffer, pH 7); (b) after 60 min. incubation with $4x10^{-6}$ M protease; (c) difference spectrum b–a; (d) signal obtained by subtracting a pure type 1 copper signal from spectrum c. Experimental settings: microwave power, 20 mW; microwave frequency, 9.112 GHz; modulation amplitude, 10 G; temperature, 100 K. Primed spectra are 10-fold amplifications.

results in an overall increase of the paramagnetic copper content in both cases, with the effect being markedly pronounced (+30 %) in the sheep protein. In this latter case, the higher content of EPR-detectable copper is clearly due to the increase of the hyperfine lines due to type 1 copper, and to the concomitant appearance of new resonances at lower magnetic field. Computer-generated difference spectra have allowed to identify the new species as a type 2-like copper center (Fig. 5), and to calculate that the extra type 1 copper and the type 2-like species equally contribute to the increase of paramagnetic copper content in plasmin-treated sheep ceruloplasmin. A straightforward explanation is that a fraction of type 1 copper

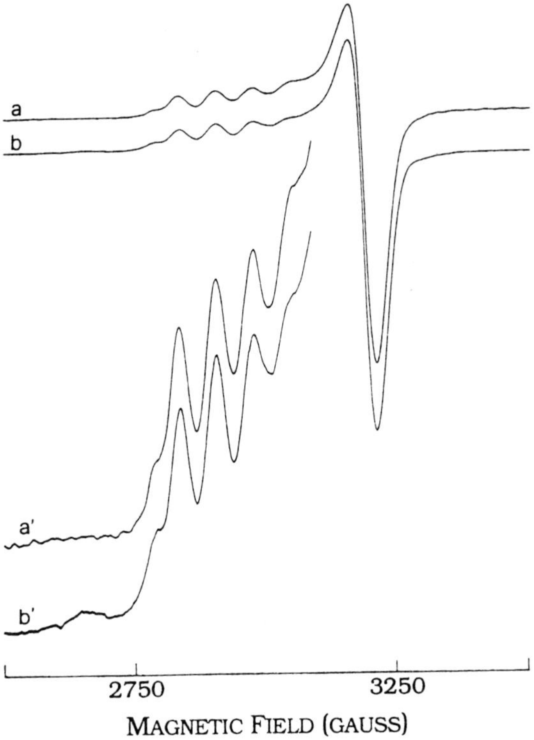

Fig. 6. EPR spectra of chicken ceruloplasmin after plasmin digestion. (a) Native protein (9×10^{-5} M in 0.1 M phosphate buffer, pH 7); (b) after 60 min. incubation with $4.5\text{x}10^{-6}$ M protease. Experimental settings as in Fig. 5.

is reduced in the "resting" protein and, upon cleavage of the polypeptide chain by plasmin, there is a structural rearrangement which results in an electron transfer from these reduced centers to the oxidized type 3 pair, which would become half-reduced. A half-met state of type 3 copper is consistent with both the decrease in the optical spectrum at 330 nm[26] and with the appearance of a type 2-like copper signal in the EPR spectrum.[27] A very similar phenomenon will be described in a later paragraph concerning Cl^- binding to human ceruloplasmin (see Sec. 4.1).

The results obtained with the chicken protein are more difficult to explain, since in this case there is no increase at either 600 nm or in the type 1 copper hyperfines, upon cleavage by plasmin, yet the chromophore at 330 nm decreases in intensity, and a type 2-like copper signal appears

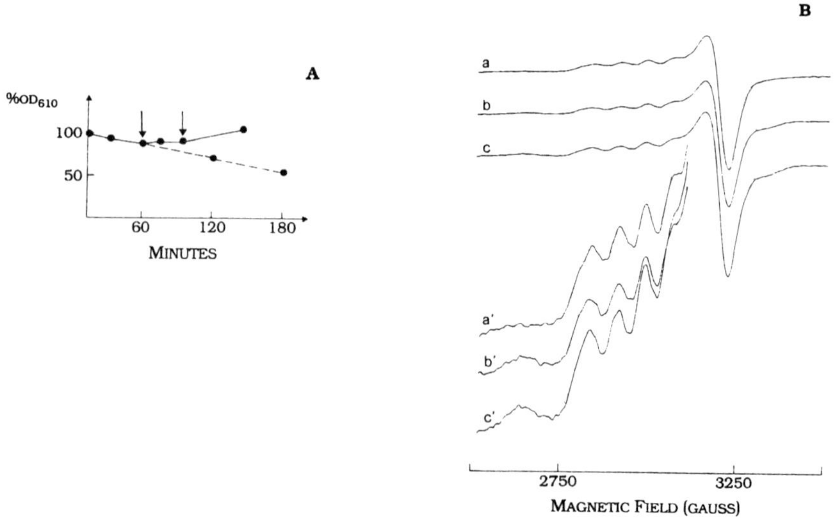

Fig. 7. Panel A: effect of temperature on the 610 nm band of trypsin-digested sheep ceruloplasmin. Incubation was carried out at 35°C for 60 min., then either it was continued at this temperature (broken line), or the temperature was lowered to 20°C for 30 min. (continuous line, first arrow) before addition of H_2O_2 (second arrow). Panel B: EPR spectra of sheep ceruloplasmin after trypsin digestion. (a) Native protein ($1x10^{-5}$ M in 0.1 M phosphate buffer, pH 7); (b) after 60 min. at 35°C with $5x10^{-6}$ M protease; (c) after treatment with 0.1 mM H_2O_2. Experimental settings as in Fig. 5.

in the EPR spectrum (Figs. 4 and 6). It can be postulated that, in this case, the electron transfer does not involve type 1 copper as the donor, but rather some unidentified protein group.

The spectroscopic response to the cleavage by trypsin is quite different. Consistent with the multiple cleavage of the ceruloplasmin chain by trypsin (see previous section), this protease induces a more extensive rearrangement of the protein structure, as it will be shown later on in this section. However, the spectroscopic properties of sheep ceruloplasmin are essentially unaltered by trypsin at room temperature, even when the 75 kDa component has been completely cleaved. On the other hand, a peculiar behavior of trypsin-treated sheep ceruloplasmin can be observed, in that the proteolytic reaction carried out at 35°C leads to the progressive decrease of the absorbance at 610 nm in the optical spectrum (Fig. 7A), down to 50 % of the original value after 3 hours. The chromophore at 330 nm is unaffected, and the EPR spectrum (Fig. 7B) confirms the decrease of the intensity of the lines due to blue copper and reveals the appearance of resonances typical of type 2 copper. This latter signal is different from that observed with plasmin, and rather resembles the type 2 copper signals which emerges upon aging of the protein, i.e. when the trinuclear cluster loses its native, tight structure.[28] That the trypsin-treated sheep protein might have shifted to the conformation of an "aged" protein is also confirmed by the observation that only half of the blue color can be restored by addition of oxygen to the ascorbate-reduced protein, a behavior typical of "aged" ceruloplasmin samples.[28] The temperature-dependent decrease of the optical and EPR features of type 1 copper stems also in this case from redox phenomena, with the blue sites being in this case, at variance with that observed with plasmin, reduced in the presence of the protease. As a matter of fact, addition of H_2O_2 restores both the A_{610} (Fig. 7A) and the type 1 copper intensity and lineshape in the EPR spectrum (Fig. 7B). As demonstrated by the behavior of sheep ceruloplasmin with plasmin, a fraction of type 1 copper is reduced in the native protein. Trypsin, at variance with plasmin, seems to magnify this phenomenon, indicating that type 1 copper can either donate or accept electrons even in the absence of exogenous substrates, depending on the conformation of the protein.

Chicken ceruloplasmin, although extensively proteolyzed by trypsin (see Fig. 2), has unaltered optical and EPR features in the presence of the protease, both at room temperature and at 35°C.

2.2. *Thermal Stability Studies*

Thermal stability studies are the other major approach, beside limited proteolysis, to the characterization of the structural relationships between different regions of a protein. The two techniques are not alternative, and can be in fact be combined for a better comprehension of the system under investigation.

2.2.1. *Heat Inactivation Measurements*

The oxidase activity of sheep, chicken and turtle ceruloplasmin is gradually lost upon heating the protein (Fig. 8B). The phenomenon is closely paralleled by the loss of the optical properties, as shown in Fig. 8A, where the variation of the absorbance at around 600 nm observed between 20°C and 80°C is reported. The changes do not depend on the heating rate in the range 0.5–2 °C/min and are essentially irreversible.

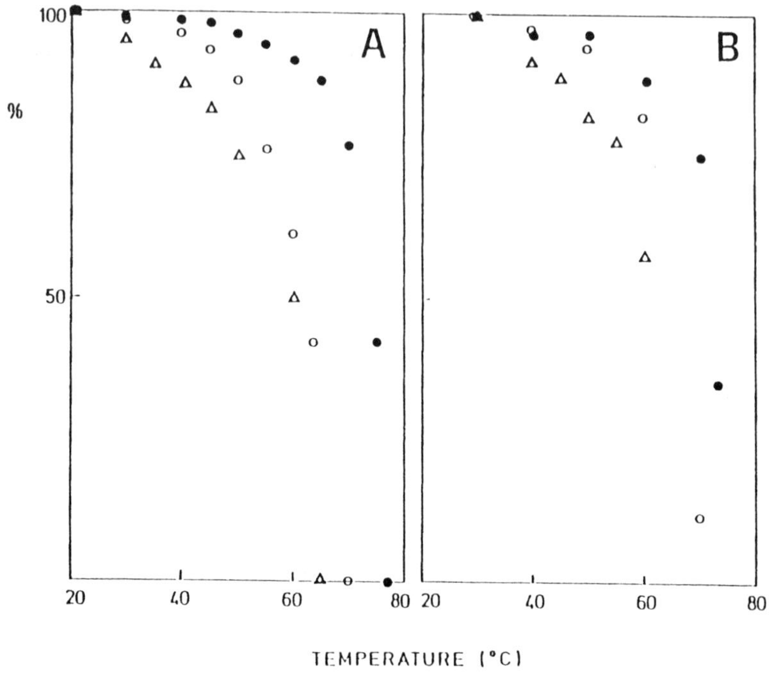

Fig. 8. Percentage variations of optical density at 600 nm (A) and *p*-phenylendiamine oxidase activity (B) of sheep (o), chicken (•) and turtle (Δ) ceruloplasmin upon heating. Heating rate was 1°C/min.

2.2.2. *Calorimetric Measurements*

Differential Scanning Calorimetry (DSC) is a powerful tool to establish the conformational stability and the domain organization of a multidomain protein. Ceruloplasmin generally has a high thermal resistance, with a temperature of maximum excess capacity, T_m, ranging from 70.6°C for the dolphin to 85.1°C for the human protein, with the noticeable exception of turtle ceruloplasmin (Table 1).[7,25,29] Denaturation is irreversible in all cases, since reheating the sample gives flat DSC curves. Despite this fact, deconvolution of the curves has been carried out in two-state transitions, for which the measured calorimetric enthalpy, $\Delta H_{\rm cal}$, equals the theoretical van't Hoff enthalpy, $\Delta H_{\rm vH}$. This approach is formally allowed only in the case of reversible denaturation, but it has been shown to be admissible even on systems having little or no reversibility.[30–32] The curve of native

Table 1. Calorimetric parameters for different ceruloplasmins.

Protein	Experimental T_m (°C)	Deconvolution[a] ΔH_{cal} (kcal/mol)	T_m (°C)	% ΔH_{cal} (kcal/mol)
sheep	71.2	550	59.1	—
			67.6	32.0
			71.3	39.0
			76.1	29.0
"aged" sheep	73.1	595	56.6	—
			67.6	26.0
			71.4	35.0
			74.3	39.0
dolphin	70.6	501	64.8	24.0
			69.8	37.0
			71.5	39.0
human	85.1	667	79.5	19.0
			83.1	35.0
			85.2	46.0
chicken	82.1	343	74.0	17.0
			78.0	27.0
			82.1	56.0
turtle	57.8	360	42.8	27.0
			57.8	47.0
			58.8	26.0
sheep+plasmin	73.2	557	68.2	25.0
			71.5	36.0
			74.2	39.0
chicken+plasmin	77.8	492	70.0	15.0
			74.7	29.0
			77.8	56.0

[a]Assuming $\Delta H_{cal} = \Delta H_{vH}$, where ΔH_{cal} and ΔH_{vH} are the calorimetric and the van't Hoff enthalpy, respectively.

sheep ceruloplasmin (Fig. 9A) can be decomposed in four transitions, with T_m values comprised between 59.1°C and 73.5°C. When copper is removed from the protein, the DSC curve shifts to considerably lower temperatures, showing an overall T_m value of 57.2°C (Fig. 9B). Moreover, only two

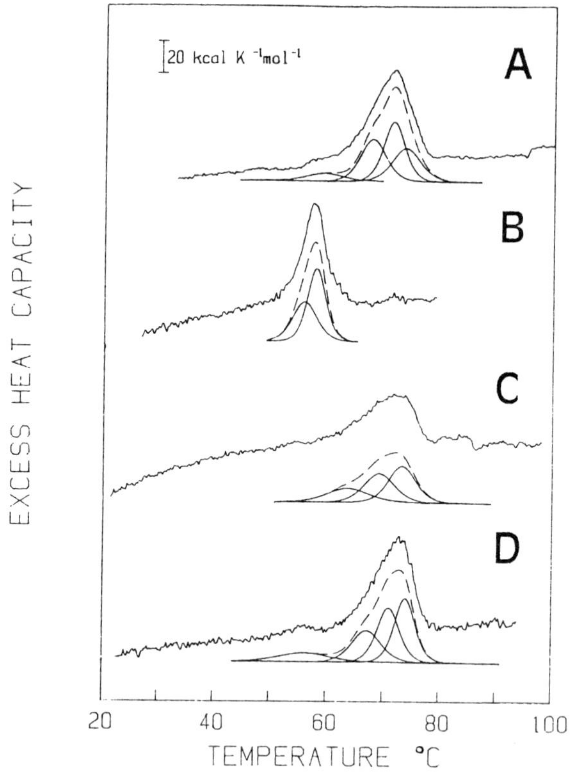

Fig. 9. Calorimetric scans of native sheep ceruloplasmin (A), aposheep ceruloplasmin (B), sheep ceruloplasmin preheated at 67°C (C) and "aged" sheep ceruloplasmin (D). The upper parts represent the experimental curves, and the lower parts are calculated curves (dashed lines) and component peaks (continuous lines) obtained by deconvolution analysis of the excess heat capacity function. Heating rate was 1°C/min.

components are needed to account for the experimental trace (Table 1). Noteworthy, the first transition observed in the DSC curve of native sheep ceruloplasmin essentially matches the position of the peak of the apoprotein, suggesting that the sample of the native protein contains a fraction of copper-free ceruloplasmin. The thermodynamic link between the three cooperative units of native ceruloplasmin has been tested by preheating the sample up to 67°C, the T_m of the first transition, and then heating again through the whole endotherm (Fig. 9C). It is evident that there is a residual ΔH_{cal} (ca. 74 %), with fractional contributions indicating that unfolding of the unit with lower melting point affects the two units at higher stability (Table 1). This result is strongly suggestive of the existence of interactions between cooperative units, which do not appear to behave as independently folded domains.

The DSC curve of a sample of "aged" sheep ceruloplasmin is shown in Fig. 9D. "Aging" here is referred to the appearance of peculiar spectroscopic properties, in particular to the detectability of type 2 copper in the EPR spectrum, reflecting the magnetic uncoupling of the trinuclear copper unit.[28] The curve looks similar to that of the native protein (cfr. Fig. 9A) and can be deconvoluted in three components with thermodynamic parameters also similar to those of native sheep ceruloplasmin (Table 1). The fractional contribution of the three transitions is however different, suggesting that a structural rearrangement of the protein takes place upon aging.

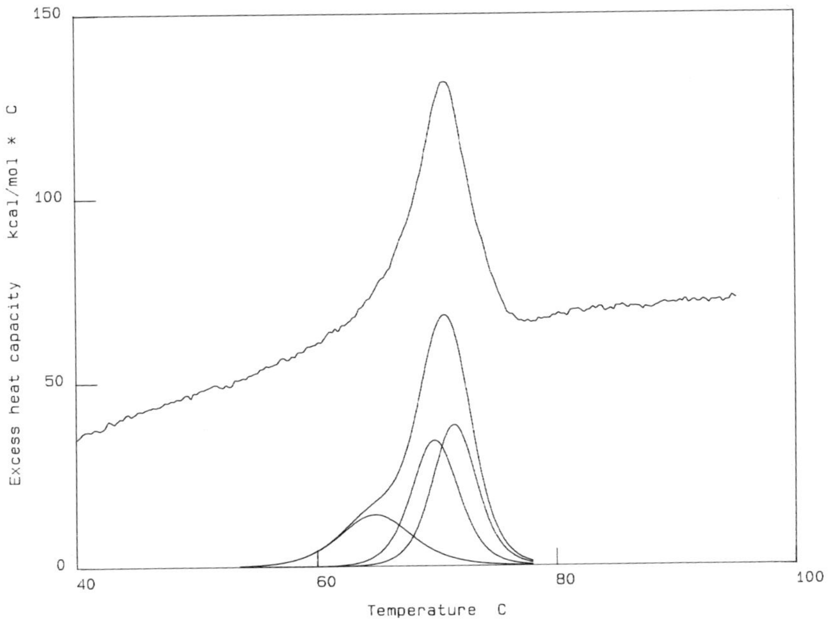

Fig. 10. Calorimetric scan of dolphin ceruloplasmin. The upper part represents the experimental curve, and the lower part is the calculated curve and component transitions. Heating rate was 1°C/min.

Interestingly, the DSC curve of dolphin ceruloplasmin (Fig. 10), a protein that has been found to have a considerable amount of EPR-detectable type 2 copper, can be deconvoluted into three units with fractional contributions very similar to those observed with the sample of "aged", rather than native sheep ceruloplasmin.[7] This result, on one hand shows that dolphin ceruloplasmin has a structural organization typical of a mammalian ceruloplasmin; on the other hand, it corroborates the idea that ceruloplasmins displaying EPR-detectable type 2 copper, though maintaining a very similar domain arrangement, have a different conformation with respect to proteins bearing an intact trinuclear cluster.

Human ceruloplasmin shows a particularly high thermostability for a mammalian ceruloplasmin, with a T_m value almost 10°C higher than that of the sheep protein.[33] Deconvolution of the DSC trace reveals again the presence of three two-state components (Table 1).

The only two non-mammalian ceruloplasmins so far examined, the chicken and the turtle proteins, behave quite differently. The DSC curve of chicken ceruloplasmin (Fig. 11A) has a T_m value of 82.1°C and presents an exothermic process revealed by the steepness of the curve on the high temperature side. The exotherm, which likely reflects aggregation of the protein, is not prevented by dilution of the sample down to 1 mg/ml (the detection limits in our experimental conditions). The deconvolution of the curve carried out by assuming the presence of three units in analogy to all other ceruloplasmins (Table 1) gives nevertheless a good fit, at least for the low-temperature side of the peak, suggesting that a similar structural organization exists in the chicken and the other proteins. It should be noted that chicken and human ceruloplasmin both share very high thermal stability (with a T_m value above 80°C) and the pattern of the fractional contributions of the three units to the curve, with the unit at highest stability contributing approximately 50 % of the total experimental ΔH_{cal}. This is interesting in view of the EPR lineshape of the blue sites of human ceruloplasmin, which is more similar to that of the chicken protein than to that of other mammalian ceruloplasmins like sheep, bovine or dolphin. It should however be noted that also turtle ceruloplasmin presents blue hyperfines in the EPR spectrum with a considerably small linewidth, yet its thermal stability is the lowest so far observed for ceruloplasmin (Fig. 11B).

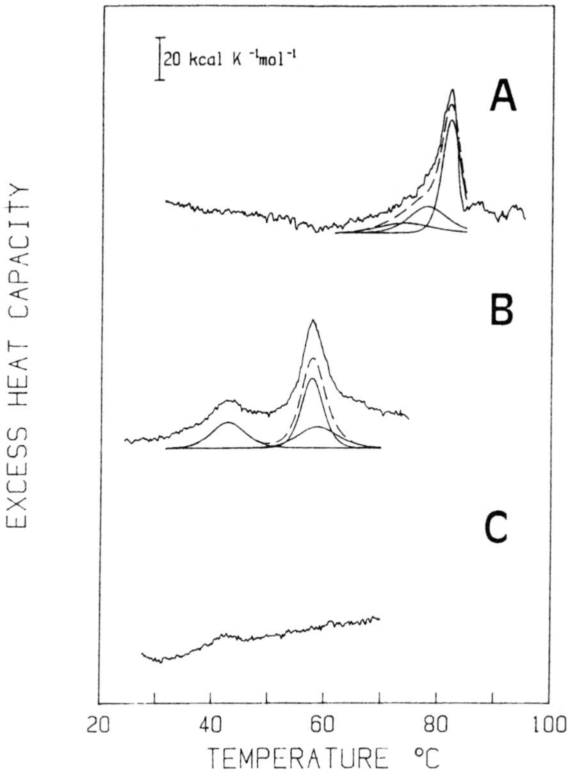

Fig. 11. Calorimetric scans of native chicken ceruloplasmin (A), native turtle ceruloplasmin (B), turtle ceruloplasmin after a complete DSC scan up to 80°C (C). The upper parts represent the experimental curves, and the lower parts are calculated curves (dashed lines) and component peaks (continuous lines) obtained by deconvolution analysis of the excess heat capacity function. Heating rate was 1°C/min.

Proteolytic cleavage by plasmin seems to have only slight effects on the overall structure of both sheep and chicken ceruloplasmin, as shown in Fig. 12. The proteins retain their melting temperatures and their three-domain structure, although the fractional contributions, at least for the sheep protein, change to indicate a structural rearrangement similar to that occurring in the "aged" sample (Table 1). It is interesting to note that the endotherm centered around 59°C in the DSC curve of native sheep ceruloplasmin vanishes upon treatment with plasmin. As discussed above, the endotherm likely represents a fraction of apoceruloplasmin, and the

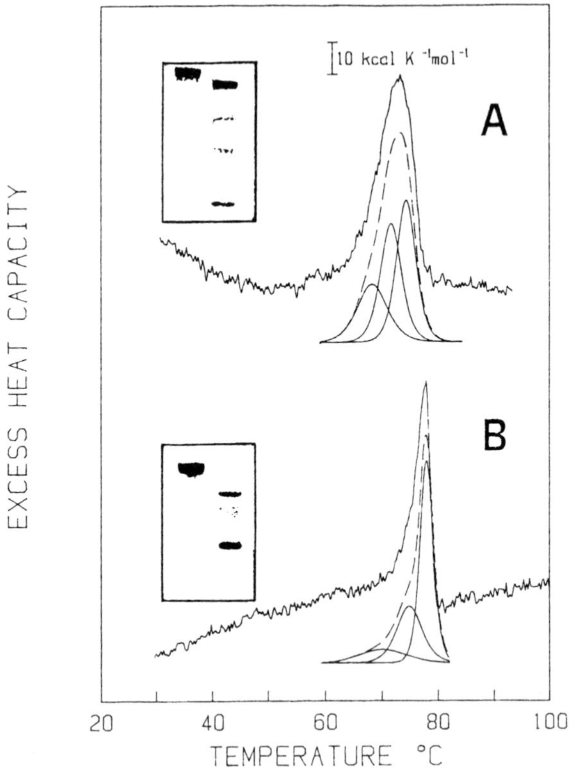

Fig. 12. Calorimetric scans of plasmin-digested (A) sheep and (B) chicken ceruloplasmin. The upper parts represent the experimental curves, and the lower parts are calculated curves (dashed lines) and component peaks (continuous lines) obtained by deconvolution analysis of the excess heat capacity function. Heating rate was 1°C/min. The insets show the SDS-PAGE patterns of the proteolyzed samples and of the parent single-chain proteins.

experiment thus indicates a higher susceptibility of apoceruloplasmin to the proteolytic attack.

Trypsin dramatically modifies the melting profiles of both sheep and chicken ceruloplasmin (Fig. 13). The number of two-state transitions increases, parallel to the number of fragments obtained in the SDS analysis. The overall values of T_m are 10°C and 15°C lower for the sheep and the chicken proteins in the presence than in the absence of the protease.

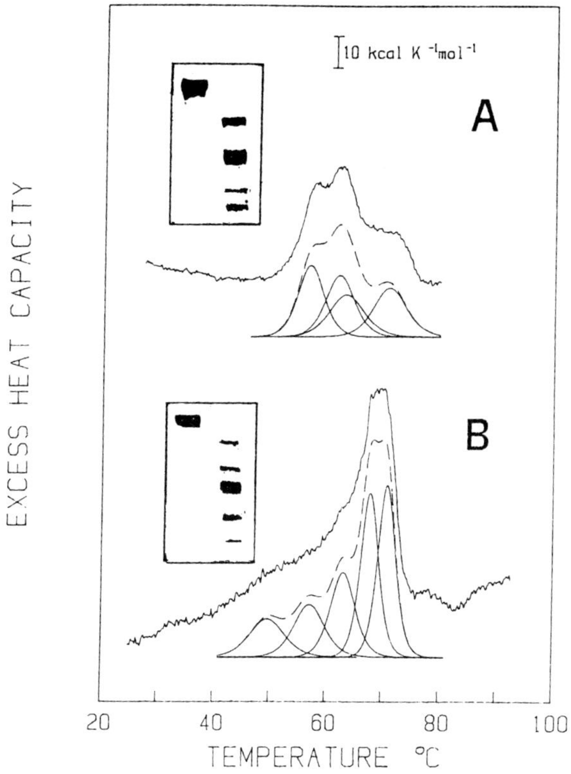

Fig. 13. Calorimetric scans of trypsin-digested (A) sheep and (B) chicken ceruloplasmin. The upper parts represent the experimental curves, and the lower parts are calculated curves (dashed lines) and component peaks (continuous lines) obtained by deconvolution analysis of the excess heat capacity function. Heating rate was 1°C/min. The insets show the SDS-PAGE patterns of the proteolyzed samples and of the parent single-chain proteins.

3. The Copper Sites of Ceruloplasmin

The exact stoichiometry of the copper sites of ceruloplasmin is, to date, not yet definitively determined. The distinction of the various copper atoms of ceruloplasmin into type 1, type 2 and type 3 sites does not need any special comment here, as it is widely explained in other parts of this book. Laccase and the monomeric unit of ascorbate oxidase invariably contain four copper sites.[34] Ceruloplasmin, on the other hand, has been reported to have as little as four[35] and as much as eight[36] copper ions per molecule. The discrepancy can be explained on the basis of three different aspects. First of all, ceruloplasmin can accommodate a number of copper ions in non-prosthetic sites.[37] If proper care is not taken to avoid metal contamination during the isolation of the protein, the stoichiometry will be overscored. This was recognized in the seventies, when it was found that the stoichiometry approached the value of six copper ions/molecule after treatment of the protein with Chelex.[38,39] A second problem is the presence of a fraction of copper-free ceruloplasmin in the plasma which possibly co-purifies with the holoprotein.[40,41] A straightforward quantitation of the amount of apoceruloplasmin in the purified samples is difficult, and this can lead to overestimate the correct stoichiometry. The third problem is that the various ceruloplasmins may have different stoichiometries, and therefore the assessment of the metal content of a particular ceruloplasmin cannot be automatically transferred to other ceruloplasmins. As a matter of fact, we now consider that the human and the dolphin proteins contain 6 tightly bound ions per molecule, while we have evidence that the chicken, the turtle and the sheep proteins have a lower content, i.e. 5 copper ions/molecule. These values, however, especially the latter ones, should be taken with caution. The experimental stoichiometry values are nearly always non-integer numbers, which more often approach the figure of six for the human and dolphin proteins, and the figure of five for the other ceruloplasmins.

While the distribution of the four copper sites in laccase and monomeric ascorbate oxidase is straightforward, with both proteins containing one type 1, one type 2 and one type 3 copper center,[34] a stoichiometry of five or six implies that one or more types of copper are present more than once. On the basis of the extinction coefficient of the "blue" band at around 600 nm, it has been generally accepted that ceruloplasmin contains two type 1 copper sites. These have been recognized to be non-equivalent, with different EPR parameters[42] and redox potentials.[43] Two type 1 sites, along with a type 2 and a type 3 site, make up a stoichiometry of five. To explain a copper content of six ions/molecule, it has been in the past hypothesized that a fourth type of copper, possibly with the metal in the reduced state, is present in ceruloplasmin.[44] The presence of three type 1 centers has been suggested on the basis of indirect evidence, such as type 1:type 2 EPR intensity ratios,[45] amino acid sequence analysis[46] or changes in the 610/280 absorbance.[47] We have recently addressed this issue and, on a purely experimental basis, we have reached the conclusion that a six copper-containing ceruloplasmin, that from humans, contains three non-equivalent type 1 copper sites.[48] The other three metal ions are from a type 2 and a type 3 site and assemble together to form a trinuclear unit.

3.1. *The Type 2-Type 3 Trinuclear Cluster*

When chicken ceruloplasmin was first purified with the single-step method involving the direct chromatography of plasma on derivatized Sepharose, the most peculiar characteristics of the isolated protein were found to be the absence of resonances from type 2 copper in the EPR spectrum and the ability of oxygen to completely restore the blue color of the protein which had anaerobically reduced with ascorbate.[49] It was later found that a mammalian ceruloplasmin also, that from the sheep, could efficiently react with oxygen, after reduction with ascorbate, only when type 2 copper was EPR-silent in the native protein.[28] Furthermore, it was found that in both cases a type 2 copper site was transiently EPR-detectable during the anaerobic reduction with ascorbate, and that the event followed the complete reduction of the type 3 copper pair. All these data were interpreted in terms of a functional and structural trinuclear unit comprising type 2 and type 3 copper. The EPR inactivity of undetectable type 2 copper would arise from unusual magnetic interactions within the cluster, although it has to be recalled that the unit cannot be diamagnetic unless type 2 copper is reduced, and that therefore the absence of any EPR resonance has to be ascribed to an instrumental inability to detect too broad signals. In a "coupled" cluster the magnetic interactions are retained, while in an "uncoupled" cluster type 2 copper becomes detectable in the EPR spectrum after breaking the structural links with the type 3 pair. According to this picture, oxygen would efficiently react only with the coupled cluster or, alternatively, the blue sites would efficiently transfer electrons only to the coupled unit. Mammalian ceruloplasmins are, in this respect, more labile, and a fraction of type 2 copper is often detected in the EPR spectrum of freshly purified samples, with the noticeable exception of ceruloplasmin from the blesbok antelope *Damiliscus dorcas phillipsi* (unpublished observations).

The confidence on the existence of a trinuclear center in ceruloplasmin, suggested by our experimental data, was strengthened by the knowledge that such a unit had been proposed to exist in laccase.[50] They had reached that conclusion on the basis of ligand-binding studies, which had shown that the anion azide, N_3^-, is able to coordinate both type 2 and type 3 copper with the same molecule at the same time. The resolution of the crystallographic structure of ascorbate oxidase[51] has strongly supported the idea that such a trinuclear unit is in place in all three major members of the multinuclear blue copper oxidases family.

A major difference exists between ceruloplasmin on one hand and laccase and ascorbate oxidase on the other, in that type 2 copper is invariably detected in the latter two proteins. Since laccase and ascorbate oxidase are fully functional systems, it can be inferred that the requirement for a "coupled" cluster is a property typical of ceruloplasmin.

3.2. *The type 1 Sites*

Concerning the issue of the blue centers of human ceruloplasmin, it should first be considered that the fraction of EPR-detectable copper varies in different batches of the purified protein between 33 % and 43 % of total copper, corresponding to 2.0 to 2.6 EPR-detectable ions on a stoichiometry of six ions/molecule.[48] To extract the contribution of the blue sites, one has to quantitate the contribution, in terms of number of spins, of the type 2 copper center. It should be recalled that this number never reaches the value of one for ceruloplasmins isolated with the single-step method. The exact estimate of the number of EPR-detectable type 2 copper centers has been made possible after the observation that the aerobic addition of a large excess ($\sim$100 e^-/Cu) of ascorbate to human ceruloplasmin causes the immediate and complete loss of the signal of the blue sites, leaving unperturbed, with unchanged intensity, the resonances of type 2 copper (Fig. 14). This has allowed to calculate that type 2 copper contributes between 0.30 and 0.80 spins per molecule in different samples of human ceruloplasmin. Knowledge of the exact magnetic parameters of type 2 copper has also allowed to exploit the procedure of Vänngård,[52] which makes use of the area underlying the $M_I = -3/2$ hyperfine line, to measure the contribution of type 2 copper to the spectrum. A striking agreement has been observed between the experimental and the theoretical method. The number of EPR-detectable blue ions per molecule has finally been calculated to be in the range 1.4–2.1. These figures, however, considerably increase when human ceruloplasmin is treated with excess potassium ferricyanide. As a representative experiment, Fig. 15 depicts the effect of the oxidant on the EPR spectrum of the protein. In this particular case, the sample initially had 2.1 EPR-active centers per molecule, 0.6 of which were from type 2

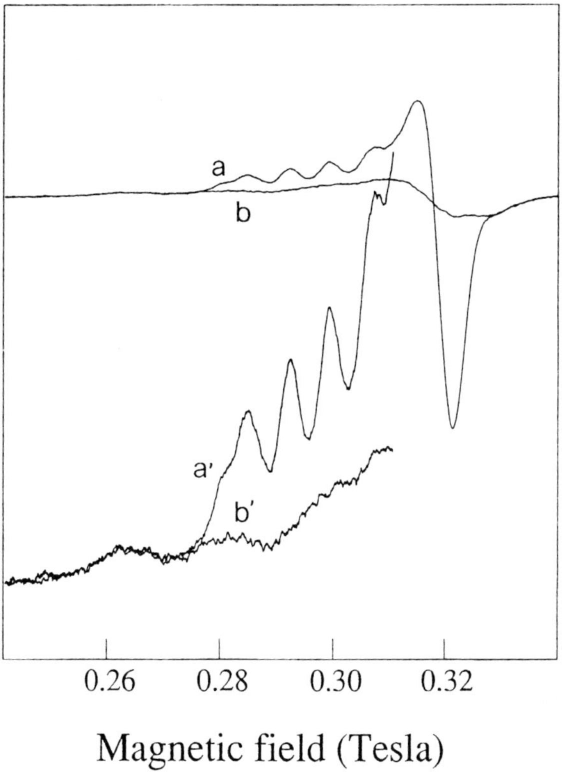

Fig. 14. X-band low temperature EPR spectra of human ceruloplasmin (4.3×10^{-5} M in 0.15 M phosphate buffer, pH 7) aerobically reacted with 17 mM sodium ascorbate, a: native protein; b: 5 min. after addition of ascorbate. Experimental settings as in Fig. 5. Primed spectra are amplified 8-fold.

(according to the methods outlined above) and 1.5 from type 1 copper sites. After reaction with ferricyanide, the paramagnetic copper accounted for 3.4 spins/molecule. Type 2 copper was apparently not affected, while the resonances from type 1 copper increased and, as evidenced by the differential spectra, a new absorption around 0.273 T appeared at longer incubation times. This novel signal accounted, in this particular experiment, for 0.35 spins per molecule. The proper calculations allowed to estimate that, out of 3.4 total spins, 2.45 were from type 1 copper centers. The EPR modifications were closely related to the optical changes, as shown in the inset of Fig. 15. These results are actually

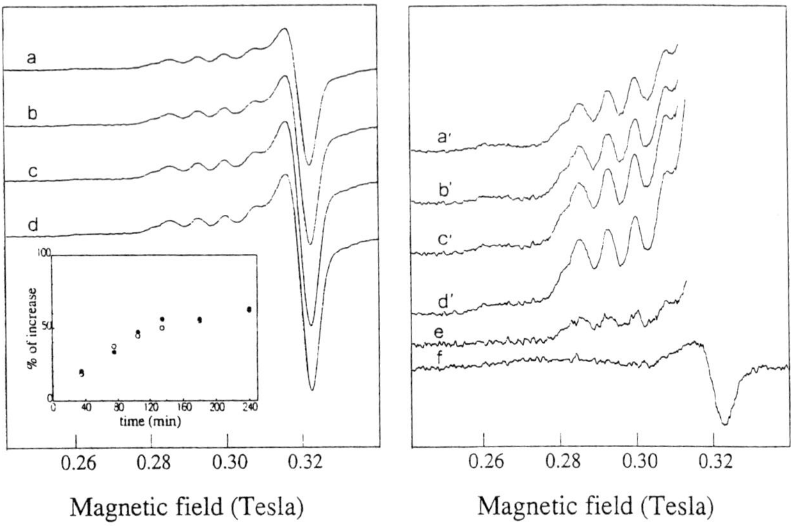

Fig. 15. X-band low temperature EPR spectra of human ceruloplasmin ($5.1x10^{-5}$ M in 0.15 M phosphate buffer, pH 7) reacted with 10 mM potassium ferricyanide for a: 0 min.; b: 35 min.; c: 105 min.; d: 200 min.. Curve e is the differential spectrum b–a; curve f is the difference d–a, to which a fraction of curve e was subtracted, in order to zero resonances from type 1 copper. Experimental conditions as in Fig. 5. Primed spectra are 5-fold amplifications. Spectra e and f are arbitrarily amplified. Inset: Time courses of the increase of the optical absorbance at 610 nm (•) and of the intensity of the EPR signal (o) of human ceruloplasmin upon reaction with ferricyanide.

quite reproducible: samples having different amounts of spectroscopically active type 1 and/or type 2 copper in their initial state all reach a common final value of ca. 2.5 EPR-detectable type 1 copper ions per molecule after reaction with ferricyanide.

Computer simulations of the EPR spectrum of human ceruloplasmin have corroborated the hypothesis that the protein contains three blue centers.[48] The simulated spectrum shown in Fig. 16 is the sum of three inequivalent blue components ($g_{//}$ = 2.262, 2.217 and 2.213; $g_{\perp}$ = 2.05, 2.05 and 2.056; $A_{//}$ = 5.5, 7.2 and 8.1 mT, respectively) and of a non-integer amount of type 2 copper ($g_{//}$ = 2.257; $g_{\perp}$ = 2.05; $A_{//}$ = 17.8 mT). Attempts to computer simulate the spectrum by assuming the presence of only two blue copper centers does not give a satisfactory fit of the experimental lineshape. This is at variance with results obtained on turtle ceruloplasmin, one of the proteins with an overall copper stoichiometry of

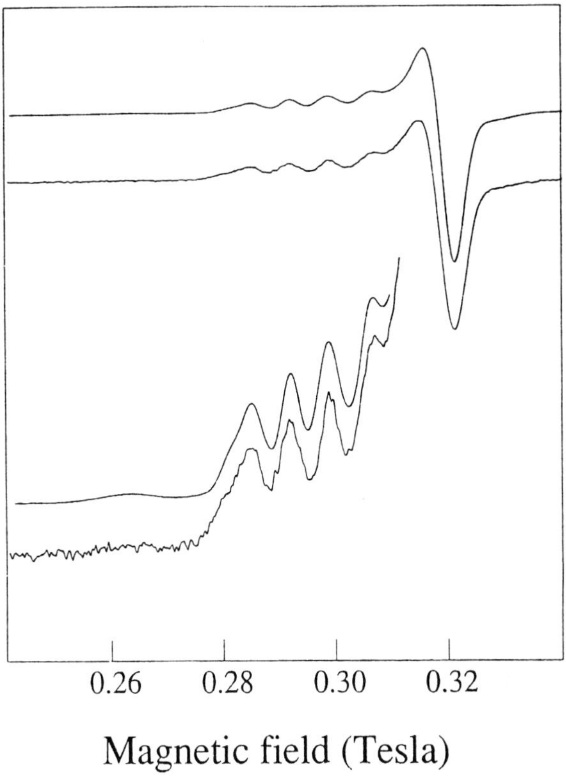

Fig. 16. Computer simulation of a typical X-band EPR spectrum of human ceruloplasmin. The experimental lineshape (lower curve) was reproduced by summing three blue components ($g_{//}$=2.262, 2.217 and 2.213; $g_{\perp}$=2.050, 2.050 and 2.056; $A_{//}$=5.5, 7.2 and 8.1 mT; LW=3.5, 3.0 and 4.5 mT, respectively) and of a non-integer amount of type 2 copper ($g_{//}$=2.257; $g_{\perp}$=2.05; $A_{//}$=17.8 mT, LW=9.0 mT).

5 ions/molecule.[6] As shown in Fig. 17, the EPR spectrum can be in this case simulated as the sum of two non-equivalent blue ions ($g_{//}$ = 2.221 and 2.217; $g_{\perp}$ = 2.055 and 2.045; $A_{//}$ = 6.8 and 9.5 mT), although a small contribution, ca. 0.3 spins/molecule, of a broad, unstructured signal centered at $g = 2.106$ has to be added to best fit the experimental lineshape.

Our results suggests that human ceruloplasmin contains three type 1 copper sites, which are partially reduced in the native protein and can be in part oxidized *in vitro* by ferricyanide. The remaining copper is present as a trinuclear cluster, with the type 2 copper exhibiting a paramagnetism lower

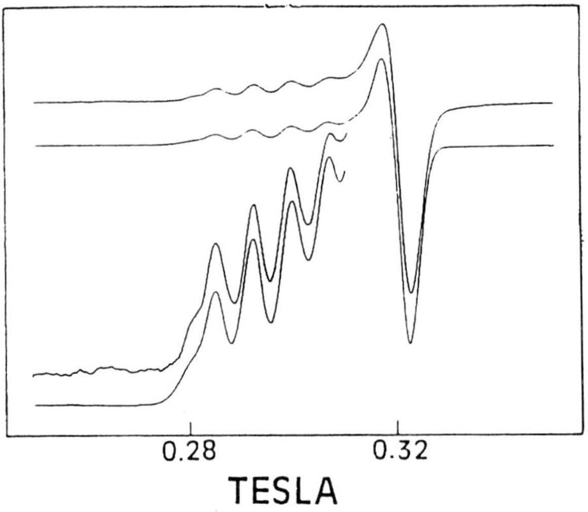

Fig. 17. Computer simulation of the EPR spectrum of turtle ceruloplasmin. Three contributions were summed to reproduce the experimental lineshape: a copper ion with $g_{\perp}$=2.055, $g_{//}$=2.221, $A_{\perp}$=0.8 mT, $A_{//}$=68.4 G, $LW_{\perp}$=30 G, $LW_{//}$=30 G; a copper ion with $g_{\perp}$=2.045, $g_{//}$=2.217, $A_{\perp}$=8 G, $A_{//}$=94.8 G, $LW_{\perp}$=40 G, $LW_{//}$=55 G; and an isotropic signal (0.3 spins/molecule) centered at g=2.106 with LW=250 G.

than expected. As stated above, the reason for this phenomenon is likely to reside in the magnetic interactions which take place inside the trinuclear unit. The experiments aimed at assessing the status and stoichiometry of type 1 copper also give evidence against the hypothesis that type 2 copper be EPR-silent in the native spectrum because it is in the reduced state. As a matter of fact, the experiment in which the protein is aerobically treated with ascorbate (Fig. 14) shows that type 2 copper has no tendency to stay reduced in the presence of oxygen. Furthermore, the intensity of its EPR signal is not affected by ferricyanide, at variance with type 1 copper, despite these latter sites are considered to be less accessible than the former one to the solvent.

The solution of the crystallographic structure of human ceruloplasmin has been very recently announced (P. Lindley, see Chap. 3). According to this preliminary communication, human ceruloplasmin would contain, as predicted by our results, three type 1 copper sites and a type 2-type 3 trinuclear unit.

4. The Ligand-Binding Properties of Ceruloplasmin

4.1. *Binding of Anions*

The ability of ceruloplasmin to interact with a number of ligands has been widely documented in over thirty years of research (Ref. 53 and references therein). Among these ligands, anions are certainly the most investigated, yet not the best understood. Reports in the sixties have shown that several anionic species, including N_3^-, Cl^-, F^-, I^-, Br^-, CN^-, OCN^-, SCN^-, act as inhibitors of the oxidase activity of ceruloplasmin, possibly by coordinating to type 2 copper. The reactivity toward anions, especially N_3^-, is on the other hand a feature common to many metalloproteins, and, as already mentioned, it has been exploited to unravel the structure of the type 2-type 3 trinuclear unit in the multinuclear blue oxidase laccase,[50] well before the existence of such a unit has been established by crystallographers in ascorbate oxidase, another member of this class of enzymes.[51]

The availability of ceruloplasmin preparations lacking appreciable amounts of EPR-detectable type 2 copper[28,29,54] has prompted us to reinvestigate the binding of chloride,[55] the anion with the better described inhibitory properties on the enzymatic activity of the protein. The results have been quite surprising, and deserve to be shown and discussed in detail.

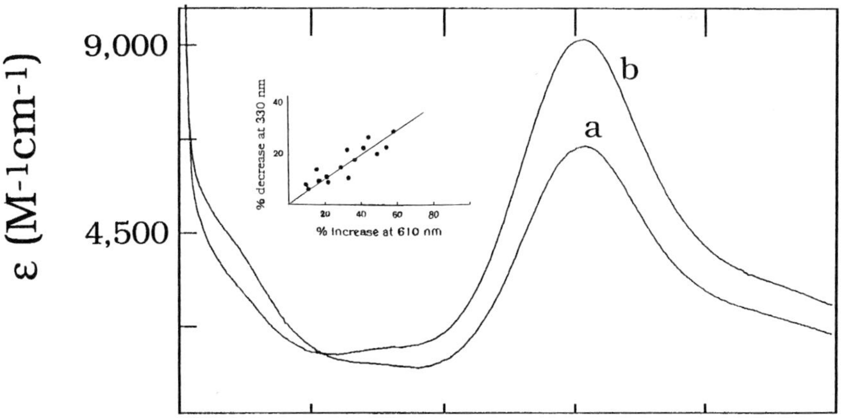

Fig. 18. Optical spectra of human ceruloplasmin (a) before and (b) after addition of 0.15 M NaCl at 25°C. The protein was 7×10^{-5} M in 0.15 M phosphate buffer pH 7.0. Inset: plot of percent enhancement of the 610 nm absorption vs. percent decrease of the 330 nm band in different samples of human ceruloplasmin treated with NaCl.

The addition of 150 mM NaCl (or KCl) to human ceruloplasmin in 150 mM phosphate buffer, pH 7.0 induces the increase of the intensity at 610 nm and the concomitant decrease of the absorbance at 330 nm (Fig. 18). The extent of the optical changes varies among different samples, ranging for the band at 610 nm between 10 % and 60 %. As shown in the inset of Fig. 18, the changes at the two chromophores are linearly related. Furthermore, the kinetics of the optical changes are similar at the two wavelengths, and an isosbestic point at 410 nm in the optical spectra of human ceruloplasmin at various concentrations of chloride can be detected. An apparent $K_d \sim$ 4 mM can be evaluated by the variation of A_{610} vs. Cl^- concentration. Addition of chloride also produced a shift of the 440 nm shoulder, present in the native protein, to higher wavelengths. It appears from these data that Cl^- elicits an increase of the extinction coefficient at 610 nm, the chromophore stemming from the type 1 copper sites.

Sheep and human ceruloplasmin behave in a similar way, while the chicken protein is essentially insensitive to chloride. This is interesting when we consider that chicken ceruloplasmin, as prepared by the single-step procedure, has one of the highest values of ε_{610nm} (ca. 9,200 $M^{-1}cm^{-1}$) among ceruloplasmins. Moreover, this is the value attained by human or sheep ceruloplasmin after incubation with chloride. It can therefore be postulated that Cl^- acts by oxidizing the fraction of reduced blue sites. Its effect would be, in this respect, similar to that exerted by ferricyanide and discussed above, with the noticeable difference that Cl^- is not, per se, a redox agent. What follows is a deeper description of the experiments carried out on human ceruloplasmin.

The effects of Cl^- are reversible, and gel-filtration on G-25 of the treated protein restores the original optical intensities at 610 and 330 nm. This latter absorbance can be selectively recovered by addition of H_2O_2 to Cl^--treated ceruloplasmin (Fig. 19), however the peroxide also induces a change of the lineshape at around 330 nm. The band is in fact now centered at higher energies, with features resembling the well-known adduct between oxidized type 3 copper and peroxide of multinuclear blue oxidases.[56–58] Furthermore, the derivative is "locked" in its spectroscopic state, since the optical features do not change upon gel-filtration, nor upon a new addition of Cl^-.

The paramagnetic copper content of ceruloplasmin, as monitored by EPR spectroscopy, increases upon addition of Cl^-. As shown in Fig. 20,

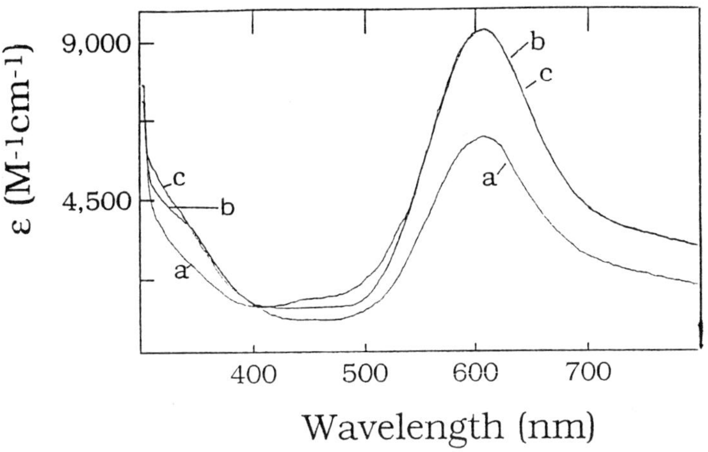

Fig. 19. Optical spectra of human ceruloplasmin treated with Cl^- and H_2O_2 at 25°C (a) ceruloplasmin; (b) ceruloplasmin plus 0.12 M NaCl; (c) sample (b) 30 min. after addition of 2.5 mM H_2O_2. The protein was $7x10^{-5}$ M in 0.15 M phosphate buffer pH 7.0.

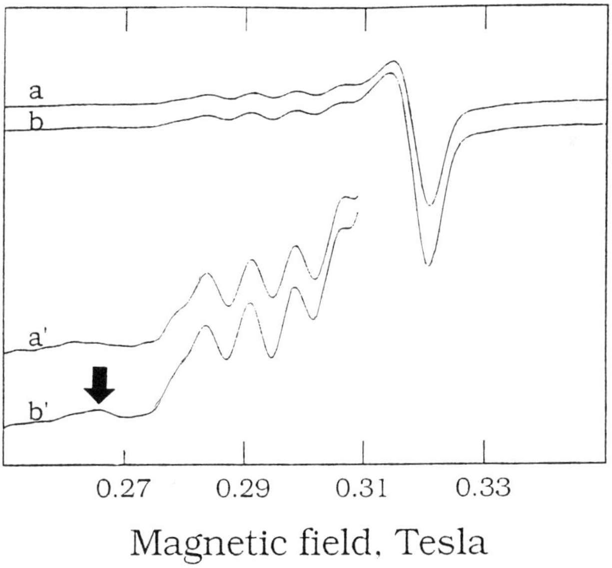

Fig. 20. EPR spectra of human ceruloplasmin (a) before and (b) after addition of 0.15 M NaCl. The protein was $5x10^{-5}$ M in 0.15 M phosphate buffer pH 7.0. Experimental settings as in Fig. 5. Primed spectra are 8-fold amplifications.

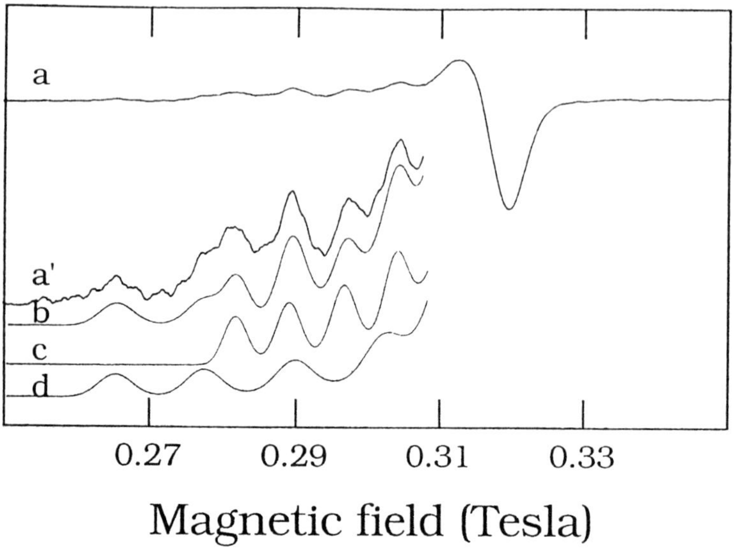

Fig. 21. (a) Difference EPR spectrum of Cl^--treated minus untreated human ceruloplasmin. Primed spectrum is an 8-fold amplification. The amplified spectrum (b) is a computer-simulated curve obtained by summing equal contributions of spectra (c) and (d). The following parameters were used for the simulations: spectrum (c), $A_{//}$=7.2 mT, $g_{//}$=2.217; spectrum (d), $A_{//}$=12.1 mT, $g_{//}$=2.289.

the increase is due to a higher intensity of the four hyperfine lines due to blue copper (consistent with the optical data), and to the appearance of a new signal with an isolated "low-field" feature (marked with an arrow in figure). Two components are present in the computer-subtracted difference EPR spectrum of NaCl-treated vs. untreated ceruloplasmin (Fig. 21, curve a). The narrow-split component (curve c) has values of $A_{//} = 7.2$ mT and $g_{//} = 2.217$ matching those of one of the three blue sites.[48] The other component (curve d) has magnetic parameters ($A_{//} =$ 12.1 mT, $g_{//} = 2.289$) suggestive of a half-met state of the type 3 copper pair, in analogy to that observed for partially reduced states of laccase or hemocyanin.[27,59,60] The two components equally contribute to the EPR intensity of the difference spectrum.

The optical and EPR data strongly suggest that Cl^- promotes a reversible electron displacement from reduced type 1 copper to the type 3 pair within the trinuclear cluster of ceruloplasmin, to generate a half-met copper pair. It is clear that, for this to occur, a fraction of type 1 copper has to be in the reduced state in the native protein. The reaction with hydrogen peroxide has been used in the past to show that 10–15 % of the blue sites are reduced in sheep and bovine, but not in chicken, turtle and dolphin ceruloplasmin.[6,7,49,57] It is not by chance that these latter proteins have the highest values of $\varepsilon_{610\ nm}$ (ca. 9,200, 10,000 and 9,000 $M^{-1}cm^{-1}$, respectively). The presence of reduced blue copper in human ceruloplasmin, on the other hand, is well documented by the experiments reported above with ferricyanide.

The decrease of the optical band at 330 nm in the presence of Cl^- corresponds to the half-reduction of the type 3 pair, in analogy to the case of laccase.[26] An electron displacement from type 2 to type 3 copper is supposed to generate two paramagnetic ions at once, consistent with the equal contribution to the EPR intensity of the 12 mT and the 7.2 mT signals in the difference spectrum.

The ability of H_2O_2 to directly reoxidize a half-reduced type 3 copper pair[26] accounts for the recovery of the 330 nm absorption when peroxide is added to Cl^--treated ceruloplasmin. The protein becomes locked in a fully oxidized state, and therefore insensitive to Cl^-.

Unexpected results have been obtained when the oxidase activity of ceruloplasmin has been measured at different pH values in the presence of Cl^-. As previously mentioned, this anion has been invariably reported to be an inhibitor of the enzymatic activity of ceruloplasmin.[61–63] Indeed, Cl^- acts as inhibitor at low pH. However, at neutral pH, where the V_{max} of the reaction dramatically drops, the anion behaves as an activator, compensating, at least in part, for the activity decline (Fig. 22). At pH values around 6, a composite situation can be observed, with Cl^- exerting an activating effect at low concentrations (i.e. 5 mM), while being an inhibitor at higher concentrations (i.e. 50 mM), suggesting the existence of multiple binding sites on ceruloplasmin under these conditions.

The effects of Cl^- on ceruloplasmin may have a physiological relevance. At the concentrations found in the human plasma (0.15 M), Cl^- should be permanently bound to ceruloplasmin. It cannot be ruled out that other factors, including other ligands or altered residues on the ceruloplasmin chain, may modulate the interaction of the protein with chloride. It is worth recalling that the type 1 copper sites of ceruloplasmin keep a higher oxidation state in old vs. young individuals, both in the plasma and in the purified protein, and that this phenomenon has been linked to the presence

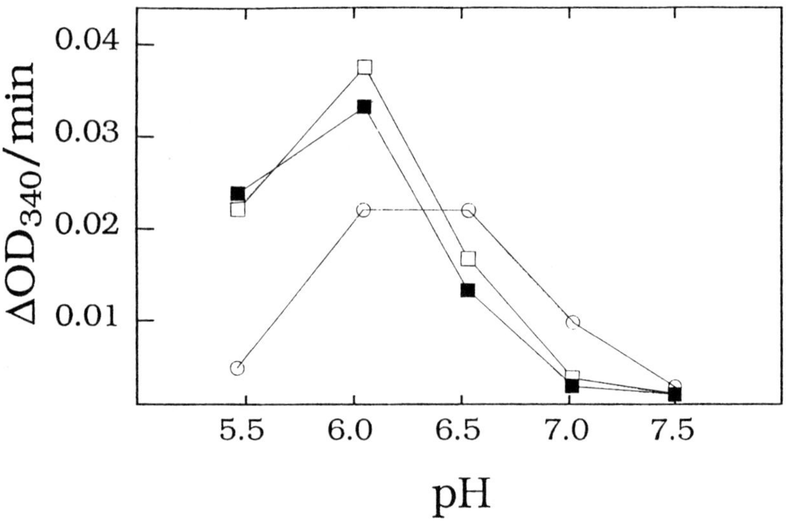

Fig. 22. Rates of oxidase activity at 25°C of human ceruloplasmin vs. *p*PD at different pH values in the absence (closed squares) or presence of 5 mM (open squares) or 50 mM (open circles) NaCl.

of oxidative modifications which occur at specific amino acid residues of the protein upon aging.[54]

The structural perturbations induced by Cl^- on human or sheep ceruloplasmin have also been found to be produced by other anions. These include, as expected, other halides like Br^- and, more surprisingly, N_3^-. This latter anion, like chloride, is a well-known inhibitor of the oxidase activity of ceruloplasmin and has been reported to have a quenching effect on the optical absorbance at 610 nm. We have preliminary data showing that azide actually behaves like chloride, showing multiple binding sites with different effects on the spectroscopic and functional properties of ceruloplasmin. In particular, in analogy with chloride, azide at slightly acidic pH enhances the oxidase activity of ceruloplasmin, which is instead repressed at higher concentrations of the anion. At variance with Cl^-, N_3^- exerts its activating effect at micromolar, rather than millimolar concentrations.

4.2. *Interaction with Nitric Oxide*

When the reaction of ceruloplasmin with nitric oxide started to be reinvestigated by our group, the importance of NO as a messenger molecule in physiological events[64] had not been fully unraveled. As a matter of fact, this simple gaseous molecule, formed by stoichiometric amounts of the two most common elements of the terrestrial atmosphere, used to be exploited as a spin probe of metal binding sites in metalloproteins, including copper- and iron-containing systems. As far as the multinuclear blue proteins were concerned, it was of general knowledge that NO could quench the spectroscopic signatures, both optical and EPR, of the type 1 copper sites, allowing the straightforward analysis of the remaining sites, in particular of type 2 copper.[65]

The reaction of NO had already been studied in the human,[45,66,67] the bovine[68] and the pig ceruloplasmin.[65] The finding that the chicken protein, purified with the single-step procedure, possessed a completely EPR-undetectable type 2 copper[49] prompted us to study the interaction of NO with such a type 2 copper-lacking protein, in the attempt to disclose the spectroscopic features of the trinuclear cluster.[69]

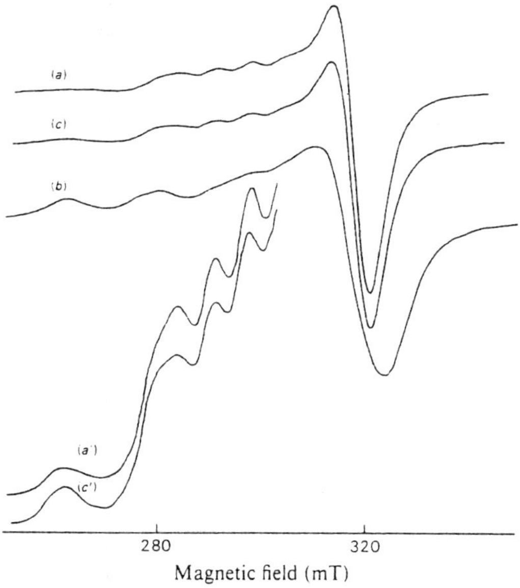

Fig. 23. Effect of NO on the EPR spectrum of bovine ceruloplasmin. (a) Native protein; (b) under 1 atm NO; (c) after removal of NO. The reaction medium contained 0.5 mM ceruloplasmin in 0.1 M phosphate buffer, pH 7. Experimental settings as in Fig. 5. Relative receiver gains: (a) and (c)–1, (b)–6, (a') and (c')–10.

The EPR spectra in Fig. 23 show the typical behavior of a type 2 copper-containing ceruloplasmin, that is the bovine protein, when anaerobically treated with gaseous NO.[68] The hyperfine lines due to type 1 copper readily disappears, and a spectrum composed of only resonances from type 2 copper remains. We later showed that the magnetic parameters of the type 2 copper center are also perturbed by the presence of NO.[48] Therefore, quantitative analyses following the NO-induced quenching of the blue copper features to establish the contribution of type 2 copper to the total EPR intensity should be reconsidered.

The EPR spectra of chicken ceruloplasmin before and after incubation with NO are shown in Fig. 24. In this case, no type 2 copper signals are evident in the native protein, and it is clear that they remain undetectable even after addition of NO. The origin of the broad, structureless

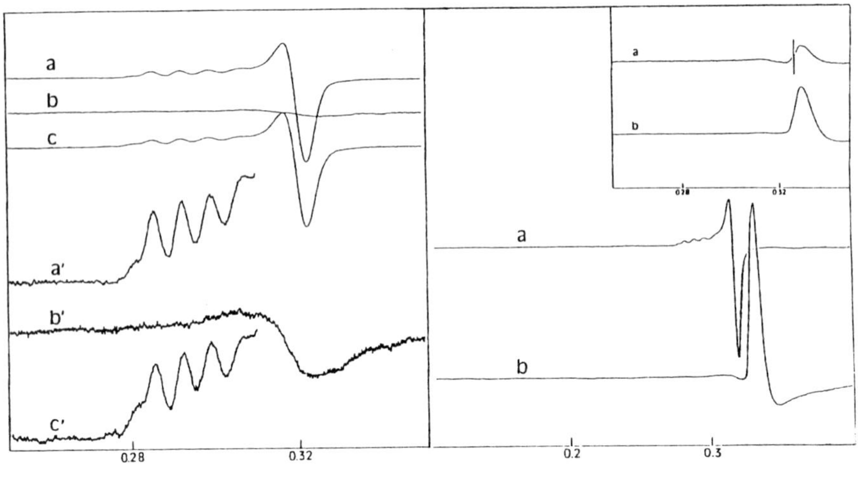

Fig. 24. EPR spectra at 100 K (panel A) or 11.5 K (panel B) of 1×10^{-4} M chicken ceruloplasmin in 0.1 M phosphate buffer, pH 7.4, incubated with 1 atm of NO: (a) native ceruloplasmin; (b) after 5 min. of incubation with NO; (c) 20 min. after anaerobic removal of NO. Experimental settings: panel A, as in Fig. 5, with primed spectra being 10-fold amplifications, except curve b', which is 20-fold; panel B, power–5 mW, frequency–9.121 Ghz, modulation amplitude–0.1 mT; panel B inset, power–(a) 0.05 mW or (b) 150 mW, frequency–9.121 GHz, and modulation amplitude–0.1 mT. The sharp line in spectrum (a) is the DPPH resonance. The arrow marks the g=2 position.

resonance which persists in the spectrum in the presence of the gas also at liquid helium temperatures (Fig. 24B) cannot be assessed with certainty. It accounts for ca. 0.4 spins/molecule and could represent a modified form of type 2 copper, broadened to the limits of hyperfine structure detection by relaxation interactions with the type 3 couple in the trinuclear cluster. It is worth recalling that a similar signal contributes to the best fit of the experimental lineshape in the computer simulated spectrum of turtle ceruloplasmin (see Fig. 17). In any case, type 1 copper is unlikely to be the source of the signal. The blue sites are fully quenched at the liquid nitrogen temperatures and are fully recovered after removal of NO (Fig. 24). This is also shown by the optical spectra reported in Fig. 25. As expected on the basis of previous findings, quenching of the blue sites is temperature-dependent, being complete only below 170 K (Fig. 25, inset). The optical

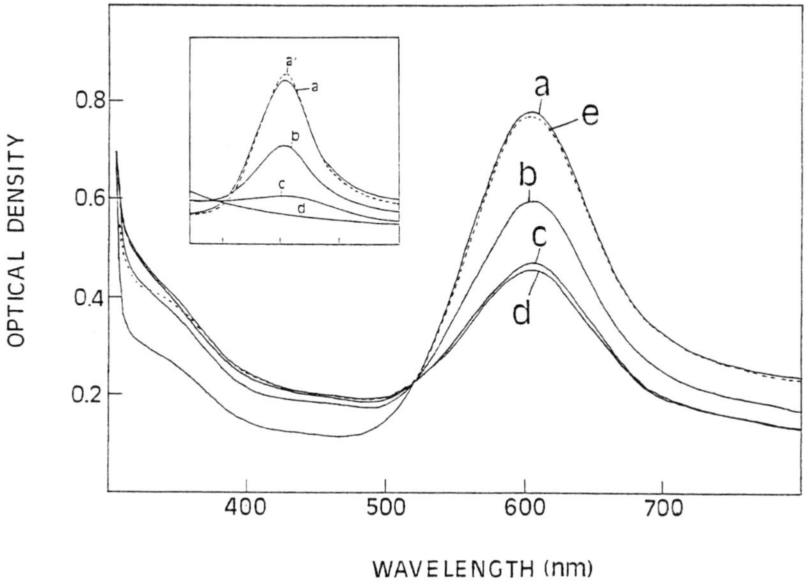

Fig. 25. Optical spectra of 9×10^{-5} M chicken ceruloplasmin in 0.1 M phosphate buffer, pH 7.4, at different times of incubation with 1 atm of NO: (a) 0'; (b) 5'; (c) 20'; (d) 40'; (e)–dashed line, after substitution of NO with argon. (Inset) Effect of temperature on the optical spectra of chicken ceruloplasmin reacted for 30 min. with 1 atm of NO: (a) native protein at 300 K; (a')–dashed line, native protein at 170 K; (b), (c) and (d) NO-treated protein at 300 K, 220 K and 170 K, respectively. The protein was in 0.1 M phosphate buffer, pH 7.4, and 60 % v/v glycerol.

spectra also indicate that part of the reaction of NO with ceruloplasmin is not reversible. As a matter of fact, the absorbance below 500 nm is not recovered when NO is removed, and remains higher than in the native sample. The new transitions have been assigned to an irreversibly modified state of the trinuclear cluster. This conclusion is mainly based on the following points. The new bands do not originate from type 1 copper, as previously proposed,[45] since they are still present after removal of NO and recovery of the optical and EPR features of the blue ions. Moreover, these bands are always formed in conjunction with changes of the 330 nm chromophore, suggesting that the copper site giving rise to this latter absorption is also responsible for the onset of the new transitions. Finally, the NO-treated protein shows an enhanced rate of reduction by ascorbate of both the 600 nm and the 330 nm chromophores with respect to the native protein (ca. 3 times faster), as shown in Fig. 26. Since it is widely accepted that, during the catalytic cycle affected by ceruloplasmin, electrons flow from the substrate to oxygen via the blue sites, which have unperturbed spectroscopic properties after removal of NO, the site with modified electron accepting ability has to be the trinuclear cluster.

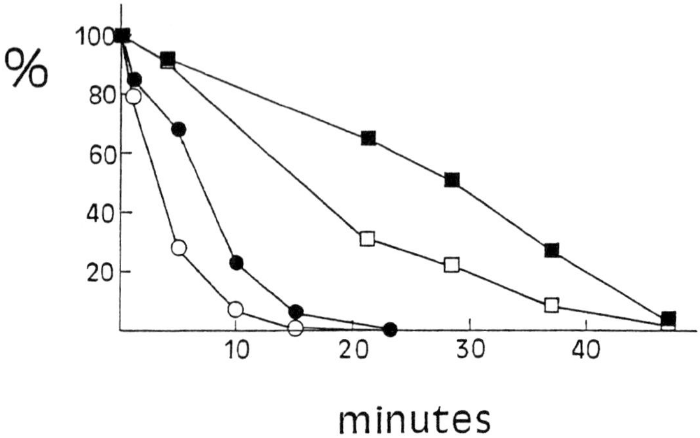

Fig. 26. Time course of the optical absorbances at 330 nm (open symbols) and 600 nm (closed symbols) during the anaerobic reduction of native (squares) and NO-treated (circles) chicken ceruloplasmin by ascorbate. Optical absorbances are expressed as percentage of the difference between the optical density at the specified wavelength of the fully oxidized and the fully reduced protein.

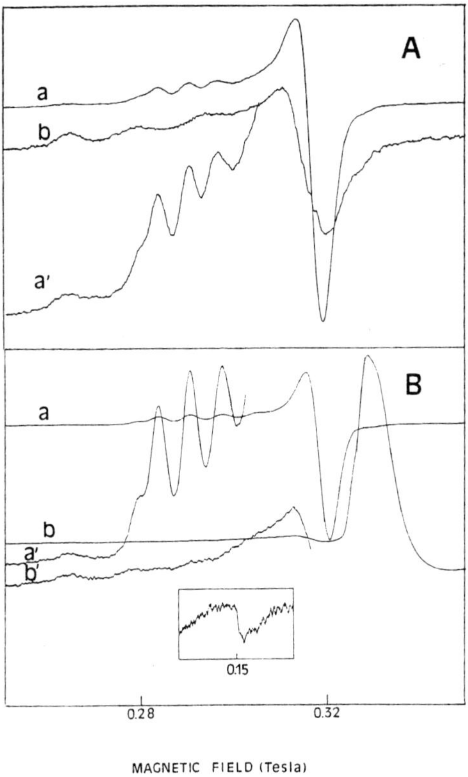

Fig. 27. EPR spectra of chicken ceruloplasmin in 0.1 M phosphate buffer, pH 7.4, incubated with 1 atm of NO taken at 100 K (panel A) or 11.5 K (panel B). Both ceruloplasmin samples were subjected to NO treatment after several months of storage at −20°C. The protein was $1x10^{-4}$ M in the experiment of panel A and $8x10^{-5}$ M in that of panel B. (a) Native protein; (b) after addition of 1 atm NO. Experimental conditions: power 20 mW (panel A), 5 mW (panel B) or 160 mW (inset of panel B); frequency 9.121 GHz, modulation amplitude, 0.1 mT. Primed spectra are 20-fold amplifications.

The state of type 2 copper clearly plays a major role in the reactivity of ceruloplasmin toward NO. With type 2 copper-containing proteins, a signal at $g=4$ in the EPR spectrum at liquid helium temperature appears in the presence of NO.[67] Chicken ceruloplasmin discloses such a signal only when "aged" samples generated by prolonged storage are used. In this case, the classical resonances of type 2 copper can also be observed (Fig. 27). A more distinctive difference between sample with and without an EPR-detectable type 2 copper is found in the ability of NO to act as an oxidant on the

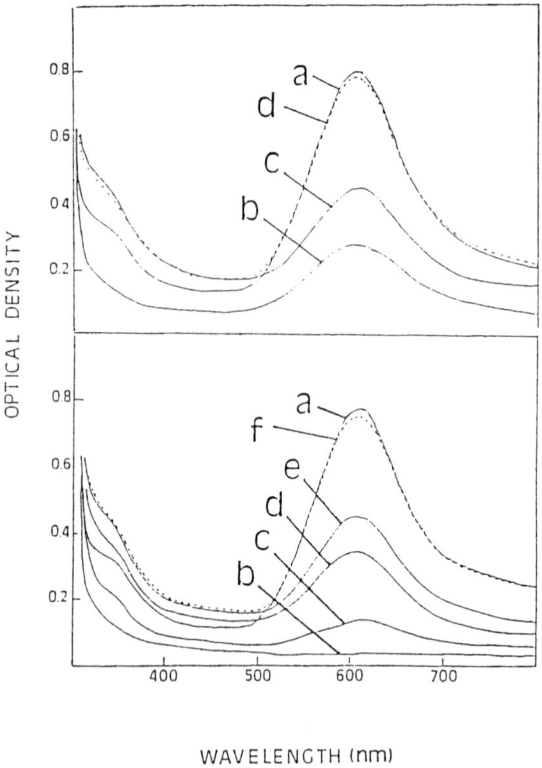

Fig. 28. (Upper panel) Optical spectra of $9x10^{-5}$ M chicken ceruloplasmin in 0.1 mM phosphate buffer, pH 7.4, incubated with 1 atm of NO after partial reduction with $2e^{-}/Cu(II)$ of ascorbate: (a) native ceruloplasmin; (b) after 60 min. of incubation with ascorbate; (c) 5 min. after addition of NO; (d) (dashed line) after evacuation of NO and subsequent readmission of oxygen. (Lower panel) Optical spectra of $9x10^{-5}$ M chicken ceruloplasmin in 0.1 M phosphate buffer, pH 7.4, incubated with 1 atm of NO after complete reduction with 2 $e^{-}/Cu(II)$ of ascorbate: (a) native ceruloplasmin; (b) after 120 min. of incubation with ascorbate; (c) 5 min. after addition of NO; (d) 10 min. after addition of NO; (e) 25 min. after addition of NO; (f)–dashed line, after evacuation of NO and readmission of oxygen.

reduced protein. As shown in Fig. 28, addition of NO to partially or fully reduced native chicken ceruloplasmin results in the increase of the optical band at 600 nm, a clear indication that a part of the blue sites can release electrons to NO. The phenomenon is absent when samples displaying type 2 copper in the EPR spectrum are used.

To unambiguously prove that the differences between chicken and other ceruloplasmins is due to the state of type 2 copper, we have examined the behavior of several samples of sheep ceruloplasmin, with a variable content of EPR-detectable type 2 copper, with NO. The sheep protein is in fact more labile than the chicken protein, being more prone to aging phenomena. Samples containing an EPR-detectable type 2 copper fully reproduce the results reported in the literature for mammalian proteins, as well as our findings on chicken ceruloplasmin with analogous spectroscopic properties (see Fig. 27). On the other hand, all these features are not observed when samples of sheep ceruloplasmin lacking an EPR-detectable type 2 copper are used, in analogy with the results obtained with native chicken ceruloplasmin.

Work in progress in our laboratories are disclosing a possible physiological implication of ceruloplasmin in NO-mediated events, as the protein has been found to significantly inhibit the NO-induced vascular relaxation of aortic vessels (unpublished results).

4.3. *Sites for Divalent Cations and Nucleotides*

The presence on ceruloplasmin of multiple non-constitutive metal-binding sites has been repeatedly suggested. It is known that the protein can interact with additional copper atoms at non-prosthetic sites,[37] and that transition metal ions, such as zinc, nickel and cobalt exert an inhibitory effect on the oxidase activity of the protein.[39]

We have recently studied the binding of various divalent cations to human and sheep ceruloplasmin[33] by following the room temperature EPR signal of Mn(II), which is virtually undetectable when the metal is bound to large molecules and therefore allows the direct measure of the free ion concentration in a mixture with a protein. A Scatchard plot obtained by titrating human ceruloplasmin with Mn(II) indicates the presence of 3 to 4 high affinity binding sites with $K_d \cong 25\ \mu M$ and more than 10 binding sites with K_d at least one order of magnitude lower (Fig. 29A). Increasing concentrations of Ca(II) or Mg(II) progressively displace the bound Mn(II), so titrations of ceruloplasmin with Mn(II) at fixed Ca(II) or Mg(II) concentrations have allowed to calculate a K_d of 1.4 mM and 0.3 mM for the high affinity sites of Ca(II) and Mg(II) respectively (Fig. 29B). The metal-binding properties of sheep ceruloplasmin are analogous to those of the human protein, with 3 to 5 high affinity ($K_d \cong 23\ \mu M$) and more than 20

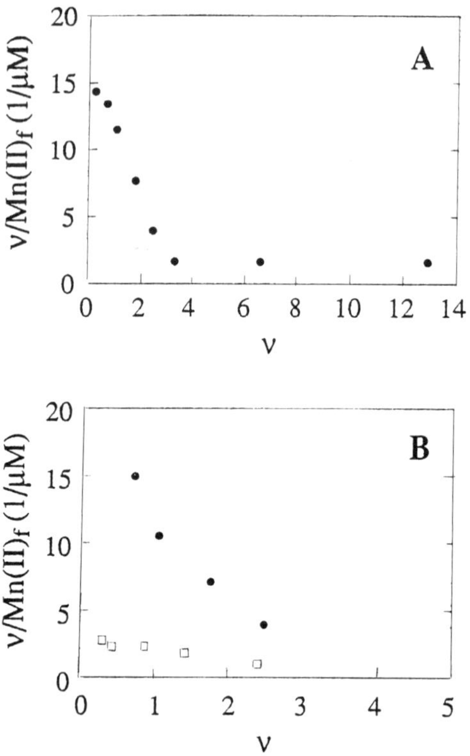

Fig. 29. Panel A. Scatchard plot of Mn(II) binding to human ceruloplasmin. 15 μM protein in 0.05 M MOPS, 0.1 M KCl pH 7.4 was titrated with $MnCl_2$ (5 μM to 1 mM). Panel B. Scatchard plot of Mn(II) binding to the high affinity sites of human ceruloplasmin in the absence (•) and presence (□) of 5 μM $CaCl_2$. 15 μM protein was titrated with 15 to 200 μM Mn(II). EPR experimental settings: frequency–9.4 GHz; power–100 mW; modulation amplitude–0.1 T; and temperature–25°C.

low affinity binding sites for Mn(II). K_d values for Ca(II) and Mg(II) are 1.4 mM and 0.7 mM, respectively.

The ability to bind divalent cations is not restricted to mammalian ceruloplasmins. As an example, the chicken protein binds 6–7 non-prosthetic metal ions with $K_d \cong 18\ \mu$M and 0.7 mM for Mn(II) and Ca(II), respectively (unpublished results from our laboratories).

The spectroscopic and catalytic properties of ceruloplasmin are not altered by the presence of Mn(II), Ca(II) or Mg(II), suggesting that no gross changes involving the prosthetic copper atoms take place upon binding. In particular, the EPR spectrum does not show signatures diagnostic of magnetic interactions between Mn(II) and constitutive Cu(II) ions, suggesting that the binding sites for divalent cation are grouped in a copper sites-free domain of the protein.

The K_d values for Ca(II) and Mg(II) are in the range of their physiological concentrations in human plasma, 2.4 mM and 0.9 mM, respectively,[70] and this suggests that binding of divalent cations to ceruloplasmin may be physiologically relevant. It needs to be recalled that the interaction of the protein with its receptor on red blood cells is mediated by calcium, but not by magnesium.[71] As a matter of fact, we have found that ceruloplasmin is more stable toward heat-induced unfolding in the presence of Ca(II), while Mg(II) exerts no effect.[33] This allows to speculate that binding of Ca(II), but not Mg(II), to ceruloplasmin favors a conformation of the protein which has a higher affinity for its receptor. This conclusion, however, cannot be generalized. In a recent study on the mechanism of reconstitution of apoceruloplasmin by Cu(I)-GSH we have found that divalent cations are required for copper to correctly reinstate in its sites, and that Mg(II) is even more efficient than Ca(II) in this respect.[25] Since the study was also aimed at elucidating the sequence of events leading to the incorporation of copper by newly synthesized ceruloplasmin, this finding, has, in turn, prompted us to investigate a possible interaction between ceruloplasmin and ATP, as most of intracellular Mg(II) is complexed to this nucleotide. Surprisingly, ceruloplasmin has been found to bind the nucleotide, with an affinity constant greater than 10^3 M^{-1} for the sheep protein. Preliminary data show that not only the site for ATP is present on other ceruloplasmins, including human and chicken, but other nucleotides can also interact with these ceruloplasmins. We are currently investigating the exact stoichiometries of nucleotide binding, and the structural relationships between the nucleotide and the copper binding sites.

It is difficult to believe that a nucleotide-binding site, with a proper binding strength and conserved in different species, exists on a protein purely by chance. As mentioned in the Introduction, the physiological role of ceruloplasmin remains to be clarified. It has been so far believed that the protein exerts its function in the extracellular compartments, i.e. in the plasma. The finding on the nucleotide-binding ability of ceruloplasmin now adds to other evidence that the protein may be involved in intracellular processes as well.

Acknowledgments

This work was supported in part by M.U.R.S.T. (40 %) and by C.N.R. grant no. 9502167.

References

1. C. G. Holmberg, *Acta Physiol. Scand.* **8**, 227 (1944).
2. C. G. Holmberg and C. B. Laurell, *Acta Chem. Scand.* **2**, 555 (1948).
3. C. G. Holmberg and C. B. Laurell, *Acta Chem. Scand.* **5**, 476 (1951).
4. U. S. Seal, *Comp. Biochem. Physiol.* **13**, 143 (1964).
5. P. J. McCosker, *Nature* **190**, 887 (1961).
6. G. Musci, M. Carbonaro, A. Adriani, A. Lania, A. Galtieri and L. Calabrese, *Arch. Biochem. Biophys.* **279**, 8 (1990).
7. M. C. Bonaccorsi di Patti, A. Galtieri, A. Giartosio, G. Musci and L. Calabrese, *Comp. Biochem. Physiol.* **103B**, 183 (1992).
8. J. B. Bingley and A. T. Dick, *Clin. Chim. Acta* **25**, 480 (1969).
9. T. Inaba and E. Frieden, *J. Biol. Chem.* **242**, 4789 (1967).
10. E. Frieden, in *Biological Roles of Copper*, Ciba Foundation Symposium 79, Excerpta Medica (Elsevier, Amsterdam/New York, 1980), pp. 93–124.
11. T. L. Ortel, N. Takahashi and F. W. Putnam, *Proc. Natl. Acad. Sci. USA* **81**, 4761 (1984).
12. L. G. Rydén and L. T. Hunt, *J. Mol. Evol.* **36**, 41 (1993).
13. C. Askwith, D. Eide, A. Van Ho, P. S. Bernard, L. Li, S. Davis-Kaplan, D. M. Sipe and J. Kaplan, *Cell* **76**, 403 (1994).
14. D. M. De Silva, C. C. Askwith, D. Eide and J. Kaplan, *J. Biol. Chem.* **270**, 1098 (1995).
15. Z. L. Harris, Y. Takahashi, H. Miyajima, M. Serizawa, R. T. A. MacGillivray and J. D. Gitlin, *Proc. Natl. Acad. Sci. USA* **92**, 2539 (1995).
16. L. Rydén, *Eur. J. Biochem.* **26**, 380 (1972).
17. M. Sato and J. D. Gitlin, *J. Biol. Chem.* **266**, 5128 (1991).
18. N. Takahashi, T. L. Ortel and F. W. Putnam, *Proc. Natl. Acad. Sci. USA* **81**, 390 (1984).
19. B. Halliwell and J. M. Gutteridge *Arch. Biochem. Biophys.* **246**, 501 (1986).
20. E. Ehrenwald, G. M. Chisolm and P. L. Fox, *J. Clin. Invest.* **93**, 1493 (1994).
21. E. Ehrenwald and P. L. Fox, *Arch. Biochem. Biophys.* **309**, 392 (1994).
22. M. C. Bonaccorsi di Patti, G. Musci and L. Calabrese, *Life Sci. Adv. (Biochemistry)* **11**, 197 (1992).
23. T. P. Ryan, T. A. Grover and S. D. Aust, *Arch. Biochem. Biophys.* **293**, 1 (1992).
24. R. E. Fleming and J. D. Gitlin, *J. Biol. Chem.* **265**, 7701 (1990).
25. G. Musci, S. Di Marco, G. C. Bellenchi and L. Calabrese, *J.Biol. Chem.* **271**, 1972 (1996).
26. D. J. Spira, M. E. Winkler and E. I. Solomon, *Biochem. Biophys. Res. Comm.* **107**, 721 (1982).

27. B. Reinhammar, R. Malkin, P. Jensen, B. Karlsson, L. -E. Andréasson, R. Aasa, T. Vänngård and B. Malmström, *J. Biol. Chem.* **255**, 5000 (1980).
28. L. Calabrese, M. Carbonaro and G. Musci, *J. Biol. Chem.* **264**, 6183 (1989).
29. M. C. Bonaccorsi di Patti, G. Musci, A. Giartosio, S. D'Alessio and L. Calabrese, *J. Biol. Chem.* **265**, 21016 (1990).
30. V. Edge, N. M. Allewell and J. M. Sturtevant, *Biochemistry* **24**, 5899 (1985).
31. V. Edge, N. M. Allewell and J. M. Sturtevant, *Biochemistry* **27**, 8081 (1988).
32. C. Q. Hu and J. M. Sturtevant, *Biochemistry* **26**, 178 (1987).
33. G. Musci, M. C. Bonaccorsi di Patti, R. Petruzzelli, A. Giartosio and L. Calabrese, *Biometals* **9**, 66 (1996).
34. J. A. Fee, in *Structure and Bonding,* eds. J. D. Dunitz, P. Hemmerich, R. H. Holm, J. A. Ibers, C. K. Jørgensen, J. B. Neilands, D. Reinen and R. J. P. Williams (Springer Verlag, Berlin, 1975), Vol 23, pp. 1–60.
35. R. A. DiSilvestro and E. D. Harris, *Arch. Biochem. Biophys.* **241**, 438 (1985).
36. C. B. Kasper and H. F. Deutsch, *J. Biol. Chem.* **238**, 2325 (1963).
37. A. Zgirski and E. Frieden, *J. Inorg. Biochem.* **39**, 137 (1990).
38. C. T. Huber and E. Frieden, *J. Biol. Chem.* **245**, 3973 (1970).
39. D. J. McKee and E. Frieden, *Biochemistry* **10**, 3880 (1971).
40. R. J. Carrico, H. F. Deutsch, H. Beinert and W. H. Orme-Johnson, *J. Biol. Chem.* **244**, 4141 (1969).
41. N. A. Holtzman and B. M. Gaumnitz, *J. Biol. Chem.* **245**, 2350 (1970).
42. P. O. Gunnarson, U. Nylen and G. Petterson, *Eur. J. Biochem.* **37**, 47 (1973).
43. J. Deinum and T. Vänngård, *Biochim. Biophys. Acta* **310**, 321 (1973).
44. L. Rydén and I. Bjork, *Biochemistry* **15**, 3411 (1976).
45. R. Wever, F. X. R. Van Leuween and B. F. Van Gelder, *Biochim. Biophys. Acta* **302**, 236 (1973).
46. A. Messerschmidt and R. Huber, *Eur. J. Biochem.* **187**, 341 (1990).
47. V. V. Rylkov, M. Y. Tarasiev and K. A. Moshkov, *Eur. J. Biochem.* **197**, 185 (1991).
48. G. Musci, M. C. Bonaccorsi di Patti and L. Calabrese, *Arch. Biochem. Biophys.* **306**, 111 (1993).
49. L. Calabrese, M. Carbonaro and G. Musci, *J. Biol. Chem.* **263**, 6480 (1988).
50. M. D. Allendorf, D. J. Spira and E. I. Solomon, *Proc. Natl. Acad. Sci. USA* **82**, 3663 (1985).
51. A. Messerschmidt, A. Rossi, R. Ladenstein, R. Huber, M. Bolognesi, G. Gatti, A. Marchesini, R. Petruzzelli and A. Finazzi-Agrò, *J. Mol. Biol.* **206**, 513 (1989).
52. T. Vänngård, in *Magnetic Resonance in Biological Systems,* eds. A. Ehrenberg, B. Malmström and T. Vänngård (Pergamon Press Ltd., Oxford, 1967), pp. 213-219.
53. L. Rydén, in *Copper Proteins and Copper Enzymes*, ed. R. Lontie (CRC Press Inc., Boca Raton, FL, 1984), Vol. 3, pp. 37–100.
54. G. Musci, M. C. Bonaccorsi di Patti, U. Fagiolo and L. Calabrese, *J. Biol. Chem.* **268**, 13388 (1993).

55. G. Musci, M. C. Bonaccorsi di Patti and L. Calabrese, *J. Protein. Chem.* **14**, 611 (1995).
56. R. Bränden, B. Malmström and T. Vänngård, *Eur. J. Biochem.* **18**, 238 (1971).
57. L. Calabrese and U. Leuzzi, *Biochem. Int.* **8**, 35 (1984).
58. J. E. Penner-Hahn, B. Hedman, K. O. Hodgson, D. J. Spira and E. I. Solomon, *Biochem. Biophys. Res. Comm.* **119**, 567 (1984).
59. R. S. Himmelwright, N. C. Eickman and E. I. Solomon, *Biochem. Biophys. Res. Comm.* **81**, 243 (1978).
60. B. Reinhammar, *J. Inorg. Biochem.* **18**, 113 (1983).
61. G. Curzon and B. E. Speyer, *Biochem. J.* **109**, 25 (1968).
62. P. O. Gunnarson, U. Nylén and G. Pettersson, *Eur. J. Biochem.* **27**, 572 (1972).
63. R. A. Løvstad and E. Frieden, *Acta Chem. Scand.* **27**, 121 (1973).
64. S. Moncada, R. M. Palmer and E. A. Higgs, *Pharmacol. Rev.* **43**, 109 (1991).
65. A. C. F. Gorren, E. de Boer and R. Wever, *Biochim. Biophys. Acta* **916**, 38 (1987).
66. F. X. R. Van Leeuwen, R. Wever and B. F. Van Gelder, *Biochim. Biophys. Acta* **315**, 200 (1973).
67. F. X. R. Van Leeuwen and B. F. Van Gelder, *Eur. J. Biochem.* **87**, 305 (1978).
68. L. Calabrese, F. Malatesta and D. Barra, *Biochem. J.* **199**, 667 (1981).
69. G. Musci, S. Di Marco, M. C. Bonaccorsi di Patti and L. Calabrese, *Biochemistry* **30**, 9866 (1991).
70. G. V. Iyengan, W. E. Kollmer and H. J. Bowen, *The Elemental Composition of Human Tissues and Body Fluids* (Verlag Chemie, New York, 1978).
71. E. L. Saenko and A. I. Yaropolov, *Biochem. Int.* **20**, 215 (1990).

CHAPTER 12

ELECTRON TRANSFER REACTIONS IN MULTI-COPPER OXIDASES

Ole Farver

Dept. of Analytical and Pharmaceutical Chemistry,
Royal Danish School of Pharmacy,
2 Universitetsparken, 2100 Copenhagen ø, Denmark
E-mail: of@charon.dfh.dk

and

Israel Pecht

Dept. of Immunology, The Weizmann Institute of Science,
76100 Rehovot, Israel

ABSTRACT

Electron transfer in proteins is the most fundamental process in biological energy conversion. Electron transfer takes place over considerable distances between active centers such as transition metal ions and organic cofactors and is generally characterized by relatively weak electronic interaction. Considerable research efforts have been invested in resolving the factors which control the rates of long distance electron transfer in proteins. These aspects are the main issue of this review which deals with mainly the blue copper containing oxidases, a family of copper proteins that have most likely undergone an evolutionary optimization in order to perform their physiological function.

Contents

1. Introduction

Electron transfer (ET)[a] reactions play a central role in biological chemistry, the most important ones being in photosynthesis and respiration.[1] In these systems cascades of electron transport processes take place within and between proteins, some of which are embedded into or attached to biological membranes while others serve as mobile electron carriers between the former immobilized proteins. The redox centers are often transition metal ions, such as manganese, iron or copper. The copper ion is thus essential in several biological systems and with the exception of the large multi-subunit oxygen-carrying Cu-protein, hemocyanin, which is present in a variety of invertebrates, all copper proteins are in one way or another involved in electron transfer processes, particularly those associated with energy conversion and substrate redox chemistry.[2] When present in excess, copper ions are highly toxic and the subtle balance of copper homeostasis is highly regulated by complex transport and storage mechanisms.[3] Understanding the copper biochemistry has advanced considerably during the last decades mainly due to intensive studies of structure-reactivity relationships among copper proteins.[4,5]

Here we will review mainly results of studies of long range electron transfer (LRET) in blue multi-copper oxidases which constitute an important family of enzymes mainly involved in coupling the one-electron oxidation of substrates with the four-electron reduction of molecular oxygen to water.[6] Blue copper oxidases have several advantageous properties for serving as more general model systems for understanding ET within the polypeptide matrix: First and foremost, the blue copper site has most probably been

[a]Abbreviations: AO–ascorbate oxidase; Az–azurin; CD–circular dichroim; ET–electron transfer; HOMO–highest occupied molecular orbital: LRET–long range electron transfer, MCD–magnetic circular dichroism; NiR–nitrite reductase; T1–type-1; T2–type-2; T3–type-3; T2D–type-2 depleted; WT–wild type.

selected by evolution in order to undergo minimal structural rearrangements upon shuttling electrons between the Cu(II) and the Cu(I) redox state (i.e. having a minimal Franck–Condon barrier).[7] Further, the copper ions are directly coordinated to amino acid residues without any intervening prosthetic groups. Thus, the protein matrix itself is the only medium separating the redox partners within these enzymes.

2. The Nature of Electron Transfer within Proteins

Electron transfer is the simplest of all chemical reactions, since in principle no chemical bonds need to be formed or broken. While electron transfer in proteins shares common features with that in small molecules, others distinguish the two systems.[8] Rates of electron transfer are expected to depend on the energy required for bond-length and -angle changes as well as solvent reorganization accompanying the ET process. In proteins, however, these processes involve the polypeptide matrix as part of the medium which is far less homogeneous than solvent molecules surrounding small molecule ET partners. Further, conformational changes preceding or succeeding ET in proteins may affect the free energy changes of the reaction. Moreover, while small molecules exchange electrons in solutions where they are in close, sometimes direct contact, in proteins the redox reaction partners are held in fixed positions by and within the polypeptide matrix and hence prevented from coming into direct inner sphere contact. Therefore, the distance between electron donor and acceptor would be one decisive parameter affecting ET rates. Both theoreticians and experimentalists have devoted considerable efforts to studies of these processes, and the comprehension of intra- and inter- protein mediated ET has advanced significantly, largely due to the determination of three-dimensional structures of an ever increasing number of redox proteins. In addition, the theoretical models for analyzing LRET have also advanced to a stage where they can more readily be employed and tested experimentally.[8–17] Some very interesting questions remain to be answered: 1. Does the ET rate depend on the nature of the medium separating the two redox partners in a protein system and how? 2. Have the structure of redox proteins undergone evolutionary selection in order to optimize their function for specificity and control of biological ET and, if so, what are the structural corollaries of this selection? These are some of the issues that we wish to address in the present chapter concentrating on blue copper proteins as models.

3. Theory of Biological Electron Transfer

Several excellent reviews on ET theory have been published in recent years,[8–17] and the reader is referred to them for a comprehensive and up-to-date discussion of this subject. In the following only an outline, which is necessary for the discussion here, is presented. LRET in proteins is characterized by a weak interaction between electron donor and acceptor, and in the non-adiabatic limit the rate constant is proportional to the square of the electronic coupling between the electronic states of reactant and product, represented in the form of a matrix element, H_{AB}. For intramolecular ET the rate constant is given by Fermi's Golden Rule[8]:

$$k = \frac{2\pi}{\hbar} H_{AB}^2 (FC) \tag{1}$$

The Franck–Condon factor (FC) for nuclear movements can for relatively low vibrational frequencies, where $kT > h\nu$, be treated classically, a condition which often applies to biological ET. The electronic motion, however, requires a quantum mechanical approach, and the semiclassical Marcus equation may be expressed as[8]:

$$k = \frac{2\pi}{\hbar} \frac{H_{AB}^2}{(4\pi\lambda RT)^{1/2}} e^{-(\Delta G^0 + \lambda)^2/4\lambda RT} \tag{2}$$

Since wave functions decay exponentially with distance, the tunneling matrix element, H_{AB} will decrease with the distance, $(r - r_0)$, as:

$$H_{AB} = H_{AB}^0 e^{-\beta(r - r_0)/2} \tag{3}$$

H_{AB}^0 is the electronic coupling at direct contact between electron donor and acceptor (where $r = r_0$) and the decay rate of electronic coupling with distance is determined by the coefficient, β.

Both theoretical and experimental studies now demonstrate that the structure and composition of the protein medium intervening between electron donor and acceptor must be included in a theoretical analysis of the electronic coupling, H_{AB}, of the redox centers. One theoretical treatment[10,11] divides the protein medium into small subunits linked by covalent bonds, hydrogen bonds and through-space contacts, and each type of link is then assigned a coupling decay factor:

$$\begin{aligned} \epsilon_C &= 0.6 \\ \epsilon_H &= \epsilon^2 \exp[(-1.7(r-2.8)] \\ \epsilon_S &= \frac{1}{2}\epsilon_C \exp[-1.7(r-1.4)] \end{aligned} \tag{4}$$

for covalent (C), hydrogen bonded (H) and van der Waals interaction (S). The ET pathways may now be identified by analyzing the bonding interactions which maximize the tunneling matrix element, H_{AB} which now can be expressed as

$$H_{AB} = P \prod_i \epsilon_{C(i)} \prod_j \epsilon_{H(j)} \prod_k \epsilon_{S(k)} \tag{5}$$

where the prefactor, P, depends on the electronic coupling of donor and acceptor with the bridging orbitals.

The above procedure can be illustrated by the system where an intramolecular LRET is induced between the single disulfide (reduced pulse radiolytically to a $RSSR^-$ radical) and the Cu(II) ion in azurin, a β-sheet structured blue single copper protein. We have investigated this ET process in a large number of wild type and single site mutated azurins.[18–22] Applying the above pathway analysis we found that the same LRET pathway applies to all of them and that the ET can suitably be described in terms of this model. From the kinetic data and activation parameters using the above theoretical expressions, a reorganization energy, $\lambda = 99 \pm 5$ kJ mol^{-1}, and electronic decay factor, $\beta = 6.7 \pm 0.3$ nm^{-1} were calculated.[22] The latter value is in excellent agreement with both the experimentally determined decay factor for electron tunneling in modified cytochrome c (7.1 nm^{-1})[23] and the theoretical value calculated for tunneling through a saturated $-(CH_2)_n$ chain (6.5 nm^{-1})[9]. In these calculations no distinction is made between systems involving σ- and π-orbitals. We assume that this rate of decrease in electronic coupling with distance, expressed by β, is also applicable to the analysis of LRET in the blue copper oxidases. Both configurational changes in the inner coordination spheres of the metal centers (λ_i) and in the surrounding medium (λ_0) contribute to the total nuclear reorganization energy (λ), i.e. $\lambda = \lambda_i + \lambda_0$. For the blue copper center λ_i has been estimated to be quite small,[24] ≤ 10 kJ mol^{-1} which is in line with the observations that very minor structural changes take place during the redox cycle of this center and hence to the notion that this copper site has been evolutionary optimized for fast and effective electron transfer.

4. Structural Aspects

We will mainly discuss three distinct groups of copper oxidases, first and foremost the blue copper oxidases which have a minimum of four copper ions pr. functional subunit. The three-dimensional structures have now been determined at high resolution for two members of this group, namely ascorbate oxidase[25] (Chapter 2) and ceruloplasmin[26] (Chapter 3). We shall only briefly consider related enzymes like cytochrome *c* oxidase[27,28] which contain three copper ions and two heme centers and a group of denitrifying enzymes, like nitrous oxide reductase and nitrite reductase of which, at the present time, only the structure of the latter has been well characterized.[29]

The minimal catalytic unit of all blue copper oxidases consists of four copper ions bound to distinct sites classified according to their unique spectroscopic properties.[6] The type 1 (T1) site is practically identical to the coordination site found in the single blue copper proteins which gives rise to the characteristic spectroscopic properties of this class of proteins. A copper(II) pair is antiferromagnetically coupled and bound to a site designated as type 3 (T3) having an intense absorption in the near UV region. Together with the proximal type 2 (T2) Cu(II) site with a rather weak optical absorption spectrum these three copper ions constitute a trinuclear copper center. Since the catalytic cycle of the blue copper oxidases proceeds by a sequential mechanism where single electrons are taken up from the reducing substrate by the T1Cu(II) ion while the trinuclear T2/T3 center serves as the dioxygen binding and reduction site during the four-electron conversion of this oxidizing substrate to two water molecules,[30] intramolecular ET must be an essential element in the enzymatic mechanism. A remarkably similar redox cycle is catalyzed by all cytochrome *c* oxidases,[31] i.e. single electron uptake from cytochrome *c* and reduction of dioxygen to water, the crucial difference being that the energy produced by this process is converted here into an electrochemical gradient, later used for ATP synthesis whereas it is lost in all other processes catalyzed by the blue oxidases.

The three-dimensional structure of ascorbate oxidase (AO) has been determined at high resolution in five distinct states (in native oxidized,[25,32] in AO depleted of the T2 copper ion (T2D),[33] and finally in the reduced, peroxide and azide forms.[34]) AO is a 140 kDa dimer, consisting of two identical subunits, and each monomer is constructed of three domains with the T1 copper bound within one domain while three copper ions constitute a novel trinuclear arrangement held together by four histidine imidazoles from one domain and four histidine imidazoles from the other one.[32] Of the three copper ions, two are coordinated to six histidines arranged in a trigonal prism, reminiscent of the hemocyanin copper site structure,[35] and are by analogy called T3 while the remaining copper ion coordinated to two histidines would then, by exclusion be T2 (Fig. 1). Significantly, none of the histidine imidazoles are involved in bridging the copper ions. The trinuclear site is accessible to exogenous ligands and may bind anions such as F^-, N_3^- and CN^-. X-ray crystallographic studies have revealed that both azide and peroxide bind to one of the T3Cu ions in a terminal end-on position.[34] The three-dimensional structure of AO has also indicated a possible pathway for the internal T1Cu(I) → T3Cu(II) ET. Thus, it is noteworthy that

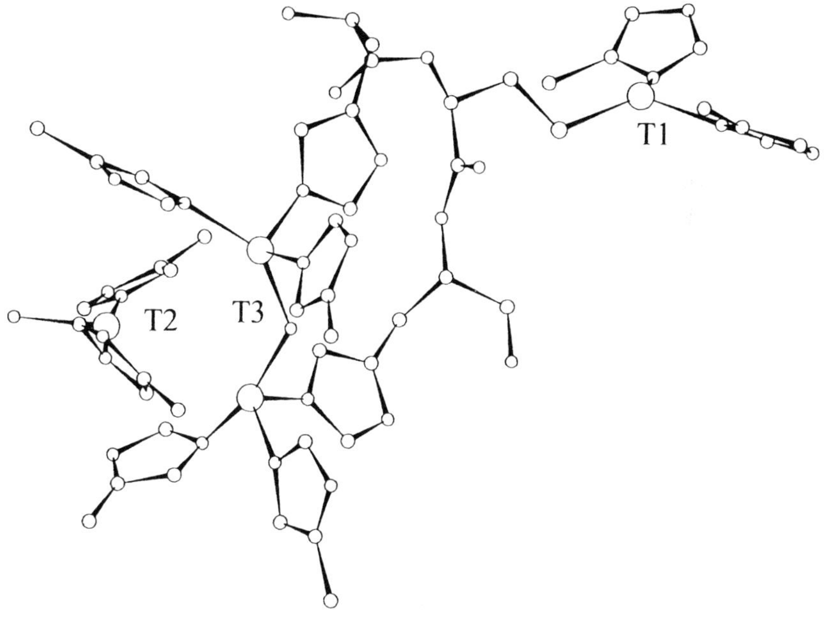

Fig. 1. Structure of the copper binding sites in ascorbate oxidase. T1Cu is shown to the right together with its two histidine and one cysteine ligands. The trinuclear T2/T3 center is shown to the left. In addition to the 8 histidines the OH^- group bridging the two T3Cu ions is shown. The coordinates were kindly provided by Dr. A. Messerschmidt.[32]

adjacent amino acid residues constitute the ligands for the T1 and T3 binding sites.[25,32,36] In this His-Cys-His sequence, the cysteine thiolate binds to T1 copper and the two histidines coordinate two separate copper ions of the trinuclear cluster (cf. Sec. 6.2).

The three-dimensional structures of two distinct cytochrome *c* oxidases have very recently been determined: namely the 200 kDa enzymes from bovine heart[27] and from the *Paracoccus denitrificans* bacteria,[28] both at a resolution of 0.28 nm. The structures and location of the different redox centers in both enzymes are similar and in remarkable agreement with predictions made on the basis of spectroscopic and mutagenesis studies. Single electrons are provided by cytochrome *c* and taken up at the Cu_A site in cytochrome *c* oxidase.[31] The above studies have finally confirmed earlier spectroscopic suggestions that in the oxidized enzyme this site is comprised of a mixed valence dinuclear Cu–Cu site[37] at a distance of 0.27 nm[27,28] with the metal ions bridged by two cysteine thiolates. The functional significance of having this type of site as the electron uptake port for this superfamily of enzymes is still an open and fascinating question. The heme a and Cu_B-heme a_3 centers are almost equidistant from the Cu_A site.[27,28] This raises another intriguing question of how preferential ET processes take place and how these are being controlled.

Very recently the three-dimensional structure of another blue copper oxidase, ceruloplasmin, has also been solved at a 0.31 nm resolution.[26] This structure is reminiscent of that of ascorbate oxidase (*vide supra*), but with two additional type 1 copper ions (cf. Chap. 3). Close sequence homology between AO, ceruloplasmin and laccase[36] suggests that essentially identical structural elements also exist in the latter enzyme serving a similar ET function.

5. Experimental Methods

Several different experimental methods have been employed in mechanistic studies of the blue oxidase activities, ranging from fast mixing (stopped-flow, rapid-freeze EPR) and T-jump to flash photolysis and pulse radiolysis. The latter technique enables production of oxidizing and reducing radicals within a fraction of a μ-second, by introducing short pulses of high-energy (~ 5 MeV) accelerated electrons into an aqueous solution.[38] Interaction of the electrons with water molecules produces a series of primary products, such as:

$$H_2O \rightarrow e^-_{aq}(2.65); OH(2.75); H(0.65); H_2(0.45); H_2O_2(0.72).$$

The numbers in parenthesis represent the G-values or the number of molecules produced pr. 100 eV absorbed energy. The first two radicals are particularly prominent, with reduction potentials at the two extreme ends of the redox scale (E^0 of OH $\approx$ 1.8 V and E^0 of e^-_{aq} ≈ -2.7 V). Radiation chemists have devised protocols which lead to the conversion of the former radicals into more selective agents with lower reaction affinity.[39] For example, in N_2O saturated solutions e^-_{aq} is converted into OH radicals by the reaction:

$$e^-_{aq} + N_2O + H_2O \rightarrow N_2 + OH^- + OH$$

with $k(e^-_{aq}) = 8.7 \times 10^9$ $M^{-1}s^{-1}$. OH and H radicals can similarly be converted into the milder reducing radical, CO_2^- ($E^0 = -1.8$ V) by reaction with formate ions[39]:

$$OH/H + HCO_2^- \rightarrow CO_2^- + H_2O/H_2$$

with $k(OH) = 3 \times 10^9$ $M^{-1}s^{-1}$; $k(H) = 3 \times 10^8$ $M^{-1}s^{-1}$. Thus, within a few hundred nanoseconds it is possible to produce an aqueous solution of reducing agents like CO_2^- which would react primarily with the redox active centers in proteins, the only exception being one type of amino acid residue, namely the disulfide group of cystine. ET reactions can then be monitored with an appropriate fast optical detection system throughout time domains from less than a μs to several minutes.

Flash photolysis, while providing an even shorter time resolution is however somewhat limited in the available spectroscopic range due to the requirement of a chromophore for photogenerating an electron donor or acceptor.

6. Intramolecular Electron Transfer in Blue Oxidases

6.1. *Intramolecular Electron Transfer Processes in Laccase*

Laccases were the first members of the blue copper oxidase family to be investigated in great detail both spectroscopically and kinetically.[6] They contain the minimum of four redox active copper ions per catalytic subunit: One of each type, T1 copper, T2 copper and a T3 copper pair. Although the three-dimensional structure is as yet not known for any laccase, ample evidence from spectroscopic studies and sequence homology with ascorbate oxidase suggests that the general structures of these two oxidases are quite similar.[36] Type 1 Cu(II) confers the intense blue color to these enzymes and is assumed to serve as the electron uptake site, while type 2 copper together with the type 3 copper pair, based on the analogy with ascorbate oxidase, constitute the trinuclear center where O_2 undergoes the 4-electron reduction to water. In AO, the through-bond distance between the T1 center and the trinuclear T2/T3 site via a His-Cys-His peptide sequence is 1.34 nm with the thiolate coordinating the T1 center while the two neighboring histidine imidazoles each bind to one of the two copper ions constituting the T3 copper pair.[32] As mentioned above, a homologous amino acid sequence motif is also found in all laccases sequenced so far and therefore a similar ET pathway may be operative in these enzymes as well.

An intramolecular ET from T1Cu(I) to the oxidized T3Cu(II) center has been proposed to be rate determining (i.e. k_{cat}) for the catalytic cycle in laccase.[40] Under anaerobic conditions, however, this T1Cu(I) to T3Cu(II) ET was found to be rather slow (*vide infra*) in contrast to the rapid reoxidation of T1 observed when the fully reduced enzyme is reacted with dioxygen.[41–44] These observations were interpreted as being a reflection of the existence of two different protein conformations, a resting and an activated state[45] (see more details below).

Some of the key questions concerning the enzymatic mechanism of laccase relate to formation of possible protein intermediates in different redox states and to what extent redox changes are induced by intramolecular ET between distant and structurally distinct sites. Some early studies addressing these questions were performed by Faraggi and Pecht, applying the pulse radiolysis technique, using fungal laccase[46] and human ceruloplasmin.[47] It was demonstrated that the hydrated electron, e^-_{aq}, reacted primarily with aromatic residues and disulfide groups on the proteins, to produce radicals, some of which decayed in a subsequent intramolecular ET step to the Cu(II) center. This behavior is most probably a result of the high reactivity of e^-_{aq} rather than an indication of physiologically significant ET pathways. However, they provided early evidence that the T1Cu(II) ion is the primary electron uptake port for these enzymes.

Quite detailed pulse radiolysis studies on *Polyporus* fungal laccase were later performed by Guissani *et al.*[48] showing that different optical transients are produced upon reaction with radicals, such as OH, e^-_{aq}, CO_2^- and O_2^-.

Laccase adducts with OH radicals or e_{aq}^- were again found to be formed very rapidly ($t_{1/2} < 100$ ns). OH reacted with aromatic amino acid residues such as tryptophan, tyrosine and histidine, as well as cystine with rate constants $\sim 10^{10}$ $M^{-1}s^{-1}$. From the absorption changes in the near UV region it was concluded that a very limited number of residues, one or two aromatics, reacts with OH radicals, yet cystine or cysteine apparently do not contribute significantly to this reaction. A more complex process takes place between *Polyporus* laccase and e_{aq}^- where again several aromatic and non-aromatic residues seem to be involved, also at diffusion controlled rates ($\sim 10^{10}$ $M^{-1}s^{-1}$).[48]

The milder reducing CO_2^- radicals do not react with aromatic residues or peptide bonds.[39] Hence, only exposed disulfide bridges react, producing the respective radical anion, $RSSR^-$, having an intense absorption band in the near UV region of the spectrum ($\varepsilon_{410} = 10,000$ $M^{-1}cm^{-1}$). Also, O_2^- forms adducts with laccase, but this radical anion is not sufficiently strongly reducing to transfer electrons to disulfide groups. A definitive site of interaction could not be identified, however.[48] Interestingly, in all cases decay of the initially formed protein radicals was found to be followed by T1Cu(II) reduction, even with strongly oxidizing radicals such as OH. The first phase observed was a fast first order reduction irrespective of the radical involved. For OH or e_{aq}^- the rate constant was 5×10^3 s^{-1} while it was slightly faster when CO_2^- were used (10^4 s^{-1}). Applying O_2^- as a reductant the rate constant was even smaller ($\sim 10^3$ s^{-1}).[48] As the detailed three-dimensional structure is not known for any laccase, the significance of these internal ET processes cannot be assessed. Still, a likely rationale could be that different radicals are originally produced with different amino acid residues on the enzyme, and electron tunneling distances could therefore differ significantly, giving rise to the range of the observed first order rate constants. No reactivity of the T3Cu(II) site, as monitored at 330 nm absorption, was observed in these studies although intramolecular T1→ T3 ET is expected to be an essential part of the enzymatic process.

Employing pulse radiolytically produced nitroaromatic radical anions as reductants, O'Neill *et al.* (1984) have shown that the T1Cu(II) sites in tree laccases (from *Rhus vernicifera* and *Rhus succedanea*) are reduced in bimolecular processes with rate constants of 2×10^6 $M^{-1}s^{-1}$ and 1.7×10^7 $M^{-1}s^{-1}$ respectively, at pH 7.4.[49] It was further found that complete reduction of T1Cu(II) in the holoproteins required two reduction equivalents.

Under similar conditions the number of electrons required for full reduction of T1Cu(II) in laccase depleted of T2 copper was found to be only one. Experimental observations using the intact protein were interpreted in terms of two alternative mechanisms: One where electrons enter the protein via T1Cu(II) with a subsequent rapid intramolecular ET to T2, and the second, that T1 and T2 are reduced independently, yet at the same rate. However, T1Cu(II) in type 2 depleted laccase is reduced stoichiometrically at the same rate as in the holoprotein which is consistent with the notion that T1 is the primary acceptor site. Subsequent intramolecular ET to the T3 site could not be monitored accurately due to the low signals at 330 nm and interference from optical absorption by the reducing radical. It was still inferred that this process is slow or does not take place at all.[49] A fast T1Cu(I) to T2Cu(II) intramolecular ET, however, is difficult to rationalize since the reduction potential for T1 is 30 mV higher than that for T2 which would shift the equilibrium towards reduced T1 copper. In addition, a fast intramolecular T1$\rightarrow$ T2 ET should show up in the time resolved absorption changes at 615 nm. This question of whether other sites can accept electrons directly from substrates and how these reduction equivalents distribute among the sites in laccase will be further discussed below.

As indicated, one of the prominent questions related to the catalytic cycle of the blue oxidases is what role does T2 copper play in intramolecular ET. This pertinent question has been addressed by Andréasson and Reinhammar[40,50] who took advantage of the possibility of independently monitoring both optical bands and the distinct paramagnetic properties of T1 and T2 Cu(II) by combining rapid-freeze EPR and stopped-flow experiments of *Rhus* laccase. At lower pH (< 7.0) the intensity of the EPR signals of T1 and T2Cu(II) were found to follow each other closely and the bimolecular reduction of T1Cu(II) was proposed to be the rate limiting step,[50] in accordance with earlier observations.[40] In the simulations of the proposed reaction scheme, this step was followed by concomitant intramolecular T1Cu(I)T2Cu(I) two-electron transfer to T3Cu(II) and at low pH (6.5) the rate constant was adjusted to 5 s^{-1} in order to yield a reasonable fit to this mechanism. However, upon increasing the pH an intramolecular process becomes rate limiting, and it was proposed that the rate of elimination of an OH^- ion coordinated to T2 copper was the limiting factor. Thus, at higher pH (> 7.5) an "inactive" form of laccase becomes dominant and it was proposed that one of the water molecules formed during the enzymatic reduction of O_2 becomes coordinated to $T2Cu^{2+}$ with a pK of 7.4.[50] This notion is in accordance with all later enzymatic studies of dioxygen reduction involving blue copper oxidases as well as cytochrome *c* oxidase. All seem to involve a relatively slow transition between an active and an inactive form of the enzyme. However, with today's knowledge that T2 and T3 copper constitute a trinuclear center it is difficult to reconcile this fact with a type-2 copper ion having distinct redox behavior from the type-3 center. It therefore seems that the mechanism of T2 copper involvement in the catalytic cycle, specifically the intramolecular ET, in the blue oxidases needs to be re-evaluated. The slow final reduction of T2Cu(II) at high pH would exclude it from entering into the catalytic process. Reoxidation by O_2 of the T1Cu(I)T2Cu(II)T3Cu(I) form of the enzyme would therefore produce a T1Cu(II)T2Cu(II)T3Cu(II)O^- intermediate by a three-electron transfer. An EPR spectrum has been obtained which indeed can be interpreted as a composite T1Cu(II), T2Cu(II) and O^- spectrum.[43]

The relationship between possible protein conformational changes and the redox changes of the copper sites was addressed by studies of anaerobic reduction of *Rhus* laccase.[51–54] It was earlier found that a significant enhancement in the intrinsic protein fluorescence took place and that this could be correlated with reduction of the T1Cu(II) site.[55] The rather detailed time-resolved analysis of the relation between the redox state of the enzyme and its absorption and fluorescence carried out by stopped-flow kinetics of *Rhus* laccase, yielded an intrinsically self-consistent model. Interpretation of the results required two separate reduction steps in order to account for the reduction linked fluorescence increase, as demonstrated by computer simulations of the experimental data.[51] According to this kinetic model, an anaerobic reduction pathway of laccase is associated with a minimum of three different fluorescence-sensitive protein states which depend on both conformational changes and redox states. One proposed step in the kinetic model is a "very fast" concomitant two-electron transfer from T1Cu(I)T2Cu(I) to the T3Cu(II) pair with a rate constant independent of the concentration of the Ru(II) used as reductant. This intramolecular ET step should follow the putative individual bimolecular reductions of T1 and T2Cu(II), and with a rate constant 10 times faster for the latter site.[51] Concomitant two-electron transfer is in agreement with the proposal of Andréasson and Reinhammar[50] (*vide supra*) although in the earlier simulations the intramolecular ET rate was proposed to be slow.[40] However, this mechanism where both T1 and T2 copper(I) synchronously transfer two electrons to T3Cu(II) is in conflict with later thermodynamic and kinetic studies which clearly established that intramolecular one-electron ET can already take place between T1Cu(I) and T3Cu(II) in molecules where just the former site has been reduced (*vide infra*). Finally, the issue of a direct, bimolecular reduction of the T2Cu(II) ion is still not clearly established.

Solvent isotope and pH effects on the catalytic activity of *Rhus* laccase under steady state conditions resolved a ratio $k_{cat}(H)/k_{cat}(D) = 2.1$.[53] It was further found that only a single proton plays a significant role in the rate limiting intramolecular ET step with a pK $\sim$ 6.0 and that the protein protic group is undissociated under optimal catalytic conditions.[53] Although the specific intramolecular ET step could not be identified, it was proposed that an oxygen radical assembly, produced in close proximity to the protein protic group might be involved. This species, a T1Cu(II)T2Cu(I)T3Cu(II)O^- could be the intermediate under steady state conditions where the oxygen radical interacts with T2 and T3 coppers of the trinuclear center[54] and though similar it still differs from the species proposed by the Gothenburg group[41–44] having type-2 copper oxidized, *vide supra*. The rate constant for intramolecular ET from T1Cu(I) to the above transient trinuclear copper-oxygen radical complex, using ascorbate as reducing substrate, was determined to be 3.4 s^{-1} similar to the rate determining step in the overall catalytic cycle at pH 6.0.[54]

A detailed analysis of the absorption and circular dichroism (CD) spectra of *Rhus vernicifera* laccase occurring upon (a) addition of one equivalent of hydrogen peroxide to oxidized laccase; (b) reduction with ascorbate under aerobic conditions; (c) reaction of partial or fully reduced enzyme with O_2; (d) reduction of peroxo-laccase; and (e) peroxide reoxidation of the reduced enzyme, have corroborated the notion that peroxo-laccase is a *bona fide* distinct and stable intermediate in the catalytic cycle of this enzyme.[56–60]

When oxidized native laccase is reacted with one equivalent of hydrogen peroxide a specific increase in the absorption band centered at 330 nm, characteristic for the T3 copper(II) pair, is observed.[56] This was assigned to the formation of a peroxo-laccase derivative. That this change is indeed due to peroxide binding and not, as has been suggested later, to oxidation of

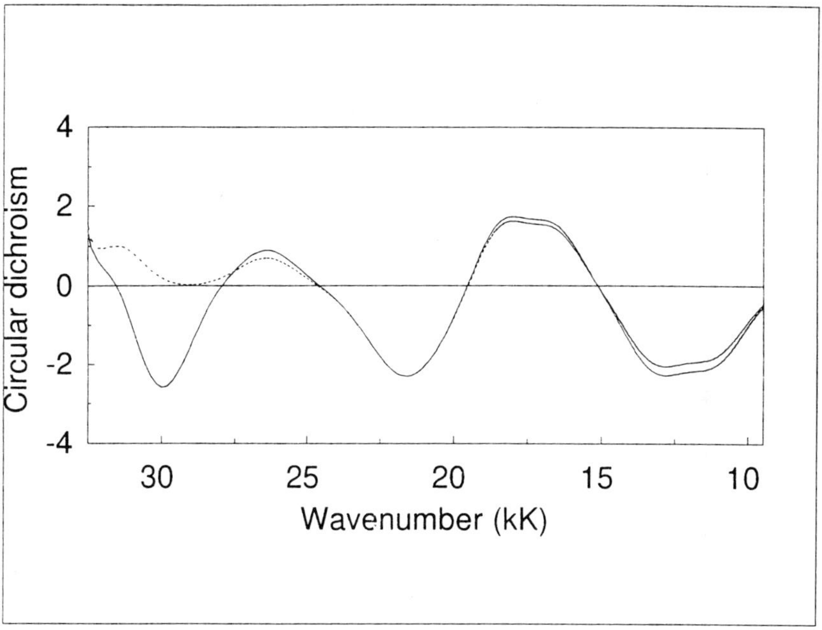

Fig. 2. CD spectra of native and peroxo-laccase.[57] Dotted line: Native *Rhus* laccase, 0.15 mM in 0.1 M phosphate buffer, pH 7.0 and 25°C. Extended line: Same laccase solution to which 0.15 mM hydrogen peroxide was added. The wavelength scale is from 300 nm (33 kK) to 1000 nm (10 kK). The absorption scale is $\Delta\varepsilon(M^{-1}cm^{-1})$.

residual reduced T3 sites has been shown by the fact that no other oxidant except H_2O_2 was able to cause this stoichiometric absorption enhancement in the near UV region. Moreover, the formation of the peroxo-laccase complex is accompanied by prominent changes in the CD spectrum (Fig. 2) specifically the formation of a new negative Cotton effect around 325 nm.[57] Reductive titration of the peroxo-laccase complex required two additional electron equivalents in order to fully reduce this species as compared with the reduction of native oxidized enzyme. Oxidative titration of reduced laccase with H_2O_2 were also performed demonstrating that while four equivalents were necessary for full reoxidation, two additional equivalents led to the formation of spectral changes in the near UV region identical to those seen upon adding peroxide to native oxidized laccase. Other strong oxidants like $PtCl_6^{2-}$, $S_2O_8^{2-}$, $Mo(CN)_8^{3-}$ and *t*-butyl peroxide do reoxidize reduced laccase quantitatively, but even large excess does not produce the spectral changes observed upon reaction with peroxide.[56] In the recently determined three-dimensional structure of the peroxo derivative of AO it was found that it binds terminally as hydroperoxide end-on to one of the two copper ions constituting the T3 center.[34] The copper–copper distance increases from 0.37 in oxidized AO to 0.48 nm and the bridging oxygen present in the oxidized native enzyme is lost. In view of the close sequence similarity between laccase and AO the notion of a T3 copper peroxide complex formation by laccase gains further support.

The T3 copper(II) pair in laccase is antiferromagnetically coupled and in the AO structure the two copper ions are bridged by OH^- or O^{2-}. We have performed magnetic susceptibility measurements on native oxidized and peroxo-laccase at room temperature in order to examine whether the exchange coupling between these two T3Cu(II) ions is affected by the interaction with peroxide.[61] The coupling between the two copper ions can be expressed by the spin Hamiltonian, H:

$$H = -2J\hat{S}_1\hat{S}_2 \tag{6}$$

and the lower limit for the exchange integral, $-J$ is 250–300 cm^{-1}. The antiferromagnetic coupling was found to decrease from the above value to $-J = 120 \pm 10$ cm^{-1} upon addition of 1 equivalent of H_2O_2. This demonstrates that peroxide coordination to the T3 site reduces the coupling between the two Cu(II) ions.[61] This is in very good agreement with the structural changes observed upon adding peroxide to AO crystals discussed above.

As stated earlier CD measurements are useful for monitoring the chemical and conformational changes in laccase also in reactions with dioxygen. When partially or fully reduced laccase reacts with O_2, specific long-lived changes are observed in the optical spectra, characterized by extra absorption bands, mainly around 330 nm.[58,59] The observed negative Cotton effects suggest that laccase-oxygen intermediates are formed, most likely equivalent to peroxo-laccase (Fig. 3). The largest change is observed when half-reduced laccase is reoxidized with O_2 leading to a CD spectrum (Fig. 3, dotted line) quite reminiscent of peroxo-laccase produced directly by reaction of H_2O_2 with the oxidized enzyme as described above (Fig. 2). Also, reductive titration of this species support the notion of formation of a peroxo intermediate.[59] The optical spectrum of laccase during turnover with ascorbate and excess dioxygen differs significantly from that of the

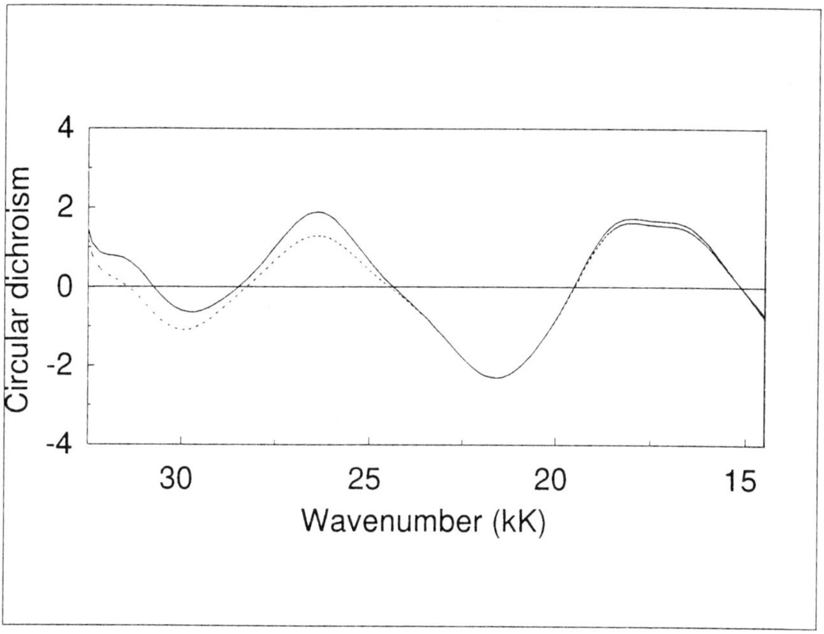

Fig. 3. CD spectra of laccase-dioxygen intermediates. Dotted line: 0.10 mM laccase reduced with 0.10 mM ascorbate followed by reoxidation with O_2. Extended line: 0.05 mM laccase after turnover with 0.25 mM ascorbate in an air saturated solution (0.25 mM O_2). All other conditions are as in Fig. 2.

resting enzyme, and remains unchanged for more than an hour after substrate exhaustion, clearly suggesting that under these conditions, the enzyme undergoes a conformational change to produce a relatively long-lived intermediate.[58] The CD spectrum of this intermediate is shown in Fig. 3 (extended line). In the presence of excess dioxygen the rate of decay of this species is $5 \times 10^{-4} s^{-1}$ at 25°C. The nature of this transient chromophore may be a free radical localized at the T3 site. The O^- radical identified using liquid He temperature EPR of the rapid reaction product of fully reduced laccase with O_2[43] (*vide supra*) carries some resemblance to this transient, only that in the latter T2 copper was found to be reduced. An attractive hypothesis is that the unpaired electron of the free radical is delocalized on both the trinuclear copper center and the oxyl radical. The rate of disappearance of this transient is two orders of magnitude slower than that of the species produced by rapid-freezing, which is not surprising considering that the transient in the former case has an oxidized T2Cu(II) site with no reduction equivalents left. Based on the above results we propose that the rate limiting step for T1Cu(I) reoxidation is an intramolecular ET and assuming

$$V(\text{steady state}) = k_{\text{ox}}[\text{T1Cu(I)}] = 0.5 \times k_2[\text{Red}][\text{T1Cu(II)}],$$

where Red refers to the reducing substrate and k_2 is the appropriate bimolecular rate constant for anaerobic reduction of laccase, k_{ox} could be calculated from a large series of experiments.[59] We found $k_{ox} \sim 3\ s^{-1}$ at room temperature, independent of the nature or concentration of the employed reductant. Interestingly, this value is very similar to that obtained for T1Cu(I)→ T3Cu(II) ET using the pulse radiolysis technique.

Results of the above studies provided further support to the formation of enzyme species produced from the resting oxidized state during the first redox cycle which is catalytically active and hence named C-state.[58] This observation probably resolves an apparent dilemma caused by the slow internal T1 to T3 ET rates observed in anaerobic reduction studies. Independent evidence for such a species has also emerged from studies performed in several other laboratories and though laccase is fully oxidized at the resting state as usually isolated, it is different from the former in both kinetic and some subtle spectroscopic properties. The coordination of an OH^- ion to the T2Cu(II) site has been suggested to be a major structural difference and its displacement is supposed to bring about the transition into the active enzyme form.[50] It is noteworthy that at least formally, a similar behavior has been reported for bovine cytochrome *c* oxidase ("a pulse state").[62]

The extensive studies carried out by Ettinger and colleagues[51–54] (*vide supra*) yielded a detailed kinetic analysis of the catalytic cycle of *Rhus* laccase where the central role of the T2 redox conversions has been confirmed.[51–54] Moreover, structural changes accompanying the catalytic redox cycle and expressed in the intrinsic fluorescence properties of the enzyme were analyzed and correlated with specific states for the different copper sites.[51–53] Further insights into the catalytic cycle were also gained by examination of the solvent isotope effect and pH dependence of the enzymatic activity.[54] Concentrating on resolving steps involving proton transfer, at least four distinct groups were proposed to be involved in inter- and intramolecular steps of laccase catalysis. The problem of proton consumption upon O_2 reduction has also been addressed in another study where a pH indicator was employed to monitor transient changes during a single catalytic cycle.[45] It was found that two protons are consumed in the same time range as the EPR detectable oxygen intermediate is formed, while two more protons are taken up by the protein concomitantly with the T2Cu(I) reoxidation. This could indicate that one water molecule is produced in the fast phase, leaving the second oxygen as an O^- radical coordinated to the T3 copper pair.

More recently, stopped-flow measurements of the reaction of native and Hg(II) derivatized T1 laccase, T1Hg(II) laccase, with dioxygen have been undertaken by Solomon *et al.*, extending earlier studies.[63–65] T1Hg(II) laccase containing a fully reduced trinuclear copper center can be reoxidized by O_2 producing a peroxide intermediate with the characteristic near UV absorption spectrum.[64,65] It was suggested that peroxide is coordinated as a μ-1,1 hydroperoxide bridging one of the oxidized T3Cu(II) ions with a reduced T2Cu(I) ion.[65] However, as stated above, the three-dimensional structure of a peroxide complex of ascorbate oxidase, clearly demonstrates a coordination of the peroxide end-on to one of the T3Cu(II) ions,[34] which must also provide a structural rationale for the analogous species formed by *Rhus* laccase. MCD measurements have further been employed in order to probe the electronic properties of the intermediate in native *Rhus* laccase and T1Hg(II) laccase.[66] The T3-peroxo intermediate was found to be paramagnetic in contrast to the native oxidized enzyme where the T3Cu(II) pair is antiferromagnetically coupled. This is in line with our earlier observations that peroxide coordinating to native oxidized laccase caused a decrease in coupling between the two Cu(II) ions.[61] Fully reduced T1Hg(II) laccase is expected eventually to form a three-electron reduced oxygen intermediate, much like the one obtained in rapid-freeze EPR upon reacting fully reduced native laccase with O_2.[43] Using $^{17}O_2$ in the above reaction of T1Hg(II) labelled laccase with dioxygen it could be deduced that a considerable degree of electronic delocalization from the HO radical onto T2Cu(II)[66] which may be the reason for the apparent relatively large stability of such an intermediate.

The central, though still unresolved role of the T2 copper center in the catalytic cycle of laccase has been further addressed by Koudelka and Ettinger[54] who carried out a detailed study of this issue using fluoride ions, known to coordinate the latter site, as their main probe. Fluoride was clearly shown to cause the reduction rate of the T3 site to become limited by a reductant independent step. These and other results were incorporated into a model where F^- ions inhibit the reduction of enzyme bound oxygen radical intermediate, also under steady-state conditions.[54] Significantly, these results may have direct implications for the role of the T2Cu(II) coordinated hydroxyl ion and to the resultant pH dependence of the catalytic activity.

In conclusion, clear evidence is now available for the reactivity of *Rhus* laccase molecules containing 1, 2 and 3 reduction equivalents with dioxygen yielding intermediates of different stabilities. This might, however, not be the case for fungal laccase where the properties of both T3 and T1 differ considerably as compared with *Rhus* laccase, hence leading to marked differences in the catalytic mechanism.

6.2. *Intramolecular Electron Transfer in Ascorbate Oxidase*

Earlier studies of the blue oxidases, laccase and ceruloplasmin have focused, in addition to oxygen intermediate formation[41–44,58–60] on the kinetics of T1Cu(II) reduction and have contributed to the notion that the T1 site acts as the electron uptake site by these enzymes.[46,47] A major breakthrough in the investigation of this group of enzymes occurred with the determination of the three-dimensional structure of ascorbate oxidase at high resolution.[25,32,36] This provided insights into the detailed structure and spatial relationship among the copper coordinating sites, cf. Fig. 1.

The reduction potentials of the T1- and T3 Cu(II)/Cu(I) couples in resting AO are identical, 350 mV at pH 7.0[67] while the T2 potential is considerably lower (< 300 mV). The catalytic cycle of this enzyme proceeds by a sequential mechanism where single electrons are transferred from the organic reducing substrates to T1Cu(II) while the trinuclear T2–T3 center serves as the dioxygen binding and reducing site. Intramolecular ET from T1Cu(I) to T3Cu(II) therefore seems to be an essential step required for the transfer of the 4 electrons (and 4 protons) necessary for reduction of O_2 to two H_2O molecules.

The pioneering study by pulse radiolysis of AO has been carried out by O'Neill *et al.*[68] examining the enzyme reactivity towards hydrated electrons as well as the CO_2^- and reduced nitroaromatic radicals. Rather low yields of T1Cu(I) were observed using the former two reagents while a stoichiometric reduction of this site was observed using the latter. Also, the reduction mechanism by the former agents was indirect, i.e. it proceeded via protein radical intermediates. No clear evidence for intramolecular ET was presented though this was proposed to take place.[68] This point gained further support from later studies by the same group[69] where the reduction of AO and T2D AO by $ArNO_2^-$, CO_2^- or O_2^- radicals has been pursued. The reduction stoichiometry of native AO by the former agents required 6–7 equivalents, for T2D AO only 3 equivalents were necessary (all based on 610 nm absorption). Reduction by O_2^- radicals, performed in excess O_2 containing solutions yielded rather limited net reduction (as expected from this experimental protocol). In contrast, T2D AO underwent full reduction of the T1Cu(II), clearly showing the requirement of an intact trinuclear center for attaining the catalytic cycle.[69]

Determination of the three-dimensional structure of ascorbate oxidase prompted three independent groups to investigate the intramolecular ET from T1Cu(I) to T3Cu(II) in AO under anaerobic conditions.[70–74] Photochemically produced lumiflavin semiquinone reduces T1Cu(II) in a fast second order process ($k = 2.7 \times 10^7\ M^{-1}s^{-1}$ at pH 7.0) followed by partial reoxidation of the T1Cu(I) site with a rate constant of 160 s^{-1}.[70] The flavin absorption in the near UV region prevented monitoring the changes at 330 nm where T3Cu(II) absorbs, but the latter process was still interpreted as intramolecular ET from T1 → T3. This interpretation gained further support by results obtained with AO depleted of type 2 copper (T2D AO), where no difference in the intramolecular ET rate was observed.[70] Using pulse radiolytically produced CO_2^- radicals as the primary electron donor, AO reactivity was monitored at both 610 nm (T1) and 330 nm (T3)[71] (Fig. 4): A bimolecular, diffusion controlled T1Cu(II) reduction ($1.2 \times 10^9\ M^{-1}s^{-1}$ at pH 7.0) was observed (Fig. 4A) followed by biphasic T1Cu(I) reoxidation (Fig. 4A and B) and T3Cu(II) reduction (Fig. 4C) proceeding with identical and concentration independent rate constants, establishing the assignment of the latter processes as intramolecular ET between the T1 and T3 sites. The observation that the intramolecular ET process proceeds in several distinct phases is very interesting. Two phases were observed at both 610 nm and 330 nm and though their respective amplitudes varied as the protein got reduced, the amplitude ratio remained constant. The phases could be resolved and gave rate constants of 200 s^{-1} for the fastest phase and 2 s^{-1} for the slowest.[71] A rationale for the multiplicity of phases in the internal ET could be that they are the result of the presence of several forms of the active sites varying in reactivity.[71]

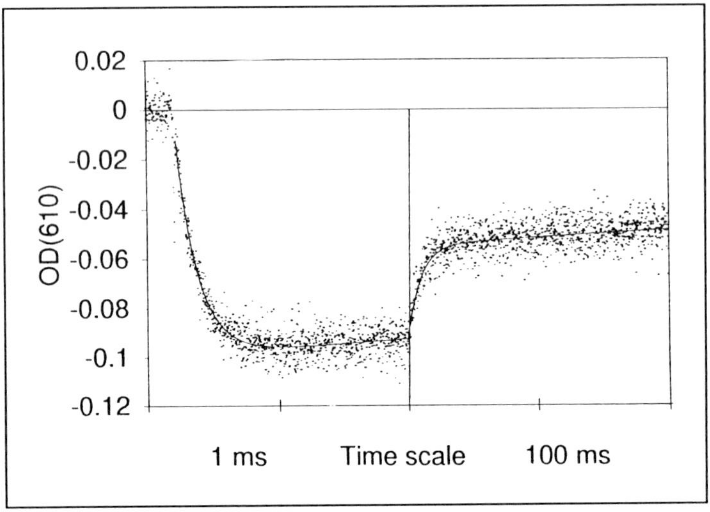

(A)

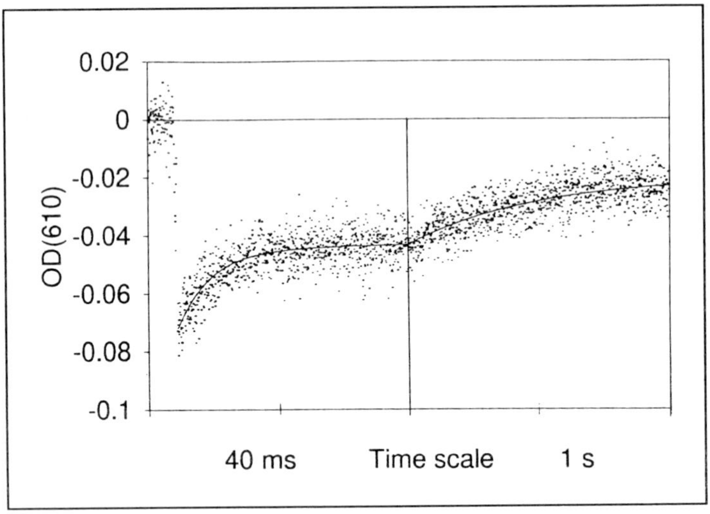

(B)

Fig. 4. Kinetics of AO reduction by CO_2^- radicals under *anaerobic* conditions. A: Time-resolved absorption changes at 610 nm of a 5.5 μM AO solution in N_2O saturated 0.1 M sodium formate/0.01 M sodium phosphate at pH 5.5 and 25°C. Pulse width 0.75 μs equivalent to $\sim$ 4μM of CO_2^-. B: Conditions are as above, but at different time scales. C: Conditions are as above, only monitored at 330 nm.

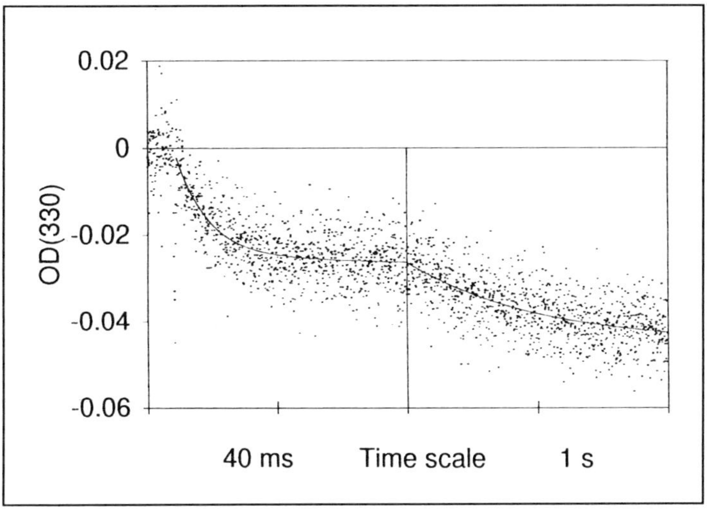

(C)

Fig. 4. (*Continued*)

This notion is further supported by the result of the high resolution structure of AO where two oxygen ligands are found at the trinuclear center.[32] Protonation equilibrium of these ligands could yield states with distinct reactivities. Current experiments aiming at examining the pH dependence of the internal ET are in progress (Farver and Pecht, unpublished).

In a later pulse radiolysis study using different organic radicals as reductants similar ET rates were obtained by monitoring only the 610 nm chromophore.[72] It is noteworthy that a single phase intramolecular ET from T1 $\rightarrow$ T3 with a rate constant of 2 s^{-1} was observed in the related blue copper oxidase, laccase using pulse radiolytically produced reducing radicals, CO_2^- and O_2^- (cf. Sec. 6.1).[75] However, in this protein no other phases of intramolecular ET process were resolved.

An underlying assumption in our studies of LRET in AO is that the structure of the blue oxidases has most probably been evolutionarily optimized for intramolecular ET in the catalytic dioxygen reduction, and AO therefore represents a relevant and very interesting system for investigating the parameters that control intraprotein ET. Several important questions then arise: Is there a control of the intramolecular ET in AO during the multielectron reduction and oxidation? In other words, does the internal ET rate depend on the number of reduction equivalents taken up by the molecule? How does the rate of electron transfer relate to the conformational changes the enzyme undergoes upon reduction which were resolved by the three-dimensional structure determination? Does the presence of reducing substrates or dioxygen affect the internal ET rates (i.e. by an allosteric mechanism)? These questions gain further significance when it is noted that steady-state kinetic measurements of AO activity[67,76] yield turnover numbers of 12,000–15,000 s^{-1} which are considerably faster than the above observed rate constants for intramolecular T1 to T3 ET.

The notion that AO molecules reduced to different degree may exhibit distinct reactivities has been ruled out by the finding that identical intramolecular ET rate constants are obtained in experiments starting with the fully oxidized enzyme and ending with less than 5 % oxidized T1 sites.[71] Also, no difference in the internal ET rate constants could be observed with enzyme preparations that were "activated" or "pulsed" by turning over 1 mM ascorbate in the presence of 0.25 mM O_2 prior to the determination of the intramolecular ET processes. However, Tollin *et al.* have recently found that freezing and thawing AO solutions may cause them to exhibit biphasic kinetics which are also correlated with a lower enzymatic activity of such samples. Turnover, could nevertheless restore the "high activity" conformation.[73] These conflicting reports need obviously more studies in order to resolve the issue.

In contrast to the above experiments which were all performed under anaerobic conditions, when studies were carried out on AO solutions containing low and controlled concentrations of O_2 (15–65 μM) quite conspicuous differences were observed[77] (Fig. 5): An additional fast intramolecular ET was discovered which strictly depended on the presence of dioxygen. While under anaerobic conditions electrons are taken up by T1Cu(II) followed by internal redistribution (Fig. 6A), with O_2 present in the solution, a lag phase is observed which depends on the amount of dioxygen present (Fig. 6B). During this phase electrons are indeed taken up by T1Cu(II) (Fig. 5A) and transferred on to T3Cu(II) (Fig. 5B and C), but these reduction equivalents eventually end up in production of an enzyme-bound oxygen intermediate, probably coordinated to T3 copper. This is clearly seen in both the time resolved spectra at 610 nm and 330 nm (Fig. 5) and in amplitude changes observed in the transient phase of the 330 nm signal. Large transient absorption formation in the near UV range was observed following the intramolecular ET from T1Cu(I) to T3Cu(II) (Fig. 5B), and most probably reflect interaction between the partially reduced trinuclear site and dioxygen. Transient absorptions in this spectral range were first observed for ceruloplasmin[78] and then studied in detail in *Rhus* laccase.[41–44,56–60] The rate constant of the fast intramolecular ET phase in AO in the presence of O_2, monitored at both 610 nm (T1) and 330 nm (T3), was 1100 s^{-1} (293 K, pH 5.8). It was maintained at this value as long as dioxygen remained in the solution.[77] Oxygen intermediates coordinated to the trinuclear center will increase the driving force of the intramolecular ET significantly and this will in turn result in an accelerated ET rate. Calculations show that a 100 mV increase in the reduction potential of the T3 center would lead to the observed enhancement in the intramolecular T1Cu(I) $\rightarrow$ T3Cu(II) ET rate.[77] A similar behavior has been observed for cytochrome *c* oxidase in its catalytic reduction of O_2.[79]

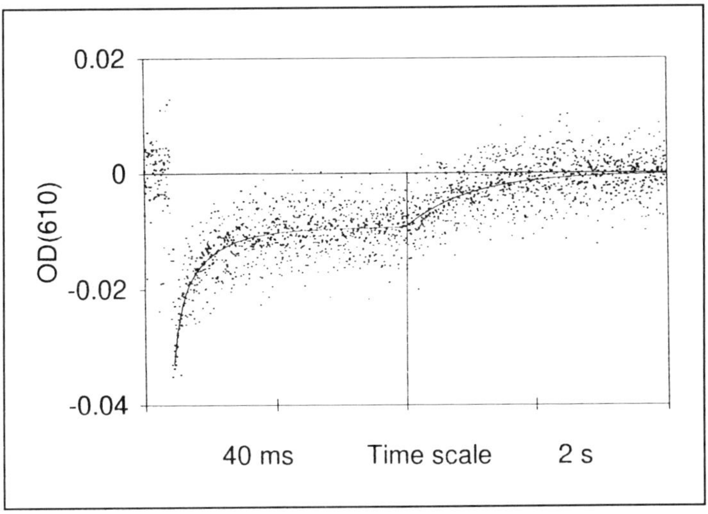

(A)

Fig. 5. Kinetics of AO reduction by CO_2^- radicals under *aerobic* conditions. A: Time-resolved absorption changes at 610 nm of a 5.5 μM AO solution in a N_2O solution containing 65 μM O2. 0.1 M sodium formate/0.01 M sodium phosphate at pH 5.5 and 20°C. Pulse width 0.5 μs $\sim$ 2.5 μM of CO_2^-. B: Conditions are as above, but monitored at 330 nm. Note the formation of a transient absorption increase in the presence of excess O_2 in the solution. C: Conditions are as before, but after consumption of free O_2 in the solution.

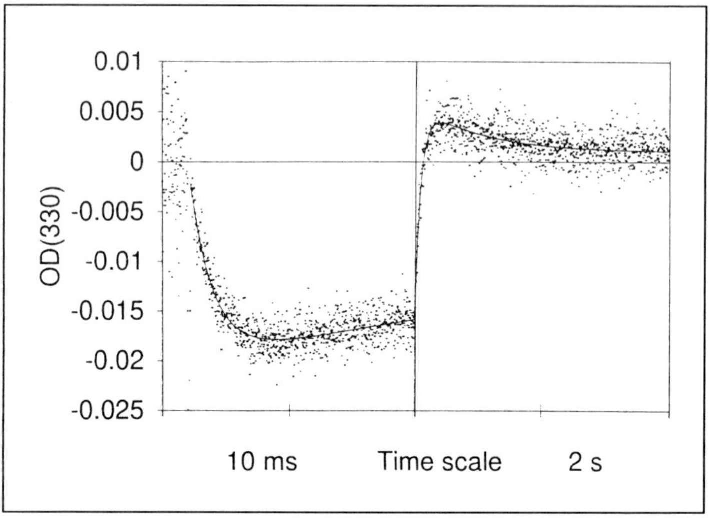

(B)

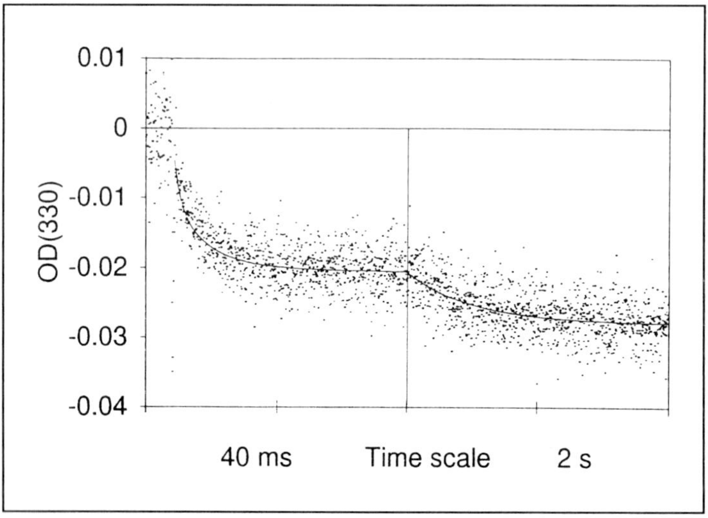

(C)

Fig. 5. (*Continued*)

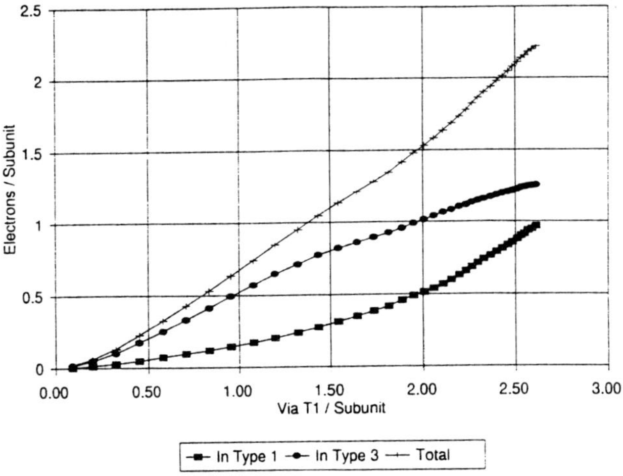

(A)

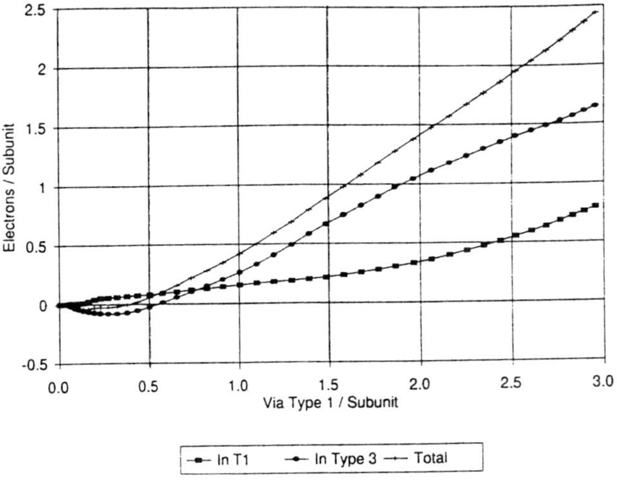

(B)

Fig. 6. Intramolecular electron distribution in AO during the time-resolved reductive titrations by CO_2^- radicals. 5.5 μM AO solution in N_2O containing 0.1 M sodium formate/0.01 M sodium phosphate solutions at pH 5.5 and 20°C. The abcissa shows the number of electrons entering the AO molecules via T1Cu(II). The ordinate shows the number of electrons accumulated in T1 and T3, as determined from the observed absorption changes at 610 nm and 330 nm. (A) Strictly anaerobic conditions. (B) Same solution as in A, except for an initial 15 μM O_2 concentration.

Applying the equations presented in Sec. 3 above to the determined activation parameters of the intramolecular ET processes in AO, the reorganization energy, λ, for this LRET could be calculated. Using a through-bond distance of 1.34 nm between the T1-Cu ligating thiolate and one of the imidazole nitrogens coordinated to the T3-copper pair together with the previously determined electronic decay factor, $\beta = 6.7\ \text{nm}^{-1}$ for ET in β-sheeted proteins (cf. Sec. 3) a calculated $\lambda = 153\ \text{kJ mol}^{-1}$ is obtained.[71] This value for the reorganization energy in AO is significantly larger than that calculated for intramolecular ET in azurin, $\lambda = 99\ \text{kJ M}^{-1}$ [22] (cf. Sec. 3). This difference can be rationalized in terms of the marked changes that take place in the three-dimensional structure of the enzyme: Upon reduction of the T3 site the copper–copper distance increases from 0.37 nm to 0.51 nm[34] and the antiferromagnetic coupling between the copper pair is disrupted along with a considerable local conformational change around the trinuclear site.

The observed maximal rate constant of 1100 s^{-1} for intramolecular ET in AO is still considerably smaller than the turnover number of ~14,000 s^{-1}.[67,76] Therefore, the rate enhancement by dioxygen coordinated to the trinuclear site is not sufficient to explain the difference between the observed maximal enzymatic activity and the T1 $\rightarrow$ T3 LRET rate. Under optimal conditions, the concentration of reducing substrate (e.g. ascorbate) is adequate to maintain a steady state of fully reduced copper sites. In this case it is obviously impossible to initiate ET using reducing radicals. Thus, an opposite approach was taken by Tollin and coworkers who have examined the reoxidation of fully reduced AO by a laser generated triplet state of 5-deaza riboflavin.[74] Subsequent to the one-electron oxidation presumably of the reduced T2–T3 cluster a rapid biphasic increase in 610 nm absorption was observed which was interpreted as intramolecular ET from T1Cu(I) to the oxidized trinuclear center. The rate constant of the faster of the two observed processes (9500 and 1400 s^{-1}, respectively) is comparable to the turnover number for AO under steady-state conditions and renders it very likely to be the rate limiting step in catalysis.

In the three-dimensional model of AO, one of the T1-copper ligands is the thiolate Cys 507-(S_γ) while imidazoles of the two neighboring histidines (His 506 and His 508) coordinate to the T3 copper ions. This led to the proposal that the shortest ET pathway from T1Cu(I) to T3Cu(II) takes place via Cys 507 and either His 506 or His 508,[25] cf. Fig. 7. Both pathways consist of nine covalent bonds yielding a total distance of 1.34 nm. Pathway calculations as developed by Beratan and Onuchic[10,11] (cf. Sec. 3) support this notion and further resolve an alternative path via a H-bond between the carbonyl oxygen of Cys 507 and His 506(N_δ). The electronic coupling calculated for the two covalent pathways is 0.010 while for the latter it is slightly higher, 0.014.[71] Another noteworthy point is the similarity between the ET pathways in AO and the one proposed for LRET between the Cu ion and Tyr-83 in plastocyanin[80]: In both cases the thiolate ligand of T1-copper is assumed to be utilized in a "remote path" intramolecular ET. A high degree of anisotropic covalency was calculated for the Cu(II) center in plastocyanin resulting in a very favorable electronic coupling via the cysteine thiolate ligand.[81] A similar anisotropic effect is presumably present in AO as indicated by the structural arrangement of T1–T3. Moreover, the same degree of anisotropic covalency is expected to exist for the more pertinent **reduced** T1Cu(I) site from which the electrons are transferred intramolecularly in AO since ET can be considered as a hole superexchange in the HOMO.[82] One question remains, namely why it is not the thiolate that provides the ET pathway in most of the electron mediating single blue copper proteins, such as azurin, amicyanin, etc., but rather an exposed imidazole ligand.

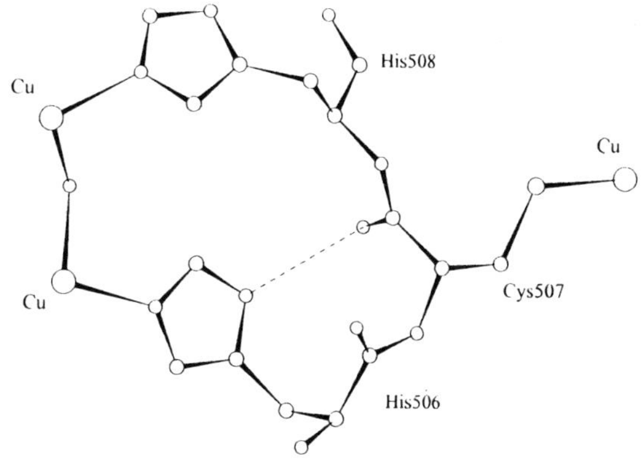

Fig. 7. Proposed intramolecular ET pathway from T1Cu(I) (right) to T3Cu(II) (left) in ascorbate oxidase. In addition to the two covalent pathways, a third one was found involving a hydrogen bond between O of Cys 507 and N_δ of His 506. The coordinates were kindly provided by Dr. A. Messerschmidt.[32]

6.3. *Intramolecular Electron Transfer in other Multi-Copper Oxidases*

In cytochrome *c* oxidase the primary step in the enzymatic process is ET from reduced cytochrome *c* to the binuclear Cu_A center.[31] The question of how and where electrons are transferred intramolecularly in this enzyme have now gained considerable impulse by the determination of the three-dimensional structure of both the mammalian and the bacterial enzymes.[27,28] These interesting problems have also been addressed using pulse radiolytically produced N-methylnicotinamide radicals as primary electron donor in reaction with bovine cytochrome *c* oxidase.[83] It was found that the Cu_A center was reduced in a fast bimolecular ET reaction with a rate constant of 1.5×10^9 $M^{-1}s^{-1}$. Following this fast phase, reoxidation of the Cu_A center was observed concomitantly with reduction of the heme-a site at a first order rate constant of 1.8×10^4 s^{-1}. This suggests that electrons are transferred intramolecularly from reduced Cu_A to oxidized heme-a. From experiments using the mixed valence CO-complex of cytochrome *c* oxidase it could be shown that heme-a_3 did not contribute to the reaction since the same kinetics were observed under these conditions.[83] The three-dimensional structure of bovine cytochrome *c* oxidase shows a direct distance from Cu_A to heme-a of ~1.5 nm while the distance from Cu_A to heme-a_3 is 1.9 nm.[27] This may explain the preferential ET between Cu_A and heme-a. It would, however, be very interesting to pursue pathway calculations as those discussed above in Sec. 6.2 for ascorbate oxidase, in order to better understand the reasons for the preferential intramolecular ET to heme-a rather than to the Cu_B-heme-a_3 center.

Nitrite reductase (NiR) represents a distinct type of blue copper enzymes which participate in the denitrification process in prokaryotic organisms by reducing the nitrite ion to nitrogen oxide. NiR from *Achromobacter cycloclastes* contains two copper ions per subunit, one of which has typical T1 characteristics while the other is classified as T2 copper. The three-dimensional structure has recently been determined to a 0.23 nm resolution and shows that the two copper ions are separated by a distance of 1.25 nm and coordinated to adjacent residues in the sequence: T2Cu - His 135 - Cys 136 - T1Cu.[29] Pulse radiolysis studies have been performed by Suzuki *et al.*[84] which show that T1Cu(II) is reduced by N-methylnicotinamide in a rapid bimolecular process ($k = 3.4 \times 10^8$ $M^{-1}s^{-1}$). Following this initial phase a slow partial recovery in T1 absorption was observed with a rate constant independent of the protein concentration. This supported the notion that an intramolecular ET takes place between T1Cu(I) and T2Cu(II). The extent of recovery of T1Cu(II) apparently depends on the amount of T2 copper present in the employed enzyme preparation, with usually a ratio of T1/T2 copper less than 1.[84] Extrapolation to the intact 1:1 T1/T2 NIR gives an apparent equilibrium constant of 2 for the

$$\mathrm{T1Cu(I)T2Cu(II)} = \mathrm{T1Cu(II)T2Cu(I)}$$

ET equilibrium. The rate constants for intramolecular ET in the above reversible reaction were estimated to be 980 s^{-1} and 420 s^{-1}, respectively, corresponding to an equilibrium constant of 2.3.[84] The former rate constant is approximately 5 times faster than that observed for the fastest phase in AO (200 s^{-1})[70–72] via a structurally related pathway discussed above in Sec. 6.2. This could be attributed to the slightly more favorable driving force for intramolecular ET in NiR.

7. Conclusion

Laccase, ascorbate oxidase and ceruloplasmin which are prominent members of the family of blue copper containing oxidases catalyze single electron uptake from specific substrates by the T1 copper site while the reduction of dioxygen to two water molecules is carried out by the trinuclear center constituted of the T2 and T3 sites. Determination of the three-dimensional structures of the latter two enzymes supports earlier spectroscopic assignments of the distinct metal binding sites, and it provides a structural rationale for some aspects of their function.[25,26] However, that significant differences do exist in the detailed reaction mechanisms has already emerged from early investigations and was further underscored by more recent ones. A consensus emerging from both kinetic and equilibrium measurements is that for all these enzymes, the T1Cu(II) center functions as the primary electron acceptor delivering the electrons intramolecularly to the trinuclear center. This is in line with both the reduction potentials of the different copper centers in these copper oxidases and moreover in the case of AO, an ascorbate docking site has been identified proximal to the T1 site.[32] Still, in some cases a bimolecular reduction of the T3 chromophore[71] and of the T2Cu(II) EPR signal[50] have also been observed. These observations may either be due to non-physiological processes, proceeding when relatively small reductants (eg. CO_2^- radicals) can access the latter sites, or a result of the bimolecular T1Cu(II) reduction rate being slower than the internal T1Cu(I) $\rightarrow$ T3Cu(II) ET step, hence yielding an apparent bimolecular character to the reduction rate of the latter site.

As elaborated above, an obvious ET pathway between the T1 and T3 copper sites has also emerged from examination of the three-dimensional AO structure, suggesting that this is the one operating in the catalytic cycle of this enzyme.[25] Moreover, deducing from the sequence homologies it is probably also the one operating in all other blue oxidases.[36] The electron distribution amongst the copper ions constituting the trinuclear center is still an open question as usually different methods have to be employed for monitoring the T2 and T3 Cu(II) centers (eg. rapid-freeze EPR and optical spectroscopy, respectively). Marked differences in this respect exist among the above enzymes although the trinuclear center is assumed to perform an identical chemical transformation of O_2 to H_2O in all these enzymes! Indeed, far less well understood are the steps where dioxygen is reduced and to what extent they affect the electron transfer rate from T1Cu(I) to T3Cu(II). For example, both the reduction potentials and stoichiometry of electron uptake as determined by redox titration revealed marked differences among the better characterized members of the family, i.e. fungal and tree laccase.[85]

Acknowledgments

The authors wish to thank Dr. Lars K. Skov for producing the protein figures for us. Part of this work was supported by grants from the German-Israeli Foundation and by the Danish Natural Science Research Council.

References

1. M. K. Johnson *et al.*, eds., *Electron Transfer in Biology, Adv. Chem. Ser.* **226**, (1990).
2. K. D. Karlin and Z. Tyeklar, eds., *Bioinorganic Chemistry of Copper* (Chapman & Hall, New York, London, 1993).
3. H. Sigel, ed., *Metal Ions in Biological Systems*, Vols. 13 & 14, (Marcel Dekker, 1981).
4. E. T. Adman, in *Topics in Molecular and Structural Biology: Metalloproteins*, ed. P. M. Harrison (Chemie Verlag, 1985), p. 1.
5. E. T. Adman, *Adv. Protein Chem.* **42**, 145 (1991).

6. O. Farver and I. Pecht, in *Copper Proteins and Copper Enzymes*, Vol. 1, ed. R. Lontie (CRC, Boca Raton, 1984), p. 183.
7. H. B. Gray and B. G. Malmström, *Comments Inorg. Chem.* **2**, 203 (1983).
8. R. A. Marcus and N. Sutin, *Biochim. Biophys. Acta* **811**, 265 (1985).
9. A. Broo and S. Larsson, *Chem. Phys.* **148**, 103 (1990).
10. D. N. Beratan, J. N. Betts and J. N. Onuchic, *Science* **252**, 1285 (1991).
11. D. N. Beratan, J. N. Onuchic, J. R. Winkler and H. B. Gray, *Science* **258**, 1740 (1992).
12. M. van de Kamp and G. W. Canters, *Curr. Opin. Struct. Biol.* **2**, 859 (1992).
13. J. Jortner and M. Bixon, *Mol. Cryst. Liquid. Cryst.* **234**, 29, (1993).
14. J. J. Regan, S. M. Risser, D. N. Beratan and J. N. Onuchic, *J. Phys. Chem.* **97**, 13083 (1993).
15. P. Siddarth and R. A. Marcus, *J. Phys. Chem.* **97**, 13078 (1993).
16. D. S. Wuttke and H. B. Gray, *Curr. Opin. Struct. Biol.* **3**, 555 (1993).
17. S. S. Skourtis, J. J. Regan and J. N. Onuchic, *J. Phys.Chem.* **98**, 3379 (1994).
18. O. Farver and I. Pecht, *Proc. Natl. Acad. Sci. USA* **86**, 6968 (1989).
19. O. Farver and I. Pecht, *J. Am. Chem. Soc.* **114**, 5764 (1992).
20. O. Farver, L. K. Skov, M. van de Kamp, G. W. Canters and I. Pecht, *Eur. J. Biochem.* **210**, 399 (1992).
21. O. Farver, L. K. Skov, T. Pascher, G. B. Karlsson, M. Nordling, L. G. Lundberg, T. Vänngård and I. Pecht, *Biochemistry* **32**, 7317 (1993).
22. O. Farver, L. K. Skov, G. Gilardi, G. van Pouderoyen, G. W. Canters, S. Wherland and I. Pecht, *Chem. Phys.* **204**, 271 (1996).
23. D. S. Wuttke, M. J. Bjerrum, J. R. Winklerand and H. B. Gray, *Science* **256**, 1007 (1992).
24. R. Margalit, N. M. Kostic, C.-M. Che, D. F. Blair, H.-J. Chiang, I. Pecht, J. B. Shelton, J. R. Shelton, W. A. Schroeder and H. B. Gray, *Proc. Natl. Acad. Sci. USA* **81**, 6554 (1984).
25. A. Messerschmidt, A. Rossi, R. Ladenstein, R. Huber, M. Bolognesi, G. Gatti, A. Marchesini, R. Petruzzelli and A. Finazzi-Agró, *J. Mol. Biol.* **206**, 513 (1989).
26. V. Saitzev, I. Saitzeva, G. Card, A. Ralph, B. Bax and P. Lindley, *J. Inorg. Biochem.* **59**, 719 (1995).
27. T. Tsukihara, H. Aoyama, E. Yamashita, T. Tomizaki, H Yamaguchi, K. Shinzawa-Itoh, R. Nakashima, R. Yaono and S. Yoshikawa, *Science* **269**, 1669 (1995).
28. S. Iwata, C. Ostermeier, B. Ludwig and H. Michel, *Nature* **376**, 660 (1995).
29. J. W. Godden, S. Turley, D. C. Teller, E. T. Adman, M. Y. Liu, W. J. Payne and J. LeGall, *Science* **253**, 438 (1991).
30. B. Reinhammar, in *The Coordination Chemistry of Metalloenzymes*, eds. I. Bertini *et al.* (D. Reidel Publ. Comp., 1983), p. 177.
31. G. T. Babcock and M. Wikström, *Nature* **356**, 301 (1992).
32. A. Messerschmidt, R. Ladenstein, R. Huber, M. Bolognesi, L. Avigliano, A. Marchesini, R. Petruzzelli, A. Rossi and A. Finazzi-Agró, *J. Mol. Biol.* **224**,

179 (1992).
33. A. Messerschmidt, W. Steigemann, R. Huber, G. Lang and P. M. H. Kroneck, *Eur. J. Biochem.* **209**, 597 (1992).
34. A. Messerschmidt, H. Luecke and R. Huber, *J. Mol. Biol.* **230**, 997 (1993).
35. K. A. Magnus, H. Ton-That and J. E. Carpenter, in *Bioinorganic Chemistry of Copper*, eds. K. D. Karlin and Z. Tyeklar (Chapman & Hall, New York, London, 1993), p. 143.
36. A. Messerschmidt and R. Huber, *Eur. J. Biochem.* **187**, 341 (1990).
37. P. M. H. Kroneck, W. A. Antholine, J. Rister and W. G. Zumft, *FEBS Lett.* **247**, 70 (1988).
38. M. Anbar, Solvated Electron, *Adv. Chem. Ser.* **50**, 55 (1965).
39. M. H. Klapper and M. Faraggi, *Quart. Rev. Biophys.* **12**, 465 (1979).
40. L.-E. Andréasson and B. Reinhammar, *Biochim. Biophys. Acta* **445**, 579 (1976).
41. L.-E..Andréasson, R. Brändén and B. Reinhammar, *Biochim. Biophys. Acta* **483**, 370 (1976).
42. R. Aasa, R. Brändén, J. Deinum, B. G. Malmström, B. Reinhammar and T. Vänngaard, *FEBS Lett.* **61**, 115 (1976).
43. R. Brändén and J. Deinum, *Biochim. Biophys. Acta* **524**, 297, (1978).
44. R. Aasa, R. Brändén, J. Deinum, B. G. Malmström, B. Reinhammar and T. Vänngaard, *Biochem. Biophys. Res. Comm.* **70**, 1204 (1976).
45. B. Reinhammar, *Chemica Scripta* **25**, 172 (1985).
46. M. Faraggi and I. Pecht, *Nature New Biology* **233**, 116 (1971).
47. M. Faraggi and I. Pecht, *Biol. Chem.* **248**, 3146 (1973).
48. A. Guissani, Y. Henry and L. Gilles, *Biophys. Chem.* **15**, 117 (1982).
49. P. O'Neill, E. M. Fielden, L. Morpurgo and E. Agostinelli, *Biochem. J.* **222**, 71 (1984).
50. L. -E. Andréasson and B. Reinhammar, *Biochim. Biophys. Acta* **568**, 145 (1979).
51. F. B. Hansen, R. W. Noble and M. J. Ettinger, *Biochemistry*, **23**, 2049 (1984).
52. F. B. Hansen, G. B. Koudelka, R. W. Noble and M. J. Ettinger, *Biochemistry* **23**, 2057 (1984).
53. G. B. Koudelka, F. B. Hansen and M. J. Ettinger, *J. Biol. Chem.* **260**, 15561 (1985).
54. G. B. Koudelka and M. J. Ettinger, *J. Biol. Chem.* **263**, 3698 (1988).
55. M. Goldberg and I. Pecht, *Proc. Natl. Acad. Sci. USA* **71**, 4684 (1974).
56. O. Farver, M. Goldberg, D. Lancet and I. Pecht, *Biochem. Biophys. Res. Comm.* **73**, 494 (1976).
57. O. Farver, M. Goldberg and I. Pecht, *FEBS Lett.* **94**, 383 (1978).
58. O. Farver, M. Goldberg and I. Pecht, *Eur. J. Biochem.* **104**, 71 (1980).
59. M. Goldberg, O. Farver and I. Pecht, *J. Biol. Chem.* **255**, 7353 (1980).
60. O. Farver, P. Frank and I. Pecht, *Biochem. Biophys. Res. Comm.* **108**, 273 (1982).
61. O. Farver and I. Pecht, *FEBS Lett.* **108**, 436 (1979).

62. F. Malatesta, P. Sarti, G. Antonini, B. Vallone and M. Brunori, *Proc. Natl. Acad. Sci. USA* **87**, 7410 (1990).
63. C. D. LuBien, M. E. Winkler, T. J. Thamann, R. A. Scott, M. S. Co, K. O. Hodgson and E. I. Solomon, *J. Am. Chem. Soc.* **103**, 7014 (1981).
64. J. L. Cole, G. O. Tan, E. K. Yang, K. O. Hodgson and E. I. Solomon, *J. Am. Chem. Soc.* **112**, 2243 (1990).
65. J. L. Cole, D. P. Ballou and E. I. Solomon, *J. Am. Chem. Soc.* **113**, 8544 (1991).
66. P. A. Clark and E. I. Solomon, *J. Am. Chem. Soc.* **114**, 1108 (1992).
67. A. Marchesini and P. M. H. Kroneck, *Eur. J. Biochem.* **101**, 65 (1979).
68. P. O'Neill, E. M. Fielden, A. Finazzi-Agró and L. Avigliano, *Biochem. J.* **209**, 167 (1983).
69. P. O'Neill, E. M. Fielden, L. Avigliano, G. Marcozzi, A. Ballini and A. Finazzi-Agró, *Biochem. J.* **222**, 65 (1984).
70. T. E. Meyer, A. Marchesini, M. A. Cusanovich and G. Tollin, *Biochemistry* **30**, 4619 (1991).
71. O. Farver and I. Pecht, *Proc. Natl. Acad. Sci. USA* **89**, 8283 (1992).
72. P. Kyritsis, A. Messerschmidt, R. Huber, G. A. Salmon and A. G. Sykes, *J. Chem. Soc. Dalton Trans.* 731 (1993).
73. G. Tollin, T. E. Meyer, M. A. Cusanovich, P. Curir and A. Marchesini, *Biochim. Biophys. Acta* **1183**, 309 (1993).
74. J. T. Hazzard, A. Marchesini, P. Curir and G. Tollin, *Biochim. Biophys. Acta* **1208**, 166 (1994).
75. O. Farver and I. Pecht, *FASEB J.* **5**, 2554 (1991).
76. P. M. H. Kroneck, F. A. Armstrong, H. Merkle and A. Marchesini, *Adv. Chem. Ser.* **220**, 223 (1982).
77. O. Farver, S. Wherland and I. Pecht, *J. Biol. Chem.* **269**, 22933 (1994).
78. T. Manabe, H. Manabe, K. Hiromi and H. Hatano, *FEBS Lett.* **23**, 268 (1972).
79. M. Brunori, G. Antonini, F. Malatesta, P. Sarti and M. T. Wilson, *FEBS Lett.* **314**, 191 (1992).
80. J. M. Guss and H. C. Freeman, *J. Am. Chem. Soc.* **169**, 521 (1983).
81. M. D. Lowery, J. A. Guckert, M. S. Gebhard and E. I. Solomon, *J. Am. Chem. Soc.* **115**, 3012 (1993).
82. J. A. Guckert, M. D. Lowery and E. I. Solomon, *J. Am. Chem. Soc.* **117**, 2817 (1995).
83. K. Kobayashi, H. Une and K. Hayashi, *J. Biol. Chem.* **264**, 7976 (1989).
84. S. Suzuki, T. Kohzuma, K. Yamaguchi, N. Nakamura, S. Shidara, K. Kobayashi and S. Tagawa, *J. Am. Chem. Soc.* **116**, 11145 (1994).
85. O. Farver, M. Goldberg, S. Wherland and I. Pecht, *Proc. Natl. Acad. Sci. USA* **75**, 5245 (1978).

CHAPTER 13

REDOX PROPERTIES OF BLUE MULTI-COPPER OXIDASES

Peter M. H. Kroneck

Fakultät Biologie, Universität Konstanz

D-78434 Konstanz, Germany

E-mail: Peter.Kroneck@uni-konstanz.de

ABSTRACT

In this chapter redox properties of the multi-copper enzymes ascorbate oxidase, ceruloplasmin and laccase have been reviewed. The reactions of the different Cu centers with biological and artificial redox substrates are discussed. Furthermore, electrochemical properties and redox potentials of the blue oxidases including aspects of the electron distribution between the individual Cu sites are described.

Contents

1. Multi-Copper Enzymes: Powerful and Versatile Catalysts

Copper, similar to iron, is an essential constituent of every form of life from bacteria to man, because of its role as a prosthetic group in a number of vital proteins and enzymes.[1] As a transition metal ion, Cu has the capacity to complex effectively a variety of protein side chains and can vary its oxidation state from Cu(I) to Cu(II), and Cu(III). In aqueous solution and in ionic compounds, despite the favorable $3d^{10}$ configuration of univalent Cu, the divalent state appears more stable as a result of the higher positive charge and the larger heat of hydration (Eq. (1)).[2]

$$Cu^{2+} + e^- \rightarrow Cu^{1+}, \quad E_0\ (Cu^{2+}/Cu^+)\ 0.1682\ V. \tag{1}$$

Thus, hydrated Cu^+ tends to disproportionate in "pure" water to give Cu^{2+} and metallic Cu. Apparently, the protein provides effective stabilization of Cu^+ and does not allow disproportionation. Yet even when strongly coordinated by a protein matrix, Cu retains it capability to bind both an organic electron donor, such as L-ascorbate, and small electron acceptor molecules like dioxygen as shown for ascorbate oxidase.[3,4]

Copper proteins are involved in the activation and transformation of kinetically inert molecules ranging from nitrous oxide and dioxygen to ammonia and methane (Eqs. (2)–(5))[5–9]:

$$N_2O + 2H^+ + 2e^- \rightarrow N_2 + H_2O \tag{2}$$

$$O_2 + 4H^+ + 4e^- \rightarrow H_2O + H_2O \tag{3}$$

$$NH_3 + O_2 + 2H^+ + 2e^- \rightarrow NH_2OH + H_2O \tag{4}$$

$$CH_4 + O_2 + 2H^+ + 2e^- \rightarrow CH_3OH + H_2O\,. \tag{5}$$

No eukaryotes and only a few prokaryotes are able to survive without Cu and Fe enzymes. These metalloenzymes are involved in the adapation of cells to aerobic life as protective proteins (e.g. superoxide dismutase, ceruloplasmin), fulfill the function of terminal oxidases (e.g. ceruloplasmin, laccase, ascorbate oxidase, cytochrome oxidase) and respiratory proteins (e.g. nitrite reductase, nitrous oxide reductase).[1,5]

In this chapter, I will focus on the redox properties of the multi-copper enzymes ascorbate oxidase, ceruloplasmin and laccase. The reaction of the Cu centers with biological and artificial redox substrates will be discussed including aspects of intramolecular electron transfer between the individual Cu sites. In view of the complexity of the topic and the enormous number of publications accumulating in the field, I recommend the following references for introduction and further reading.[4,7,8,10–18]

2. Redox Reactions of Blue Multi-Copper Oxidases

Within the class of copper enzymes which contain the blue type-1 Cu center, ascorbate oxidase, ceruloplasmin and laccase are the most prominent and best investigated representatives.[4] For ascorbate oxidase and ceruloplasmin high-resolution X-ray structures have become available[19–22] documenting the presence of the "classical" blue type-1 Cu site, $Cu2N_{His}S_{Cys}S_{Met}$, and a novel trinuclear Cu center coordinated solely by histidine nitrogens. This novel Cu site arises from an interaction of the EPR-detectable type-2 Cu and the antiferromagnetically coupled type-3 Cu pair.[19–24]

Laccase and ascorbate oxidase are widely distributed in plants whereas ceruloplasmin is found in blood plasma from various mammalian species. Laccase is also produced by many fungi and is particularly abundant in fungi which are involved in lignin degradation. Recently, the first bacterial laccase from *Azospirillum liperum* was described which is required for melanin production.[25,26] All these enzymes exhibit a rather low specificity and can oxidize a number of quite different substrates including one-electron reductants (A^-, e.g. hexacyanoferrate(II)) and two-electron reductants (AH_2, e.g. L-ascorbate) as shown in Eqs. (6) and (7):

$$2AH_2 + O_2 \rightarrow 2A + 2H_2O \quad (2 \times 2e^- \ \text{transfer}) \tag{6}$$

$$4A^- + O_2 + 4H^+ \rightarrow 4A + 2H_2O \quad (4 \times 1e^- \ \text{transfer})\,. \tag{7}$$

There are other multi-copper oxidases which will be mentioned briefly. Sulochrin oxidase is a blue Cu-containing glycoenzyme from *Penicillum frequentans* and *Oospora sulphurea-ochracea* that catalyzes a stereospecific formation of bisdechlorogeodin from sulochrin. The enzyme has probably a dimeric structure containing 6 Cu atoms.[27] Dihydrogeodin oxidase from the fungus *Aspergillus terreus* catalyzes the regio- and stereo-specific phenol oxidative coupling reaction of dihydrogeodin to (+)-geodin.[28] It is reported to contain 16 Cu/M_R 153,000 and shows the characteristic optical properties of type-1 Cu. Bilirubin oxidase catalyzes the oxidation of bilirubin to biliverdine and appears to have spectral properties very similar to those of fungal laccase which does not have bilirubin oxidase activity.[29,30] Phenoxazinone synthase has been isolated and purified from the bacterium *Streptomyces antibioticus* and appears to belong into the blue multi-copper oxidase family. The enzyme requires 4–5 Cu atoms/monomer for full activity, EPR spectroscopy demonstrates both the presence of type-1 and type-2 Cu sites.[31] The enzyme catalyzes the oxidative coupling of two molecules of a substituted *o*-aminophenol to the phenoxazinone chromophore in the final step in the biosynthesis of the antineoplastic agent actinomycin D.[32]

Interestingly, Cu-resistant strains of *Xanthomonas campestris* carry histidine-rich polypeptide regions. These regions on the copper resistance genes strongly resemble the copper-binding motifs of small blue copper proteins and the multi-copper oxidases fungal laccase, plant ascorbate oxidase, and human ceruloplasmin.[33] Nitrite reductase, which transforms nitrite to nitric oxide in denitrifying bacteria, contains both the type-1 and the type-2 Cu center but lacks the binuclear type-3 site.[34] Yet, its arrangement of Cu atoms within the protein indicates a strong structural relationship with the "classical" blue multi-copper oxidases.[4] Along these lines, the multi-copper enzymes nitrous oxide reductase and cytochrome *c* oxidase carry both the binuclear mixed-valence Cu_A center[5,35–39] which can be thought of as a dimer of type-1 Cu centers (F. Neese, W. G. Zumft, W. E. Antholine and P. M. H. Kroneck, unpublished data). Furthermore, the catalytic center of nitrous oxide reductase, Cu_Z may consist of a type-3 Cu pair similar to the one found in the multi-copper oxidases.[5,40,41] In contrast to the regular type-3 Cu pair observed in the blue multi-copper oxidases, this site shows unusual redox properties which may be explained by the formation of a second mixed-valence Cu site in nitrous oxide reductase (W. E. Antholine, F. Neese, W. G. Zumft and P. M. H. Kroneck, unpublished data).

2.1. *Ascorbate Oxidase*

The copper-dependent enzyme ascorbate oxidase (L-ascorbate:oxygen oxidoreductase, EC 1.10.3.3) is generally isolated from squash or cucumber. The Refs. 3, 4, 42 and 43 are recommended for review of the literature. Considerable caution must be used when evaluating ascorbate oxidase activity of biological extracts. Many metal-containing proteins, or free Cu^{2+} ions in solution, can undergo oxidoreduction in the presence of ascorbate, generally producing hydrogen peroxide instead of water. Some earlier reports about bacterial ascorbate oxidases could not be confirmed.[3]

As its name implies, ascorbate oxidase displays its greatest specificity towards L-ascorbic acid (vitamin C), but the D-isomer and reductic acid, a substrate analog, are also oxidized at high rates.[3,44] Mechanistic investigations with the substrates reductate, D-araboascorbate and L-ascorbate gave an oxygen reactivity of 1:20:300 which was attributed to the "enzymic memory" linked to a substrate-induced conformation of the enzyme.[45]

Highest ascorbate oxidase activity is usually associated with those parts of plants that exhibit the fastest growth. The close association with the cell-wall material[3,46] lends some support to the theory that the enzyme might be important for plant growth and ripening of the fruit.[47] The oxidation of various natural compounds present in cell walls has been studied.[48] Catechols, flavanoids and hydroxycinnamic acids have been sucessfully employed. Ascorbate oxidase exhibited a secondary catechol oxidase activity, which was tested against natural and synthetic compounds having a catechol-like structure. These findings suggest that the enzyme is involved in the process of darkening at least in the fruits and vegetables which contain ascorbate oxidase and *o*-diphenolic compounds.[48] Activity was also reported towards hydroquinones, aminophenols, and polyhydric phenols. Hydroquinone and 4-aminophenol did not react, but several chloroderivatives were oxidized at low rates.[49] More recently, the oxidation of 2-aminoascorbic acid to red 2,2'-nitrilodi-2(2')-deoxy-L-ascorbic acid has been used as a sensitive assay for ascorbate oxidase.[50] The activity of the enzyme was also used for its electrophoretic detection by photoreduction of the dye nitroblue tetrazolium.[51]

In the absence of dioxygen, both the type-1 and the trinuclear center of ascorbate oxidase seem to react specifically with inorganic reagents, such as hexacyanoferrate(II), tetrathiomolybdate(VI) and nitric oxide. Cucumber ascorbate oxidase reacted with an excess of hexacyanoferrate(II), leading to a partially reduced enzyme.[52–54] The reaction rate of the Fe(II) complex with ascorbate oxidase plus dioxygen corresponded to approximately 20 % of the rate measured for L-ascorbate. Removal of the "type-2 Cu" (T2D-ascorbate oxidase) lowered the hexacyanoferrate(II) activity to 3 % of that of native enzyme. Parallel to a new band around 400–500 nm in the electronic spectrum, an unusual EPR signal was observed lacking the Cu hyperfine structure in the g_{II}-region (g_{II} 2.28, g_{I} 2.08) which was not observed for T2D-ascorbate oxidase. From its intensity and microwave power behavior the EPR signal was assigned to an adduct of the type-3 Cu pair and hexacyanoferrate(II).[54]

Potassium tetrathiomolybdate(VI) reduced both native and T2D ascorbate oxidase. Whereas no change of the absorbance at 610 nm was observed, the EPR spectroscopy clearly revealed that close to 100 % of the type-2 Cu had been reduced. Upon further addition of thiomolybdate, the type-3 center was reduced (determined optically at 330 nm) prior to the type-1 site. In the case of the T2D enzyme, electrons were immediately transferred to type-1 Cu.[55]

Concerning the reactivity of ascorbate oxidase towards hexacyanoferrate(II) and tetrathiomolybdate(VI) note that X-ray structure analysis of T2D-ascorbate oxidase had revealed the removal of about 1.3 Cu atoms/ ascorbate oxidase monomer. The Cu was lost from the trinuclear site whereby the EPR-active type-2 Cu was depleted most in the crystals.[56] This suggests a preferential formation of a 1-Cu-depleted form, with the hole equally distributed over the trinuclear site. Most likely, the "so-called" T2D enzyme represents a mixture of species with an antiferromagnetically coupled Cu pair.

Upon incubation of oxidized ascorbate oxidase with nitric oxide in the absence of dioxygen, the characteristic EPR signal due to type-1 Cu disappeared concomitantly with a partial bleaching of the blue chromophore.[57] The specific spectral changes induced by NO were reversible, and were explained by an electron transfer from NO to the type-1 Cu.

2.2. *Ceruloplasmin*

Ceruloplasmin (ferro-O_2-oxidoreductase, EC 1.16.3.1), the copper protein isolated from plasma, still represents a puzzle concerning its physiological role. At least four functions have been attributed to ceruloplasmin: (i) copper transport, (ii) oxidase activity acting on Fe(II), (iii) oxidase activity acting on a number of organic substrates, and (iv) protective antioxidant activity. References[1,7,58] are recommended for introduction to the field.

Among the organic substrates of ceruloplasmin *p*-phenylenediamines seem to be the best,[59] but catechol derivatives, aminophenols, indoles and other compounds related to the biogenic phenylalkylamines also served as substrates.[7,60–62] The first step in the oxidation of N,N,N',N'-tetramethyl-*p*-phenylenediamine (TPD) is a one-electron transfer to produce the radical cation $TPD^{+\cdot}$, as reported for the reaction of reductate, or L-ascorbate, with ascorbate oxidase.[3] The reaction between ceruloplasmin and substrate may well resemble an outer sphere redox reaction, i.e. any compound with the suitable energy of ionization and redox potential might thus function as a substrate. The catalytic power of the multi-copper enzyme would then be mainly a result of the high redox potential of the accepting type-1 Cu center.[58]

The great efficiency of ceruloplasmin to catalyze the oxidation of Fe(II) to Fe(III) coupled to the reduction of dioxygen to water led to the suggestion that ceruloplasmin is a ferroxidase[1,63,64] (Eq. (8)):

$$\mathrm{CP} - [\mathrm{Cu(II)}]_4 + 4\mathrm{Fe(II)} \rightarrow \mathrm{CP} - [\mathrm{Cu(I)}]_4 + 4\mathrm{Fe(III)} \quad (8)$$

$$\mathrm{CP} - [\mathrm{Cu(I)}]_4 + \mathrm{O}_2 + 4\mathrm{H}^+ \rightarrow \mathrm{CP} - [\mathrm{Cu(II)}]_4 + 2\mathrm{H_2O}\,. \quad (9)$$

Ceruloplasmin was found to inhibit xanthine oxidase and ferritin-dependent peroxidation of phospholipid liposomes, and was also shown to inhibit superoxide-mediated mobilization of iron from ferritin.[65] It was suggested that this inhibition functions largely via the ability of ceruloplasmin to reincorporate reductively mobilized iron back into ferritin.

Sodium tetrathiomolybdate (VI) inhibited the diamine oxidase activity of ceruloplasmin in sheep.[66,67] EPR studies on the interaction between thiomolybdate and ceruloplasmin confirmed that Cu(II) was reduced, but no specific binding of the molybdenum compound to type-1 or type-2 Cu was observed. Tetrathiomolybdate did not remove Cu from ceruloplasmin but seemed to bind to the reduced Cu centers through its sulfide moiety.[68]

Nitric oxide was used as a spin probe for the Cu sites in ceruloplasmin.[57,69–74] Obviously, NO formed a reversible charge-transfer with oxidized type-1 Cu, and no redox reactions were involved. The type-2 Cu center seemed unaffected, at least in short times of incubation. The interaction of NO with EPR-silent reduced human ceruloplasmin[70] led to EPR signals indicative for dipolar coupled binuclear Cu(II) sites. The reaction of NO with a ceruloplasmin lacking the EPR-detectable type-2 Cu site in the native state was also investigated.[74] The interaction with NO can lead to complex reaction pathways as described for laccase,[75] cytochrome oxidase,[76] or ascorbate oxidase.[57,73] Nitric oxide complexed to the type-1 Cu center could be photodissociated at 77 K in the case of azurin, ceruloplasmin and ascorbate oxidase.[73]

2.3. *Laccase*

Laccase (benzenediol:oxygen oxidoreductase, EC 1.10.3.2) has a rather low specificity towards reducing substrates. Good substrates are mono-, di- and polyphenols.[77–81] Fungal laccase oxidized steroid hormones with a phenolic group in a two-phase system,[82] and was also active in the oxidative metabolism of natural antitumor agents with chlorpromazine as redox mediator.[83] More recently, laccase was shown to oxidize dopamine to decarboxy dopachrome, an intermediate of melanin synthesis in the pathogenic fungus *Cryptococcus neoformans*.[84] The extracellular laccase of the white-rot fungus *Pleurotus ostreatus* exhibited phenol oxidase activity towards a number of lignin-related compounds, such as 3,5-dimethoxy-5-hydroxyacetophenone which was oxidized to the corresponding phenoxy radical.[85] Laccase was the first enzyme used to polymerize a synthetic lignin *in vitro*, and was proposed to perform this function also *in vivo*, which has been questioned lately.[86]

Like ascorbate oxidase and ceruloplasmin, both fungal and tree laccase reacted with NO in the absence of dioxygen.[75] A half met-nitrite derivative of T2D laccase from *Rhus vernicifera* could be prepared.[87]

3. Electrochemistry, Oxidation/Reduction Potentials of the Cu Centers, and Electron Transfer in Blue Multi-Copper Oxidases

Electron transfer reactions of the Cu centers in blue oxidases have been studied by direct electrochemical methods (polarography, cyclic voltammetry), potentiometric titrations (spectroelectrochemistry), stopped-flow kinetics with organic and inorganic reductants (hydroquinones, L-ascorbate and structural analogs, Cr^{2+} ion, Fe(II) complexes), and pulse radiolysis.[88–92] For a review on pulse radiolysis studies of electron transfer in blue copper proteins and blue oxidases, Refs. 93 and 94 are recommended.

The low temperature (77 K) irradiation of oxidized ceruloplasmin and *Rhus* laccase at the 330 nm absorption which arises from type-3 Cu, led to the reduction of the type-1 Cu sites. Type-2 Cu remained unaffected under these conditions.[95] Rate constants have been determined using laser flash photolysis for the electron-transfer reactions between reduced flavins and flavodoxin semiquinone and blue copper proteins including nitrite reductase, *Rhus* laccase and ascorbate oxidase.[96,97]

3.1. *Direct Electrochemistry*

Compared to small blue copper proteins, such as plastocyanin, azurin, or rusticyanin,[98–104] attempts to study the direct electrochemistry of multi-copper proteins have not been very successful so far. Laccase from *Polyporus versicolor* which had been adsorbed irreversibly on the surface of a pyrolytic graphite electrode, catalyzed the electro-reduction of dioxygen. In the absence of dioxygen, an electrochemical response from the laccase-coated electrode was achieved by adding 2,9-dimethylphenanthroline to the supporting electrolyte.[105] Polarographic responses were observed for the metal centers in Cu(II)-substituted horse liver alcohol dehydrogenase, ceruloplasmin and hemocyanin, by differential pulse polarography in the presence of EDTA.[106] Nitrite reductase from *Achromobacter cycloclastes* gave a direct electrochemical response at a di-4-pyridyl disulfide modified gold electrode in the presence of apopseudoazurin.[107] One well-defined cyclic voltammetric response was observed for the type-1 Cu center with a mid-point potential, $E_{1/2} = 240$ mV (vs. NHE), which corresponded pretty well to the value obtained for nitrite reductase from *Alcaligenes* by potentiometric titration in the presence of the $Fe(CN)_6^{3-}/Fe(CN)_6^{4-}$ couple.[108]

3.2. *Oxidation/Reduction Potentials and Aspects of Electron Transfer*

Oxidation/Reduction Potentials for the different Cu centers in the blue oxidases have been obtained through potentiometric titrations in the presence of charged and neutral mediator redox couples, such as $Fe(CN)_6^{3-}/Fe(CN)_6^{4-}$, $Mo(CN)_8^{3-}/Mo(CN)_8^{4-}$, and ferrocenium compounds (Refs. 4, 7, 42, 58, 81, 91 and 92). Most recently, potentials of a series of recombinant fungal laccases and bilirubin oxidase were determined using the couples $Fe(dipyridyl)_2Cl_2/Fe(dipyridyl)_2Cl_3$ or NaI_3/NaI.[109]

The redox potentials, $E°$, of the type-1 Cu in tree laccase and ascorbate oxidase from two sources are in the range 330–400 mV at around pH 7 and 25°C.[3,4,7,42;58,81] Similar values have been determined for small blue Cu proteins carrying only the type-1 center.[7,110] Spectroelectrochemical titrations of the type-1 Cu in native and T2D *Rhus* laccase (pH 7.0, 25°C) at varying concentrations of (hydroxyethyl)ferrocene gave a mediator-independent value of $E° = 434 \pm 5$ mV[92] which corresponds closely to the value determined earlier at high concentrations of $Fe(CN)_6^{4-}$.[111] One of the type-1 centers of human ceruloplasmin ($E°$ 580 mV) and the type-1 Cu in *Polyporus* laccase ($E°$ 785 mV) both exhibit significantly higher potentials, which has been attributed to changes in the Cu coordination.[4] Earlier, based on the temperature dependence (7–41°C) of the reduction potential determined by thin-layer spectroelectrochemistry, it was suggested that the high $E°$ value of the blue copper center in *Polyporus* laccase is attributable mainly to stabilization of the Cu(I) center by enhanced ligand binding interactions, and that protein solvation effects play a lesser role.[112] Note that rusticyanin from *Thiobacillus ferrooxidans*, which contains a single type-1 Cu, has also a very high $E° = 680$ mV (pH 3.2).[13] In this case it is suggested, that the Cu is less accessible to the solvent. Furthermore, a shorter Cu-S_{Met} bond has been assumed.[113] For the type-1 Cu protein pseudoazurin from *Alcaligenes faecalis*, exchange of alanine for the proline residue adjacent to a coordinated histidine (His 81) caused a marked increase in reduction potential by 139 mV. In this case, a water molecule occupied the pocket created by the exchange Pro80Ala.[114]

Most recently, four new fungal laccases have been purified, cloned and expressed in an *Aspergillus oryzae* system.[109] The laccases have been studied with regard to their spectroscopic properties, redox potential, specificity, and stability. Two of the four laccases showed potentials around 0.7–0.8 V, two had potentials near 0.5 V. It was observed that the higher redox potential correlated with higher activity which led to the speculation that structural differences in the substrate-activation site, i.e. the type-1 Cu center, controlled the redox potential range as well as the substrate specificity.[109] For further discussion of the connection between structure, spectroscopic properties and electron transfer properties of the type-1 Cu center, the Refs. 17 and 115 are recommended.

The redox potential for the type-3 Cu center (determined at 330 nm) have been reported to be equal or slightly higher than the potentials of the corresponding type-1 Cu sites.[3,4,81] For *Rhus* laccase, $E° = 394$ mV (determined by EPR) was reported for the type-2 Cu.[111] Potentiometric titrations of ascorbate oxidase (mediator $Fe(CN)_6^{4-}$, pH 7.0, 25°C) gave a midpoint potential $E°_{610\ nm}$ of 344 mV, with a Nernst factor $n_{610\ nm} = 1.1$,[3] vs 327 ± 3 mV, $n_{610\ nm} = 0.97 \pm 0.03$ at 10°C (W. Jakob and P. M. H. Kroneck, unpublished data). A double Nernst plot of the results of a simultaneous spectrophotometric titration at 330 and 610 nm gave an $E°_{330\ nm}$ that was nearly identical to $E°_{610\ nm}$. At 10°C, however, a value of 30 ± 10 mV was measured for the difference $E°_{330\ nm} - E°_{610\ nm}$. Thus, at 25°C, where this difference is considerably diminished in ascorbate oxidase, thermodynamic control for the occupancy of the individual Cu sites is less pronounced.[3] The Nernst coefficient ratio $n_{330\ nm}/n_{610\ nm}$ did not approach the expected value of 2 as expected for cooperative two-electron acceptor.[3,116] Similar double Nernst plots for spectrophotometric titrations of tree laccase produced slopes of 1.0–2.0, depending on the titrant used. Compounds such as hydroquinone or ferrocyanide gave a ratio of 2.0, whereas $Ru(NH_3)_6^{2+}$ gave a ratio of 1.0[117] indicative for uncoupling of the type-3 Cu pair in the case of strong reductants. Note that in both ascorbate oxidase and ceruloplasmin, and most likely in laccase, the type-3 Cu pair interacts with the so-called EPR detectable type-2 Cu center.[4,22] Thus, any discussion of the redox potentials and electron distribution among the different Cu sites in blue oxidases must take into account the properties of this coupled trinuclear Cu center which is the site for the four-electron reduction of dioxygen.

4. Conclusions

Multi-copper proteins are involved in many vital processes and represent powerful redox catalysts in the conversion of small inorganic molecules, e.g. nitrite, nitrous oxide, dioxygen, methane and ammonia. The blue oxidases ascorbate oxidase, ceruloplasmin and laccase catalyze the four-electron reduction of dioxygen without the release of toxic oxygen intermediates, such as H_2O_2 or hydroxyl radicals. They accommodate a wide range of one-electron and two-electron reductants, and can carry out stereospecific and regiospecific reactions *in vivo* and *in vitro*.

Type-1 Cu centers, or 'modified' type-1 Cu centers, i.e. the dinuclear mixed-valence Cu_A site, have evolved as potent electron acceptors. Subsequently, electrons are transferred to dinuclear Cu (Fe, Cu in the terminal oxidases) and trinuclear Cu centers which bind the electron-accepting substrate. The efficient catalytic function of the blue oxidases has been attributed to the synergistic use of the blue type-1 Cu site and a trinuclear Cu cluster.[4,118,119] In several laccases the type-1 Cu appeared to influence the specificity and activity of the enzyme.[109] Intramolecular electron transfer between type-1 Cu and the trinuclear cluster provides the electrons needed for the transformation of dioxygen to water.[4,21]

In the multi-copper enzymes nitrous oxide reductase and cytochrome *c* oxidase Cu_A, a mixed-valence, $S=1/2$ [Cu(1.5+)...Cu(1.5+)] site, acts as the electron acceptor, with a redox potential $E°$ around 260–290 mV (Refs. 38, 39, 120–122). CD spectra of the Cu_A domain in cytochrome *c* oxidase[123] documented the presence of the cupredoxin-fold, a Greek key

β-barrel and common structural motif in small type-1 Cu proteins of plants and bacteria.[13] In cytochrome *c* oxidase, reduction of dioxygen is carried out at the dinuclear heme a_3 – Cu_B site.[38,39,124] In nitrous oxide reductase, N_2O is split into dinitrogen and water at the Cu_Z site, a presumably dinuclear type-3 Cu center of so far unknown structure with unusual chemical and spectroscopic properties.[5] In the case of the particulate methane monooxygenase from *Methylococcus capsulatus* (Bath) a trinuclear Cu cluster has been suggested to constitute the active site.[9]

Acknowledgments

The author thanks E. I. Solomon for providing results before publishing it. The collaboration with O. Farver, A. Messerschmidt, I. Pecht and W. G. Zumft is gratefully acknowledged.

References

1. E. Frieden, in *Metal Ions In Biological Systems*, ed. H. Sigel (Marcel Dekker, Inc., New York and Basel, 1981), Vol. 13, p. 1.
2. N. N. Greenwood and A. Earnshaw, *Chemie der Elemente* (Verlag Chemie, Weinheim, 1988), p. 1530.
3. P. M. H. Kroneck, F. A. Armstrong, H. Merkle and A. Marchesini, in *Ascorbic Acid: Chemistry, Metabolism and Uses*, eds. P. A. Seib and B. M. Tolbert (American Chemical Society, Washington, 1982).
4. A. Messerschmidt, *Adv. Inorg. Chem.* **40**, 121 (1993).
5. W. G. Zumft and P. M. H. Kroneck, *Adv. Inorg. Biochem.* **11**, 193 (1995).
6. P. M. H. Kroneck, J. Beuerle and W. Schumacher, in *Metal Ions in Biological Systems*, eds. H. Sigel and A. Sigel (Marcel Dekker, Inc., New York, 1992), Vol. 28, p. 455.
7. J. A. Fee, *Structure and Bonding* **23**, 1 (1975).
8. *Copper Proteins and Copper Enzymes*, ed. R. Lontie (CRC Press, Inc., Boca Raton, 1984), Vol. I-III.
9. H. -H. T. Nguyen, A. K. Shiemke, S. J. Jacobs, B. J. Hales, M. E. Lidstrom, and S. I. Chan, *J. Biol. Chem.* **269**, 14995 (1993).
10. A. M. Mayer and E. Harel, *Phytochemistry* **18**, 193 (1979).
11. R. Huber, *Chem. Scripta* **29**, 103 (1989).
12. A. Messerschmidt and R. Huber, *Eur. J. Biochem.* **187**, 341 (1990).
13. E. T. Adman, *Adv. Prot. Chem.* **42**, 145 (1991).
14. E. I. Solomon, M. J. Baldwin and M. D. Lowery, *Chem. Rev.* **92**, 521 (1992).
15. E. I. Solomon and M. D. Lowery, *Science* **259**, 1575 (1993).
16. *Bioinorganic Chemistry of Copper*, eds. K. D. Karlin and Z. Tyeklár (Chapman & Hall, New York and London, 1993).

17. B. G. Malmström, *Eur. J. Biochem.* **223**, 711 (1994).
18. R. Basosi, W. E. Antholine and J. S. Hyde, *Biological Magnetic Resonance* **13**, 103 (1993).
19. A. Messerschmidt, A. Rossi, R. Ladenstein, R. Huber, M. Bolognesi, G. Gatti, A. Marchesini, R. Petruzzelli and A. Finazzi-Agrò, *J. Mol. Biol.* **206**, 513 (1989).
20. A. Messerschmidt, R. Ladenstein, R. Huber, M. Bolognesi, L. Avigliano, R. Petruzzelli, A. Rossi and A. Finazzi-Agrò, *J. Mol. Biol.* **224**, 179 (1992).
21. A. Messerschmidt, H. Luecke and R. Huber, *J. Mol. Biol.* **230**, 997 (1993).
22. P. Lindley *et al.*, in *Multi-Copper Oxidases*, ed. A. Messerschmidt (World Scientific, Singapore, 1996) pp. 81–102.
23. D. Spira-Solomon, M. D. Allendorf and E. I. Solomon, *J. Am. Chem. Soc.* **108**, 5318 (1986).
24. J. L. Cole, L. Avigliano, L. Morpurgo and E. I. Solomon, *J. Am. Chem. Soc.* **113**, 9080 (1991).
25. A. Givaudon, A. Effosse, D. Faure, P. Potier, M. L. Bouillant and R. Bally, *FEMS Microbiol. Lett.* **108**, 205 (1993).
26. D. Faure, M. L. Bouillant and R. Bally, *Appl. Environm. Microbiol.* **60**, 3413 (1994).
27. H. Nordlöv and S. Gatenbeck, *Arch. Microbiol.* **131**, 208 (1982).
28. I. Fujii, H. Iijiima, Y. Ebizuka and U. Sankawa, *Chem. Pharm. Bull.* **31**, 337 (1983).
29. C. V. Rene, O. Yokosuka, and H. H. Barbara, *Biochem. J.* **236**, 625 (1986).
30. K. Hiromi, Y. Yamaguchi, Y. Sugiura, H. Iwamoto and J. Hirose, *Biosci. Biotech. Biochem.* **56**, 1349 (1992).
31. J. C. Freeman, P. G. Naygar, T. P. Begley and J. J. Villafranca, *Biochemistry* **32**, 4826 (1993).
32. C. E. Barry and T. P. Begley, *Biochemistry* **28**, 6323 (1989).
33. Y. -A. Lee, M. Hendson, N. J. Panopoulos and M. N. Schroth, *J. Bacteriol.* **176**, 173 (1994).
34. J. W. Godden, S. Turley, D. C. Teller, E. T. Adman, M. Y. Liu, W. J. Payne and J. LeGall, *Science* **253**, 438 (1991).
35. P. M. H. Kroneck, W. E. Antholine, J. Riester and W. G. Zumft, *FEBS Lett.* **242**, 70 (1988).
36. P. M. H. Kroneck, W. E. Antholine, D. H. W. Kastrau, G. Buse, G. C. M. Steffens and W. G. Zumft, *FEBS Lett.* **268**, 274 (1990).
37. W. E. Antholine, D. H. W. Kastrau, G. C. M. Steffens, W. G. Zumft and P. M. H. Kroneck, *Eur. J. Biochem.* **209**, 875 (1992).
38. S. Iwata, C. Ostermeier, B. Ludwig and H. Michel, *Nature* **376**, 660 (1995).
39. T. Tsukihara, H. Aoyama, E. Yamashita, T. Tomizaki, H. Yamaguchi, K. Shinzawa-Itoh, R. Nakashima, R. Yaono and S. Yoshikawa, *Science* **269**, 1069 (1995).
40. J. Riester, W. G. Zumft and P. M. H. Kroneck, *Eur. J. Biochem.* **178**, 751 (1989).

41. J. A. Farrar, A. J. Thomson, M. R. Cheesman, D. M. Dooley and W. G. Zumft, *FEBS Lett.* **294**, 11 (1991).
42. B. Mondovi and L. Avigliano, in *Copper Proteins and Copper Enzymes*, ed. R. Lontie (CRC Press, Inc., Boca Raton, 1984), Vol. III, p. 101.
43. A. Finazzi-Agrò, *Life Chem. Rep.* **5**, 199 (1987).
44. B. Gerwin, S. R. Burstein and J. Westley, *J. Biol. Chem.* **249**, 2005 (1974).
45. M. Katz and J. Westley, *J. Biol. Chem.* **254**, 9142 (1979).
46. M. Hallaway, P. D. Phethean and J. Taggart, *Phytochem.* **9** , 935 (1970).
47. G. R. Stark and C.R. Dawson, in *The Enzymes*, eds. P. D. Boyer, L. Hardy and K. Myrbäck (Academic Press, New York, 1963).
48. A. Marchesini, P. Capelletti, L. Canonica, B. Danieli, and S. Tollari, *Biochim. Biophys. Acta* **484**, 290 (1977).
49. J. Dayan and C. R. Dawson, *Biochem. Biophys. Res. Comm.* **73**, 451 (1976).
50. K. Wimalasena and S. Dharmasena, *Anal. Biochem.* **210**, 58 (1993).
51. M. Maccarrone, A. Rossi, G. D'Andrea, G. Amicosante and L. Avigliano, *Anal. Biochem.* **188**, 101 (1990).
52. K. Kawahara, T. Sakurai, S. Suzuki and A. Nakahara, *Inorg. Chim. Acta* **92**, L33 (1984).
53. K. Kawahara, S. Suzuki, T. Sakurai, and A. Nakahara, *Arch. Biochem. Biophys.* **241**, 179 (1985).
54. T. Sakurai, *Inorg. Chim. Acta* **195**, 255 (1992).
55. P. M. H. Kroneck and W. Jakob, *Rev. Port. Quim.* **27**, 47 (1985).
56. A. Messerschmidt, W. Steigemann, R. Huber, G. Lang and P. M. H. Kroneck, *Eur. J. Biochem.* **209**, 597 (1992).
57. F. X. R. van Leeuwen, R. Wever, B. F. van Gelder, L. Avigliano and B. Mondovì, *Biochim. Biophys. Acta* **403**, 285 (1975).
58. L. Rydèn, in *Copper Proteins and Copper Enzymes*, ed. R. Lontie (CRC Press, Inc., Boca Raton, 1984), Vol. III, p. 37.
59. C. G. Holmberg and C. -B. Laurell, *Acta Chem. Scand.* **5**, 476 (1951).
60. S. N. Young and G. Curzon, *Biochem. J.* **129**, 273 (1972).
61. B. C. Barrass, D. B. Coult, R. M. Pinder and M. Skeels, *Biochem. Biopharmacol.* **22**, 2891 (1973).
62. B. C. Barrass, D. B. Coult, P. Rich and K. J. Tutt, *Biochem. Biopharmacol.* **23**, 47 (1974).
63. S. Osaki, D. A. Johnson and E. Frieden, *J. Biol. Chem.* **241**, 2746 (1966).
64. E. Frieden, *TIBS* **1**, 273 (1976).
65. V. M. Samokyszyn, D. M. Miller, D. W. Reif and S. D. Aust, *J. Biol. Chem.* **264**, 21 (1989).
66. J. Mason, M. Lamand and C. A. Kelleher, *Br. J. Nutr.* **43**, 515 (1980).
67. B. Lannon and J. Mason, *J. Inorg. Biochem.* **26**, 107 (1986).
68. M. V. Chidambaram, G. Barnes and E. Frieden, *J. Inorg. Biochem.* **22**, 231 (1984).
69. R. Wever, F. X. R. van Leeuwen and B. F. van Gelder, *Biochim. Biophys. Acta* **302**, 326 (1973).

70. F. X. F. van Leeuwen, R. Wever and B. F. van Gelder, *Biochim. Biophys. Acta* **315**, 200 (1973).
71. F. X. R. van Leeuwen and B. F. van Gelder, *Eur. J. Biochem.* **87**, 327 (1978).
72. L. Calabrese, F. Malatesta and D. Barra, *Biochem. J.* **199**, 667 (1981).
73. A. C. F. Gorren, E. de Boer and R. Wever, *Biochim. Biophys. Acta* **916**, 38 (1987).
74. G. Musci, S. Di Marco, M. C. Bonaccorsi di Patti and L. Calabrese, *Biochemistry* **30**, 9866 (1991).
75. C. T. Martin, R. H. Morse, R. M. Kanne, H. B. Gray, B. G. Malmström and S. I. Chan, *Biochemistry* **20**, 5147 (1991).
76. R. Boelens, H. Rademaker, R. Wever and B. F. van Gelder, *Biochim. Biophys. Acta* **765**, 196 (1984).
77. R. A. Holwerda and H. B. Gray, *J. Am. Chem. Soc.* **96**, 6008 (1974).
78. L. E. Andréasson and B. Reinhammar, *Biochim. Biophys. Acta* **445**, 579 (1976).
79. L. E. Andréasson and B. Reinhammar, *Biochim. Biophys. Acta* **568**, 145 (1979).
80. J. D. Clemmer, B. L. Gilliland, R. A. Bartsch and R. A. Holwerda, *Biochim. Biophys. Acta* **568**, 307 (1979).
81. B. Reinhammar, in *Copper Proteins and Copper Enzymes*, ed. R. Lontie (CRC Press, Inc., Boca Raton, 1984) Vol. III, p. 1.
82. G. Lugaro, G. Carrea, P. Cremonesi, M. M. Casellato and E. Antonini, *Arch. Biochim. Biophys.* **159**, 1 (1973).
83. F. Eckenrode, W. Peczynska-Czoch and J.P. Prosazza, *J. Pharm. Sci.* **71**, 1246 (1982).
84. P. R. Williamson, *J. Bacteriol.* **176**, 656 (1994).
85. H. -D. Youn, K. -J. Kim, J. -S. Maeng, Y. -H. Han, I. -B. Jeong, G. Jeong, S. -O. Kang and Y. C. Hah, *Microbiology* **141**, 393 (1995).
86. J. F. Dean and K. -E. L. Eriksson, *Holzforschung* **48**, 21 (1994).
87. D. J. Spira and E. I. Solomon, *Biochem. Biophys. Res. Comm.* **112**, 729 (1983).
88. J. W. Dawson, H. B. Gray, R. A. Holwerda, and E. W. Westhead, *Proc. Natl. Acad. Sci. USA* **69**, 30 (1972).
89. S. Wherland, R. A. Holwerda, R. C. Rosenberg and H. B. Gray, *J. Am. Chem. Soc.* **97**, 5260 (1975).
90. R. C. Rosenberg, S. Wherland, R. A. Holwerda and H. B. Gray, *J. Am. Chem. Soc.* **98**, 6364 (1976).
91. R. A. Holwerda, H. K. Baek, R. M. Wynn and D. B. Knaff, *Biochim. Biophys. Acta* **791**, 112 (1984).
92. R. M. Wynn, D. B. Knaff and R. A. Holwerda, *Biochemistry* **23**, 241 (1984).
93. G. V. Buxton, *Adv. Inorg. Bioinorg. Mech.* **3**, 131 (1984).
94. O. Farver and I. Pecht, *Biophys. Chem.* **50**, 203 (1994).
95. Y. Henry and J. Peisach, *J. Biol. Chem.* **253**, 7751 (1978).

96. G. Tollin, T. E. Meyer, G. Cheddar, E. D. Getzoff and M. A. Cusanovich, *Biochemistry* **25**, 3363 (1986).
97. T. E. Meyer, A. Marchesini, M. A. Cusanovich and G. T. Tollin, *Biochemistry*, **30**, 4619 (1991).
98. G. Lappin, C. A. Lewis and W. J. Ingledew, *Inorg. Chem.* **24**, 1446 (1985).
99. F. A. Armstrong, P. A. Cox, H. A. O. Hill, B. N. Oliver and A. A. Williams, *J. Chem. Soc. Chem. Commun.*, 1236 (1985).
100. F. A. Armstrong, P. A. Cox, H. A. O. Hill, V. J. Lowe and B. N. Oliver, *J. Electroanal. Chem.* **217**, 331 (1987).
101. J. E. Frew and H. A. O. Hill, *Eur. J. Biochem.* **172**, 261 (1988).
102. Z. Salamon and G. Tollin, *Arch. Biochem. Biophys.* **294**, 382 (1992).
103. J. Haladjian, M. Bruschi, F. Nunzi and P. Bianco, *J. Electroanal. Chem.* **352**, 329 (1993).
104. J. W. Furbee Jr., C. R. Thomas and R. S. Kelly, *Anal. Chem.* **65**, 1654 (1993).
105. C. -W. Lee, H. B. Gray and F. C. Anson, *J. Electroanal. Chem.* **172**, 289 (1984).
106. M. Stunicková, J. Pirincová and J. Kovár, *Bioelectrochem. and Bioenerg.* **25**, 109 (1991).
107. T. Kohzuma, S. Shidara and S. Suzuki, *Bull. Chem. Soc. Jpn.* **67**, 138 (1994).
108. M. Masuko, H. Iwasaki, T. Sakurai, S. Suzuki and A. Nakahara, *J. Biochem.* (Tokyo) **96**, 447 (1984).
109. F. Xu, W. Shin, S. H. Brown, J. Wahleithner, U. M. Sundaram and E. I. Solomon, *Biochim. Biophys. Acta* **1292**, 303 (1996).
110. A. G. Sykes, *Adv. Inorg. Chem.* **36**, 377 (1991).
111. B. Reinhammar, *Biochim. Biophys. Acta* **275**, 245 (1972).
112. V. T. Taneguchi, B. G. Malmström, F. C. Anson and H. B. Gray, *Proc. Natl. Acad. Sci. USA* **79**, 3387 (1982).
113. F. Nunzi, F. Guerlesquin, W. Shepard, B. Guigliarelli and M. Bruschi, *Biochem. Biophys. Res. Commun.* **203**, 1655 (1994).
114. M. Nishiyama, J. Suzuki, T. Ohnuki, H. C. Chang, S. Horinouchi, S. Turley, E. T. Adman and T. Beppu, *Prot. Eng.* **5**, 177 (1992).
115. S. Larsson, A. Bro and L. Sjölin, *J. Phys. Chem.* **99**, 4860 (1995).
116. B. G. Malmström, L. -E. Andréasson and B. Reinhammar, in *The Enzymes*, eds. P. D. Boyer, H. Lardy and K. Myrbäck (Academic Press, New York, 1975).
117. O. Farver, M. Goldberg, S. Wherland and I. Pecht, *Proc. Natl. Acad. Sci. USA* **75**, 5245 (1978).
118. O. Farver and I. Pecht, *Proc. Natl. Acad. Sci. USA* **89**, 8283 (1992).
119. P. Kyritsis, A. Messerschmidt, R. Huber, G. A. Salmon and A. G. Sykes, *J. Chem. Soc. Dalton. Trans.*, 731 (1993).
120. C. L. Coyle, W. G. Zumft, P. M. H. Kroneck, H. Körner and W. Jakob, *Eur. J. Biochem.* **153**, 459 (1985).

121. H. Wang, D. F. Blair, W. R. Ellis, H. B. Gray and S. I. Chan, *Biochemistry* **25**, 167 (1986).
122. S. M. Musser and S. I. Chan, *Biophys. J.* **68**, 2543 (1995).
123. P. Wittung, B. Källebring and B. G. Malmström, *FEBS Lett.* **349**, 286 (1994).
124. O. Einarsdóttir, *Biochim. Biophys. Acta* **1229**, 129 (1995).

CHAPTER 14

MODEL COMPOUNDS FOR MULTI-COPPER OXIDASES

Phalguni Chaudhuri
Max-Planck-Institut für Strahlenchemie
P.O. Box 101365, D-45413 Mülheim an der Ruhr
Germany
E-mail: Chaudh@mpi-muelheim.mpg.de

ABSTRACT

Triangular tricopper complexes as models for the "blue oxidases" are reviewed, with particular emphasis on the magneto-structural problems. As the structure of the metalloenzyme ascorbate oxidase isolated from green zucchini is known at 1.9 Å resolution, the synthetic analogues described here are of a corroborative nature. The trinuclear Cu(II) complexes with the $Cu_3(\mu_3\text{-}X)$-core have been briefly discussed to show that they are not very good models for the trinuclear site of the "blue oxidases". On the other hand, the copper compounds without the μ_3-X core are only few in number; structural analogues of the blue oxidases have proved to be elusive. A representative survey of the Cu(I) or Cu(II) complexes with the Cu_3-core has been presented, including a short description of the phenomenon "spin frustration", operative in the Cu(II) complexes.

Contents

1. Introduction

Enzymes known as oxidases use molecular oxygen as simply a reservoir for electrons. Oxidases can be either "non-blue" two-electron-(reduction of dioxygen to hydrogen peroxide, $O_2 + 2e^- + 2H^+ \rightarrow H_2O_2$) or four-electron-(reduction to water, $O_2 + 4e^- + 4H^+ \rightarrow 2H_2O$) transferring enzymes. Laccase, ascorbate oxidase and ceruloplasmin belong to the "blue" oxidases.[1–7] These are multi-copper enzymes catalyzing the four-electron reduction of dioxygen to water with concomitant one-electron oxidation of a variety of substrates, such as polyphenols, aromatic polyamines and L-ascorbate. These enzymes contain three types of biological copper,[2–11] according to the Malmström-classification: Type 1 or blue copper exhibits an intense absorption ($\epsilon \approx 5000$ l mol^{-1}cm^{-1}) due to a charge transfer band ($RS^- \rightarrow Cu^{2+}$) near 600 nm and an EPR spectrum with $A_{\|} < 95 \times 10^{-4}$cm^{-1}; Type 2, or normal copper has an EPR spectrum typical of low molecular weight Cu(II) complexes ($A_{\|} > 140 \times 10^{-4}$cm^{-1}) and limited optical absorption; Type 3 is characterized by a strong absorption in the near UV-region ($\lambda_{max} = 330$ nm, $\epsilon \approx 2800$ l mol^{-1}cm^{-1}) and is EPR silent. Type 3 copper is a pair of copper ions which are strongly antiferromagnetically coupled so that the Type 3 is diamagnetic, even at room temperature. The super-exchange interaction is mediated by an endogenous bridging ligand, which in most cases is a hydroxo-group. In full conformity with the expectation these spectroscopic signals disappear upon reduction of Cu(II) with d^9 electronic configuration to Cu(I). The intense blue color of these oxidases owes its origin to the presence of Type 1 copper ions. Since copper normally functions as a one-electron redox unit [$Cu(II) + e^- \rightarrow Cu(I)$] and the reduction of molecular oxygen to water requires four electrons, it is not too difficult to envisage that the minimum catalytic unit within these enzymes involves four copper ions, one of each biological copper types.

2. Scope

This chapter will deal only with the coordination complexes that might be of importance in modeling structurally, both geometrically and electronically, the bioinorganic active sites of the oxidases catalyzing the reduction of molecular oxygen to water. Main emphasis will be given to ascorbate oxidase, whose X-ray structure (1.9 Å resolution) has recently been reported.[12] The results obtained in the last years have not only deepened our knowledge about the enzymes, but also raised several important questions about the electronic structures of these enzymes, as will be demonstrated in this article. This article is not intended to provide a fully comprehensive coverage of the literature for the model complexes, because, at least partly, the determination of what constitutes a model complex is both subjective and dependent on the current state of knowledge about the metalloenzyme in question. This article aims to give a substantial and representative survey. Thus we concentrate here mainly on aspects of coordination chemistry of the copper ion, with particular emphasis on the magneto-structural problems.

Some information concerning the structures and functions of the metalloenzyme active sites is required for any discussion of model complexes. A brief description of the natural sites are therefore included and more detailed accounts may be found in the references or articles elsewhere in this volume.

3. Copper Proteins

3.1. *The Oxidized Forms*

No crystal structure of laccase, the simplest member of the family "Blue Oxidases", is available. Early spectroscopic work[13] suggested that it contains one Type 1 ("blue" copper), one Type 2 ("normal" copper) and a Type 3 (antiferromagnetically coupled dinuclear copper), i.e. four copper ions in three distinct sites, a (2+1+1) arrangement. However, the crystal structure (1.9 Å resolution) of ascorbate oxidase, which is thought to be a dimer of laccase-like subunits, has recently been reported[12] and confirmed the presence of four copper ions in each subunit of ascorbate oxidase. The functional unit with four copper ions consists of mononuclear and trinuclear moieties, as shown below in Fig. 1. The isolated Type 1, "blue" copper is bound to two histidine-nitrogens, a cysteine-thiolate sulfur and a methionine-thioether sulfur, and has a coordination geometry similar to the copper site in plastocyanin. Spectroscopic[14–16] and crystallographic data[12] showed that the catalytic centre should be interpreted as a trinuclear centre, rather than an antiferromagnetically coupled dinuclear copper centre (Type 3) and an independent EPR active Type 2 copper centre due to the proximity of the latter. The trinuclear cluster has a biologically unprecedented structure with three copper ions arranged in an almost equilateral

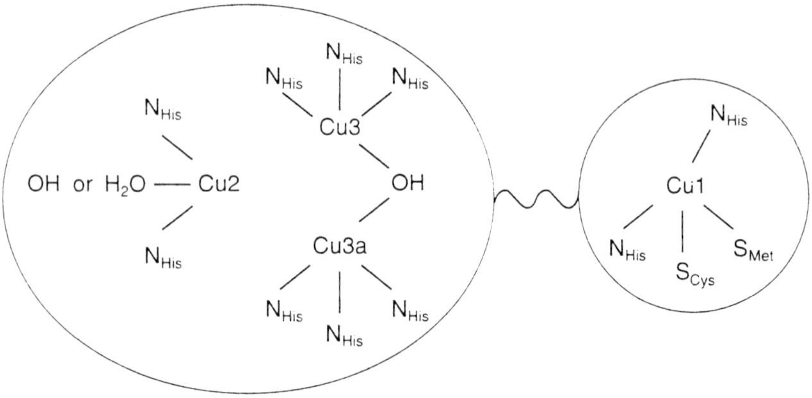

Fig. 1. A schematic view of the copper sites in ascorbate oxidase.

triangle of sides 3.74 Å ($\sigma = 0.08$ Å) and the individual distances do not deviate by more than 0.16 AA from this mean value. The shortest and the longest distances between the mononuclear Type 1 copper center and the trinuclear copper center are 12.2 and 14.9 Å, respectively. The Cu3 and Cu3a pair is bridged by an OH^-, which leads probably to a strong antiparallel exchange coupling, and make up the EPR silent Type 3 copper pair; each copper is coordinated to three histidine ligands in a trigonal prismatic arrangement. In contrast to the hemocyanin active site, each Cu3 and Cu3a ion has a distorted tetrahedral coordination geometry with a shared vertex. The remaining copper, Cu2, of the trinuclear site is coordinated to three ligands, two histidine nitrogens and one oxygen from either a hydroxide or a water ligand, thus providing a T-shaped geometry probably representing the spectroscopic Type 2 copper. It is worth noting here that a T-shaped coordination geometry is well documented for the Cu(I) oxidation state, but not known for the complexes with Cu(II) ions.[17,18]

Although EPR studies on laccase[15] suggested the presence of a central water molelcule that could point towards the Type 3 pair and bridges with the Type 2 copper, there is no evidence from the X-ray structure[12] of ascorbate oxidase for such a μ_3-OH/O donor at the centre of the trinuclear copper site. In other words, no bridging ligand between the Type 2 copper and the Type 3 pair is seen in the X-ray structure. The Cu-N and Cu-O distances, 2.06 ± 0.06 and 2.02 ± Å respectively, are unexceptional and all comparable to those of Cu(II) compounds with oxygen and nitrogen donor ligands.[17,18]

An alignment of the amino acid sequences of the related blue oxidases, laccase and ceruloplasmin, has been made on the basis of the spatial structure of ascorbate oxidase. These model studies have led to the proposal[11] that laccase has one Type 1 copper ion and a trinuclear copper cluster; ceruloplasmin has three Type 1 copper ions and one trinuclear centre, which has been recently confirmed by the crystal structure determination of human ceruloplasmin by Lindley *et al.*[19]

The Type 2 and Type 3 copper centers are fundamentally different from those in the single center enzymes in that they are not independent and form a trinuclear copper center, which is the minimum structural unit required for the multielectron reduction of dioxygen to water. That the trinuclear centers in laccase, ascorbate oxidase and ceruloplasmin are essentially identical has been suggested both by the spectroscopic[20] and crystallographic[21] evidences.

The binding of azide to laccase as well as to ascorbate oxidase has been studied extensively.[15] The data were interpreted as the binding of at least one azide as a μ-1,3-N_3 bridging Type 2 Cu and one copper of the coupled Cu3 site. Solomon and coworkers proposed from their spectroscopic data[22] the same type of bridging mode between Type 2 and Type 3 copper for dioxygen which binds as μ-1,1-hydroperoxide. Recently the X-ray structures of the azide and peroxide forms of ascorbate oxidase have been published by Messerschmidt *et al.*[23] In both of the structures only terminal binding of azide and oxygen has been observed. These results show that the two copper ions are not equivalent. Messerschmidt *et al.* proposed that the binding site and mode determined for the peroxide derivative of ascorbate oxidase is representative of all reaction intermediates of dioxygen in all blue oxidases. A tentative mechanism for the reduction of dioxygen through to water has also been proposed.[21]

3.2. *The Reduced Form*

At 2.2 Å resolution the structure of reduced ascorbate oxidase was solved by Messerschmidt *et al.*[21,23] The structure of the Type 1 copper changes very little on reduction. Similar results have also been obtained for the reduced forms of plastocyanin[24] and azurin.[25] On the contrary there are

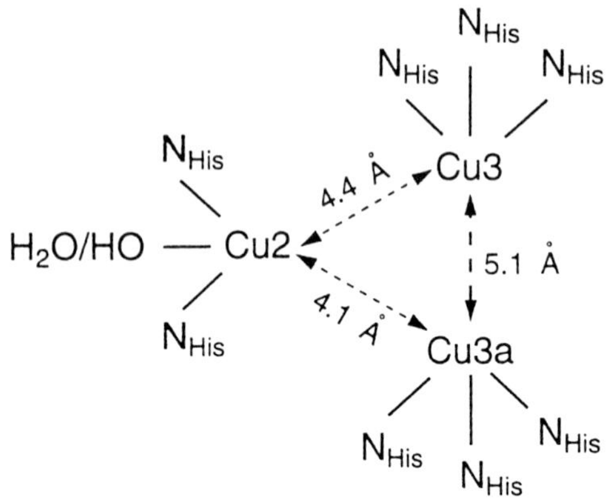

Fig. 2. The trinuclear copper site in the reduced form of ascorbate oxidase (cf. Fig. 1).

considerable structural changes at the trinuclear copper site. There is no bridging ligand between the copper pair of the Type 3, and the coordination geometry at each of the Type 3 together with the Type 2 becomes trigonal planar (three coordinated) in the reduced form of the ascorbate oxidase. The Cu$\cdots$Cu distance increases from an average of 3.7 Å to 5.1, 4.4 and 4.1 Å, as shown in Fig. 2. The Type 2 copper remains virtually unchanged between the reduced and oxidized forms. The only significant change in the coordination geometry of the Type 2 copper is the elongation of the Cu-O bond from 2.02 Å in the oxidized form to 2.21 Å in the reduced form.

4. Low Molecular Weight Models

As a number of good reviews[26] are available on model compounds for Type 1, or blue copper proteins, we are refraining from surveying the model compounds for this class of proteins.

4.1. *Trinuclear Cu(II) Complexes with the $Cu_3(\mu_3$-X)-Core*

The existence of an M_3O-core held by peripheral bridging ligands is well documented in the chemistry of trivalent metal ions, particularly for Fe(III), Mn(III), and Cr(III).[27] In the case of Cu(II) the fully characterized examples are based mostly on oxime ligands and involve usually a central μ_3-OH group.[28–37] Recently compounds with $Cu_3(\mu_3$-OH) core have also been documented with peripheral pyrazolate[29,37] and Schiff base[30] ligands. Thus there are quite a number of examples for the triangular tricopper(II) complexes in the literature with the $Cu_3(\mu_3\text{-X})$ core, where X=OH^- or O^{2-}. The general structure contains a triangular Cu_3X core maintained by peripherally disposed ligand molecules which also function as copper–copper bridging groups. In most of the cases the copper atoms are arranged at the corners of an equilateral triangle with the sides close to 3.24 Å. The central oxygen atom is not located on the plane containing the copper ions and possesses a pseudotetrahedral environment. Thus the oxygen is able to function as a tripod-like ligand due to the favorable orientation of its three lone pairs towards the Cu_3 triad.

The primary interest in the structure of these complexes with oxime ligands arose from their low magnetic moments.[28–37] The compounds containing the $Cu_3(\mu_3\text{-OH})$ core exhibit a varying degree of intramolecular antiferromagnetic interactions, −J lying between 15 and 300 cm^{-1}($\hat{H} = -2J\vec{S}_1 \cdot \vec{S}_2$. The antiferromagnetic interaction within the compound with the Cu_3O[32] unit is so strong that only a single unpaired electron per Cu_3-unit is observed even at high temperatures. This indicates that only the

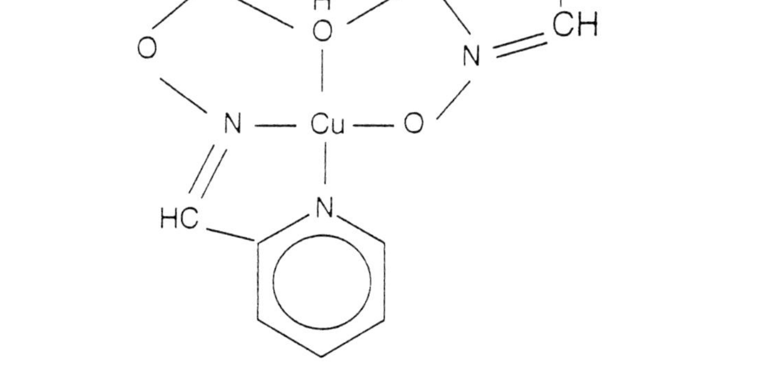

Fig. 3. The structure of $[Cu_3L_3(OH)]^{2+}$ where HL is pyridine-2-aldoxime.

twofold orbitally degenerate term ^{2}E is thermally populated and the excited spin quartet is located at more than 1000 cm^{-1} above. The structure of such a compound with pyridine aldoxime[38] as the ligand is shown in Fig. 3.

The triangular mixed-valence $Cu^{III}Cu_2^{II}O$-core circumscribed by oxime ligands has also been reported.[39] However, the complexes discussed so far cannot be considered as synthetic models for the trinuclear site of the blue oxidases, as there is no evidence for a μ_3-OH or μ_3-O donor at the centre of the cluster in the X-ray structure of ascorbate oxidase. The Cu$\cdots$Cu distance, av. 3.74 Å, is too long to support such a bridge.

The report of the 2.5 Å-resolution X-ray structure of oxidized ascorbate oxidase from green zucchini by Messerschmidt and coworkers[40] in 1989, which unambiguously demonstrated the existence of the biologically unprecedented trinuclear site, sparked efforts to synthesize triangular tricopper complexes without the μ_3-X group so that they model the structure of the metalloenzyme active site. Since the metalloenzyme ascorbate oxidase is structurally well characterized, so any model will be of a corroborative nature.[41] Biomimetic studies to elucidate the function of trinuclear copper cluster are few in number. Structural analogues have proved to be elusive.

4.2. *Trinuclear Cu(I) Complexes, the Cu_3^I-Core*

Trinuclear Cu(I) complexes are not uncommon,[17,18] but with either μ_2-X, μ_3-X bridging or terminal anionic ligands, and threefold symmetry. The structure of an unsymmetrical trinuclear Cu(I) cluster $[Cu_3L^1(PPh_3)_2](PF_6)_3 \cdot 2CH_3CN$ containing aromatic nitrogen ligands has been reported by Karlin *et al.*[42] and the cation of the complex is shown schematically in Fig. 4. Cu(1) possesses a distorted trigonal pyramidal coordination, whereas Cu(2) and Cu(3) are found in a distorted tetrahedral environment with coordination to the phosphorous atoms of the triphenylphosphine ligand. In comparison to the metal$\cdots$metal separations in the reduced ascorbate oxidase, the intramolecular Cu(I)$\cdots$Cu(I) distances with Cu(1)$\cdots$Cu(2) = 6.06 Å, Cu(2)$\cdots$Cu(3) = 7.67 Å and Cu(1)$\cdots$Cu(3) = 7.43 Å are rather large. As a model for the reduced state of the enzyme the compound may prove to be limited owing to the presence of cuprous-state stabilizing phosphine ligands; the coordination spheres of the metal ions differ both in terms of the nature and the geometric arrangement of the donor atoms. The trinuclear Cu(I) complex does, however, demonstrate the feasibility of generating unsymmetrical trinuclear copper clusters.

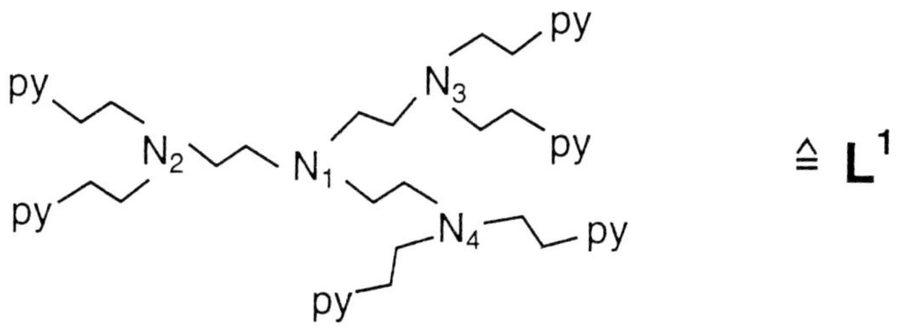

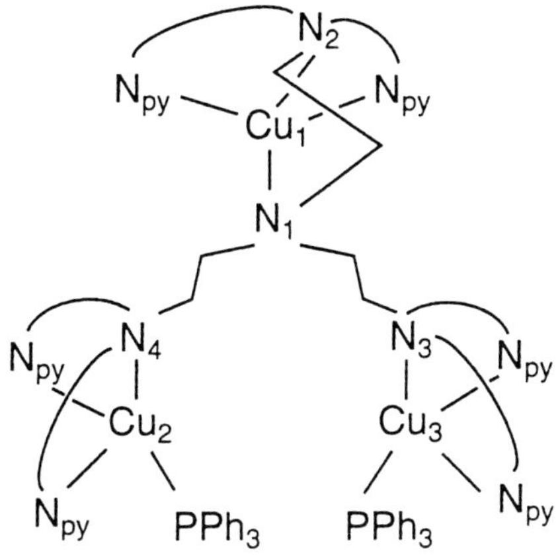

Fig. 4. The tricopper(I) complex $[Cu_3L^1(PPh_3)_2]^{3+}$.

Recently the trinuclear Cu(I) complex $[Cu(\mu\text{-}L^2)_2\ \{Cu(2\text{-cyanoguanidine})\ (CH_3CN)\}_2]\ (BF_4)_3\cdot\ CH_3CN$ where L^2 is a tetradentate Bu^t-substituted 3,6-bis(N-pyrazolyl)pyridazine ligand, has been structurally characterized.[43] For the sake of clarity a schematical drawing, together with the heterocyclic ligand L^2, is shown in Fig. 5. The trinuclear cation, with an approximate twofold symmetry, is based on a near isosceles triangle of Cu(I) ions (Cu$\cdots$Cu = 3.62, 3.63 and 5.01 Å) held together by a heterocyclic nitrogen ligand L^2. All three copper ions are coordinated by four nitrogen donor ligands with pseudotetrahedral geometry. Like the reduced ascorbate oxidase, this complex contains two different types of copper ions and has approximate twofold symmetry; the copper$\cdots$copper distances are

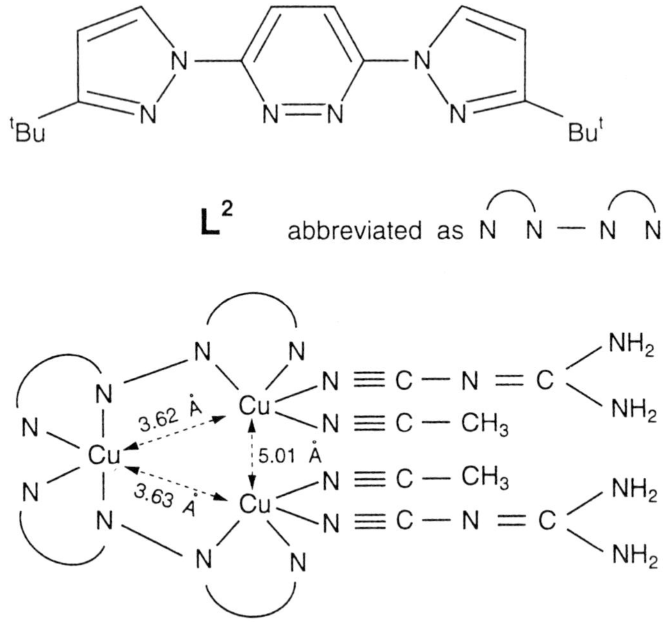

Fig. 5. A schematic drawing of the structure of trinuclear Cu(I) complex with the ligand L^2.

also similar (cf. Fig. 2). However, unlike the enzyme case, the copper sites are four-coordinated, bridged by a complicated pyridazine ligand and presumably inert to oxygen.

4.3. *Trinuclear Cu(II) Complexes, the Cu_3^{II}-Core*

The trinucleating ligand L^3 yielded the trinuclear Cu(I) complex, $[Cu_3^I(L^3)](PF_6)_3$, which upon oxygenation at low temperature gave a hexanuclear Cu(II) complex[44] (Fig. 6), in which the polypodal ligand L^3 had been hydroxylated to L-O-. The Cu(II) ions are pentacoordinated and the centrosymmetric molecule consists of two types of copper ion environments.

There are two isosceles triangular arrays of Cu(II) ions present. Cu(1) and Cu(2) exist as a copper pair with μ-1,1-azido or μ-hydroxo and an endogenous bridging phenoxo group, derived from the ligand via hydroxylation of the m-xylyl dicopper(I) precursor and are separated by 3.11 Å from each other. The third copper Cu(3) is at a distance of 7.78 and 7.46 Å from

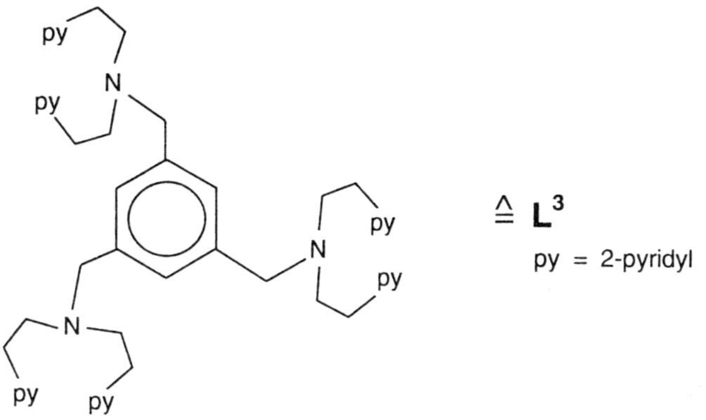

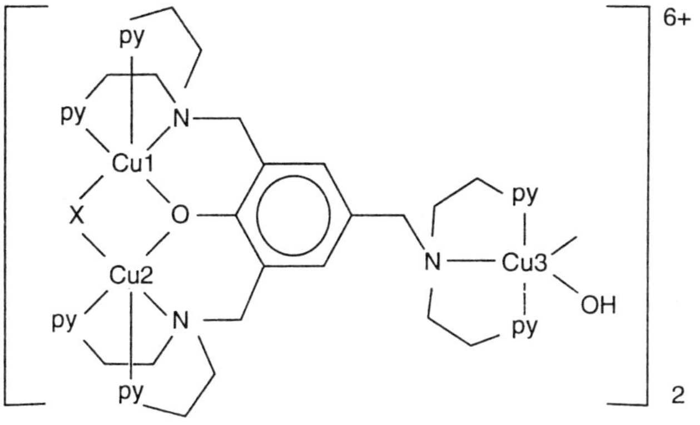

Fig. 6. Structure of the hexanuclear Cu(II) complex $[Cu_3(L^3\text{-}O)(X)(OH)]_2^{6+}$.

the pair. The dinuclear unit, Cu(1)/Cu(2) pair, with exogenous hydroxo-bridge is EPR silent, indicating that this pair can be envisaged as the Type 3 copper centre in the trinuclear site of the blue oxidases.

The presence of Cu-O_2-Cu adduct as an intermediate has been deduced from the low temperature UV-VIS spectroscopy. Although the trinuclear unit contains two different types of copper environments, Cu(1)/Cu(2) pair and Cu(3), as in the copper oxidases, but unlike the enzymes the Type

3-like copper pair in the complex is bridged by two groups, a hydroxo/or azido and a phenoxo. The distance between the Type 2-like and Type 3-like copper centres of the complex is longer (7.78 and 7.46 Å) than that in the oxidized enzyme.

During the studies of ligating properties of the very large polyaza-macrocycles in solution, two solid products with formulae $Cu_3([33]ane\ N_{11})(ClO_4)_6 \cdot 2H_2O$ and $Cu_3([36]ane\ N_{12})(ClO_4)_6 \cdot 3H_2O$ have been isolated.[45] For both macrocycles many other species are present in the equilibrium solution. No X-ray structure has been determined, but considering the molecular topology two structures, a linear and a triangular, have been proposed; the proposed triangular structure together with the large polyaza-cycloalkane ligands are shown below in Fig. 7.

Fenton and coworkers[46] have reported the aimed synthesis and crystal structure of a triangular tricopper(II) compound, based on a functionalized bibracchial tetraimine Schiff-base macrocycle, with Cu···Cu separations of 3.6, 4.9 and 5.9 Å. The copper compound was synthesized by functionalization of the amine pendant arms in a disilver complex $[Ag_2L](BF_4)_2$ with salicylaldehyde followed by transmetallation with Cu^{II}. The crystal structure of $[Cu_3(OH)(L^4)(OH_2)]^{3+}$ is shown in Fig. 8. The geometry of the

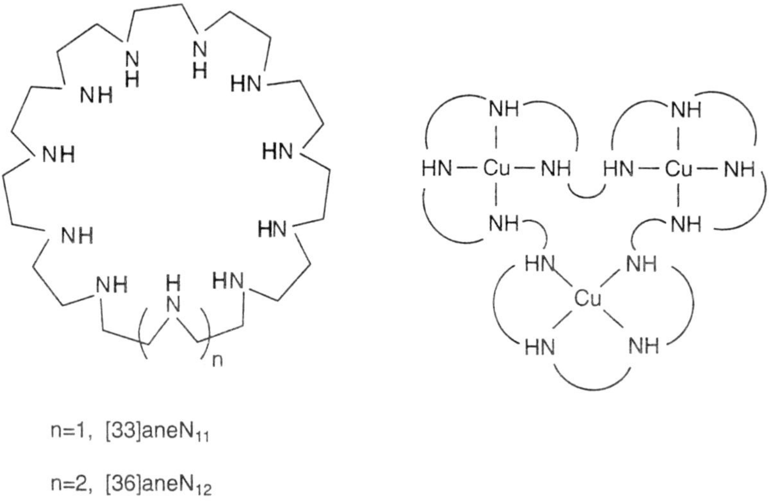

Fig. 7. The proposed structure of trinuclear $[Cu_3(macro)]^{6+}$ complex and two large polyaza macrocycles.

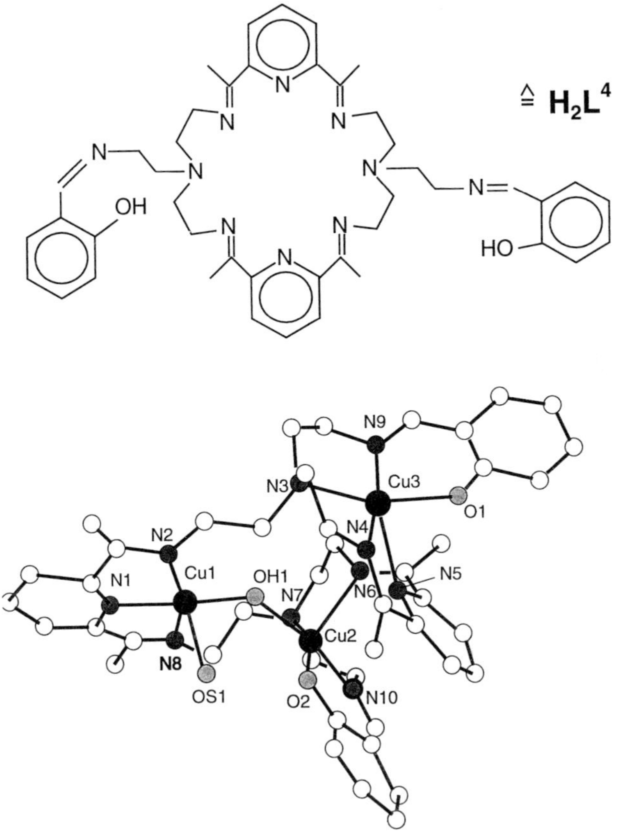

Fig. 8. The crystal structure of $[Cu_3(OH)(L^4)(OH_2)]^{3+}$ and the ligand H_2L^4 (taken from Ref. 46).

cation is very unsymmetrical; each copper has a unique donor set and approximately square pyramidal geometry. The compound consists of Type 3-like center, Cu(1) and Cu(2), and a Type 2-like atom Cu(3). The two copper ions Cu(1) and Cu(2) are bridged by a hydroxy group originated apparently from water of the reaction medium and are 3.6 Å apart, as expected for a $Cu_2(\mu\text{-OH})$ unit. The major feature of the structure in relation to the copper oxidases is that there is a μ_2-hydroxo bridge linking Cu(1) and Cu(2), but no direct bridge involving Cu(3).

The oxidized Type 3 site in oxidases displays a broad near UV strong absorption band maximizing at 330 nm ($\epsilon \sim$ 2700–3050 l $mol^{-1}cm^{-1}$) and a *d–d* band at 740–745 nm. The trinuclear Cu(II) complex described above exhibits a *d–d* band at λ_{max} = 627 nm with an extinction coefficent of 440 l $mol^{-1}cm^{-1}$, and a LMCT band at λ_{max} = 376 nm ($\epsilon \sim 7.4 \times 10^3$ l $mol^{-1}cm^{-1}$), thus obscuring the spectral region of most relevance to the Type 3 copper proteins. The X-band EPR spectrum at 3.8 K recorded on a solid sample yields a quasi-isotropic g value of 2.107 and does not show any fine structure. The EPR data can be interpreted as a system with an isolated $S = 1/2$ with the unpaired electron present in a $d_{x^2-y^2}$ orbital. It is noteworthy in this connection that the trinuclear sites of the blue oxidases show axial EPR spectra with $g_{\|} \sim$ 2.19–2.24, $g_{\perp} \sim$ 2.03–2.06 and $A_{\|}$ (185–200) $\times 10^{-4}cm^{-1}$.

Variable temperature (5–300 K) magnetic susceptibility data of the Cu_3-compound indicate clearly an $S = 1/2$ ground state and have been interpreted considering the trinuclear unit as composed of a magnetically isolated mononuclear site and a moderately antiferromagnetically coupled copper pair with the singlet-triplet separation of only 202 cm^{-1}. This moderate strength of coupling has been explained due to the long Cu(2)-μ(OH) distance of 1.995 (± 20) Å. It must be pointed out that the Cu-(μ-OH) distance in the trinuclear site of the enzyme ascorbate oxidase is also very similar, 2.00 Å, although the susceptibility data[15] on Type 2 depleted (T2D) laccase give a lower limit of the singlet-triplet energy splitting of $-2J >$ 400 cm^{-1}. In summary, although the coordination spheres of the copper ions differ appreciably both in terms of the nature and the geometric arrangement of the donor atoms from the trinuclear site in ascorbate oxidase, this trinuclear complex reproduces features of the biosite to a greater extent than any synthetic complex reported until now.

Very recently the crystal structure of the trinuclear cation

$$[\{Cu(\mu\text{-}HL^5)Cu(OH_2)(\mu\text{-}L^5)Cu\}\mu\text{-}Cl]^{2+}$$

with C_{2v} symmetry has been reported[47] together with its magnetic properties. The three Cu(II) ions are arranged at the corners of an isosceles triangle with the Cu(2)$\cdots$Cu(1) and Cu(2)$\cdots$Cu(2′) distances of 3.37 and 4.77 Å, respectively (Fig. 9). This is another example of a triangular Cu_3 unit, devoiding of any μ_3-X group, which is maintained by peripherally disposed oximato groups and a chloride anion that function also

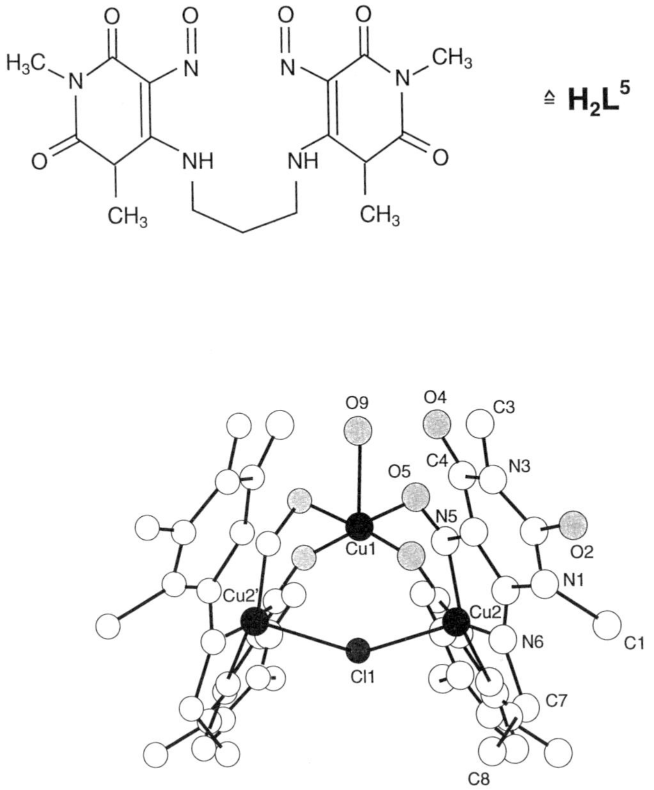

Fig. 9. A schematic drawing of the ligand H_2L^5 and the crystal structure of the trinuclear cation $[\{Cu(\mu\text{-}HL^5)Cu(OH_2)(\mu\text{-}L^5)Cu\}(\mu\text{-}Cl)]^{2+}$ (taken from Ref. 47).

simultaneously as copper–copper bridging groups. The ensemble and copper–copper distances within the tricopper unit are close to those found in the oxidized ascorbate oxidase. The polycrystalline EPR spectrum at 4.2 K exhibits a quasi-isotropic resonance at $g = 2.07$ and is devoid of any half-field ($\Delta M_s = 2$) transition or fine structure so that it resembles that of an axial doublet associated with a mononuclear $S = 1/2$ system with the unpaired electron occupying the $d_{x^2-y^2}$ orbital. Magnetic susceptibility data in the temperature range 5–300 K have been analyzed by assuming the spin exchange via the chloride group to be effectively zero and the exchange coupling constant J is determined only by the exchange pathway involving the oximato groups. A strong antiferromagnetic interaction between Cu(1) and

Cu(2)/Cu(2′) has been found, $J = -189\ \text{cm}^{-1}$. The strong spin interaction in this compound has been explained by a σ-superexchange mechanism, as suggested by Chaudhuri *et al.*[48] for the oximato-bridged butterfly-shaped trinuclear Cu(II) complexes.

Recently Chaudhuri *et al.*[49] reported the X-ray structure and magnetic properties of the imidazolate-bridged trinuclear Cu(II) complex $[L^6_3Cu_3(Im)_3]^{3+}$, whose structure is illustrated in Fig. 10. Each of the three L^6Cu-units is coordinated via two imidazolate anions in the equatorial positions to two remaining L^6Cu-units, yielding three distorted square pyramidal CuN_5-polyhedra. The three copper ions are arranged at the corners of an equilateral triangle with the Cu··· Cu separation of 5.894(5) Å for molecule 1. There are two independent molecules per asymmetric unit

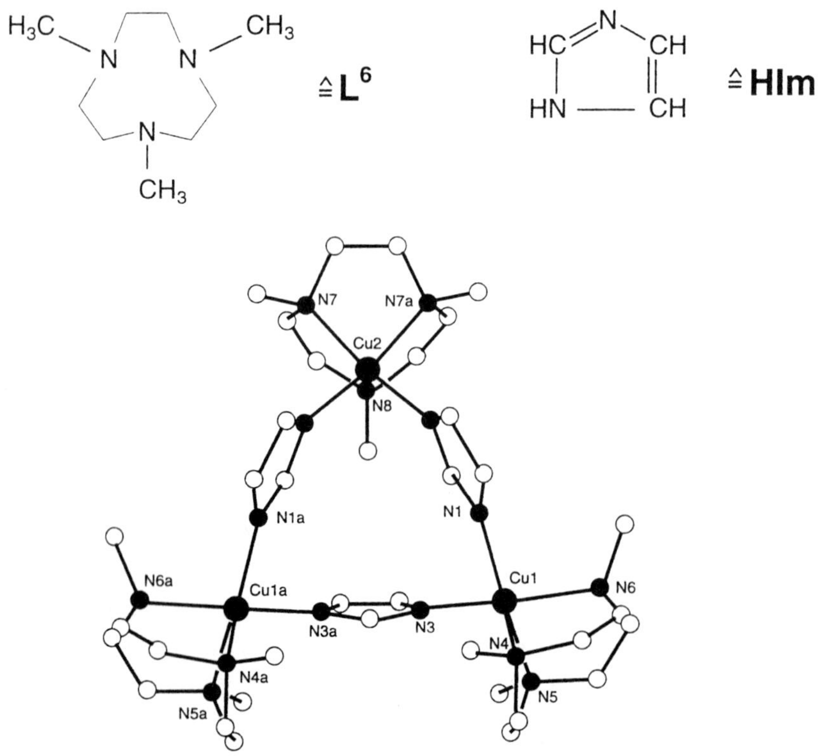

Fig. 10. The structure of the cation in $[L^6_3\ Cu_3(Im)_3]^{3+}$ and the ligands L^6 and HIm (taken from Ref. 49).

with distinctly different configurations caused by the different orientations of the imidazolate as well as the L^6 ligand for each molecule. In molecule 2 there are two long, 5.968(4) Å and one short 5.877(5) Å Cu$\cdots$Cu distances, which are considerably longer than that in the metalloenzyme sites and unlike the enzyme, the three copper sites are nearly identical. These small differences in the Cu$\cdots$Cu distances within the two molecules in the unit cell might be not only due to the crystal packing effect, but also due to the probable presence of the so-called magnetic Jahn–Teller effect,[50] according to which the antiferromagnetically exchange coupled trinuclear systems can be subject to an intrinsic structural distortion.

Variable temperature (2–295 K) magnetic susceptibility data yielded the exchange coupling constant $J = -37.5$ cm^{-1}, with the doublet-quartet separation of 112.5 cm^{-1}, i.e. the magnetic interaction between the Cu(II) centers is antiferromagnetic, causing spin frustration.[51] If the exchange interaction between the copper ions, Cu(1) and Cu(2), are antiferromagnetic, then Cu(3) senses simultaneously spin parallel and antiparallel, thereby causing spin frustration. An $S = 1/2$ ground state is the inescapable consequence of a triangular structure for Cu(II) complexes.

A ground state of degeneracy greater than two is the invariable result of an antiferromagnetically coupled trinuclear complex in which the metal ions with odd number of electrons occupy the corners of an equilateral triangle. The degeneracy is a direct consequence of spin frustration. In other words, it is impossible for all possible pairings of spins to be simultaneously antiparallel in an equilateral triangle. The degeneracy of the ground state is removed if the symmetry of the triangular ensemble is lowered by any structural distortion — distortion into an isosceles or a scalene triangle, thus making at least two exchange coupling constants necessary to describe the spin coupling of the system. By removing the degeneracy of the ground state the system gains energy for the spontaneous distortion to occur; this has been named "magnetic Jahn–Teller effect" by Murao.[50]

The single crystal and polycrystalline powder X-band EPR spectra[54,49] of $[L^6_3Cu_3(Im)_3](ClO_4)_3)$ (Fig. 10) between 4.2 and 300 K were recorded with the aim to have additional confirmation of the model. No half-field ($\Delta M_s = 2$) transition was detected for the polcrystalline powder sample at various temperatures, indicating a weak zero-field splitting. The g values for the polcrystalline powder increases gradually within the experimental error from a minimum of 2.08 at 4.2 K up to a maximum of 2.11 at room temperature. Despite the fact that the number of magnetically non-equivalent sites is four, only one signal could be detected at all the crystal orientations with respect to the static magnetic field. The experimental g values are almost independent of temperature, showing that they correspond to the same values for both the quartets and the doublets. The spectra are explained based on the spin-frustration, the low-symmetry, and the intercluster exchange interactions.

The study with this trinuclear Cu(II) complex raises the question whether the trinuclear site in blue oxidases should be viewed as a spin-frustrated trinuclear system or as a strongly antiferromagnetically coupled Type 3 pair with a peripheral Type 2 centre. The implications of Type 2-Type 3 exchange interactions for the magnetic properties of the trinuclear site in the multi-copper oxidases have, however, been considered in detail by Solomon and coworkers,[15] showing that the antiferromagnetic coupling within the Type 3 centres is the dominant magnetic interaction among the coppers, resulting in an $S = 1/2$ ground state in which the paramagnetism is localized on the Type 2 site. However, a moderate exchange coupling between the Type 2 site and Type 3 copper pair may also exist, as indicated by the same authors.

5. Conclusions and Perspectives

Research aimed at modelling the trinuclear site of the blue copper oxidases, laccase, ascorbate oxidase and ceruloplasmin, will continue. There have been certain exciting results that indicate that is is possible to prepare synthetic analogues that bind and activate dioxygen as well as mimic qualitatively the magnetic and spectroscopic properties of the metalloenzymes.

The use of multidentate cyclic and acyclic ligands for modelling the trinuclear copper site has been shown to be of great value, because of their ability to promote or impose nonclassical geometries or donor sets on the copper ions and of their relative stability. It has already been pointed out by Sorrell[52] that ligands that incorporate imidazole residues are especially desirable and still largely unexplored, and might lead to the synthesis of better biomimics.

The triangular Cu_3-core, both with and without μ_3-X donor atom, is a stable fragment that assembles readily in the presence of appropriate ligands, so its presence in the different blue oxidases is not especially surprising. Available spectroscopic and crystallographic data demonstrate the inequivalence of the Type 3 coppers. No report is known to model the interactions of the trinuclear site with exogenous ligands that might provide more insights about the nature of the trinuclear site. According to the crystallographic data on oxidized ascorbate oxidase there is no bridging ligand linking the Type 2 and Type 3 pair. Without a bridging group capable of acting as superexchange pathway it is difficult to conceive any interaction between the Type 2 and the Type 3 pair, separated by a distance of ca. 3.7 Å. So it is very tempting to contemplate that there may be a donor atom present that is not resolved in the crystal structure!

The structural unit containing a 3+1 arrangement of the Cu(II) ions has not yet been realized in the low molecular weight copper compounds. Magnetic susceptibility data for met T2D laccase, which contains an oxidized Type 3 site and an oxidized Type 1 site, exhibit linear Curie law behavior over the temperature range 8 to 200 K, showing that no exchange interaction is present between the sites that are ca. 13 Å apart from each other. In this context it is noteworthy that a binuclear Cu(II) complex has been described where the superexchange mechanism is operative between the copper centres separated by a distance of as large as 11.25 Å.[53]

As the trinuclear cluster is thought to represent the active site for the binding and irreversible multielectron reduction of dioxygen, and the Type 1 site can be considered as an isolated entity, our main focus in the future would be to model the geometric and electronic structure of the trinuclear site, with an aim to answer the fundamental questions concerning the electronic and geometric structure contributions to reactivity.

Acknowledgments

This work was partially funded by the German Research Council (DFG). It is a pleasure to acknowledge the coworkers and collaborators, who are listed as co-authors in the literature cited for their contributions. The author wishes to express his sincere appreciation to Prof. K. Wieghardt; this work would not have been possible without his help and encouragement.

References

1. R. Malkin and B. G. Malmström, *Adv. Enzymol.* **33**, 177 (1970).
2. B. G. Malmström, L. -E. Andréasson and B. Reinhammer, in *"The Enzymes"*, Vol. XII, ed. P. D. Boyer (Academic, New York, 1975).
3. A. J. Fee, *Struct. Bonding* (Berlin) **23**, 1 (1975).

4. E. I. Solomon, K. W. Penfield and D. E. Wilcox, *Struct. Bonding* (Berlin) **53**, 1 (1983).
5. B. G. Malmström, *Ann. Rev. Biochem.* **51**, 21 (1982).
6. K. G. Strothkamp and C. R. Dawson, *Biochemistry* **13**, 434 (1974).
7. H. Beinert, *Coord. Chem. Rev.* **33**, 55 (1980).
8. D. M. Dooley, R. A. Scott, J. Ellinghaus, E. I. Solomon and H. B. Gray, *Proc. Natl. Acad. Sci. USA* **75**, 3019 (1978).
9. A. A. Gewirth and E. I. Solomon, *J. Am. Chem. Soc.* **110**, 3811 (1988).
10. B. R. M. Reinhammer, *Biochim. Biophys. Acta* **275**, 245 (1972).
11. A. Messerschmidt and R. Huber, *Eur. J. Biochem.* **187**, 341 (1990).
12. A. Messerchmidt, R. Ladenstein, R. Huber, M. Bolognesi, L. Avigliano, R. Petruzzelli, A. Rossi and F. Finazzi-Agro, *J. Mol. Biol.* **224**, 179 (1992).
13. D. J. Spira-Solomon, M. D. Allendorf and E. I. Solomon, *J Am. Chem. Soc.* **108**, 5318 (1986).
14. J. L. Cole, G. O. Tan, E. K. Yang, K. O. Hodgson and E. I. Solomon, *J. Am. Chem. Soc.* **112**, 2243 (1990).
15. J. L. Cole, P. A. Clark and E. I. Solomon, *J. Am. Chem. Soc.* **112**, 9534 (1990).
16. J. L. Cole, L. Avigliano, L. Morpurgno and E. I. Solomon, *J. Am. Chem. Soc.* **113**, 9080 (1991).
17. B. J. Hathaway, in *Comprehensive Coordination Chemistry*, Vol. 5, ed. G. Wilkonson (Pergamon, Oxford, 1987), p. 533.
18. M. Melnik, L. Macaskova and C. E. Holloway, *Coord. Chem. Rev.* **126**, 71 (1993).
19. P. Lindley *et al.*, in *Multi-Copper Oxidases*, ed. A. Messerschmidt (World Scientific, Singapore, 1996), pp. 81–102.
20. (a) E. I. Solomon, M. J. Baldwin and M. D. Lowery, *Chem. Rev.* **92** 521 (1992); (b) E. I. Solomon and M. D. Lowery, *Science* **259**, 1575 (1993).
21. A. Messerschmidt, *Adv. Inorg. Chem.* **40**, 121 (1994).
22. J. L. Cole, D. P. Ballou and E. I. Solomon, *J. Am. Chem. Soc.* **113**, 8544 (1991).
23. A. Messerschmidt, H. Luecke and R. Huber, *J. Mol. Biol.* **230**, 997 (1993).
24. J. M. Guss, P. R. Harrowell, M. Murata, V. A. Norris and H. C. Freeman, *J. Mol. Biol.* **192**, 361 (1986).
25. W. E. B. Shepard, B. F. Anderson, D. H. Lewandowski, G. E. Norris and E. N. Baker, *J. Am. Chem. Soc.* **112**, 7817 (1990).
26. See for example, (a) E. N. Baker, in *Encyclopedia of Inorganic Chemistry* Vol. 2 (John Wiley, New York, 1994), p. 883; (b) H. Toflund, J. Becher, P. H. Olesen, J. Z. Pedersen, *Isr. J. Chem.* **25**, 56 (1985); (c) E. Bouwman, W. L. Driessen and J. Reedijk, *Coord. Chem. Rev.* **104**, 143 (1990); (d) *Bioinorganic Chemistry of Copper*, ed. K. D. Karlin and Z. Tyeklár (Chapman & Hall, New York, 1993).
27. R. D. Cannon and R. P. White, *Progr. Inorg. Chem.* **36**, 195 (1988).

28. Y. Agnus, R. Louis, B. Metz, C. Boudon, J. P. Gisselbrecht and M. Gross, *Inorg. Chem.* **30**, 3155 (1991); and references therein.
29. M. Angaroni, G. A. Ardizzoia, T. Beringhelli, G. La Monia, D. Getteschi, N. Masciocchi and M. Moret, *J. Chem. Soc. Dalton Trans.*, 3305 (1990).
30. (a) N. A. Bailey, D. E. Fenton, R. Moody, P. J. Scrimshire, E. Beloritzky, P. H. Fries and J. M. Latour, *J. Chem. Soc. Dalton Trans.* 2817 (1988); (b) E. Kwiatkowski, M. Kwiatkowski, U. Dettlaff-Weglikowska and D. M. Ho, *J. Cryst. Spectrosc. Res.* **22**, 411 (1992).
31. R. Beckett, R. Colton, B. F. Hoskins, R. L. Martin and D. G. Vince, *Aust. J. Chem.* **22**, 2527 (1969).
32. R. J. Butcher, C. J. O'Connor and E. Sinn, *Inorg. Chem.* **20**, 537 (1981).
33. J. -P. Coses, F. Dahan and J. -P. Laurent, *Inorg. Chem.* **25**, 413 (1986).
34. S. Baral and A. Chakravorty, *Inorg. Chem. Acta* **39**, 1 (1980).
35. P. F. Ross, R. K. Murmann and E. O. Schlemper, *Acta Crystallogr.* **B30**, 1120 (1974).
36. R. W. Green. and M. C. K. Svasti, *Aust. J. Chem.* **16**, 356 (1963).
37. F. B. Hulsbergen, R. W. M. Ten Hoedt, G. C. Verschoor, J. Reedijk and A. L. Spek, *J. Chem. Dalton Trans.*, 539 (1983).
38. R. Beckett and B. F. Hoskins, *J. Chem. Soc. Dalton. Trans.*, 291 (1972).
39. D. Datta and A. Chakravorty, *Inorg. Chem.* **22**, 1611 (1983).
40. A. Messerschmidt, A. Rossi, R. Ladenstein, R. Huber, M. Bolognesi, G. Gatti, A. Marchesini, R. Petruzzelli and A. Finazzi-Agró, *J. Mol. Biol.* **206**, 513 (1989).
41. H. A. Hill, *Chem. Br.* **12**, 119 (1976).
42. K. D. Karlin, Q. Gan, A. Farooq, S. Liu and J. Zubieta, *Inorg. Chem. Acta.* **165**, 37 (1989).
43. P. Hubberstey and C. E. Russell, *J. Chem. Soc. Chem. Comm.*, 959 (1995).
44. K. D. Karlin, Q. -F. Gan, A. Farooq, S. Liu and J. Zubieta, *Inorg. Chem.* **29**, 2549 (1990).
45. A. Bencini, A. Bianchi, E. Garcia-Espana, M. Micheloni and P. Paoletti, *Inorg. Chem.* **27**, 176 (1988).
46. H. Adams, N. A. Bailey, M. J. S. Dwyer, D. E. Fenton, P. C. Hellier, P. D. Hempsted and J. M. Latour, *J. Chem. Soc. Dalton Trans*, 1207 (1993).
47. E. Colacio, J. M. Dominquez-Vera, A. Escuer, M. Klinga, R. Kivekäs and A. Romerosa, *J. Chem. Soc. Dalton Trans.* 343 (1995).
48. P. Chaudhuri, M. Winter, B. P. C. Della Védova, E. Bill, A. Trautwein, S. Gehring, P. Fleischhauer, B. Nuber and J. Weiss, *Inorg. Chem.* **30**, 2148 (1991).
49. (a) P. Chaudhuri, I. Karpenstein, M. Winter, C. Butzlaff, E. Bill, A. X. Trautwein, U. Flörke and H. -J. Haupt, *J. Chem. Soc. Chem. Commun.*, 321 (1992); (b) P. Chaudhuri, M. Winter, U. Flörke and H. -J. Haupt, unpublished results.
50. T. Murao, *Phys. Lett.* **49A**, 33 (1974).

51. S. Ghose, A. W. Hewat and M. Pinkney, *Solid State Commun.* **74**, 413 (1990).
52. T. N. Sorrell, *Tetrahedron* **45**, 3 (1989).
53. (a) P. Chaudhuri, K. Oder, K. Wieghardt, S. Gehring, W. Haase, B. Nuber and J. Weiss, *J. Am. Chem. Soc.* **110**, 3657 (1988); (b) K. -S. Bürger, P. Chaudhuri, K. Wieghardt and B. Nuber, *Chem. Eur. J.* **1**, 583 (1995).
54. J. Padilla, D. Gatteschi, U. Flörke and P. Chaudhuri, unpublished results.

CHAPTER 15

^{199m}HG-DERIVATIVES OF LACCASE AND ASCORBATE OXIDASE: A HYPERFINE SPECTROSCOPIC STUDY

T. Butz and W. Tröger

Fakultät für Physik und Geowissenschaften, Universität Leipzig, Linnéstr. 5, 04103 Leipzig, Germany

ABSTRACT

Apo-laccase and -ascorbate oxidase were partially and fully reconstituted with radioactive ^{199m}Hg. The nuclear quadrupole interaction (NQI) of this isotope was determined by time differential perturbed angular correlation (TDPAC) of γ-rays. Up to three different NQI's were observed and assigned to Hg residing on type-1, type-2, and type-3 Cu-sites. Site assignments were based on additional experiments with type-2 depleted enzymes (T2D), on comparisons with small blue Cu-proteins like azurin and stellacyanin containing type-1 Cu-sites only, as well as on blocking studies with non-radioactive Hg. Under aerobic conditions, Hg(II) seems to be preferentially introduced onto type-2 Cu-sites, contrary to expectations based on chemical affinities.

Laccase exhibits great similarities with ascorbate oxidase with respect to observed NQI's as well as Hg-reconstitution probabilities.

Contents

Abbreviations: NQI–nuclear quadrupole interaction; TDPAC–time differential perturbed angular correlation; T2D–type 2 depleted; EPR–electron paramagnetic resonance; CD–circular dichroism; XAS–X-ray absorption fine structure; NMR–nuclear magnetic resonance; ENDOR–electron-nuclear double resonance; FWHM–full width at half maximum; AO–ascorbate oxidase; LAC–laccase; DMG–dimethylglyoxine; EDTA–ethylenediaminetetraacetate.

1. Introduction

Laccase and ascorbate oxidase are multi-copper enzymes which catalyze the four-electron reduction of dioxygen to water with concomitant one-electron oxidation of the substrate.[1] They are glyco-proteins and particularly laccase is notoriously resistant to all attempts to crystallize it. The crystal structure determination of ascorbate oxidase (*Cucurbita pepo medullosa*) by Messerschmidt *et al.*[2,3] revealed that the monomeric unit contains one type-1 Cu, responsible for the intense blue colour, and one trinuclear Cu-center with one EPR-active type-2 Cu and one EPR-silent antiferromagnetically coupled type-3 Cu-pair. Both type-3 Cu-atoms are nearly equivalent.

All four Cu-atoms can be removed under suitable dialysis conditions and a partial or complete reconstitution with Cu[4,5] and other metals[2,5] is possible. In addition, the selective depletion of the type-2 Cu-site (T2D) is reported[6,7] which offers the possibility to perform spectroscopy on T2D-systems without interference from type-2 Cu or even the preparation of "type-2-only" substituted derivatives. McMillin *et al.*[8,9] report on laccase with Hg blocking selectively the type-1 Cu-site. The success of such metal manipulations is — in most cases — not verified by crystal structure determinations but rather by spectroscopic techniques such as optical absorption, electron paramagnetic resonance (EPR), circular dichroism (CD), X-ray absorption (XAS), and others. None of these techniques gives a clear, well separated signature of a given metal residing on a specific Cu-site for all sites simultaneously. In this respect, the nuclear quadrupole interaction of the metal atom is expected to yield a clear signature for the local environment of the metal, preferably combined with a chemical shift determination. Unfortunately, nuclear magnetic resonance (NMR), nuclear quadrupole resonance (NQR) and electron-nuclear double resonance (ENDOR) are rather difficult to perform on such "dilute systems" and require large quantities of protein. However, there is another hyperfine technique which is extremely sensitive because it is based on the observation of γ-rays emitted from radioactive nuclei: time differential perturbed angular correlation (TDPAC) of γ-rays. A suitable isotope is ^{199m}Hg ($t_{1/2}$ = 43 min). The "m" indicates longlived (with respect to sample preparation 3/4 of an hour is short!) isomeric states, i.e. there is no nuclear transmutation involved. Unfortunately, there are no suitable isomeric states in any of the existing Cu-isotopes. The only alternative could be ^{62}Zn which decays to ^{62}Cu. The decay properties are not unfavorable, but the nuclear quadrupole moment of the I = 5/2 state is apparently very small. Hence we concentrated on ^{199m}Hg.

The theory of TDPAC with emphasis on NQI's is briefly introduced in Sec. 2. Experimental details concerning the production of ^{199m}Hg, the preparation of holo-, apo-, and T2D-enzymes, their reconstitution with ^{199m}Hg, the spectrometer and the data analysis are described in Sec. 3. In Sec. 4 all experimental results are presented. This section also contains all efforts to assign individual NQI-signals to specific Cu-sites. This includes experiments with T2D-enzymes, comparisons with small blue copper proteins containing type-1 Cu only, and blocking studies with non-radioactive Hg. In Sec. 5 we discuss the proposed site assignments, the Hg-reconstitution probabilities, and similarities between laccase and ascorbate oxidase. Finally, Sec. 6 deals with open problems like metal exchange in the trinuclear cluster or recent observations of specific Hg-binding to holo-enzymes.

2. Theory of TDPAC and NQI

The TDPAC technique of γ-rays is described in detail in Refs. 10–15 and with special emphasis on chemical and biological applications in

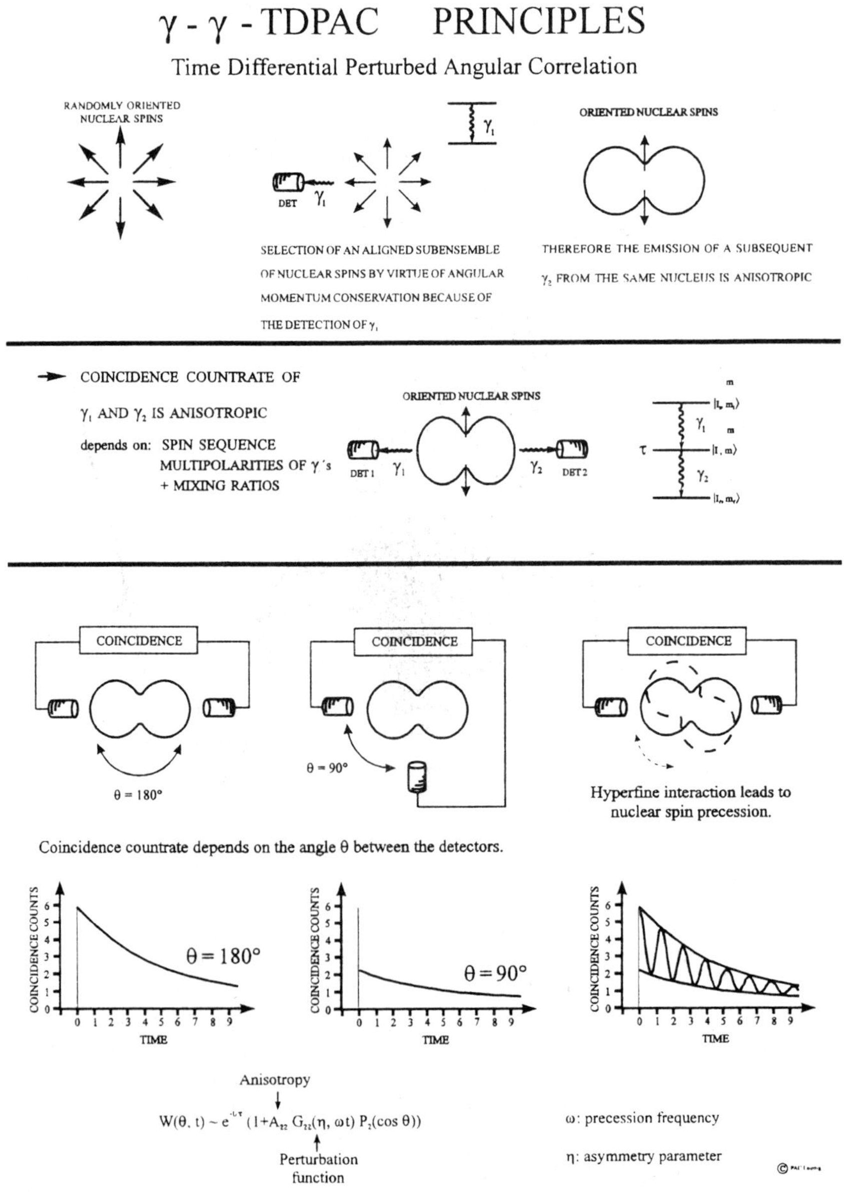

$$W(\theta, t) \sim e^{-t/\tau}\,(1 + A_{22}\, G_{22}(\eta, \omega t)\, P_2(\cos\theta))$$

Fig. 1. Schematic illustration of the γ-γ-TDPAC principles.

Refs. 16–18. Here, only a very short introduction will be given. In Fig. 1 the γ-γ-TDPAC principles are illustrated in a very schematic way. γ-γ-TDPAC requires a nucleus in an excited state which decays via a γ-γ-cascade. Due to angular momentum conservation, the emission probability of the second quantum is correlated to the emission direction of the first γ-quantum, i.e. the coincidence countrate $W(\theta)$ of γ_1 and γ_2 depends on the angle θ between both γ–quanta (detectors):

$$W(\theta) \sim 1 + A_{22}P_2(\cos\theta) + \cdots + A_{kk}P_k(\cos\theta) + \cdots \tag{1}$$

The anisotropy A_{22} is a constant depending on the nuclear spins, multipolarities and mixing ratios of the γ-γ-cascade. $P_2(\cos\theta)$ denotes a Legendre polynomial. In general there could be higher-order terms $A_{kk}P_k(\cos\theta)$, which are usually small. If the intermediate nuclear state has a finite life time τ_n, the time differential angular correlation can be recorded:

$$W(\theta,t) \sim e^{-(t/\tau_n)}(1 + A_{22}P_2\cos\theta + \cdots + A_{kk}P_k(\cos\theta) + \cdots) \tag{2}$$

with t denoting the time elapsed between the detection of γ_1 and γ_2. The angular correlation (Eq. 2) can be perturbed under the influence of extranuclear fields while the nucleus is in the intermediate state: the nucleus can interact via its magnetic dipole moment with magnetic fields or via its electric quadrupole moment with the electric field gradient, produced by extranuclear charges. Both effects belong to hyperfine interactions and lead to a precession of the nuclear spin

$$W(\theta,t) \sim e^{-(t/\tau_n)}(1 + A_{22}G_{22}(t)P_2(\cos\theta) + \cdots A_{kk}G_k^{(+)}P_k(\cos\theta) + \cdots) \tag{3}$$

This equation is applicable to a randomly orientated ensemble of nuclear spins, i.e. powder samples or precipitated protein. The perturbation function $G_{22}(t)$ contains all information about the hyperfine interaction. For a pure static nuclear quadrupole interaction with axial symmetry and nuclear spin I $= 5/2$ for the intermediate state in a powder sample we have:

$$G_{22}(t) = 0.2 + 0.371\cos(\omega t) + 0.286\cos(2\omega t) + 0.143\cos(3\omega t) \tag{4}$$

with $\hbar\omega = E_{\pm 3/2} - E_{\pm 1/2}$ denoting the smallest energy difference of the quadrupole split state. A deviation from axial symmetry modifies the splitting of the energy levels and, hence, the precession frequencies. For non-axial symmetry the perturbation function $G_{22}(t)$ remains a sum of 3 cosine functions and a time independent term, the so-called hardcore, but the intensities $f_i(\eta)$ and the precession frequencies $\omega_i(\eta)$ depend on the asymmetry parameter η:

$$G_{22} = \sum_{i=0}^{3} f_i(\eta)\cos(\omega_i t) \qquad (\omega_0 = 0)\,. \tag{5}$$

The asymmetry parameter η of the nuclear quadrupole interaction is defined as follows:

$$\eta = \frac{V_{yy} - V_{xx}}{V_{zz}} \tag{6}$$

where V_{xx}, V_{yy}, V_{zz} denote the principal elements of the traceless electric field gradient tensor:

$$\begin{pmatrix} V_{xx} & 0 & 0 \\ 0 & V_{yy} & 0 \\ 0 & 0 & V_{zz} \end{pmatrix} \tag{7}$$

with $V_{xx} + V_{yy} + V_{zz} = 0$ and $|V_{zz}| \geq |V_{yy}| \geq |V_{xx}|$. If $V_{xx} = V_{yy}$ then $\eta = 0$ and there is axial symmetry. Otherwise one obtains $0 < \eta \leq 1$. Instead of V_{zz} the nuclear quadrupole interaction frequency

$$\nu_Q = \frac{eQV_{zz}}{h} \tag{8}$$

is used, with Q denoting the electric quadrupole moment of the intermediate state. Analytic expressions for perturbation functions were derived by Butz.[19]

A few extensions of the above equations are necessary in practice:

(i) The finite solid angle of the detectors yields a correction factor < 1 which reduces the effective anisotropy: $A_{22}^{\mathrm{eff}} < A_{22}$.

(ii) Inequivalent probe sites lead to a superposition of perturbation functions, e.g. for three different sites:

$$G_{22}(t) = \alpha_{\mathrm{I}} G_{22}^{\mathrm{I}}(t) + \alpha_{\mathrm{II}} G_{22}^{\mathrm{II}}(t) + \alpha_{\mathrm{III}} G_{22}^{\mathrm{III}}(t)\,. \tag{9}$$

The amplitudes α_j (j = I, II, III) give the populations of the different sites.

(iii) The finite time resolution τ_{FWHM} of the spectrometer reduces the amplitude of the cosine terms approximately as $e^{-0.09\omega^2\tau_{FWHM}^2}$, i.e. very high frequencies become increasingly more difficult to detect.

(iv) A distribution of the NQI, in V_{zz} as well as in η, leads to an exponential decay of the cosine intensities which reads for a Lorentzian distribution of ω_i with a width δ : $e^{-\delta\omega_i t}$. Thus, for sufficiently long times, all oscillations vanish and only the time independent hardcore survives.

(v) For fluctuating NQI's all intensities including the time independent hardcore decay with time. The details of the perturbation function for dynamic interactions depend on the type of motion and can be found in Refs. 20 and 21.

3. Experimental

3.1. *Holo-, Apo-, T2D-Enzyme Preparation*

Acetone powder of the latex Japanese lacquer tree was obtained from Saito & Co. Ltd. (Tokyo, Japan), and laccase was extracted by the method of Reinhammar.[22] All chemicals were reagent grade and were used without further purification. All preparation steps mentioned in this section were carried out at 5°C. We used ion exchange resins CM-Sephadex C50 and DEAE-Sephadex A50 from Pharmacia. The enzyme was isolated, purified and stored in potassium phosphate buffer (0.1 M pH = 6) at −80°C. Ascorbate oxidase was obtained from Boehringer/Mannheim. As buffer TRIS 0.1 M/20 mM NaCl with and without FICOLL was used. No further purification steps were performed with the protein. The FICOLL was removed by applying the enzyme to a DEAE-Sephadex A50 column. The column loaded with the enzyme was washed with TRIS 0.1 M/20 mM NaCl. Two to three times the column volume was used to be sure that all FICOLL was removed. The enzyme was eluted from the column with 0.1 TRIS/180 mM NaCl and then dialysed against 0.1 M TRIS/20 mM NaCl. The type-2 depleted forms of ascorbate oxidase (T2D-AO) and laccase (T2D-LAC) were prepared by the method of Avigliano and Graziani.[6,7] Buffer solutions were prepared from deionized water (Millipore Q). All the glassware was washed several times with buffer solution to avoid any metal contamination. The concentrations of the holo-proteins were 2–24 μM. For the selective removal of type-2 Cu ascorbate oxidase was dialyzed for 12 hours under strictly anaerobic conditions (Ar atmosphere with slight excess pressure) against acetate buffer (0.05 M, pH = 5.0) with 0.002 M DMG (dimethylglyoxine),

0.001 M EDTA (ethylenediaminetetraacetate), and 0.05 M $K_4[Fe(CN)_6]$ (ferrocyanide) under reducing conditions. The T2D-laccase was prepared by an anaerobic dialysis against the same buffer as ascorbate oxidase but without EDTA, followed by an aerobic dialysis against 0.001 M EDTA in the same buffer. Subsequently the enzymes were dialyzed 2 to 3 times in air against the storing buffer (potassium phosphate buffer, 0.1 M, pH = 6). Apo-ascorbate oxidase (apo-AO) was prepared according to the method of Savini *et al.*[4] The enzyme (2-24 μM) was anaerobically dialyzed under reducing conditions for 12 hours against potassium phosphate buffer (0.1 M, pH = 7.2) with 0.025 M KCN (potassium cyanide) and subsequently for 12 hours in air against potassium phosphate buffer (0.08 M, pH = 7.2) with 0.001 M EDTA. Two or three dialysises against the pure storing buffer (potassium phosphate buffer 0.01 M, pH = 6) followed to remove all EDTA. In the case of apo-laccase (apo-LAC) 0.01 M ascorbate and 0.05 M KCN were used for the anaerobic dialysis.[5] The T2D-and apo-forms of the enzymes were frozen in aliquots of 50 ml and 2 ml at −80°C and stored for several days to several weeks at the same temperature.

To reconstitute the enzymes with the shortlived radioactive isotope ^{199m}Hg ($\tau_{1/2}$ = 43 min) the samples were transported on dry ice to Geneva where ^{199m}Hg is produced by the mass separator ISOLDE.

3.2. *Isotope Production at ISOLDE*

The short-lived radio-isotope ^{199m}Hg was produced by the on line isotope separator ISOLDE at the European Organization for Nuclear Research CERN in Geneva, Switzerland.[23] The basic principle of the ISOLDE facility is that a thick target of molten lead is placed in the 1 GeV proton beam from a proton accelerator, the proton synchrotron booster (PSB). In the target a variety of radioactive isotopes is produced by spallation, fission or fragmentation of the lead atoms. The radioactive isotopes — among them ^{199m}Hg — are ionized and accelerated to 60 keV before they are separated in different beams according to their masses in an analyzing magnet and delivered to the ISOLDE experimental hall. Here, the ^{199m}Hg beam is directly implanted into ice from deionized water at about 150 K. Typical collection times for the ^{199m}Hg activity are 10 to 60 minutes. Melting the ice gives a pure aqueous ^{199m}Hg(II)-solution of 50–150 μl, referred to as "carrier-free".

3.3. *Hg-Reconstitution of Enzymes*

Immediately before the Hg-reconstitution the enzymes are quickly thawed. The enzymes were incubated at 0°C for 5–60 min either with carrier-free ^{199m}Hg or with stoichiometric amounts of ^{199m}Hg/Hg, i.e. non-radioactive Hg (0.44 mM $HgCl_2$ dissolved in potassium phosphate buffer 0.1 M, pH = 6.0) was added to the carrier-free ^{199m}Hg solution. The enzyme concentration was 2–24 μM, for the Hg-reconstitution usually 0.2–7.7 nmol enzyme was employed. After the incubation the enzymes were either immediately precipitated with cold ethanol or washed with buffer solution and concentrated further with AMICON-concentrators (cut-off: 30.000) and then precipitated with cold ethanol. The precipitate formed a massive pellet at the bottom of the tube during centrifugation (3.000 rpm – 6.000 rpm for 5–10 minutes), the supernatant was carefully removed. The binding of Hg to the enzymes was checked by monitoring the ^{199m}Hg-activity in the filtrate or in the supernatant. Typically, only 10 % of the used activity was found in the filtrate or in the supernatant in experiments with 4 equivalents of ^{199m}Hg/Hg or less per 4 empty Cu-sites.

In the blocking experiments we first added non-radioactive Hg (0.44 mg/ml $HgCl_2$ dissolved in potassium phosphate buffer 0.1 M, pH = 6) in order to block the binding sites with the highest reconstitution probabilities for Hg. After an incubation time of 5 to 60 min the remaining unoccupied sites were filled up partially or completely by radioactive Hg. The TDPAC measurements were performed either at −70°C or around 0°C.

3.4. *TDPAC Spectrometer and Data Reduction*

The TDPAC measurements were performed on the 375 keV–158 keV γ-γ-cascade of ^{199m}Hg. The nuclear quadrupole moment of the intermediate state is Q = 0.674(77) barn.[24] Due to the short half-life $\tau_{1/2}$ = 2.4 ns of the intermediate state and the high precession frequencies ($\omega_1 \sim$ 1 Grad/s) an excellent time resolution of the TDPAC spectrometer is required. Furthermore the 375 keV-transition is highly converted (total conversion coefficient $\alpha \sim 5$) which requires a high efficiency of the TDPAC spectrometer in order to gain TDPAC spectra of a sufficient statistical quality. For the TDPAC studies on ascorbate oxidase a conventional 4-detector setup[13] and the so-called TDPAC camera,[25] a 6-detector setup, was used. The time resolutions of the spectrometers were 450 ps and 550 ps for the 4- and

6-detector setup, respectively. In the conventional 4-detector setup the sample is surrounded by 4 detectors in a plane at 90° intervals. In the TDPAC-camera the 6 detectors are arranged in a cube such that there are three pairs facing each other. Each detector consists of a conically shaped BaF_2 scintillator mounted on a XP2020Q photomultiplier tube with a quartz window. Each detector covers about 10 % of 4π solid angle. The dynode (energy) signal is amplified and selected in single channel analyzers for γ_1 and γ_2. The anode (timing) signal is delayed until the energy signals are available. A routing/coincidence unit selects the start (γ_1) and stop (γ_2) signals of a detector combination W_{ij} and creates an address for the multichannel analyzer subgroups corresponding to W_{ij}.

The time difference between start and stop pulses is determined by a combination of a time to amplitude converter and an amplitude to digital converter. In the case of a 4-detector spectrometer for technical reasons only 8 out of 12 subgroups (4 combinations under 180° and 4 combinations under 90°) were used. With the TDPAC-camera 30 subgroups (6 × 180°, 24 × 90°) can be recorded simultaneously, which results in a 2.4 times higher efficiency of the TDPAC-camera compared to the conventional 4-detector setup. The recorded 8 or 30 detector combinations can be combined (after substraction of accidental background and shift to a common origin):

$$R(t) = 2\frac{W(180°) - W(90°)}{W(180°) + 2W(90°)} \tag{10}$$

$$W(180°) = \begin{cases} \sqrt[4]{W_{13}W_{31}W_{24}W_{42}} & \text{for the 4-detector spectrometer} \\ \sqrt[6]{W_{13}W_{31}W_{24}W_{42}W_{56}W_{65}} & \text{for the 6-detector spectrometer} \end{cases}$$

$$W(90°) = \begin{cases} \sqrt[4]{W_{14}W_{41}W_{23}W_{32}} & \text{for the 4-detector spectrometer} \\ \sqrt[24]{P} & \text{for the 6-detector spectrometer} \end{cases}$$

$$\text{with} \quad P = \prod_{i,j=1, i\neq j}^{6} W_{ij} \quad \text{without} \quad W_{13}, W_{31}, W_{24}, W_{42}, W_{56}, W_{65}\,.$$

In the ratio $R(t)$, the exponential decay factors of the intermediate state cancel and the detector efficiencies are eliminated to first order. For powder samples and $k = 2$, $R(t)$ reduces to $R(t) = A_{22}G_{22}(t)$. This is usually called the TDPAC time spectrum which is presented together with its Fourier transform.

4. Results and Site Assignments

4.1. *Hg-Reconstitution of T2D-Enzymes*

Hg-reconstitution studies of T2D-enzymes were carried out with ^{199m}Hg under aerobic conditions. Figure 2 shows TDPAC spectra (left) and their Fourier-transforms (right) for ^{199m}Hg-T2D-ascorbate oxidase (top) and -laccase (bottom). In both cases a unique Hg-site with well-defined NQI parameters is observed. For ascorbate oxidase, a nearly equidistant three-line pattern is observed which clearly shows that the asymmetry parameter η is small. We required a small line-broadening of $\delta = 3$ % (Lorentzian line shape) in order to fit the ascorbate oxidase data whose origin is not yet clear.

The situation is less clear for laccase. Here, an excess line-broadening of $\delta = 7$ % is required to fit the data. Alternatively, two closely lying NQI's

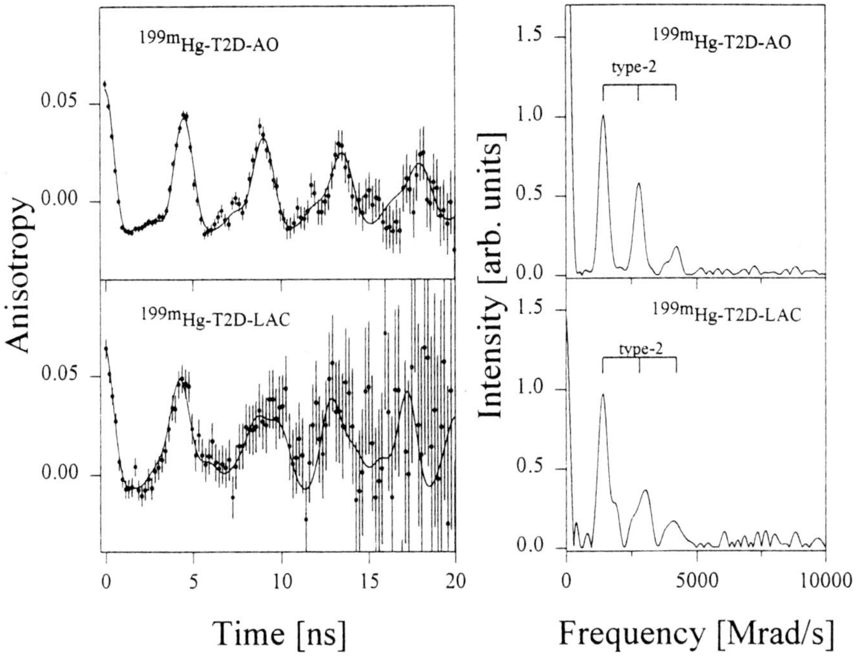

Fig. 2. TDPAC spectra (left) and their Fourier transforms (right) for ^{199m}Hg-T2D-ascorbate oxidase (top) and ^{199m}Hg-T2D-laccase (bottom).

Table 1. The NQI parameters for ^{199m}Hg on type-1, -2, -3 sites in ascorbate oxidase (*Cucurbita pepo medullosa*) and laccase (*Rhus vernicifera*).

	ω [Grad/s]	η	ν_Q [GHz]	$\|V_{xx}\|$	$\|V_{yy}\|$ [10^{21}V/m^2]	$\|V_{zz}\|$
AO						
type-1	1.57(2)	0.88(2)	1.02	3.8	59.0	62.8
type-2	1.44(1)	0.17(2)	1.48	37.7	53.2	90.9
type-3	1.12(1)	0.83(2)	0.756	3.9	42.5	46.4
LAC						
type-1	1.42(3)	0.91(4)	0.907	2.5	53.2	55.7
type-2	1.35(1)	0.16(2)	1.39	35.9	49.6	85.6
type-3	0.95(1)	0.66(3)	0.724	7.6	36.9	44.4

would also fit the data. In Table 1, these NQI-parameters for Hg on type-2 Cu-sites are listed together with the NQI-parameters for Hg on the other two sites whose assignment will become clear below.

4.2. *Partial Hg-Reconstitution of Apo-Enzymes*

First, experiments with carrier-free activity of ^{199m}Hg were performed, i.e. the radioactive probes are added to the apo-enzymes in "infinite dilution". This procedure has the advantage that denaturation upon addition of heavy metals is certainly impossible. However, trace amounts of unwanted binding sites can compete with the vast excess of empty Cu-binding sites provided their affinity is much higher. In addition, experiments were performed with one equivalent of Hg with respect to 4 empty Cu-sites per monomeric subunit by adding the appropriate amount of non-radioactive Hg in the form of $HgCl_2$-solution. No significant difference was observed compared to the "infinite dilution" case.

Figure 3 shows TDPAC spectra (left) and their Fourier-transforms (right) for ^{199m}Hg-ascorbate oxidase (top) and laccase (bottom) for the case of "infinite dilution" (AO) and 1Hg/4Cu (LAC). It is immediately clear, that Hg does not reconstitute one single site only. The least squares fitting analysis of these data together with all further data obtained for

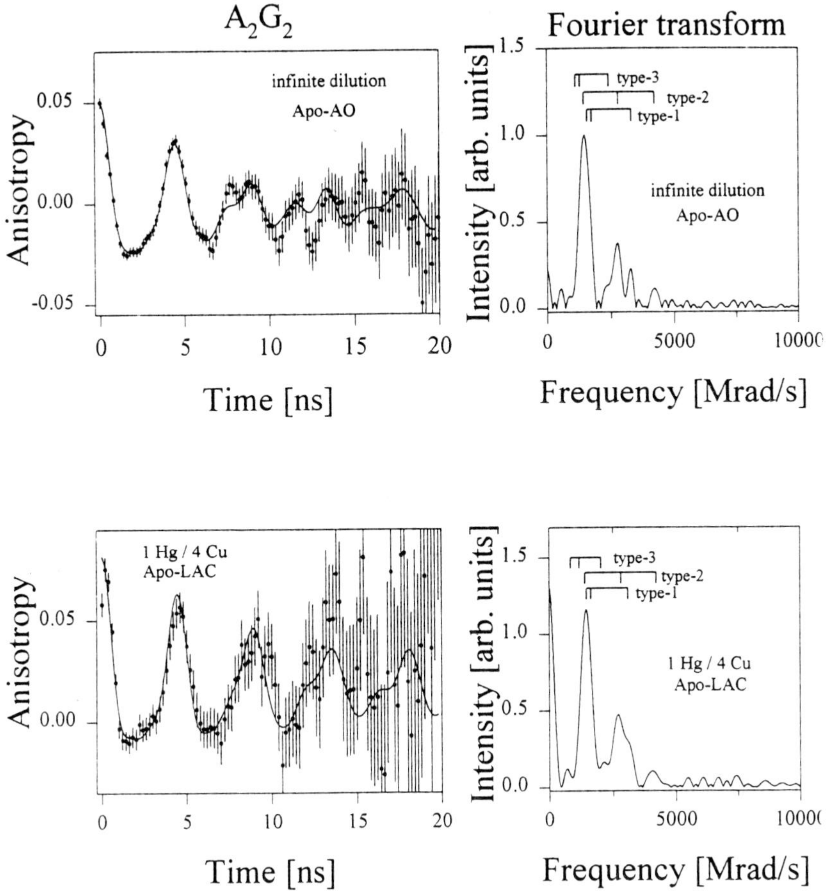

Fig. 3. TDPAC spectra (left) and their Fourier transforms (right) for apo-ascorbate oxidase at "infinite dilution" and for apo-laccase at 1Hg/4Cu stoichiometry.

partial Hg-reconstitution yields the following results: all data can be fitted assuming three inequivalent Hg-NQI's, one of them being identical in frequency ω, asymmetry parameter η, and line broadening δ to that observed for the T2D-enzymes. There is a further NQI with a large asymmetry parameter η and a fundamental frequency ω close to that of the type-2 site and a third site with a large asymmetry parameter η but much lower fundamental frequency. All fitted NQI-parameters for the ^{199m}Hg-data are listed in Table 1.

Table 2. The NQI parameters for ^{199m}Hg-derivatives of azurin (*Pseudomonas aeruginosa*)[29] and stellacyanin (*Rhus vernicifera*),[29] which contain the type-1 site only.

	ω [Grad/s]	η	ν_Q [GHz]	$\lvert V_{xx}\rvert$	$\lvert V_{yy}\rvert$ [10^{21}V/m^2]	$\lvert V_{zz}\rvert$
azurin	1.325(6)	0.698(7)	0.983	9.1	51.2	60.3
stellacyanin	1.287(8)	0.656(10)	0.984	10.4	50.0	60.4

The metal site identification for the two new sites apart from the type-2 site is essentially based on the comparison of the enzyme data with those obtained for the small blue Cu-proteins azurin and stellacyanin. Table 2 lists the observed ^{199m}Hg-parameters for both mono-Cu-proteins which contain a type-1 Cu-site only. It is immediately evident that the high-frequency site has to be identified with the type-1 Cu-site for both enzymes. Hence, by exclusion, the low frequency site has to be assigned to the type-3 site. This will be further corroborated by reconstitution studies with higher Hg/4Cu-ratios. Figure 4 shows the NQI-parameters for the three sites of ascorbate oxidase in the form of a Czjzek plot[26] using a cross correlation analysis.[27] In the Czjzek-Plot η = constant lines start from the origin and have slopes ranging from 60° to 0°. At the upper boundary we have $\eta = 0$ while $\eta = 1$ at the lower horizontal boundary. Lines of constant V_{zz} start at right angles from the $\eta = 0$ boundary and are forming a sort of herring-bone. Since both the abscissa and the ordinate are multiplied by eQ/h, ν_Q can be read off by, first, extrapolating along a V_{zz} = constant or ν_q = constant line towards the $\eta = 0$ boundary and, second, going horizontally to the ordinate. We first analyzed the T2D-data. Then we subtracted the appropriate amount of the type-2 signal from our data and used linear combinations of spectra to emphasize the type-1 or type-3 spectral contributions, respectively. In this way, the three "spots" in Fig. 4 are obtained sequentially.

From these reconstitution studies with low Hg/4Cu ratios we obtain the Hg-reconstitution probabilities for the three different sites. A striking result is the preferential reconstitution of the type-2 site both in ascorbate oxidase and laccase by Hg(II). The carrier-free and the 1Hg/4Cu-data for ascorbate oxidase both yield about 20 % : 60 % : 20 % for type-1,-2,-3 sites. The 1Hg/4Cu-data for laccase yield (with less accuracy) about

40 % : 60 % : 0 %. In all further data analyses the NQI-parameters listed in Table 1 were kept fixed and the only adjustable parameters were the site populations.

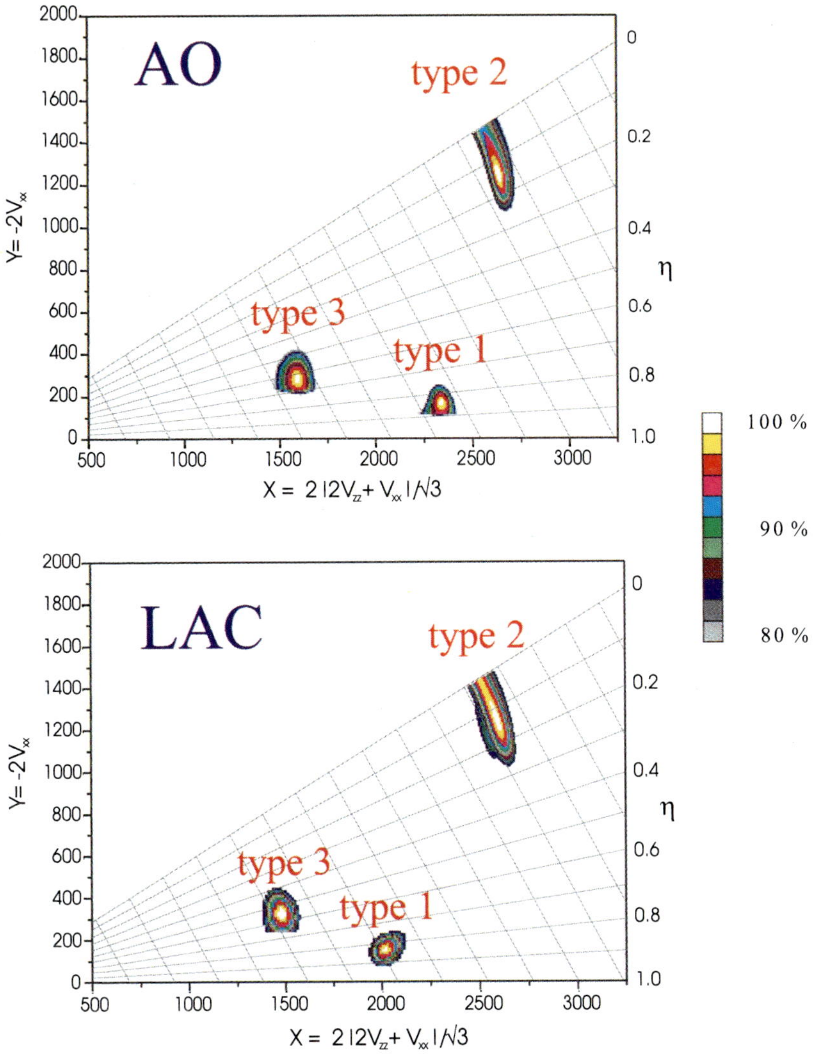

Fig. 4. Cross correlation analysis of the ^{199m}Hg-NQI's of ascorbate oxidase and laccase showing three different metal sites.

4.3. *Nearly Full or Full Hg-Reconstitution of Apo-Enzymes*

Experiments were performed with 2Hg/4Cu and with 3Hg/4Cu for ascorbate oxidase. Based on the Hg-reconstitution probabilities discussed in the preceding section and under the assumption that no metal exchange between Hg on type-2 and on type-3 sites takes place we would expect a population of type-1,-2,-3 sites of 25 % : 50 % : 25 % for the case of 2Hg/4Cu, i.e. the type-2 site is fully reconstituted and the remaining Hg is equally distributed on type-1 and type-3 sites. The experimental observation was 19 % : 49 % : 32 % which is consistent with our expectations within an estimated uncertainty of about 10 % in the populations. For the case of 3Hg/4Cu we would expect populations of 33 % : 33 % : 33 %. The experimental result is 23 % : 35 % : 42 % which gives again a slight preference for the type-3 site compared to the type-1 site, but still is in agreement with expectations within error limits. Unfortunately, there are no data with full Hg-reconstitution of ascorbate oxidase yet.

For laccase, an experiment with full Hg-reconstitution was performed.[28] We obtained populations of 21 % : 30 % : 49 %, indicating that all three sites are fully reconstituted by Hg(II) within experimental uncertainties.

4.4. *Blocking Studies*

Blocking studies with ascorbate oxidase (see Fig. 5) were carried out with the following quantities: Per 4 Cu sites 1/2 equivalent of non-radioactive Hg followed by 3/2 equivalents of radioactive Hg; 1 equivalent of non-radioactive Hg followed by 1 equivalent of radioactive Hg. For the first case we expect — disregarding a possible metal exchange, i.e. the blocked sites are not accessible in the subsequent reconstitution procedure — populations of 27 % : 46 % : 27 %, which is roughly what is observed experimentally: 14 % : 60 % : 26 %, and in a second experiment 20 % : 54 % : 26 %. For the second case we expect 25 % : 50 % : 25 %. We observed 19 % : 48 % : 33 %. Thus, these blocking studies with ascorbate oxidase corroborate the site assignments as discussed in Sec. 4.2.

With laccase, 1 equivalent of non-radioactive Hg was used to block essentially the type-2 site, followed by 3 equivalents of radioactive Hg. In this case, we would expect the following populations: 20 % : 13 % : 67 %, provided there is no metal exchange. We observed 17 % : 32 % : 51 %, in partial disagreement with expectations as far as type-2 is concerned. The predominance of the type-3 population, however, is clearly visible as the

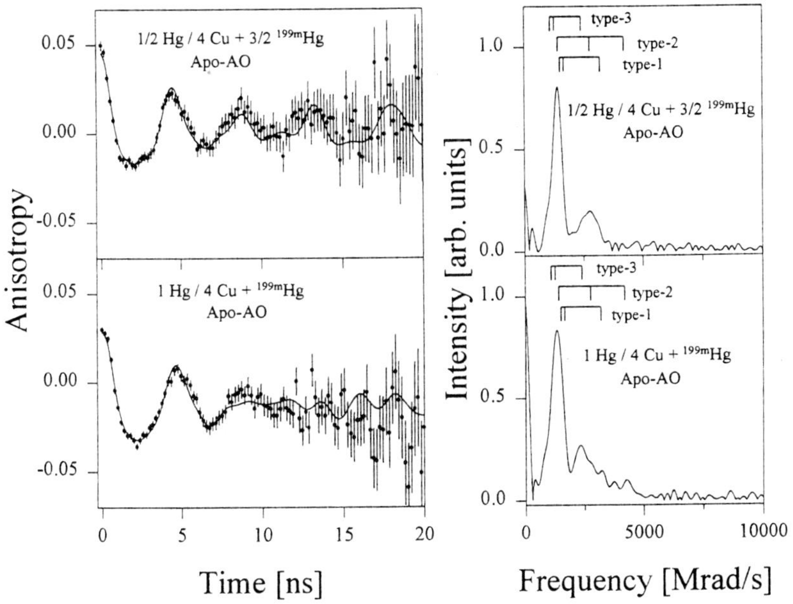

Fig. 5. TDPAC spectra (left) and their Fourier transforms (right) for the blocking experiment with apo-ascorbate oxidase: top — blocking with 1/2 equivalent of non-radioactive Hg per 4 Cu sites plus subsequent filling up with 3/2 equivalents of radioactive ^{199m}Hg per 4 Cu sites; bottom — blocking with 1 equivalent of non-radioactive Hg per 4 Cu sites plus subsequent filling up with 1 equivalent of radioactive ^{199m}Hg per 4 Cu sites.

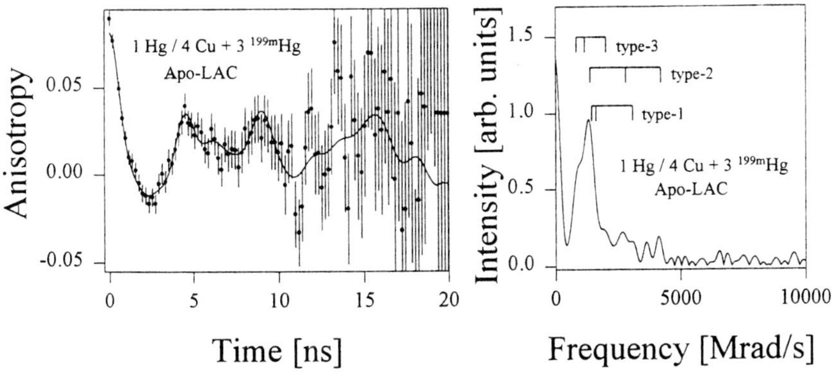

Fig. 6. TDPAC spectra (left) and their Fourier transforms (right) for the blocking experiment with apo-laccase: blocking with 1 equivalent of non-radioactive Hg per 4 Cu sites plus subsequent filling up with 3 equivalents of radioactive Hg per 4 Cu sites.

appearance of the low-frequency shoulder in the Fourier-spectra shown in Fig. 6 (right). The same is true, although less pronounced, for ascorbate oxidase (see Fig. 5. right). Hence, this blocking study also supports the site assignments made in Sec. 4.2, although the agreement with expectations is less convincing.

5. Discussion

5.1. *Site Assignments*

Our site assignments (see Table 1) were based on T2D-enzyme data, on comparisons with small blue Cu-proteins (see Table 2), and on blocking studies. There is further corroborating evidence for this assignment from considerations of strength and symmetry of the NQI concerning the coordination geometries derived from the ascorbate oxidase crystal structure. Type-2 Cu is essentially two-fold coordinated by two histidines (N_ε-bonds) with an additional (weak) bond to an OH-group. Hence, for an almost linear N-Hg-N bond we expect a large NQI with a small asymmetry parameter, as is indeed observed. The type-3 Cu atoms have a trigonal prismatic coordination by six histidines with a bridging OH-group. Therefore, each type-3 Cu atom is four-fold coordinated in a distorted tetrahedral geometry. Such a metal coordination would cause a relatively low NQI and a significant asymmetry parameter. This holds for the assigned type-3-Hg-NQI that exhibits the lowest frequency and a substantial asymmetry parameter for ascorbate oxidase as well as for laccase. Finally, the type-1 Cu has two histidine N_ε-bonds, one cysteine S-bond and a distant methionine S-bond. Our ^{199m}Hg studies with azurin mutants, where methionine is replaced by several other amino acids (Asn, Asp, Gln, Leu, Ala, Glu) showed that methionine is not influencing the NQI to a large extent, although it might be relevant to stabilize the metal center.[29] Thus — as far as the NQI-data are concerned — Hg is threefold coordinated by the equatorial ligands N(His), N(His), S(Cys). The same situation prevails for the type-1 site in AO and LAC.

In conclusion, it should be pointed out that the coordination geometries for all three Cu-sites are conserved in the Hg-derivatives both for ascorbate oxidase and laccase.

5.2. *Hg-Reconstitution Probabilities*

The biggest surprise is certainly the observation that the type-2 site has the highest reconstitution probability for Hg both for ascorbate oxidase and laccase, contrary to expectations based on affinities. This is also in sharp disagreement with reports on laccase (from the Chinese lacquer tree *Rhus vernicifera*) by McMillin and coworkers[8,9] who describe a preparation of a type-1 Hg-blocked heterometal enzyme. However, McMillin *et al.* reconstituted Hg as Hg(I) under reducing conditions in keeping with their observations, that laccase is fully reconstituted by Cu(I) under reducing conditions only.[30] The same holds true for ascorbate oxidase. Further experiments under these conditions with Hg(I) are needed to reconcile the apparently conflicting results. A simple test was carried out to corroborate our findings: after incubation of 1 equivalent of Hg (non-radioactive) with apo-ascorbate oxidase, Cu(I) was reconstituted and the reappearance of the absorption band at 610 nm after oxidation demonstrated that the type-1 site was essentially unblocked by Hg(II), in agreement with the TDPAC results.

5.3. *Comparison between Ascorbate Oxidase and Laccase*

The ^{199m}Hg-data for ascorbate oxidase and laccase clearly demonstrate the similarities between both enzymes with respect to the number of metal binding sites, the strength and asymmetry parameters for the ^{199m}Hg-NQI on all three sites, and — to a lesser extent — also with respect to the preference for the type-2 reconstitution. All frequencies tend to be a little lower in laccase than in ascorbate oxidase. There is no explanation for this observation at present.

Summing up, our TDPAC-results suggest that the coordination geometries for laccase should be rather similar to those of ascorbate oxidase for all three metal sites.

6. Open Questions

6.1. *What is the Origin of the ^{199m}Hg Excess Linewidth at Type-2 Sites?*

The instrumental linewidth for ^{199m}Hg-TDPAC is roughly 250 Mrad/s and limited by the rather short lifetime of the I = 5/2 state. Hence, an excess linewidth is difficult to detect unless it is large. The observed linewidth of $\delta \sim 3$ % for ascorbate oxidase corresponds to an additional ~90 Mrad/s (full width at half maximum) and is clearly detectable. The observed $\delta \sim 7$ % for laccase is considered to be rather large already. What is the origin of this linebroadening? In the T2D-enzymes Hg is in the vicinity of the antiferromagnetically coupled type-3 Cu-pair. This could certainly affect the linewidth of the Hg on type-2 sites. However, we do observe the same linebroadening for the partially Hg-reconstituted enzymes, too. This and the fact that we require the same NQI-parameters in the data analysis for the T2D- and the partially reconstituted enzymes rather suggests that the NQI at the type-2 site is not affected at all by the occupancy (Cu–Cu, Hg–Hg, Hg-vacancy, vacancy–vacancy) of the two type-3 sites. We could invoke intramolecular flexibility at this site, but thus far we have no data over a wide temperature range to support this assumption. Hence, we do not have any explanation at the moment for the linewidth at the type-2 sites. We return to this question in Sec. 6.3.

6.2. *Metal Exchange in the Trinuclear Cluster?*

Our data on T2D-enzymes reveal a unique signal and, hence, do not support the idea of Hg-Cu exchange in the trinuclear cluster. It should be kept in mind that for T2D-enzymes under mild dialysis conditions all type-3 sites should be occupied by Cu and in some molecules even the type-2 site might contain Cu. These molecules could not offer binding sites for Hg at all. Whenever Hg binds to an empty type-2 site in the T2D enzyme, the trinuclear cluster has all metal binding sites occupied. Thus a possible exchange mechanism would imply the creation of an empty site, a rearrangement of metals within the trinuclear cluster and a final rebinding of the — temporarily — unbound metal ion. This is rather unlikely to occur.

The situation is, however, different for the partially Hg-reconstituted apo-enzymes. Here, the trinuclear cluster is never fully occupied by 3 Hg atoms. Hence, a Hg-Hg exchange between type-2 and type-3 sites may proceed via an empty type-2 or type-3 site. If this happens, our blocking studies would suffer in the sense, that type-2 sites blocked with non-radioactive Hg become accessible for radioactive Hg upon subsequent incubation due to metal exchange. Our blocking studies do suggest such a metal exchange, but the experimental uncertainties are still too large to proof it unequivocally.

6.3. *Are there Further Specific Binding Sites for Hg?*

Recently, we performed binding studies with 1 equivalent of Hg per 4 Cu on holo-enzymes. To our surprise, an appreciable fraction (50–70 %) of Hg did bind to the enzymes. The observed NQI-signals were rather similar to those obtained previously for the type-2 sites. There are two obvious routes which have to be followed: (i) either Hg is able to kick out Cu on type-2 sites; this Cu should be detectable by the 2,2'bichinolin reaction,[31] or (ii) there are further specific binding sites for Hg other than the empty Cu-sites which closely resemble the type-2 coordination (essentially two-fold). In this latter case all populations derived from partially and fully reconstituted enzymes would suffer from the inclusion of this as yet unidentified new site into the count for type-2 sites. At present, we cannot exclude this latter possibility, but we prefer the idea that Hg is indeed competing with Cu on type-2 sites. It should be mentioned that the Hg-concentrations applied to the holo-enzymes (and also apo-enzymes) were in the same range as protein concentrations (nM) or even significantly smaller ("infinite dilution") and that the incubation times were at most 60 minutes. These are much milder conditions than those used for replacing type-1 Cu by Hg in holo-plastocyanin crystals.[32] Yet another mechanism has to considered: a locally high concentration of Hg(II) can induce denaturation which creates new metal binding sites which might resemble type-2 binding sites. Hence, all samples should be tested for possible denatured fractions after completion of the TDPAC experiments. This was thus far carried out only occasionally. Further experiments are needed to decide which of the three (or more) possibilities is correct.

Acknowledgments

The cooperation with A. Messerschmidt, U. Thoenes, Th. Schweins, R. Huber, N. Ctortecka, C. Lippert, X. Ni, U. Schmidt, P. Schmidt, M. Ceolín, M. A. Taylor, D. Forkel-Wirth, S. G. Jahn and the ISOLDE-Collaboration is gratefully acknowledged. We thank Prof. D. R. McMillin for valuable discussions. We thank Boehringer/Mannheim for their generous gift of ascorbate oxidase. This work was supported by the Deutsche Forschungsgemeinschaft (DFG) and by the Bundesministerium für Bildung und Wissenschaft, Forschung und Technologie (BMBF).

References

1. D. McMillin, C. S. Peyratout and C. Miller, in *Encyclopedia of Inorganic Chemistry,* Vol. 2 (Wiley, Surrey, New York, 1994), p. 869.

2. A. Messerschmidt, A. Rossi, R. Ladenstein, R. Huber, M. Bolognesi, G. Gatti, A. Marchesini, R. Petruzzelli and A. Finazzi–Agró, *J. Mol. Biol.* **206**, 513 (1989).
3. A. Messerschmidt, R. Ladenstein, R. Huber, M. Bolognesi, L. Avigliano, R. Petruzzelli, A. Rossi and A. Finazzi–Agró, *J. Mol. Biol.* **224**, 179 (1992).
4. I. Savini, L. Morpurgo and L. Avigliano, *Biochem. Biophys. Res. Comm.* **131**, 1251 (1985).
5. K. Ando, *J. Biochem.* (Tokyo) **68**, 501 (1970).
6. L. Avigliano, A. Desideri, S. Urbanelli, B. Mondoví and A. Marchesini, *FEBS Lett.* **100**, 318 (1979).
7. M. T. Graziani, L. Morpurgo, G. Rotilio and B. Mondoví, *FEBS Lett.* **70**, 87 (1976).
8. M. M. Morie–Bebel, M. C. Morris, J. L. Menzie and D. R. McMillin, *J. Am. Chem. Soc.* **106**, 3677 (1984).
9. A. Schmidt Klemens, D. R. McMillin, H. T. Tsang and J. E. Penner–Hahn, *J. Am. Chem. Soc.* **111**, 6398 (1989).
10. R. M. Steffen and H. Frauenfelder, in *Perturbed Angular Correlations*, eds. E. Karlsson, E. Matthias and K. Siegbahn (North Holland, Amsterdam, 1964).
11. H. Frauenfelder and R. M. Steffen, in *Alpha-, Beta-, Gamma-Ray Spectroscopy, Vol. 2*, ed. K. Siegbahn (North Holland, Amsterdam, 1965).
12. R. M. Steffen and K. Alder, in *The Electromagnetic Interaction in Nuclear Spectroscopy*, ed. W. D. Hamilton (North Holland, Amsterdam, 1975).
13. A. Lerf and T. Butz, *Angew. Chem. Int. Ed. Engl.* **26**, 110 (1987).
14. T. Butz, in *The Time Domain in Surface and Structural Dynamics*, eds. G. J. Long and F. Grandjean (Kluwer Academic Publishers, NATO ASI Series, 1988).
15. T. Butz, accepted for publication in Zeitschrift f. Naturf.
16. R. Bauer, *Quart. Rev. Biophys.* **18**, 1 (1985).
17. T. Butz, *Hyp. Int.* **80**, 1079 (1993).
18. T. Butz, *Hyp. Int.* **84**, 47 (1993).
19. T. Butz, *Hyp. Int.* **52**, 189 (1989); erratum: *Hyp. Int.* **73**, 387 (1992).
20. S. Dattagupta, *Hyp. Int.* **11**, 77 (1981).
21. D. C. Lupascu, PhD thesis, Georg-August-Universität Göttingen, 1995.
22. B. Reinhammar, *Biochim. Biophys. Acta* **205**, 35 (1970).
23. E. Kugler, *Nucl. Instr. Meth.* **B79**, 322 (1993),
24. W. Tröger, T. Butz, P. Blaha and K. Schwarz, *Hyp. Int.* **80**, 1109 (1993).
25. T. Butz, S. Saibene, Th. Fraenzke and M. Weber, *Nucl. Instr. Meth.* **A284**, 417 (1989).
26. G. Czjzek, *Hyp. Int.* **14**, 189 (1983).
27. T. Butz, M. Ceolín, P. Ganal, P. Schmidt, M. A. Taylor and W. Tröger, Physica Scripta, in press.
28. T. Butz, W. Tröger, A. Messerschmidt, U. Thoenes and R. Huber, *Hyp. Int.* **80**, 1127 (1993).
29. W. Tröger, PhD thesis, TU München, 1994.

30. A. Schmidt Klemens and D. R. McMillin, *J. Inorg. Biochem.* **38**, 107 (1990).
31. P. E. Brumby and V. Massey, *Methods Enzymol.* **10**, 473 (1967).
32. W. B. Church, J. M. Guss, J. J. Potter and H. C. Freeman, *J. Biol. Chem.* **261**, 234 (1985).

INDEX